Environment
SCIENCE, ISSUES, SOLUTIONS

Environment
SCIENCE, ISSUES, SOLUTIONS

Manuel Molles
University of New Mexico

Brendan Borrell

w.h.freeman
Macmillan Learning
NEW YORK

PUBLISHER: Katherine Parker

SENIOR ACQUISITIONS EDITOR: Bill Minick

SENIOR DEVELOPMENTAL EDITORS: Andrea Gawrylewski and Beth Marsh

MARKETING MANAGER: Maureen Rachford

SENIOR MEDIA AND SUPPLEMENTS EDITOR: Amy Thorne

SENIOR MEDIA PRODUCER: Chris Efstratiou

MEDIA PRODUCER: Jenny Chiu

EDITORIAL ASSISTANTS: Shannon Moloney and Allison Greco

DIRECTOR OF DESIGN, CONTENT MANAGEMENT: Diana Blume

COVER AND TEXT DESIGN: Dirk Kaufman

SENIOR PROJECT EDITOR: Vivien Weiss

PRODUCTION MANAGER: Susan Wein

ILLUSTRATIONS COORDINATOR: Janice Donnola

ILLUSTRATIONS: Tommy Moorman

PHOTO EDITORS: Robin Fadool and Jennifer Atkins

PHOTO RESEARCHERS: Elyse Rieder, Jennifer Atkins, and Stephanie Heimann-Roland

ART MANAGER: Matthew McAdams

COMPOSITION: Sheridan Sellers

PRINTING AND BINDING: RR Donnelley

COVER AND TITLE PAGE: Jim Richardson/National Geographic Creative

Library of Congress Control Number: 2015953764

ISBN-13: 978-0-7167-6187-7
ISBN-10: 0-7167-6187-4

W. H. Freeman and Company
One New York Plaza
Suite 4500
New York, NY 10004-1562

SUSTAINABILITY PLEDGE Macmillan is committed to lessening our company's impact on the environment. The Macmillan family of publishing houses intends to reduce our 2020 CO_2 emissions by 64% against a 2009 baseline.

To all people everywhere, and the web of life that sustains us

ABOUT THE AUTHORS

(Courtesy of Manuel Molles)

Manuel Molles

Manuel Molles is Professor Emeritus of Biology at the University of New Mexico, where he has been a member of the faculty and Curator for the Museum of Southwestern Biology since 1975. Presently, he and his wife Mary Anne live in a cabin in the mountains of La Veta, Colorado, where he writes full time and manages his 100-acre property. He received his Bachelor of Science degree in fisheries from Humboldt State University in 1971, and his Ph.D. in zoology from the University of Arizona in 1976. His dissertation topic was "Fish Species Diversity on Model and Natural Patch Reefs: Experimental Insular Biogeography." Manuel has taught and conducted ecological research in Latin America, the Caribbean, and Europe. He was awarded a Fulbright Research Fellowship to do research on river ecology in Portugal, and has been a visiting professor at the University of Coimbra, Portugal, at the Polytechnic University of Madrid, Spain, and at the University of Montana. Most recently, in 2014 Manuel was awarded the Ecological Society of America Eugene P. Odum Award for "Excellence in Ecology Education."

(Courtesy of Brendan Borrell)

Brendan Borrell

Brendan Borrell is a biologist and journalist who has written about science and the environment for dozens of outlets, including *Bloomberg Businessweek*, *Outside*, *Nature*, *New York Times*, *Scientific American*, and *Smithsonian*. His reporting at home and abroad has given him a firsthand view of some of the most pressing environmental issues of today. He has visited the phosphate mines of Morocco, followed a rhino hunt in South Africa, and taken a road trip through the expanding soy plantations of central Brazil. Brendan received his Ph.D. in Integrative Biology from the University of California, Berkeley, in 2006. For his dissertation research, he studied the evolution, ecology, and physiology of nectar feeding in the orchid bees of Costa Rica and Panama. His articles have received awards from the American Society for Journalists and Authors, and his reporting has been funded by the Alicia Patterson Foundation, the Pulitzer Center on Crisis Reporting, and the Mongabay Special Reporting Initiative.

CONTENTS

WHY I WROTE THIS BOOK

I wrote this book because I am concerned about the future of wild places and the welfare of humanity, particularly the welfare of the next few generations who will inherit the world we leave.

I am motivated by a sense of urgency and mounting evidence that the time to establish a sustainable relationship with Earth is fast running out. The roots of these concerns about the environment developed early. I grew up on a family farm, where, from childhood, I was responsible for growing irrigated crops and raising a wide variety of livestock. There, husbanding animals and tilling soil, I grew to appreciate a well-run farm. However, my focus was not entirely on farming. There were wild places nearby where I was free to roam when my farm chores and schoolwork were done. Our farm overlooked the Merced River in central California at the transition between the flats of the Central Valley and the foothills of the Sierra Nevada. The headwaters of the Merced River drain Yosemite Valley, that long-ago haunt of John Muir.

My father trained me to do all the farm chores, but he also taught me to appreciate wild nature, especially the habits of birds—his first love. Likely because of these early influences, I would spend every available moment on or in the Merced River. However, my knowledge of the place where I grew up was not limited by what I saw in my ramblings, since my family had lived in the area since the mid-1800s. The stories of two great uncles who arrived in northern California as young boys in 1865, three years before Muir began living in Yosemite, were particularly exciting. Incredibly, one of them, Uncle Jim, was still active when I was a child. Those early days were, he said, a time of extensive wetlands and abundant wildlife, of rivers teeming with salmon, the ocean thick with whales, and most of the redwood forests still uncut. I never tired of those tales of what once was, but they also filled me with a deep sense of what had been lost in less than a century. However, I was also encouraged by the survival of unspoiled ecosystems near our farm, just an hour and a half drive from San Francisco, which we called The City and where I learned to value culturally rich urban environments.

My hope is that through this text, I can contribute in some small way to a sustainable balance between wild ecosystems, ecosystems managed for resource extraction, and urban ecosystems. It is my belief that a healthy future for humanity depends on achieving such a balance.

The core of what appears on these pages—the organization, topics, tone, and language—is inspired by what I have learned from the more than 10,000 students who attended my classes during my decades of teaching. Whether in the field, laboratory, or lecture hall, it was these students who taught me what in a subject is significant and how to communicate it. Through this text I hope to share a vision for sustainability with a new generation of students who will be the keepers of humanity's future.

I am also motivated by the feeling that my career would be incomplete without reaching out beyond my academic publications to write this textbook, which I have written while living in mountains surrounded by old growth, mixed conifer forest, abundant wildlife, and fishing for trout when I have a spare moment.

Manuel C. Molles
La Veta, Colorado

BRIEF CONTENTS

CONTENTS

(Kelli-Ann Bliss/NOAA)

Chapter 1   Introduction   1

(Jean Michel Labat/Ardea.com)

Chapter 2   Ecosystems and Economic Systems   31

(Jim Peaco, Yellowstone National Park, NPS)

Chapter 3   Conservation of Endangered Species   59

(Cheryl Jaworowski/USGS)

Chapter 4   Species and Ecosystem

(Jorg Hackemann/Shutterstock)

Chapter 5   Human Populations   125

(NASA)

Chapter 6   Sustaining Water Supplies   155

(Dudarev Mikhail/Shutterstock)

Chapter 7   Sustaining Terrestrial Resources   187

(Bill Dewey, Taylor Shellfish Farms)

Chapter 8   Sustaining Aquatic Resources   227

(Dado Galdieri/Bloomberg via Getty Images)

Chapter 9   Fossil Fuels and Nuclear Energy   261

(Andrew Henderson/National Geographic Creative)

Chapter 10   Renewable Energy   295

(fotog/Getty Images)

Chapter 11   Environmental Health, Risk, and Toxicology   329

(USFWS photo by Susan White)

Chapter 12   Solid and Hazardous Waste Management   357

(Geoff Liesik/The Deseret News via AP)

(Jean-Louis Klein & Marie-Luce Hubert/ Science Source)

A UNIQUE CHAPTER STRUCTURE

Each chapter is divided into three sections: Science, Issues, and Solutions.

"It clearly distinguishes between the science and political, social and economic choices required by the problems. It is more congenial to my teaching than any of my current or recent texts, period."

–Brian Mooney,
Johnson & Whales University

"I love this [science-issues-solutions] approach. Science is the tool that lays the foundation for what follows."

–Barry Perlmutter,
College of Southern Nevada

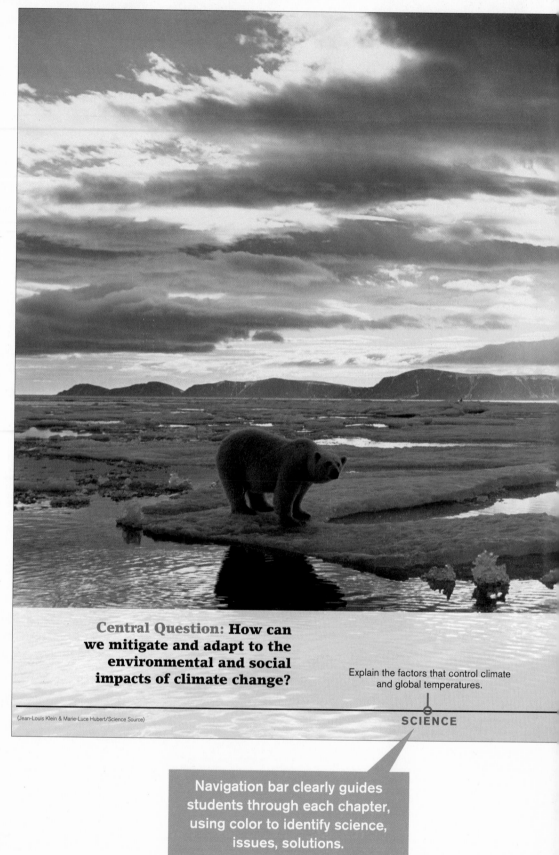

(Jean-Louis Klein & Marie-Luce Hubert/Science Source)

Central Question: How can we mitigate and adapt to the environmental and social impacts of climate change?

Explain the factors that control climate and global temperatures.

SCIENCE

Navigation bar clearly guides students through each chapter, using color to identify science, issues, solutions.

CHAPTER 14

Global Climate Change

Analyze the causes and impacts of
a warming global climate.

Discuss the
that could m

ISSUES　　　　　　　**SOLUTIONS**

14.1–14.4 Science

Each chapter begins by explaining the basic
science relevant to the chapter's topic, as a
foundation for the coverage to follow.

14.5–14.8 Issues

Students draw upon the science coverage
to get a better understanding of current
environmental issues.

14.9–14.11 Solutions

Each chapter concludes by asking students to
evaluate the success or failure of solutions (either
implemented or proposed) for environmental
problems in different parts of the world.

A CENTRAL QUESTION SETS THE LEARNING GOAL FOR THE CHAPTER

SOME CONSEQUENCES OF A WARMER EARTH

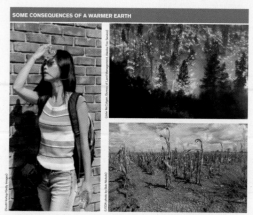

Heat waves are setting temperature records and impacting larger and larger areas around the world. High temperatures combined with drought have been conducive to large wildfires of unprecedented magnitude. Drought has had severe impacts on agricultural production in regions such as the midwestern United States.

Tracking Wildfires in the West

Raging fires and extreme weather events could become more common with a changing global climate

At 7 A.M. on June 23, 2012, a jogger was running along the Waldo Canyon Trail in the mountains above Colorado Springs, Colorado, when he smelled smoke. He veered off the trail to investigate and found a smoldering fire in the woods. After he reported the fire to the local sheriff's department, high winds and drought conditions in the forest caused the fire to spread over 600 acres in several hours' time, leading to evacuations of several nearby communities. By the time firefighters finally contained the Waldo Canyon Fire, two and a half weeks later, it had burned 7,384 hectares (18,247 acres) and 346 homes, killing two people. It ranked as the most destructive fire

in Colorado's history, resulting in insurance claims of more than $450 million. Although the fire may have been started by an arsonist, another suspect has been singled out for its rapid spread and devastating impact: climate change.

That year, the wildfire season in the West came on the heels of a period of unrelenting heat. During the 12 months from August 2011 to July 2012, land temperatures in the 48 contiguous United States were the warmest in 117 years of record-keeping. Across Colorado, wildfires blackened nearly 67,000 hectares (165,000 acres) and destroyed over 600 homes. In Montana and New Mexico, they consumed another 529 homes. In Utah and Wyoming, they forced the

shutdown of natural gas fields, interrupting the flow of critical energy supplies. All told, wildfires in the United States in 2012 burned more than 1.7 million hectares (4.1 million acres).

Abnormally high temperatures in the United States had other impacts as well. For instance, cattle had so little healthy pasture that the USDA allowed ranchers to graze their cattle on conservation lands set aside for erosion control and wildlife habitat. Approximately half of the nation's corn crop and one-third of the soybean crop had failed or were near failing—an episode that would play out in the global economy as an increase in food prices. Reduced farm income would hurt a wide range of businesses located in agricultural regions.

> "Preservation of our environment is not a liberal or conservative challenge, it's common sense."
>
> President Ronald Reagan, State of the Union address, January 1984)

Climate scientists modeling future climates believe that the summer of 2012 may provide a preview of some

of the environmental and economic consequences of climate change. In fact, they have concluded that by mid-century, if present trends continue, the western United States would be subject to droughts worse than any occurring in the previous 1,000 years. Human action has played a significant role in changing Earth's climate, particularly by increasing the concentrations of gases in the atmosphere that trap the Sun's energy, leading to a temperature increase of almost 1°C since 1880. Climate scientists predict that climate change will include a higher frequency of heat waves, droughts, and other weather extremes along with the loss of the polar ice caps and a rise in sea level.

By the end of the 21st century, climate models suggest that the temperature of Earth's surface will rise another 2 to 3°C. "Warming of the climate system is unequivocal, and since the 1950s, many of the observed changes are unprecedented over decades to millennia," wrote the authors of the fifth assessment of the Intergovernmental Panel for Climate Change (IPCC), published in 2014. "It is extremely likely that human influence has been the dominant cause of the observed warming."

The good news is that once we recognize that we are significant contributors to climate change, there are steps we can take to reduce the problem. However, as we address this issue, we will need to avoid causing other forms of disruption, both environmental and economic.

Central Question

How can we mitigate and adapt to the environmental and social impacts of climate change?

SCIENCE   ISSUES   SOLUTIONS

In each chapter, a case study introduces the student to the topic and establishes the overall learning goal for the chapter. This learning goal is called the **Central Question**.

"Using the Central Question as a theme through the chapter allows students to keep a focus on a thesis statement, tying together the supporting information. I find the Central Question very helpful in connecting concepts throughout the chapter."

—Terri Matiella, University of Texas, San Antonio

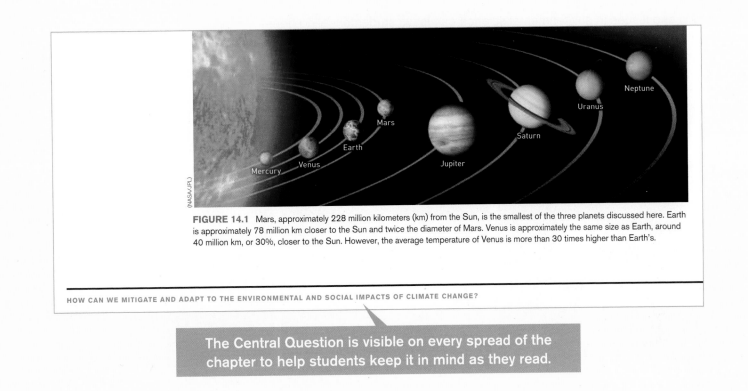

FIGURE 14.1 Mars, approximately 228 million kilometers (km) from the Sun, is the smallest of the three planets discussed here. Earth is approximately 78 million km closer to the Sun and twice the diameter of Mars. Venus is approximately the same size as Earth, around 40 million km, or 30%, closer to the Sun. However, the average temperature of Venus is more than 30 times higher than Earth's.

HOW CAN WE MITIGATE AND ADAPT TO THE ENVIRONMENTAL AND SOCIAL IMPACTS OF CLIMATE CHANGE?

The Central Question is visible on every spread of the chapter to help students keep it in mind as they read.

Central Question: How can we mitigate and adapt to the environmental and social impacts of climate change?

14.1–14.4 Science	14.5–14.8 Issues	14.9–14.11 Solutions	Answer the Central Question:
• What affect does the atmosphere have on planetary temperatures? • How did scientists learn about the greenhouse effect and its role on Earth? • How do global temperatures and CO_2 concentrations vary over time? • Which atmospheric factor exerts the most control over global temperatures and how do we know?	• What is the primary cause of increased CO_2 levels and how do we know? • What global physical effect results from rising CO_2 levels? • What types of changes on Earth have accompanied rising global temperatures? • What societal costs have resulted from climate change?	• What tactics can we take to reducing carbon emissions? • What new economic opportunities may arise from reducing greenhouse gas emissions? • What role do carbon sinks play in balancing the carbon budget?	_____ _____ _____ _____ _____ _____ _____

At the end of each chapter, students create an Active Summary as a recap of the Science, Issues, and Solutions sections presented in the chapter; it also prepares them to answer the Central Question.

"This layout has great value in terms of encouraging students to read, and it also requires the student to answer questions along the way that feed back into the Central Question. This lends itself to a curriculum based more on concepts and discussion rather than simple fact recitation."

–Megan Lahti, Arizona Western College

A Focus on Solutions

The topics and issues in environmental science can leave students feeling hopeless and powerless about environmental issues. Because of the unique chapter structure, this text emphasizes solutions—what has been done (and how well it worked) and what more can be done (and how science can help us implement it).

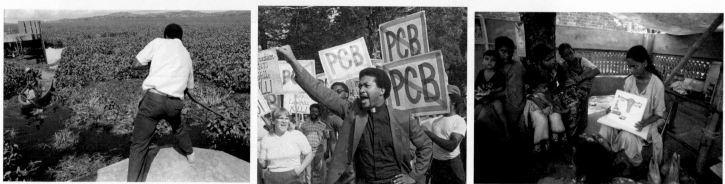

(AP Photo/Danny Wilcox Frazier)

(Greg Gibson/AP Photo)

(Mark Henley/Panos Pictures)

Empowering Students

Following the Solutions section of each chapter, students work through a list of activities they might try in order to directly engage with environmental science issues and feel that their experience counts.

Climate Change and You

Many consider climate and atmospheric change to be the most serious environmental challenge that our species has ever faced. Massive releases of greenhouse gases resulting from the activity of a growing human population have already warmed Earth and threaten to radically disrupt the entire biosphere. The challenges posed by climate change put our collective life and economic support systems at risk. In the face of such a challenge, what can an individual do?

☐ *Follow the science.*

Although climate scientists are in overwhelming agreement on climate change and its causes, the deniers of climate change science present competing conclusions on the present state and dynamics of Earth's climate, as well as the societal and environmental stakes. The best way to sort your way through these competing narratives is to build on what you have learned in this course by following developments in published science, paying particular attention to data associated with global temperatures, storm intensities, depth and frequencies of drought, sea level rise, and so forth.

☐ *Conserve energy.*

Collectively, we can alter the amount of energy produced simply by conserving energy. Energy utilities report that conservation by consumers has already reduced energy demand in both the United States and Europe. A first step is to make sure that your residence is well insulated. If possible, you can also set your thermostat to reduce energy used for heating in winter (no higher than 68° F) and cooling in summer (no cooler than 78° F). Save energy by walking or bicycling whenever practical and safe, or use public mass transport. If you operate a motor vehicle, you can try to maximize fuel economy by choosing a fuel-efficient one and keeping it well maintained.

☐ *Support efforts to reduce greenhouse gas emissions.*

As a citizen, you can use your voice and vote to support transitioning to renewable energy sources and reducing greenhouse gas production. You can support local, regional, and national programs fostering conservation agriculture and forestry practices that help sustain these natural carbon sinks. You can also support legislation that levels a cost on carbon emissions associated with power production and other industrial activity. As a consumer, you can go one step further and support clean energy initiatives offered by your local electrical utility.

☐ *Become involved.*

In ways large and small, we can all be a force for constructive change. After completing this course in environmental science, you should have a broader understanding of the science, issues, and potential solutions to today's environmental challenges. More important, you are better prepared to expand that base of knowledge far beyond where it is now. As you do so, let your informed voice be heard where appropriate and become involved individually and with organizations that reflect your knowledge and understanding of the most pressing environmental issues, whether they be related to climate change or the many other issues surveyed in this text. In the end, because these issues reflect what we do or have done to the environment of our planet, all are interrelated.

"This [science-issues-solutions framework] allows students to understand the basis for the issues, and then helps them look toward the future with a sense of hopefulness and optimism [that] these issues can be addressed, instead of leaving them with a sense of 'doom and gloom.'"

—Terri Matiella,
University of Texas, San Antonio

Critical Thinking and Problem Solving

Think About It questions after each chapter section ask students to analyze what they've just read and apply it to new situations.

Margin questions throughout the chapter help students engage with the issues and can serve as lecture or discussion prompts.

Critical Analysis

Critical Analysis questions at the end of each chapter require students to apply higher-level Bloom's skills to environmental issues and solutions.

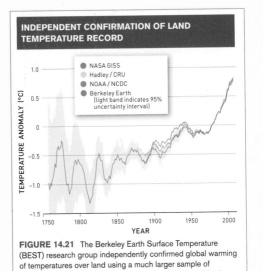

FIGURE 14.21 The Berkeley Earth Surface Temperature (BEST) research group independently confirmed global warming of temperatures over land using a much larger sample of meteorological stations and controlling for urban heat island effects. (Data from BEST, http://berkeleyearth.org/)

A focus on data in each chapter builds quantitative skills and mathematical reasoning.

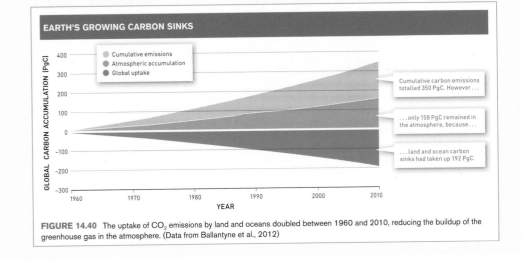

FIGURE 14.40 The uptake of CO_2 emissions by land and oceans doubled between 1960 and 2010, reducing the buildup of the greenhouse gas in the atmosphere. (Data from Ballantyne et al., 2012)

LaunchPad

LaunchPad gives instructors everything they need to quickly set up a course, shape the content of their syllabus, craft presentations and lectures, assign and assess homework, and guide the progress of individual students and the class as a whole. Meanwhile, LaunchPad is the students' one-stop shop for class preparation, homework, and exam prep.

Instructor Resources

 LaunchPad macmillan learning   The new standard in online course management, LaunchPad makes it easier than ever to create interactive assignments, track online homework, and access a wealth of extraordinary teaching and learning tools. Fully loaded with our customizable e-Book and all student and instructor resources, the LaunchPad is organized around a series of prebuilt LaunchPad units—carefully curated, ready-to-use collections of material for each chapter of *Environment: Science, Issues, and Solutions*.

LECTURE TOOLS

Lecture Slides

These slides combine art, classroom discussion questions, and descriptions of key concepts from the book for classroom presentation.

Layered Slides

Slides for select figures deconstruct key concepts, sequences, and processes in a step-by-step format, allowing instructors to present complex ideas in clear, manageable parts.

Optimized Art (Jpegs and layered slides)

Infographics are optimized for projection in large lecture halls and split apart for effective presentation.

Clicker Questions

Designed as interactive in-class exercises, these questions reinforce core concepts and uncover misconceptions.

ASSESSMENT

LearningCurve macmillan learning   Activities use a game-like interface to guide students through a series of questions tailored to their individual level of understanding.

Videos

Videos from an array of trusted sources bring the stories of the book to life and make the material meaningful to students. Each video includes assessment questions to gauge student understanding.

Test Bank

A collection of questions, organized by chapter, presented in a sortable, searchable platform. The Test Bank features multiple-choice questions and uses infographics and graphs from the book.

Student Resources

(alexsl/istockphoto)

Student resources reinforce chapter concepts and give students the tools they need to succeed in the course. All student resources are organized and can be found in the LaunchPad.

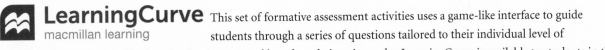

LaunchPad Students have access to a variety of study tools in the LaunchPad, along with a complete online version of the textbook. Carefully curated LaunchPad Units provide suggested learning paths for each chapter in the text.

LearningCurve This set of formative assessment activities uses a game-like interface to guide students through a series of questions tailored to their individual level of understanding. A personalized study plan is generated based on their quiz results. LearningCurve is available to students in the LaunchPad.

Graphing Tutorials

Students build and analyze graphs, using their critical thinking skills to predict trends, identify bias, and make cause-and-effect connections.

Video Case Studies

Videos from an array of trusted sources bring the stories of the book to life and allow students to apply their environmental, scientific, and information literacy skills. Each video includes questions that engage students in the critical thinking process.

Key Term Flashcards

Interactive flashcards can help students drill and learn the most important terms in each chapter.

Critical Thinking Activities

Assignable activities engage students in the material and inspire critical thinking based on content from the textbook.

Environment and Your Activities

Activities prompt students to get directly involved in environmental science issues in their lives and communities.

Reviewers

We extend our deep appreciation to the following instructors who reviewed, tested, and advised on the book manuscript at various stages.

Matthew Abbott, *Des Moines Area Community College–Newton campus*
David Aborn, *University of Tennessee at Chattanooga*
Michael Adams, *Pasco-Hernando Community College*
Loretta Adoghe, *Miami Dade College*
Shamim Ahsan, *Metropolitan State University of Denver*
Steve Ailstock, *Anne Arundel Community College*
Marc Albrecht, *University of Nebraska–Kearney*
Thomas Algeo, *University of Cincinnati*
John Aliff, *Georgia Perimeter College*
Keith Allen, *Bluegrass Community and Technical College*
Albert Allong, *Houston Community College*
Brannon Andersen, *Furman University*
Matt Anderson, *Broward College*
Dean Anson, *Southern New Hampshire University, and Lakes Region Community College*
Clay Arango, *Central Washington University*
Walter Arenstein, *San Jose State University*
Felicia Armstrong, *Youngstown State University*
Paul Arriola, *Elmhurst College*
Tom Arsuffi, *Texas Tech University*
Augustine Avwunudiogba, *California State University, Stanislaus*
Sonia Aziz, *Moravian College*
Abbed Babaei, *Cleveland State University*
Daphne Babcock, *Collin College*
Nancy Bain, *Ohio University*
Jack Baker, *Evergreen Valley College*
James Baldwin, *Boston University*
Becky Ball, *Arizona State University at the West Campus*
Deniz Ballero, *Georgia Perimeter College*
Teri Balser, *University of Wisconsin–Madison*
Barry Barker, *Nova Southeastern University*
Morgan Barrows, *Saddleback College*
Brad Basehore, *Harrisburg Area Community College*
Damon Bassett, *Missouri State University*
David Baumgardner, *Texas A&M University*
Ray Beiersdorfer, *Youngstown State University*
Timothy Bell, *Chicago State University*
Tracy Benning, *University of San Francisco*
David Berg, *Miami University*
Leonard Bernstein, *Temple University*
David Berry, *California State Polytechnic University*
Susan Berta, *Indiana State University*
Joe Beuchel, *Triton College*
Cecilia Bianchi-Hall, *Lenoir Community College*
Jennifer Biederman, *Winona State University*
Andrea Bixler, *Clarke University*
Kim Bjorgo-Thorne, *West Virginia Wesleyan College*
Brian Black, *Penn State Altoona*
Brent Blair, *Xavier University*
Steve Blumenshine, *California State University, Fresno*
Ralph Bonati, *Pima Community College*
Emily Boone, *University of Richmond*
Polly Bouker, *Georgia Perimeter College*
Michael Bourne, *Wright State University*

Richard Bowden, *Allegheny College*
Anne Bower, *Philadelphia University*
Scott Brame, *Clemson University*
Susan Brantley, *Gainesville State College*
Susan Bratton, *Baylor University*
Beth Braun, *City Colleges of Chicago*
Randi Brazeau, *MSU Denver*
James Brenneman, *University of Evansville*
Mary Brown, *Western Michigan University*
Robert Bruck, *North Carolina State University*
Susan Buck, *University North Carolina Greensboro*
Amy Buechel, *Gannon University*
Robert Buerger, *University of North Carolina Wilmington*
Bonnie Burgess, *Loyola Marymount University*
Rebecca Burton, *Alverno College*
Willodean Burton, *Austin Peay State University*
Peter Busher, *Boston University*
Nancy Butler, *Kutztown University*
Anya Butt, *Central College*
Elena Cainas, *Broward College*
John Campbell, *Northwest College*
Daniel Capuano, *Hudson Valley Community College*
Heidi Carlson, *Harrisburg Area Community College*
Deborah Carr, *Texas Tech University*
Margaret Carroll, *Framingham State University*
Kelly Cartwright, *College of Lake County*
Mary Kay Cassani, *Florida Gulf Coast University*
Michelle Cawthorn, *Georgia Southern University*
Dominic Chaloner, *University of Notre Dame*
Linda Chamberlain, *Lansing Community College*
Karen Champ, *College of Central Florida*
Fu-Hsian Chang, *Bemidji State University*
Ron Cisar, *Iowa Western Community College*
Lu Anne Clark, *Lansing Community College*
Reggie Cobb, *Nash Community College*
Marlene Cole, *Boston College*
Elena Colicelli, *College of Saint Elizabeth*
Beth Collins, *Iowa Central Community College*
David Corey, *Midlands Technical College*
Douglas Crawford-Brown, *University of North Carolina at Chapel Hill*
Joan Curry, *University of Arizona College of Agriculture*
Angela Cuthbert, *Millersville University*
Sanhita Datta, *San Jose City College*
James Dauray, *College of Lake County*
Tom Davinroy, *Metropolitan State University of Denver*
Elizabeth Davis-Berg, *Columbia College Chicago*
Robert Dennison, *Heartland Community College*
Michael Denniston, *Georgia Perimeter College*
Frank Dirrigl, *The University of Texas–Pan American*
Jan Dizard, *Amherst College*
Melinda Donnelly, *University of Central Florida*

Michael Draney, *University of Wisconsin–Green Bay*
Daniel Druckenbrod, *Longwood University*
Dani DuCharme, *Waubonsee Community College*
John Duff, *University of Massachusetts Boston*
George Duggan, *Middlesex Community College*
Don Duke, *Florida Gulf Coast University*
Robert Dundas, *California State University, Fresno*
John Dunning, *Purdue University*
Karen Duston, *San Jacinto College*
James Eames, *DePaul University*
Robert East, *Washington & Jefferson College*
Nelson Eby, *University of Massachusetts*
Kenneth Ede, *Oklahoma State University–Tulsa*
Matthew Eick, *Virginia Tech*
Diana Elder, *Northern Arizona University*
Catherine Etter, *Cape Cod Community College*
Luca Fedele, *Virginia Tech*
Jeff Fennell, *Everett Community College*
Fleur Ferro, *Community College of Denver*
Steven Fields, *Winthrop University*
Brad Fiero, *Pima County Community College*
Jonathan Fingerut, *Saint Joseph*
Ken Finkelstein, *Suffolk University Boston*
Geremea Fioravanti, *Harrisburg Area Community College*
Linda Fitzhugh, *Gulf Coast Community College*
Stephan Fitzpatrick, *Georgia Perimeter College*
Margi Flood, *Gainesville State College*
April Ann Fong, *Portland Community College, Sylvania Campus*
Nicholas Frankovits, *University of Akron*
Sabrina Fu, *UMUC*
Elyse Fuller, *Rockland Community College*
Karen Gaines, *Eastern Illinois University*
Danielle Garneau, *SUNY Plattsburgh*
Carri Gerber, *OSU-ATI*
Phil Gibson, *University of Oklahoma*
Paul Gier, *Huntingdon College*
Kristin Gogolen-Wylie, *Macomb Community College*
Michael Golden, *Grossmont College*
Julie Gonzalez, *Des Moines Area Community College*
Rachel Goodman, *Hamdpen-Sydney College*
Pamela Gore, *Georgia Perimeter College*
Karl Gould, *Webber International Univ.*
Gail Grabowsky, *Chaminade University*
Ann Gunkel, *Cincinnati State College*
Maureen Gutzweiler, *Harrisburg Area Community College*
Edward Guy, *Lakeland Community College*
Sue Habeck, *Tacoma Community College*
Charles Hall, *State University of New York College of Environmental Science and Forestry*
Robert Hamilton, *Kent State University*
Robert Harrison, *University of Washington, Seattle*
Stephanie Hart, *Lansing Community College*
Susan Hartley, *University of Minnesota Duluth*
Alyssa Haygood, *Arizona Western College*

Stephen Hecnar, *Lakehead*
Rod Heisey, *Penn State University*
Keith Hench, Ph.D., *Kirkwood Community College*
Carl Herzig, *St. Ambrose University*
Crystal Heshmat, *Hudson Valley Community College*
Crystal Heshmat, *Mildred Elley and Hudson Valley Community College*
Jeffery Hill, *University of North Carolina Wilmington*
Jason Hlebakos, *Mt. San Jacinto College*
Carol Hoban, *Kennesaw State University*
Melissa Hobbs, *Williams Baptist College*
Jeffrey Matthew Hoch, *Nova Southeastern University*
Kelley Hodges, *Gulf Coast State College*
Robert Hollister, *Grand Valley State University*
Joey Holmes, *Rock Valley College*
Claus Holzapfel, *Rutgers University Newark*
Barbara Holzman, *San Francisco State University*
Aixin Hou, *Louisiana State University*
Phillip Hudson, *Southern Illinois University Edwardsville*
LeRoy Humphries, *Fayetteville Technical Community College*
Todd Hunsinger, *Hudson Valley Community College*
Andrew Hunt, *University of Texas at Arlington*
Jodee Hunt, *Grand Valley State University*
Catherine Hurlbut, *Florida State College at Jacksonville*
Lilia Illes, *University of California, Los Angeles*
Emmanuel Iyiegbuniwe, *Western Kentucky University*
Kazi Jaced, *Kentucky State University*
Morteza Javadi, *Columbus State Community College*
Richard Jensen, *Hofstra University*
Mintesinot Jiru, *Coppin State University*
Alan Johnson, *Clemson University*
Kevin Johnson, *Florida Institute of Technology*
Gina Johnston, *California State University, Chico*
Seth Jones, *University of Kentucky*
Elizabeth Jordan, *Santa Monica College*
Stan Kabala, *Duquesne University*
Charles Kaminski, *Middlesex Community College*
Ghassan Karam, *Pace University*
John Kasmer, *Northeastern Illinois University*
Jennifer Katcher, *Pima Community College*
Dawn Kaufman, *St. Lawrence*
Jerry Kavouras, *Lewis University*
Reuben Keller, *Loyola University Chicago*
Kiho Kim, *American University*
Myung-Hoon Kim, *Georgia Perimeter College*
Andrea Kirk, *Tarrant County College*
Elroy Klaviter, *Lansing Community College*
Kristie Klose, *University of California, Santa Barbara*
Leah Knapp, *Olivet College*
Ned Knight, *Linfield College*

Miriam Kodl, *California State University, Monterey Bay*

John Koprowski, *University of Arizona*

Janet Kotash, *Moraine Valley Community College*

Elaine Kotler, *Manchester Community College*

Jean Kowal, *University of Wisconsin–Whitewater*

George Kraemer, *Purchase College*

Paul Kramer, *Farmingdale State College*

William Kroll, *Loyola University of Chicago*

Beth Ann Krueger, *Central Arizona College–Aravaipa Campus*

James Kubicki, *The Pennsylvania State University*

Katherine LaCommare, *Lansing Community College*

Troy Ladine, *East Texas Baptist University*

Diane Lahaise, *Georgia Perimeter College*

Megan Lahti, *Arizona Western College (Adjunct)/ NAU–Yuma (FT)*

Kate Lajtha, *Oregon State University*

Susan Lamont, *Anne Arundel Community College*

Gaytha Langlois, *Bryant University*

Andrew Lapinski, *Reading Area Community College*

Kim Largen, *George Mason University*

Grace Lasker, *Lake Washington Institute of Technology*

Joyce Ellen Lathrop-Davis, *Community College of Baltimore County*

Jennifer Latimer, *Indiana State University*

Kathy Lauckner, *Community College of Southern Nevada*

George Leddy, *Los Angeles Valley College*

Hugh Lefcort, *Gonzaga University*

Marcie Lehman, *Shippensburg University*

Norman Leonard, *University of North Georgia*

Jennifer Lepper, *Minnesota State University Moorhead*

Kurt Leuschner, *College of the Desert–Applied Sciences*

Stephen Lewis, *California State University, Fresno*

J. D. Lewis, *Fordham University*

Yanna Liang, *Southern Illinois University*

Matt Liebman, *Suffolk University Boston*

Theo Light, *Shippensburg University*

Tatyana Lobova, *Old Dominion University*

Eric Lovely, *Arkansas Tech University*

Jia Lu, *Valdosta State University*

Anthony Lupo, *University of Missouri*

Quen, Lupton, *Craven Community College*

Jonathan Lyon, *Merrimack College*

Jeffrey Mahr, *Georgia Perimeter College*

Steven Manis, *MGCCC*

Nancy Mann, *Cuesta College*

Heidi Marcum, *Baylor University*

Nilo Marin, *Broward College*

Tamara Marsh, *Elmhurst College*

Rob Martin, *Florida State College*

Patrick Mathews, *Friends University*

Terri Matiella, *The University of Texas San Antonio*

Eric Maurer, *University of Cincinnati*

Costa Mazidji, *Collin College*

DeWayne McAllister, *JCCC*

Charles McClaugherty, *University of Mount Union*

James McEwan, *Lansing Community College*

Dale McGinnis, *Eastern Florida State College*

Colleen McLean, *Youngstown State University*

Dan McNally, *Bryant University*

Karen McReynolds, *Hope International University*

Patricia Menchaca, *Mount San Jacinto Community College: Menifee Campus*

Michael Mendel, *Mount Vernon Nazarene University*

Heather Miceli, *Johnson and Wales University*

Chris Migliaccio, *Miami Dade College*

Donald Miles, *Ohio University*

William Miller, *Temple University*

Dale Miller, *University of Colorado–Boulder*

Kiran Misra, *Edinboro University of Pennsylvania*

Mark Mitch, *New England College*

Scott Mittman, *Essex County College*

Brian Mooney, *Johnson and Wales University*

David Moore, *Miami Dade College*

Elizabeth Morgan, *College of the Desert*

Sherri Morris, *Bradley University*

John Mugg, *Michigan State University*

Kathleen Murphy, *Daemen College*

Courtney Murren, *College of Charleston*

Carole Neidich-Ryder, *Nassau Community College*

Douglas Nesmith, *Baylor University*

Todd Nims, *Georgia Perimeter College*

Ken Nolte, *Shasta College*

Fran Norflus, *Clayton State University*

Leslie North, *Western Kentucky University*

Kathleen Nuckolls, *University of Kansas*

Kathleen O'Reilly, *Houston Community College*

Mary O'Sullivan, *Elgin Community College*

Mark Oemke, *Alma College*

Victor Okereke, *Morrisville State College*

John Ophus, *University of Northern Iowa*

Natalie Osterhoudt, *Broward Community College*

William Otto, *University of Maine at Machias*

Wendy Owens, *Anne Arundel Community College*

Phil Pack, *Woodbury University*

Raymond Pacovsky, *Palm Beach State College*

Chris Paradise, *Davidson College*

William Parker, *Florida State University*

Denise Lani Pascual, *Indiana University–Purdue University Indianapolis*

Ginger Pasley, *Wake Technical Community College*

Elli Pauli, *George Washington University*

Daniel Pavuk, *Bowling Green State University*

Clayton Penniman, *Central Connecticut State University*

Barry Perlmutter, *College of Southern Nevada*

Joy Perry, *University of Wisconsin Colleges*

Dan Petersen, *University of Cincinnati*

Chris Petrie, *Eastern Florida State College*

Linda Pezzolesi, *Hudson Valley Community College*

Craig Phelps, *Rutgers, The State University of New Jersey*

Neal Phillip, *Bronx Community College*

Frank Phillips, *McNeese State University*

Linda Phipps, *Lipscomb University*

Scott Pike, *Willamette University*

Greg Pillar, *Queens University of Charlotte*

Thomas Pliske, *Florida International University*

Gerald Pollack, *Georgia Perimeter College*

Gary Poon, *Erie Community College, City Campus*

Shaun Prince, *Lake Region State College*

Carol Prombo, *Washington University in St. Louis*

Mary Puglia, *Central Arizona College*

Jennifer Purrenhage, *University of New Hampshire*

Ann Quinn, *Penn State Erie, The Behrend College*

Jodie Ramsay, *Northern State University*

Dan Ratcliff, *Rose State College*

James Reede, *California State University, Sacramento*

Daniel Ressler, *Susquehanna University*

Marsha Richmond, *Wayne State University*

Jennifer Richter, *University of New Mexico*

Melanie Riedinger-Whitmore, *University of South Florida St. Petersburg*

Lisa Rodrigues, *Villanova University*

William Rogers, *West Texas A&M University*

Thomas Rohrer, *Central Michigan University*

Scott Rollins, *Spokane Falls Community College*

Charles Rose, *St. Cloud State University*

Judy Rosovsky, *Johnson State College*

William Roy, *University of Illinois at Urbana–Champaign*

John Rueter, *Portland State University*

Dennis Ruez, *University of Illinois at Springfield*

Jim Sadd, *Occidental College*

Eric Sanden, *University of Wisconsin–River Falls*

Shamili Sandiford, *College of DuPage*

Robert Sanford, *University of Southern Maine*

Karen Savage, *California State University, Northridge*

Timothy Savisky, *University of Pittsburgh at Greensburg*

Debora Scheidemantel, *Pima Community College*

Douglas Schmid, *Nassau Community College*

Nan Schmidt, *Pima Community College*

Jeffery Schneider, *SUNY Oswego*

Andrew Scholl, *Kent State University at Stark*

Kimberly Schulte, *Georgia Perimeter College*

Bruce Schulte, *Western Kentucky University*

Joel Schwartz, *California State University, Sacramento*

Peter Schwartzman, *Knox College*

Andrew Sensenig, *Tabor College*

Lindsay Seward, *University of Maine*

Cindy Seymour, *Craven Community College*

Rich Sheibley, *Edmonds Community College*

Brian Shmaefsky, *Lone Star College–Kingwood*

Kent Short, *Bellevue College*

Joseph Shostell, *Penn State University–Fayette*

William Shoults-Wilson, *Roosevelt University*

Abert Shulley, *CCBC*

Douglas Sims, *College of Southern Nevada*

David Skelly, *Yale University*

Sherilyn Smith, *Le Moyne College*

Rolf Sohn, *Eastern Florida State College*

Douglas Spieles, *Denison University*

Dale Splinter, *University of Wisconsin–Whitewater*

Clint Springer, *Saint Joseph's University*

Alan Stam, *Capital University*

Craig Steele, *Edinboro University*

David Steffy, *Jacksonville State University*

Michelle Stewart, *Mesa Community College*

Julie Stoughton, *University of Nevada Reno*

Peter Strom, *Rutgers University*

Robyn Stroup, *Tulsa Community College*

Andrew Suarez, *University of Illinois*

Keith Summerville, *Drake University*

Karen Swanson, *William Paterson University of New Jersey*

Melanie Szulczewski, *University of Mary Washington*

Ryan Tainsh, *Johnson & Wales University*

Michael Tarrant, *University of Georgia*

Franklyn Te, *Miami Dade College*

Melisa Terlecki, *Cabrini College*

David Terrell, *Warner Pacific College*

William Teska, *Pacific Lutheran University*

Donald Thieme, *Valdosta State University*

Nathan Thomas, *Shippensburg University*

Jamey Thompson, *Hudson Valley Community College*

Heather Throop, *New Mexico State University*

Tim Tibbetts, *Monmouth College*

Ravindra Tipnis, *Houston Community College SW*

Conrad Toepfer, *Brescia University*

Gail Tompkins, *Wake Technical Community College*

Tak Yung (Susanna) Tong, *University of Cincinnati*

Brant Touchette, *Elon University*

Jonah Triebwasser, *Marist and Vassar Colleges*

Chris Tripler, *Endicott College in Massachusetts*

Mike Tveten, *Pima Community College–Northwest Campus*

Richard Tyre, *Valdosta State University*

Janice Uchida, *University of Hawaii*

Lauren Umek, *DePaul University College of Health and Science*

Shalini Upadhyaya, *Reynolds Community College*

Quentin van Ginhoven, *Vanier College*

Thomas Vaughn, *Middlesex Community College*

Robin Verble, *Texas Tech University*

Elisheva Verdi, *Sacramento City College*

Nicole Vermillion, *Georgia Perimeter College*

Eric Vetter, *Hawaii Pacific University*

Paul Vincent, *Valdosta State University*

Caryl Waggett, *Allegheny College*

Daniel Wagner, *Eastern Florida State College*

Meredith Wagner, *Lansing Community College*

Xianzhong Wang, *Indiana University–Purdue University Indianapolis*

Deena Wassenberg, *University of Minnesota*

John Weishampel, *University of Central Florida*

Edward Wells, *Wilson College*

Nancy Wheat, *Hartnell College*

Van Wheat, *South Texas College*

Deborah Williams, *Johnson County Community College*

Frank Williams, *Langara College*

Justin Williams, *Sam Houston State University*

Kay Williams, *Shippensburg University*

Shaun Willson, *East Carolina University*

Angela Witmer, *Georgia Southern University*

Mosheh Wolf, *University of Illinois at Chicago*

Janet Wolkenstein, *Hudson Valley Community College*

Kerry Workman Ford, *California State University, Fresno*

David Wyatt, *Sacramento City College*

Joseph Yavitt, *Cornell University*

Marcy Yeager, *Northern Essex Community College*

Jeff Yule, *Louisiana Tech University*

Natalie Zayas, *CSU Monterey Bay*

Caralyn Zehnder, *Georgia College & State University*

Lynn Zeigler, *Georgia Perimeter College*

Michael Zito, *Nassau Community College*

Central Question: How do science and values help address environmental issues?

Explain what makes up the environment, what science is, and how science can address uncertainty.

SCIENCE

Introduction

ISSUES

Analyze the global environmental impact of humans.

SOLUTIONS

Discuss how personal views affect how we address environmental problems and the goal of sustainability.

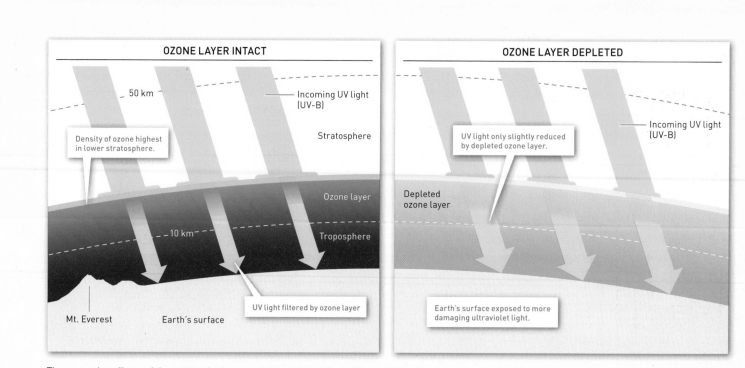

The protective effects of the stratospheric ozone layer and the effects of ozone depletion.

A Growing Impact

With the discovery of a hole in the ozone layer, the impact of a growing human population became more apparent than ever.

Polar bears are drowning in the Arctic Ocean! The Amazonian rain forest is being cleared for soybeans and cattle ranches! Another oil well has blown out in the Gulf of Mexico! It seems that every day a new and shocking environmental tragedy appears in the headlines. Environmental activists argue that we're one step away from apocalypse, while politicians and businessmen hem and haw about the true impact of these kinds of events and who bears responsibility.

Amid heated debates over the most pressing environmental issues of the 21st century, it's sometimes difficult to separate the science from spin. Are people who deny that humans are changing the climate honestly questioning the evidence or are they seeking to delay action? And do environmental activists ever consider the impact that restrictive environmental regulations would have on the economy and the livelihoods of people?

As we shall learn in this text, such controversies and philosophical dilemmas over environmental issues are nothing new. In fact, it may be easier to understand the current debates around climate change and offshore oil drilling by looking deeply at one of the most frightening news headlines in the recent past: "Hole Found in Earth's Atmosphere!" The year was 1985 and British researchers working in the Antarctic had measured a major reduction in ozone levels in the upper atmosphere. **Ozone,** a molecule made up of three oxygen atoms, is considered a pollutant in the lower atmosphere, but it performs a critical role in the upper atmosphere, shielding against potentially harmful **ultraviolet,** or **UV, light** from the Sun.

ozone A molecule made up of three oxygen atoms; considered a pollutant in the lower atmosphere, but in the upper atmosphere it shields against potentially harmful rays from the Sun.

ultraviolet (UV) light Shorter-wavelength, higher-energy rays from the Sun that can damage living tissue.

> "Science cannot resolve moral conflicts, but it can help to more accurately frame the debates about those conflicts."
>
> Heinz Pagels, physicist and science writer, *Dreams of Reason: The Computer and the Rise of the Sciences of Complexity* (1988)

Ultraviolet light, which has shorter wavelengths and higher energy than visible light, can damage living tissue, as anyone who has ever been sunburned knows. Consequently, an ozone hole would lead to problems in human health, such as increased incidence of skin cancer and cataracts, agricultural problems, such as damage to crops, and ecological problems, such as harm to the abundant marine life around the Antarctic. Although the evidence for the ozone hole was debated for years, the science was eventually settled, and governments took action to solve the problem. The ozone hole tapped into deeper fears about how the activities of humans may be impacting the environment, foreshadowing many of the challenges we're faced with today.

The depletion of Earth's ozone layer was not the first sign of human impact on the environment. However, it was a clear and dramatic indication that human impact had achieved truly global proportions. Immediately, questions swirled around the discovery. What had produced the hole in Earth's protective shield? How serious was the situation and could anything be done to repair the protective ozone layer? Addressing these questions would require contributions from the fields of science, medicine, communication media, politics, international diplomacy, national and international law, and many more. Addressing the unresolved environmental issues in the early 21st century will inevitably require the application of not only science, but also human values. As we explore the central question of Chapter 1, we'll return repeatedly to the example of the ozone hole because it reveals the entire process of how science shapes our societies.

Central Question

How do science and values help address environmental issues?

(Kelli-Ann Bliss/NOAA)

1.1–1.4 Science

Is food a chemical factor in the environment, a biological factor, or both? What does your answer imply about the classification of environmental factors?

environment The physical, chemical, and biological conditions that affect an organism.

biotic Living components of the environment.

abiotic Physical and chemical components of the environment.

biological environment The kinds and diversity of pathogens, predators, parasites, and competitors with which an organism interacts.

Because of our origins as hunters and subsistence farmers, humans have long been interested in the relationship between organisms and their environment. Even, today, with a larger fraction of the world's population living in cities and working in jobs as diverse as driving a taxi or programming computer software, we recognize that our impact on the environment extends to the entire planet and those historical interests assume a new urgency. But just what is "environment"?

1.1 Environment is everything

The **environment** consists of both the **biotic** and **abiotic** factors that affect an organism. Biotic factors are the living components of the environment. Abiotic factors include the physical and chemical components of the environment. In environments where humans have significant influences, we must also consider cultural components (**Figure 1.1**).

Think of the "feel" of a misty morning compared to the direct rays of the summer Sun. That's your physical, abiotic environment, which includes factors such as temperature, humidity, and cloud cover, which affects the intensity of sunlight. The physical environment also includes factors that play themselves out over time, such as seasonal changes in temperature or day length. It also includes noise, such as the cock-a-doodle-doo of a rooster, the roar of a freeway at rush hour, or the pinging of underwater sonar.

Furthermore, abiotic factors include the chemicals found in the environment. When you drink a glass of water, with its dissolved oxygen, minerals, and pollutants, you are ingesting a piece of the chemical environment. The chemical environment includes the composition of air, water, and soil. The number, kinds, and concentrations of pollutants the air may contain, as well as the odors in your surroundings, are part of your chemical environment, as are the nutrients in the food you eat (**Figure 1.2**). A plant's chemical environment includes all the nutrients in the soil or surrounding water, as well as the gases in the surrounding air and soil.

Chemical and physical factors are often closely intertwined, and these relationships are at the center of many of today's environmental problems. For example, scientists discovered that when we released refrigerant chemicals known as chlorofluorocarbons (CFCs) into the environment, we thinned stratospheric ozone. This, in turn, changed the physical environment at the Earth's surface by permitting more UV light to pass through the atmosphere. Conversely, altering a physical factor can change important aspects of the chemical environment. For instance, increasing the temperature of a pond will reduce the concentration of oxygen that the pond water can hold. Chemical and physical factors have direct and indirect influences on the biological environment.

A scientific study of the New York City subway system that began in 2013 mapped out species of bacteria found on everything from the turnstiles to the benches to the garbage cans. Pathomap, as the study is called, is a partial record of the **biological environment** faced by commuters each and every day. More generally, your biological environment will include all the viral or bacterial diseases you've contracted during your life and

RELATIONSHIPS BETWEEN ORGANISMS AND THE ENVIRONMENT

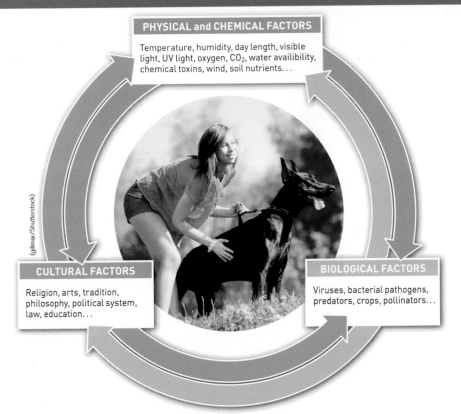

PHYSICAL and CHEMICAL FACTORS
Temperature, humidity, day length, visible light, UV light, oxygen, CO_2, water availibility, chemical toxins, wind, soil nutrients...

CULTURAL FACTORS
Religion, arts, tradition, philosophy, political system, law, education...

BIOLOGICAL FACTORS
Viruses, bacterial pathogens, predators, crops, pollinators...

(gillmar/Shutterstock)

FIGURE 1.1 The environment includes physical, chemical, biological, and cultural factors. While humans are the only species with complex cultures, other species are profoundly affected by the human cultural system in the region where they live.

A CHEMICAL FEAST OF BIOLOGICAL ORIGIN

(Joe Gough/Shutterstock)

FIGURE 1.2 Though we generally do not think of food this way, the food that we eat, including the nutrients and spices it contains along with any chemical contaminants, makes up a part of our chemical environment.

the frequency with which you are exposed to new pathogens, plants, and animals (**Figure 1.3**). The biological communities inhabiting the waters and lands from which that food is extracted, the insects that pollinate crop plants, and the microorganisms that help sustain the fertility of soils represent an extended biological environment. We respond to our biological environment as we choose to eat particular foods and avoid others, when we wash our hands, or attempt to control the insects that intrude into our lives.

Culture makes human interaction with the environment even more complex than that of other species. Cultural factors such as religion, philosophy, and the educational system influence how we view and interact

Are all environmental problems facing humans ultimately cultural?

SCIENCE　　　　　ISSUES　　　　　SOLUTIONS

ANNUAL FLU SHOT: DEFENSE IN THE FACE OF BIOLOGICAL CHALLENGE

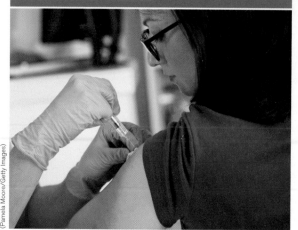

(Pamela Moore/Getty Images)

FIGURE 1.3 Some of the clearest examples of the human biological environment include disease-causing viruses, bacteria, and fungi.

with the environment. Culture affects everything from how we view wild and domestic species and the way we dispose of trash to our attitude toward birth control. The economic and political systems and the laws of the community and nation where we reside can have major effects on how environmental questions are addressed, how the environment is managed, and how citizens view environmental regulation. On a more personal scale, our cultural environment affects how many and with which people we come into contact on a daily basis and the

types of interactions we have or do not have with those individuals.

⚠ Think About It

1. In developed countries, it is often said that we are insulated from the physical and biological forces of nature. Do you think this statement is true? Why or why not?

2. How does ozone thinning demonstrate the interaction of physical, chemical, and cultural aspects of the environment?

1.2 Science uses a formal method to gather evidence about how nature works

The first step toward understanding and evaluating environmental issues is through **science,** a formal process of gathering evidence using observations, experiments, and models. **Environmental science** examines how humans have changed and are changing the environment from local to global scales. Environmental science is also a practical discipline that goes beyond academic studies to find ways of reducing human harm to the environment. Scientific knowledge is put to use through **technology,** such as developing alternatives to CFCs or improved methods of treating wastewater, and through new laws and regulations, which are developed based

science A formal process used to study nature, and the body of knowledge resulting from that process.

environmental science Study of the influence of humans on the environment and the effects of the environment on humans; also attempts to find ways of reducing human harm to the environment.

technology Practical application of scientific knowledge and methods to create products and processes.

A STRIKING SIMILARITY

(NaturePL/NaturePL/Superstock)

Przewalski's horse

(© ARCO/C Heusler/age fotostock)

Painting of horse in Chauvet Cave, France

FIGURE 1.4 Przewalski's horse (left), the last true wild horse of Eurasia, would be extinct today if some had not been captured in Mongolia and taken back to Europe in the early 20th century. Artists working in Chauvet Cave over 31,000 years ago painted this horse (right), producing an image strikingly similar to living Przewalski's horses.

SCIENTISTS STUDY NATURE AT MANY SCALES

(Steve Byland/Getty Images)

Gopher tortoise

(Eye of Science/Science Source)

Leg of a silkworm

(NASA,ESA, CXC and the University of Potsdam, JPL-Caltech, and STScI)

The Small Magellanic Cloud

FIGURE 1.5 Scientists can study some subjects, such as this tortoise (left), with unaided senses. However, the study of other subjects that are very small, such as the microanatomy of an insect (center), or very large or distant, such as stars and planets (right), require instruments that extend the human senses.

on society's value systems. You may think about science as something outside your typical experience, but it simply formalizes what people everywhere have done throughout history: learn about the world through observation and experience.

Ancient Observations

In 1994 a group of French cave explorers discovered one of the most richly painted caves in southwestern Europe, Chauvet Cave. It featured remarkably detailed images of a wild horse with a short, brushlike mane. The horse looked strikingly similar to Przewalski's horse, but that animal lives some 5,000 miles away on the grassy Mongolian Steppe. When scientists analyzed the pigments in the cave, they discovered that the paintings had been made some 31,000 years earlier, before wild horses had become extinct in Europe. Although fossil remains of wild horses had been found in Europe before, the cave paintings captured the horse's pigment and fur, giving scientists further confirmation of how similar these extinct European horses were to the living Przewalski's horse (**Figure 1.4**). Most important to our discussion, those paintings demonstrate clearly that careful observation, the basis of all science, was well developed among our Ice Age ancestors. Modern science has gone far beyond these ancient people, as it has developed as a formal process.

The Scientific Process

The domain of science includes anything that can be observed by the senses or by extensions of them, from subatomic particles to distant galaxies (**Figure 1.5**). Modern instruments such as chemical sensors, microscopes, and telescopes give scientists access to information about nature that was inaccessible to earlier humans. For example, the invention of an ozone detector by Gordon Dobson in 1924, now called the Dobsonmeter, allowed the British Antarctic Survey team to measure the ozone hole over Antarctica in 1984. However, the process of science is basically the same whether it involves technically sophisticated instruments or not. The core of science consists of gathering evidence using observations, experiments, and models to test hypotheses (**Figure 1.6**).

Hypothesis

A **hypothesis** is an explanation of an observation, or a set of relationships, based on a limited amount of information. Once a scientist or team of scientists decides on a question to study, they will propose a hypothesis or a number of alternative hypotheses as tentative answers to the question. For example, a scientist observing the resemblance of Przewalski's horse and the paintings of ancient horses in Chauvet Cave might ask, "Are living Przewalski's horses closely related genetically to the horses roaming western Eurasia during the time that the Chauvet paintings were made"? Once scientists have proposed a hypothesis, they devise procedures to test it. The methods appropriate for a particular scientific investigation—observation, experimentation, modeling, or some combination of these sources of information— depend on the nature of the hypothesis.

hypothesis An explanation of an observation, or a set of relationships, based on a limited amount of information; hypotheses are used to guide scientific experiments, observation, and modeling.

THE PROCESS OF SCIENCE

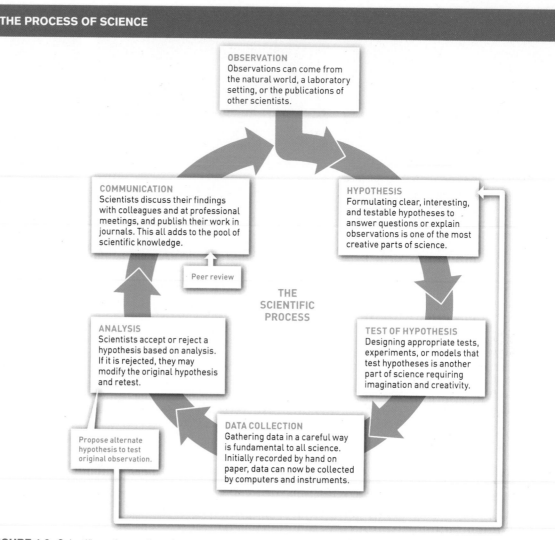

OBSERVATION
Observations can come from the natural world, a laboratory setting, or the publications of other scientists.

COMMUNICATION
Scientists discuss their findings with colleagues and at professional meetings, and publish their work in journals. This all adds to the pool of scientific knowledge.

Peer review

HYPOTHESIS
Formulating clear, interesting, and testable hypotheses to answer questions or explain observations is one of the most creative parts of science.

THE SCIENTIFIC PROCESS

ANALYSIS
Scientists accept or reject a hypothesis based on analysis. If it is rejected, they may modify the original hypothesis and retest.

Propose alternate hypothesis to test original observation.

TEST OF HYPOTHESIS
Designing appropriate tests, experiments, or models that test hypotheses is another part of science requiring imagination and creativity.

DATA COLLECTION
Gathering data in a careful way is fundamental to all science. Initially recorded by hand on paper, data can now be collected by computers and instruments.

FIGURE 1.6 Scientific understanding of nature grows out of a formal process that draws as much from intuition and creativity as it does from established procedure. Science is sometimes equated with the body of knowledge it produces. The more similar findings accumulate for a particular question, the less uncertainty there is regarding the question. However, in science, all findings or questions are open to further investigation, which is why scientists never say that something has been "proven."

Observation

As you have seen, one of the most basic sources of information about the material world is observation. **Observation** is either qualitative or quantitative information gathered systematically from the natural world. Making and recording observations can be used to test many hypotheses; in fact, observations often precede and lead to hypotheses. Modern observations, once recorded in paper notebooks, are now recorded in digital format by computers or other instruments. Familiar examples of routine observations include the temperature records made at weather stations. Daily temperatures have been recorded at some weather stations for more than a century and a half (**Figure 1.7**). Though these long-term temperature observations were not originally made to test a particular hypothesis, they have provided

important information as scientists evaluated hypotheses concerning possible global warming. Where possible, scientists will go beyond observing nature and conduct experiments.

Experiments

Experiments are studies in which one or more physical, chemical, or biological factors are controlled and other factors are allowed to vary among study systems. There are two major types of experiments important to environmental science: laboratory experiments and field experiments.

In **laboratory experiments,** scientists attempt to control, or keep constant, all factors that may influence their study system, while they vary the factor of interest and observe its effect. For instance, an environmental

observation Qualitative or quantitative information gathered systematically from the natural world.

laboratory experiments Experiments in which scientists attempt to control, or keep constant, all factors that may influence their study system, while they vary the factor of interest and observe the effect of the variation on the study system.

LONG-TERM GLOBAL TEMPERATURE RECORD

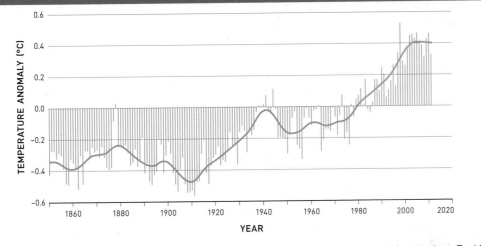

FIGURE 1.7 Scientific observations, such as carefully made temperature measurements, have revealed much about Earth's environment. Long-term records indicate that the climate has warmed during the past 150 years. Temperature anomalies are relative to the average global temperatures during the period of 1961 to 1990. (Data from Jones, 2012)

What are some other questions and hypotheses related to how increased atmospheric CO_2 might affect forests?

scientist studying the accumulation of a pesticide in a population of fish may start with the hypothesis that the rate at which fish absorb the pesticide increases with water temperature. A laboratory experiment designed to test the influence of temperature on uptake of the pesticide would establish test populations of fish in the laboratory, all living in water from the same source containing the same concentration of the pesticide, equal concentrations of oxygen, the same rates of water circulation, the same number of hours of darkness and light of the same intensity, and so forth. The one factor that the laboratory scientist would vary across study populations would be temperature.

In **field experiments,** there is much less control over the environment than in laboratory experiments. As a consequence, the experimenter generally manipulates a single factor, the factor of interest, while allowing all other factors to vary normally. For instance, scientists at Duke University were interested in how the increasing atmospheric concentration of carbon dioxide, CO_2, might affect temperate forests. Because they could not take several large tracts of intact forest into a laboratory, they designed a field experiment: They brought in large tanks of CO_2 and continuously released the gas in their experimental plots, increasing its concentration in the air. All other factors, such as temperature, precipitation, wind speed, and wind direction, were outside the control of the experimenters and varied normally.

The experimental sites were compared to similar sites in which CO_2 was not elevated. These sites served as a **control group,** providing a baseline for comparisons. One of the questions the Duke study was designed to answer was: Will increased atmospheric CO_2 concentrations

increase tree growth? The hypothesis associated with this question was, "Tree growth will be higher on plots with elevated CO_2." (**Figure 1.8**). Another research question was: Will increased atmospheric CO_2 concentrations increase the amount of carbon stored in forest soils?

Over the 10 years of the Duke field experiment, the researchers answered some of their questions. They found that the production of organic matter—a measure of tree growth—was 22% to 30% higher on experimental

field experiments
Experiments in which the experimenter generally controls or manipulates a single factor, the factor of interest, while allowing all other factors to vary normally.

control group A baseline for comparisons.

LARGE-SCALE FIELD EXPERIMENT

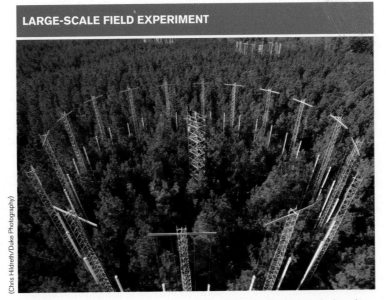

(Chris Hildreth/Duke Photography)

FIGURE 1.8 A research project at the Duke University Forest was designed to explore the response of a forest to increased levels of atmospheric CO_2. During the experiment, carbon dioxide concentration was raised in forest study plots by releasing CO_2-rich air from towers arranged in 30-meter-diameter rings.

plots, compared with control plots. They also found an increase in plant litter (fallen leaves, twigs, etc.) and root production, which meant that more carbon was stored in the soils of the experimental plots. However, most forest trees live for a very long time, some for many centuries, meaning that even a decade-long experiment is only a snapshot of the natural world. To make predictions about long-term processes, scientists often turn to models.

Models

Models are simplified representations of reality that scientists use to better understand or explain a natural phenomenon. For instance, a map of your hometown is a simplified representation of the street system, scaled down in size. Conversely, one might build a scaled-up physical model of a virus to better see how its components fit together. Models can also be more abstract in order to understand how complex processes unfold over time. For instance, a model of Earth's climate might consist of a series of mathematical equations or many lines of computer code. Scientists can then run simulations with these models to test hypotheses where experiments with the actual system would be impossible, impractical, or not permitted for legal or ethical reasons.

A decade before the discovery of the ozone hole over the Antarctic, the chemists Mario Molina and Sherwood Rowland developed a model predicting that the release of CFCs into the atmosphere would lead to the breakdown of ozone in the **stratosphere,** a layer of Earth's atmosphere extending from 10 to 50 kilometers (6.2 to 31.1 miles) above sea level. Their model proposed that chlorine released from CFCs would strip an oxygen atom from ozone, producing oxygen and forming chlorine monoxide in the process (**Figure 1.9**). Chlorine monoxide would then break down, free chlorine atoms would go on to convert more ozone into oxygen, and the process would run again. A decade later, theoretical and observational studies by Susan Solomon showed that the key chemical reactions involved in ozone depletion occur on the surface of stratospheric clouds. Molina and Roland cautioned that if the production of CFCs were not reduced, Earth's ozone shield could be lost. While models do not perfectly match the actual system, information gathered from them can yield important insights.

⚠ Think About It

1. How are the observations of the artists of Chauvet Cave and those of modern scientists similar? How are they different?

2. Why are hypotheses critical to the improvement of scientific understanding of nature? (Hint: Imagine observations and experiments in the absence of hypotheses.)

models In science, simplified representations of a system, constructed on a scale more convenient for study than the actual system of interest.

stratosphere The layer of Earth's atmosphere beginning at an elevation of 10 kilometers and extending outward to 50 kilometers (6.2 to 31.1 miles) above sea level.

MODEL OF HOW CHLORINE FROM CFCs BREAKS DOWN OZONE IN THE STRATOSPHERE

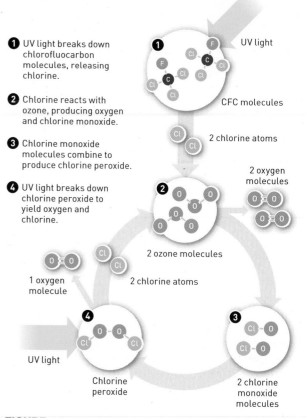

❶ UV light breaks down chlorofluocarbon molecules, releasing chlorine.

❷ Chlorine reacts with ozone, producing oxygen and chlorine monoxide.

❸ Chlorine monoxide molecules combine to produce chlorine peroxide.

❹ UV light breaks down chlorine peroxide to yield oxygen and chlorine.

UV light

CFC molecules

2 chlorine atoms

2 oxygen molecules

2 ozone molecules

2 chlorine atoms

1 oxygen molecule

UV light

Chlorine peroxide

2 chlorine monoxide molecules

FIGURE 1.9 In 1974, Mario Molina and Sherwood Rowland published a detailed model illustrating how releasing CFCs (chlorofluorocarbons) would lead to the breakdown of stratospheric ozone.

3. What are some natural phenomena that would be best studied by laboratory experiments, by field experiments, or by models? What phenomena might be realistically studied using all three approaches?

1.3 Scientific evidence can reduce uncertainty about natural phenomena

Even after conducting a successful experiment to test a hypothesis, scientists may still have an incomplete picture of a natural phenomenon. For example, the scientists at the Duke experimental forest proposed that higher levels of atmospheric CO_2 would increase growth of forest vegetation. Indeed, they found that root growth increased in their experimental plots. Case closed, right? Not exactly. While one well-conducted experiment can provide scientific evidence in support of or against a hypothesis, uncertainty will remain in

EMERGENCE OF A THEORY FROM A BODY OF SCIENTIFIC RESEARCH

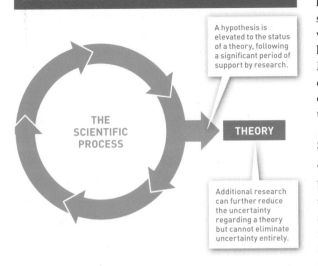

A hypothesis is elevated to the status of a theory, following a significant period of support by research.

THE SCIENTIFIC PROCESS

THEORY

Additional research can further reduce the uncertainty regarding a theory but cannot eliminate uncertainty entirely.

FIGURE 1.10 What constitutes sufficient research to elevate a hypothesis to a theory varies widely among fields of scientific inquiry.

regard to its generality. For example, would an increase in atmospheric CO_2 result in increased root growth at other forests? With additional research using different types of experiments under different conditions, the information supporting a particular hypothesis may increase.

At some point in the history of the investigation of a phenomenon, scientists may develop a theory (**Figure 1.10**). A **theory** is a general explanation of a phenomenon that has been supported by repeated observation,

experimentation, and modeling of specific hypotheses. Some of the most famous and well-established theories have names such as the theory of evolution by natural selection or the germ theory of disease, both of which have been developed over the course of testing hypotheses about a wide variety of species and diseases. More recently, when we talk about human-caused climate change as a scientific theory, we mean that the overwhelming majority of evidence supports the view that Earth's climate is changing due to human activity.

Sources of Scientific Uncertainty

There are many sources of scientific uncertainty. Looking back at the story of the discovery of the ozone hole, we can see how science gradually reduced uncertainty about this phenomenon. The first source of scientific uncertainty is simply how much we already know about a system. In building their model of CFCs, the chemists Molina and Rowland acknowledged a number of areas of uncertainty, including a lack of knowledge of the rates at which molecules moved through the stratosphere and the rate at which chlorine produced in the stratosphere would be removed. Because of these uncertainties and others, Molina and Rowland could not predict when environmental problems due to depletion of stratospheric ozone might begin.

A second potential source of uncertainty is measurement error, as a result of either human or instrument error. For instance, when the British Antarctic Survey first began recording low levels of stratospheric ozone with their Dobsonmeter (**Figure 1.11**), they considered the possibility that the machine was malfunctioning. To eliminate this possibility and reduce

theory A scientific hypothesis that has withstood sufficient testing—through observation, experimentation, and modeling—so it has a high probability of being correct.

INSTRUMENTS FOR MEASURING STRATOSPHERIC OZONE

Dobsonmeter

Total Ozone Mapping Spectrometer (TOMS)

FIGURE 1.11 The ground-based Dobsonmeter (left) is capable of making accurate measurements of stratospheric ozone concentrations—but over a limited area. In contrast, the satellite-based Total Ozone Mapping Spectrometer, or TOMS (right), is capable of measuring concentrations of stratospheric ozone over large areas.

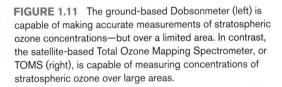

uncertainty, the team sent for a backup Dobsonmeter, which confirmed the first set of results.

The level of certainty can also be limited by the methods used in a scientific study. The original report on ozone depletion was limited to the area of the atmosphere that could be sampled by a Dobsonmeter from the Antarctic base. Consequently, the extent of the atmosphere that was depleted of ozone was uncertain until NASA conducted satellite-based measurements the following year (Figure 1.11). Because of the nature of scientific inquiry, there will always be some degree of uncertainty in our understanding of natural phenomena.

Precautionary Principle

Because uncertainty is characteristic of science, scientists must cultivate a critical view of their disciplines, their hypotheses, prevailing theories, and scientific understanding in general. This skepticism, however, is a double-edged sword in environmental science. While it drives the process of science, it can also slow action on global environmental problems. Ozone depletion, for instance, is a situation in which there is a significant probability of environmental damage. Though uncertainties remained, particularly in the early days of ozone depletion research, would it have been wise to take no action?

Science cannot eliminate uncertainty about nature entirely, which is why we often fall back on the **precautionary principle.** At a 1998 conference at the Wingspread Conference Center in Racine, Wisconsin, 32 environmental scientists expressed the principle as follows: "Where there are threats of harm to human health or the environment, precautionary measures should be taken even if some cause and effect relationships are not fully established scientifically." You may have heard the precautionary principle as a piece of folk wisdom: "Better safe than sorry."

The precautionary principle shifts the burden of proof from those supporting environmental protection to those recommending actions that may be harmful or may result in harm if no action is taken. However, because such a position may hurt businesses or other stakeholders, it often receives significant opposition. For instance, measures to control ozone-depleting CFCs were initially met with strong resistance from those in the industry.

⚠ Think About It

1. A study similar to the one at the Duke experimental forest was done at the Oak Ridge National Laboratory in Tennessee. The Oak Ridge study also resulted in higher root growth rates in response to elevated CO_2 concentrations, but another study done in Switzerland did not. Do the differences in results make any of the studies

invalid? What do scientists generally do when faced with conflicting results such as these?

2. What criteria should be used to determine when to follow the precautionary principle? Are there situations in which the precautionary principle should not be followed?

1.4 The integrity of science depends on following a strict code of ethical conduct

The scientific process depends critically on the competence and honesty of scientists. If scientists engage in unethical research practice, their findings cannot be trusted, leaving society with no sound basis for making decisions regarding environmental issues.

Data Treatment

The measurements and other information gathered during a scientific study are generally referred to as **data.** Because of their fundamental importance to science and, increasingly, to society at large, data must be gathered and managed very carefully (**Figure 1.12**). The most important obligation of the scientist is to be as accurate as possible in the recording, reporting, and sharing of data with other researchers. If research results cannot be verified, they are generally eventually discarded as "unsupported by further study."

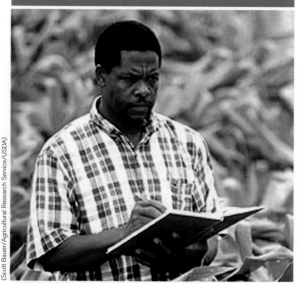

RECORDING AND MANAGING DATA: FUNDAMENTAL PROCESSES IN SCIENCE

(Scott Bauer/Agricultural Research Service/USDA)

FIGURE 1.12 No aspect of research is more important to the integrity of science than careful data gathering and management. Here, Dr. Eton Codling records data on corn growing on soil treated to reduce nutrient losses.

How should environmental regulation based on the precautionary principle be influenced by potential harm to health and environment and level of scientific uncertainty?

precautionary principle
Advises that precautionary measures should be taken to protect human or environmental health, even if some potential threats are not fully understood scientifically.

data The measurements made during a scientific study.

Research Misconduct

No violation of scientific ethics is more serious than research misconduct, which goes beyond sloppy data collection into purposeful deception. This sin generally falls into three categories: fabrication, falsification, and plagiarism. Fabrication occurs when a scientist simply makes up data or results. Falsification involves tampering with research materials or equipment or changing data, such that it alters the outcome of the research. Plagiarism is using the ideas, procedures, or research results of others without giving them proper credit. Research misconduct can destroy the careers of individual scientists and undermine the role of science in informing policy decisions. Fortunately, there are checks on the scientific process.

Publishing and Peer Review

Early scientists, such as the physicist Isaac Newton, often kept the results of their work secret, fearing that others would steal the credit. Scientific publishing came into being in the 17th century, under the sponsorship of the Royal Society of London, as a way of ensuring that scientists received credit for their discoveries. Heirs to that early effort are the many current research journals, such as *Nature,* one of the world's most prestigious scientific journals. A key element in modern scientific publishing is **peer review,** in which experts in the field of research review scientific studies prior to publication for originality, soundness of the methods and analyses, significance of the results, and accuracy of the conclusions.

Conflict of Interest

One factor that can interfere with the integrity of the scientific process is conflict of interest. **Conflict of interest** refers to competing interests, including personal, philosophical, and financial interests, which may interfere with the judgment of a researcher or reviewer of scientific work. Studies have shown that when a company funds research on a topic it has a financial interest in, such as the safety of a new drug or the environmental impact of a construction site, the results are more likely to be favorable to the company than if the research had been conducted independently. Personal relationships between researchers, both positive and negative, can encourage poor judgment and even research misconduct. Similarly, strongly held philosophical positions or beliefs can also lead to improper manipulation of the scientific process.

⚠ Think About It

1. Why is it important to avoid conflicts of interest as much as possible throughout a scientific investigation? Is it possible to avoid all conflicts of interest?

2. How does peer review help to keep researchers honest and promote research ethics?

1.1–1.4 Science: Summary

The environment consists of a great number of physical, chemical, biological, and cultural factors that interact in complex ways. Science is both a formal process used to study these interactions and the body of knowledge resulting from that process. The domain of science comprises anything in the material world that can be observed by the senses or by extensions of them, from subatomic particles to distant galaxies. The core of the scientific process consists of gathering evidence through observations, experiments, and models to test hypotheses. A theory is a scientific hypothesis that has withstood sufficient testing—through observation, experimentation, and modeling—indicating it has a high probability of being correct. However, even after extensive testing and support for a theory, a certain level of uncertainty will remain. In the face of uncertainty, the precautionary principle advises taking steps to avoid harm to humans or the environment, even if some cause-and-effect relationships are not fully established scientifically.

For science to be useful, scientists must commit to a strict code of research ethics, taking care in gathering and managing data. Scientists should make their data available to other researchers to root out conflicts of interests, fabrication of data, falsification of procedures and results, and plagiarism. The peer review process helps assure the quality of published papers and the ethical behavior of scientists.

Why is being "open-minded" such a critical quality to being successful in scientific research?

peer review As part of the process of publishing scientific papers, experts in the field of research covered by a prospective scientific paper review the research prior to publication; they check for soundness of the methods, analyses, results, and coverage of the relevant prior publications on the subject.

conflict of interest Competing interests, including personal, philosophical, and financial interests, that may interfere with an objective judgment.

1.5–1.6 Issues

All species influence the environment in which they live, but human impact on the environment dwarfs that of other species. Earth is now home to 7 billion people. Farms have replaced natural vegetation, covering nearly 40% of Earth's land surface. In the oceans, we have removed more than 90% of large fish over the last half-century. The burning of coal, oil, and natural gas spews 36 billion metric tons of climate-warming carbon dioxide per year. Some of the most pressing issues for environmental scientists are achieving a sustainable human population trajectory, maintaining the productivity of farmlands and aquatic ecosystems, protecting the biodiversity and natural ecosystems on which human welfare depends, and addressing rapid climate change (**Figure 1.13**) These impacts did not begin overnight, so it's worth turning back the clock and exploring the history.

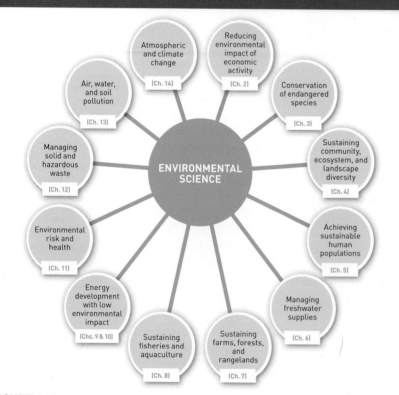

ENGAGING ENVIRONMENTAL ISSUES AND EXPLORING SOLUTIONS USING ENVIRONMENTAL SCIENCE

FIGURE 1.13 The remainder of this book will examine major contemporary environmental issues and review potential ways to address them through the lens of environmental science.

1.5 Human impact and environmental awareness began long ago

When considering the environmental issues we face today, it is important to recognize that significant, society-changing environmental impacts by human populations began long ago (**Figure 1.14**).

Because people everywhere were once connected very closely with the environment around them, it is likely they recognized environmental changes as those events happened. However, we can only know for certain where observations of environmental change have been recorded in pictures or in words. Those records reveal that the first indications of environmental awareness, the forerunner to environmental science, began thousands of years ago.

Ancient Records of Environmental Impact

The first written records of environmental impact by human populations are over 2,000 years old. The early history of Greece includes numerous references to environmental impacts and recommendations on how to avoid damage to the environment. Wood was in great demand for building ships, for cooking and heating homes, and for the charcoal used in smelting metal ores and firing pottery. Consequently, the forests around Athens, one of the main cities of ancient Greece, were gone between 500 and 400 BCE. Both Plato and Aristotle—a natural scientist as well as a philosopher—recommended regulating the use of forests and grazing lands and establishing foresters and wardens to monitor forests and woodlands and to enforce environmental regulations.

In China, the philosopher Mencius, who lived in the 4th century BCE, was a skilled observer of the world around him. He also recorded examples of deforestation: "Bull Mountain was once beautifully wooded. But, because it was close to the city, its trees all fell to the axe." Mencius recognized that the vegetation on Bull Mountain had the potential to recover, but it was prevented from doing so because of overgrazing by livestock: "However, as the days passed things grew and with the rains and the dews it was not without greenery. Then came the cattle and goats to graze. That is why, today, it has that scoured-like appearance." Severe deforestation continued in China until modern times, with devastating impacts on the environment (**Figure 1.15**). Deforestation was

ENVIRONMENT AND THE RISE AND FALL OF A CIVILIZATION

(De Agostini Picture Library/Getty Images)

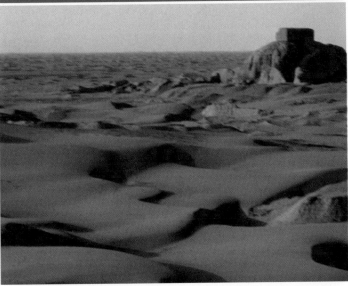

FIGURE 1.14 The Sumerian civilization was one of the first to build cities. The Sumerians depended on irrigated agriculture in the deserts of Mesopotamia, but salt buildup eventually rendered their soils unproductive. Today, the once thriving Sumerian cities are reduced to ruins surrounded by desert

DEFORESTATION IN CHINA

(Jim Richardson/National Geographic Creative)

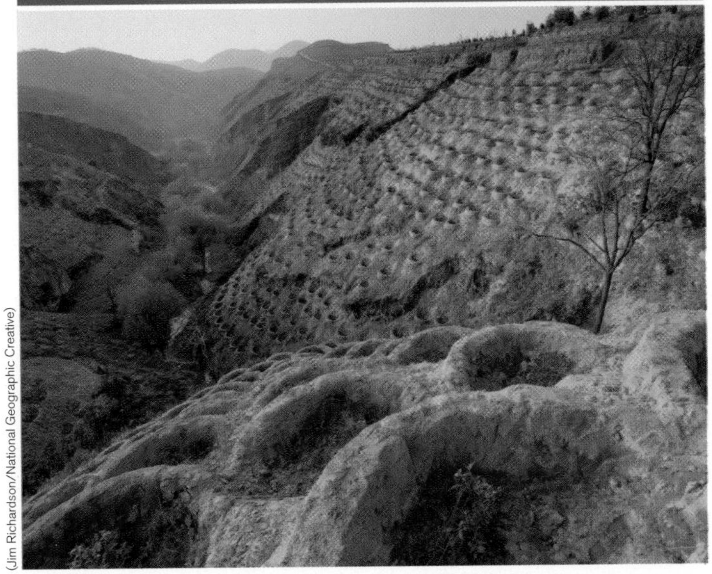

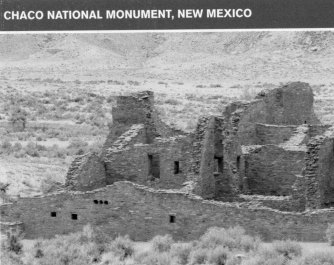

FIGURE 1.15 Deforestation and resulting soil erosion were noted and recorded by Chinese scholars over 2,000 years ago. Today, China is attempting to reverse these losses through massive campaigns to restore once forested lands.

also a problem in parts of pre-Columbian North America (**Figure 1.16**), medieval Britain, as well as North America during the Colonial period.

Growing Environmental Awareness

Concern for the health of the environment emerged wherever societies faced environmental limits—for example, in response to deforestation in ancient China and Greece. Those concerns multiplied rapidly from the 18th through 20th centuries as human populations grew and economic activity increased. The history of environmental awareness and the conflict between opposing positions over environmental issues can help us understand the challenges we face today.

Benjamin Franklin and Environmental Protection

Benjamin Franklin (1706–1790) may be best known for his political and scientific accomplishments in the 18th century, such as helping craft the United States' Declaration of Independence or studying the electrical nature of lightning. However, Franklin was also an early American environmental activist, leading the struggle, for instance, to reduce industrial pollution in his adopted home, the city of Philadelphia. In 1739 Franklin and his neighbors petitioned the local government to move leather-tanning operations, or tanneries, to locations away from the city center, where they were discharging wastes

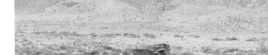

CHACO NATIONAL MONUMENT, NEW MEXICO

(Radoslaw Lecyk/Shutterstock)

FIGURE 1.16 When the Anasazi (a civilization that lived in the American Southwest from 200 CE to 1300 CE) constructed Pueblo Bonito, the building site was surrounded by productive woodland. Woodcutting and drought gradually converted the woodland into a desert shrubland, exhausting the Anasazi's sources of wood for fuel and building.

IN 1739 PHILADELPHIA USED DOCK CREEK FOR WASTE DISPOSAL

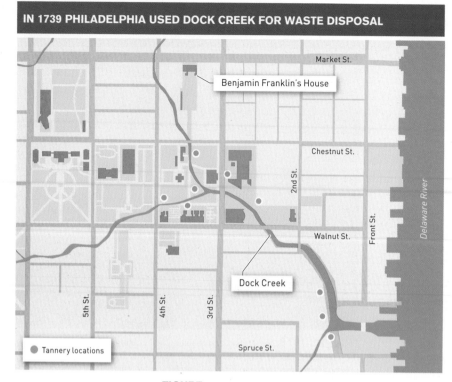

FIGURE 1.17 Leather-tanning yards were built along Dock Creek, where they disposed of their wastes, creating a stench that permeated the surrounding neighborhoods—including the one where Benjamin Franklin lived.

What are some ways in which economic and environmental interests might have found common ground in Colonial Philadelphia?

into Dock Creek (**Figure 1.17**). In addition to reducing the value of properties in the area, the presence of the tanneries made moving through the city center slow, thereby making fire-fighting difficult. During the conflict, Franklin used the paper he ran, the *Pennsylvania Gazette*, to argue that by creating a public nuisance, the activities of a few businesses were harming the daily lives and economic well-being of large numbers of Philadelphians. In the end, however, the power and influence of the tanners prevailed and Dock Creek would continue to be used as an open sewer until it was covered over in the early 19th century.

Franklin also wrote extensively about the influence of environmental conditions on disease, especially yellow fever, which killed 500 residents and visitors to Philadelphia just two years after the tannery petition. Always thinking about the welfare of his fellow Philadelphians, Franklin included in his will plans for the building of a water system that would deliver clean water to the city, as well as the money to pay for that system. In one essay, Franklin also outlined his ideas about the main controls on human population growth.

Thomas Malthus, Population Growth, and Resources

Born in England, Thomas Malthus (1766–1834), a member of the Anglican clergy, pursued a number of academic interests, including the work for which he is best known: the principles governing human population growth. Malthus was skeptical of the writings of a number of authors of his time, who predicted a future utopian world of healthy thriving human populations free from historical wants and afflictions, such as hunger, disease, and war. He had observed that there were no perfect human societies in his time, nor did historical records suggest that one had ever existed. Malthus proposed that human populations had the capacity to grow faster than society could increase its capacity to serve them, especially when it came to food supply (**Figure 1.18**). He predicted that a human population would grow past the capacity of the environment to support it even if all potential lands were converted to producing food for the population.

George Perkins Marsh: Human Populations and Nature

George Perkins Marsh (1801–1882) is mainly remembered for *Man and Nature,* the first environmental science book written by an American, which he published in 1864 (**Figure 1.19**). However, this was just one facet of Marsh's varied life work. Born in Vermont in 1801, Marsh practiced law but he also worked as a sheep farmer, businessman, lecturer, U.S. congressman, linguist (he spoke 20 languages), and diplomat. Marsh traveled extensively in the Middle East through places that were

COMPARISON OF GROWTH IN A HUMAN POPULATION WITH GROWTH IN FOOD SUPPLY

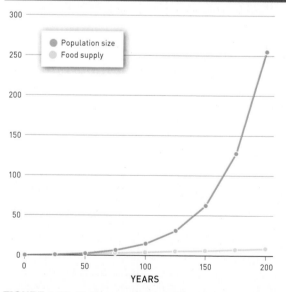

FIGURE 1.18 Malthus used mathematics to show that a human population doubling every 25 years would soon outpace growth in the food supply, even if the food supply was increased every year by the total production capacity in year 1.

FIGURE 1.19 Marsh's 1864 book, which included topics such as deforestation, extinction of animals, desertification, destruction of wetlands, and climate change, now seems eerily prophetic.

stark deserts, though they were described in the Bible as lush and productive.

These experiences, combined with the changes he had observed during his early life in Vermont, convinced him that environmental impact by humans could change productive land into desert wastes. Marsh's writings about nature and its uses and abuses were widely read and came at a critical time, when massive industrialization was occurring around the world.

Land Conservation and Preservation in the United States

Marsh's writings inspired three early leaders of land protection in the United States (**Figure 1.20**). One of those leaders was John Muir (1838–1914), a Scottish-born American naturalist. Muir's family immigrated to Wisconsin, where he eventually attended the University of Wisconsin, an experience that deepened his appreciation for nature, particularly for botany and geology. In 1868 Muir moved west to California, where he began promoting the preservation of Yosemite Valley and other lands. Muir was an early advocate of a **preservation ethic,** an environmental ethic emphasizing

the protection of natural ecosystems in their original unspoiled states. His evocative, poetic writings and his long-term advocacy for land preservation made him a central figure in the environmental movement, which was advanced by Muir's founding of the Sierra Club.

The second of the three figures linked to Marsh's book was Gifford Pinchot (1865–1946), first chief of the U.S. Forest Service. Pinchot was born to a wealthy family that had made its fortune through lumbering. His father, regretting some of the damage that lumbering was doing to the land, endowed the Yale School of Forestry and encouraged his son to pursue a career in forestry management. Pinchot's ideas centered on his **conservation ethic,** which promoted the efficient use of natural resources, a philosophy that put him on a collision course for conflict with John Muir. In Pinchot's words, "When conflicting interests must be reconciled, the question will always be decided from the standpoint of the greatest good of the greatest number in the long run."

preservation ethic
An environmental ethic emphasizing the protection of natural ecosystems in their original unspoiled states.

conservation ethic
A philosophy of resource management that promotes the efficient use of natural resources to provide the greatest good to the greatest number of people.

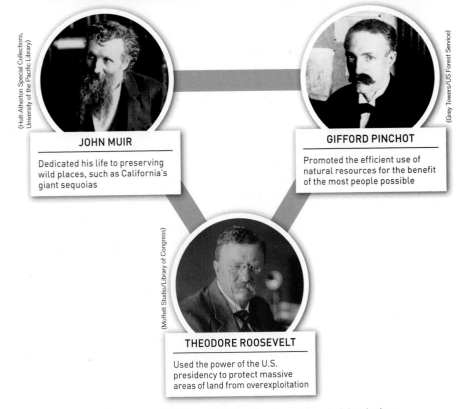

FIGURE 1.20 With the writings of George Perkins Marsh serving as their inspiration, John Muir, Gifford Pinchot, and Theodore Roosevelt worked tirelessly throughout their lives to protect the natural heritage of the United States.

This philosophy guides the management of the U.S. Forest Service to this day. During the course of his professional life, Pinchot worked with several U.S. presidents, including Theodore Roosevelt (1858–1919). Roosevelt, a legendary outdoorsman who was president from 1901 to 1909, was deeply moved by Marsh's book. In a speech on October 4, 1907, he asserted, "The conservation of natural resources is the fundamental problem. Unless we solve that problem, it will avail us little to solve all others." **Conservation** includes the preservation, wise use, or restoration of species, ecosystems, or natural resources. He created 150 national forests, 5 national parks, and 18 national monuments, along with numerous other protected areas. In total, Roosevelt protected over 930,000 square kilometers (360,000 square miles), an area equal to the combined territories of California, Oregon, Washington, and Ohio. While the environmental harm that Roosevelt, Pinchot, and Muir campaigned to avoid on the landscape is visually obvious, other, less visible human threats to the environment were still on the horizon.

Rachel Carson's Warnings

Born in the Allegheny Valley, Rachel Carson (1907–1964) was a child of the countryside. Carson entered the Pennsylvania College for Women as an English major, changing her major to zoology in her junior year. After obtaining a master's degree at Johns Hopkins, she found work as a writer and editor for the U.S. Bureau of Fisheries, where she began writing magazine articles and books (**Figure 1.21**). In 1962 she published what would become her best-known book, *Silent Spring,* which warned of the potential dangers of indiscriminate use of chemical pesticides, a practice that became common following World War II. DDT (dichlorodiphenyltrichloroethane) was first widely used during the war as an insecticide to control malaria, and several other insecticides were created during the late 1940s and 1950s. Hailed as the ultimate weapon in the fight against insect pests, the application of these chemical poisons was killing far more than the targeted insects, including birds and fish.

As a result, Carson and many others became concerned about their potential to harm the environment. They were also alarmed that heavy spraying of chemical insecticides threatened humans, since some of them, including DDT, had been classified as chemical carcinogens. In her book, Carson urged caution in our application of chemical insecticides. She warned that we might inadvertently destroy the birds that for millions of years had greeted the day with a "dawn chorus," producing, in our attempts to control insects, a "silent spring."

The chemical industry that made the pesticides, and the agricultural entities that made use of the pesticides, quickly went on a campaign to vilify Carson and other concerned environmentalists. They led organized attacks on *Silent Spring,* attempting to discredit both the book and its author. Carson was portrayed as a "hysterical woman," a "liar," and a promoter of "junk science."

Her opponents also asserted that she was not qualified to write the book. Carson's position, however, was vindicated in 1963 by President John F. Kennedy's Science Advisory Committee, which agreed that pesticides had been abused and represented a threat to the environment. Carson didn't live to see the passing of the Clean Water Act in 1972 or the banning of DDT in the United States that same year. But, as a consequence of the DDT ban, the peregrine falcon recovered from the brink of extinction and the bald eagle, the symbol of the nation, is common once again across the lower 48 states and off the endangered species list (**Figure 1.22**).

⚠ Think About It

1. What do early human impacts on forests, which began more than 2,000 years ago, suggest about the historic relation of economic activity and environmental impact?

2. At the scale of the entire globe, the huge mismatch between food production and the human population that Malthus predicted has

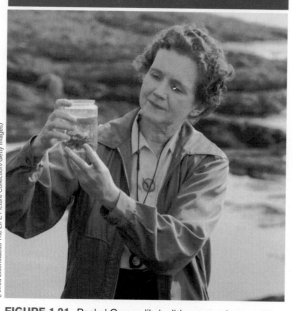

RACHEL CARSON, SCIENTIST, AUTHOR, AND ADVOCATE FOR PRECAUTION IN THE USE OF CHEMICAL PESTICIDES

(Alfred Eisenstaedt/The LIFE Picture Collection/Getty Images)

FIGURE 1.21 Rachel Carson likely did more to alert people to the potential dangers of chemical pesticides than anyone in history.

conservation The preservation, wise use, or restoration of species, ecosystems, or natural resources.

THE BALD EAGLE: SYMBOL OF A NATION AND IMPRESSIVE BENEFICIARY OF THE DDT BAN

FIGURE 1.22 Before the ban on DDT use, accumulation of DDT in the prey of bald eagles had reduced reproduction by the eagles, driving them to near extinction in the contiguous United States.

not yet occurred. What does this suggest about the assumptions made by Malthus? Since his dire predictions failed to materialize, can we safely ignore the ideas he developed?

3. How might the destruction of the great bison herds of North America and the near extirpation of many other wildlife species, all of which occurred within the lifetimes of John Muir, Gifford Pinchot, and Theodore Roosevelt, have affected the development of the conservation movement in the United States?

4. How might the precautionary principle have influenced Rachel Carson's recommendations for pesticide use?

1.6 Human impact on the environment has become a global issue

For much of human history, environmental impacts were local or regional. With the rapid growth of the human population during the past two centuries, however, our effects on the environment have extended to the entire planet.

Earth Day

The change to a broader environmental perspective is reflected in the history of Earth Day (**Figure 1.23**). Gaylord Nelson, U.S. senator from Wisconsin, began

promoting the idea of a day to celebrate the environment and inform people about environmental issues in the early 1960s. The first Earth Day celebration was held on April 22, 1970. On that day, grassroots organizations across the United States organized events in which 20 million people participated, fully 10% of the U.S. population. In his speech in Denver on the first Earth Day,

EARTH DAY: A GLOBAL EVENT

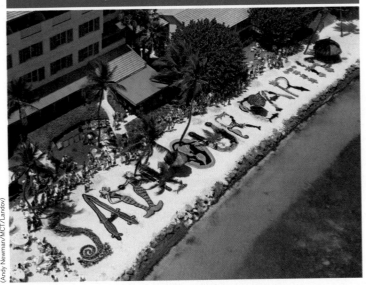

FIGURE 1.23 The Earth Day celebration, which has grown from an event centered on the United States to one including almost all nations on Earth and involving over 1 billion participants, helps sustain awareness of major environmental issues.

CONTRIBUTORS TO ECOLOGICAL FOOTPRINT

How do you explain the rapid growth in participation in Earth Day celebrations?

How do the human impacts on the environment not included in ecological footprint calculations affect our actual footprint?

CARBON
The area of forest needed to absorb and store CO₂ produced by burning of fossil fuels, minus the amount absorbed by the oceans.

CROPLAND
Area of land for growing crops for human use and for feeding domestic animals.

GRAZING LAND
Area of land used for grazing of livestock raised as sources of milk, cheese, meat, wool, and other products.

FOREST
Forest area required to produce timber, firewood, wood pulp, and other wood products.

BUILT-UP LAND
Area of land covered by houses, roads, factories, water reservoirs, and other forms of human infrastructure.

FISHING GROUNDS
Area of freshwater and marine ecosystems needed to support harvests of fish and shellfish.

ECOLOGICAL FOOTPRINT

FIGURE 1.24 The human ecological footprint, which can be scaled from individuals and national populations to the entire global population, is calculated based on six factors. (After WWF [World Wide Fund for Nature], 2012)

ecological footprint
The environmental impact of a human population as the area of land and sea needed to produce the resources it consumes and to absorb the wastes it produces.

renewable resources
Natural resources, such as wood, forage, or fish, that are replaced through natural processes on relatively short timescales and thus can last indefinitely under careful management.

nonrenewable resources
Natural resources, such as fossil fuels, that exist in a limited supply and are not renewed on timescales meaningful to humans.

Nelson revealed that he was thinking broadly about what makes an environment healthy when he said, "Our goal is not just an environment of clean air and water and scenic beauty. The objective is an environment of decency, quality and mutual respect for all other human beings and all other living creatures." By 2009 Earth Day was celebrated in 174 countries through more than 15,000 organizations, involving more than 1 billion people, or nearly 15% of the global population. As the Earth Day celebration grew, so did our sophistication in measuring our influence on the environment.

Ecological Footprint

One measure of human impact on the environment is called the **ecological footprint,** developed by Mathis Wackernagel of the Global Footprint Network and William Rees of the University of British Columbia. An ecological footprint represents the environmental impact of a human population as the area of land and sea needed to produce the resources it consumes, the area covered by infrastructure, such as buildings and roads, and the area of forest needed to absorb carbon dioxide emissions.

As shown in **Figure 1.24,** this index of environmental impact is based mainly on a population's use of **renewable resources,** such as wood, crops, forage for animals, fish for human consumption, and forests for absorbing carbon dioxide. In principle, renewable resources can last indefinitely, since they are replaced through natural processes on relatively short timescales. The ecological footprint does not take into account a population's use of **nonrenewable resources,** such as minerals or fossil fuels, except where their use impacts renewable resources and CO₂ emissions.

The Global Footprint Network has estimated that there were approximately 12 billion hectares of productive Earth surface in 2008, while the ecological footprint of the global population required 18.3 billion hectares (**Figure 1.25**). In other words, human use of natural resources was approximately 50% higher than Earth's capacity to replace them. As a result, it took one and a half years for Earth to replace the resources consumed, and to absorb the CO₂ emissions of the global population

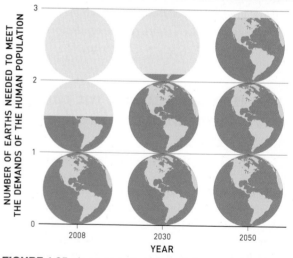

OUR GROWING ECOLOGICAL FOOTPRINT

FIGURE 1.25 At present rates of growth in resource use, the productive capacity of nearly three Earths will be required to meet the demands of human populations in 2050. (Data from WWF [World Wide Fund for Nature], 2012)

in just one year. According to the 2012 Living Planet Report, our ecological overdraft will rise to a bit over 100% by 2030 and to nearly 200% by 2050. At that point, if present trends continue, the productive capacity of three Earths will be needed to meet our single Earth's human demand.

Ecological footprints can also be calculated for individuals and countries. For instance, individuals in rich, highly developed nations, in general, have a much larger ecological footprint than do individuals in developing nations. If each of the 7 billion humans on Earth in 2012 had the ecological footprint of an average person in the United States, approximately four Earths would be required to support them. Calculations of humanity's ecological footprint demonstrate that concerns over our impact on the environment are justified. The growing impact of humans on the environment presents one of the greatest challenges ever faced by our species. What do you think we should do in the face of this challenge? How can we address the immense problems that we face?

⚠ Think About It

1. What is the value of a movement such as the Earth Day celebrations? What is the significance of the massive participation in Earth Day celebrations around the world?

2. How does the ecological footprint approach to estimating the mismatch between human needs and Earth's productive capacity differ from that of Malthus?

1.5–1.6 Issues: Summary

Human impact on the environment has gradually increased through the course of our history. Awareness of the environmental harm caused by human populations increased substantially in the 18th through 20th centuries as a result of the work of Benjamin Franklin, Thomas Malthus, George Perkins Marsh, and others. John Muir, Gifford Pinchot, and U.S. President Theodore Roosevelt were active in land preservation and conservation. Rachel Carson warned of the potential dangers of indiscriminate use of chemical pesticides in her book *Silent Spring.* Awareness of environmental issues has grown beyond local and regional perspectives to include the entire Earth. The global extent of environmental concern is reflected in the growth of the Earth Day celebration. The impact of the human population can be measured by the ecological footprint, which indicates that human pressures have grown rapidly beyond the capacity of Earth to supply renewable natural resources and absorb carbon dioxide.

1.7–1.9 Solutions

Science has taught us much about the natural world, the environment, and the challenges that we face in managing our resources. Science can also lead to technologies and policies used to avoid or repair environmental harm. However, science is silent about the environmental decisions you and I need to make as a society, which can have economic consequences, and about those things we call truth, beauty, right, and wrong. It does not instruct us on how we should treat living systems, as individuals or societies, nor does it tell us what value we

should place on the natural world. For answers to questions such as these, we need to look elsewhere.

1.7 Environmental ethics extends moral responsibilities to the environment

The area of philosophy that concerns the moral relationship of humans to the environment, including all of its human and nonhuman parts, is called **environmental ethics.** Our

environmental ethics
The branch of philosophy that concerns the moral responsibilities of humans with regard to the environment.

ENVIRONMENTAL ETHICS PERSPECTIVES AND MORAL RESPONSIBILITIES

ANTHROPOCENTRIC ENVIRONMENTAL ETHICS

Evaluates the impact of environmental change by its potential effects on humans.

BIOCENTRIC ENVIRONMENTAL ETHICS

Extends moral responsibility to all living organisms.

ECOCENTRIC ENVIRONMENTAL ETHICS

Extends moral responsibility to entire natural systems, including the organisms and nonliving parts of the system.

FIGURE 1.26 Anthropocentric, biocentric, and ecocentric ethics specify areas of moral responsibility to the environment, ranging from a human-centered perspective to one that includes all parts of the natural world, both biotic and abiotic.

anthropocentric Human-centered; for example, human-centered environmental ethics emphasizes impacts on humans.

biocentric Centered on life in all its forms; for example, biocentric environmental ethics extends moral obligation to all life.

ecocentric Centered on entire ecosystems; for example, ecocentric environmental ethics extends moral obligation to the nonliving components of the environment, emphasizing the integrity of whole natural systems.

views about the gravity of environmental problems and the need for solutions to them will emerge from ethical worldviews (**Figure 1.26**).

Worldviews

Perspectives on nature that place humans at the center are called **anthropocentric,** or human-centered. Anthropocentric perspectives generally view humans not as part of nature but as owners of nature, which exists for our benefit. With such a worldview, it may be completely permissible to cause the extinction of species if the actions associated with those extinctions directly or indirectly benefit humans. It has been suggested that anthropocentric worldviews are the cause of most environmental harm. However, some anthropocentric

philosophers have argued that causing a species extinction is wrong because it may lead to the harming of humans as well.

By contrast, the **biocentric** view emphasizes the effects of actions on all living organisms. From a biocentric perspective, cruelty to animals, needless destruction of plants, and, by extension, extinction of species are morally wrong because organisms have intrinsic value. The **ecocentric** environmental ethic extends moral obligation beyond organisms to the nonliving components of the environment and emphasizes the integrity of whole natural systems. For example, a biocentric view would posit that if a site is disturbed by mining, the native species that had occupied the site before the disturbance should be restored. By contrast, an ecocentric position would be that

HETCH HETCHY VALLEY: THE FOCUS OF MORE THAN A CENTURY OF ENVIRONMENTAL CONFLICT

Hetch Hetchy Valley before damming

Hetch Hetchy Valley after damming

FIGURE 1.27 The controversy centered on the Hetch Hetchy Valley pitted Muir's preservation ethic against Pinchot's conservation ethic.

the entire landscape should be restored to its original form, including depth and texture of topsoil, patterns of drainage channels, and so forth.

Aldo Leopold (1887–1948), a major figure in environmental science, argued specifically for an ecocentric position when he proposed a **land ethic.** In his most famous 1949 book *A Sand County Almanac,* Leopold wrote, "A thing is right when it tends to preserve the integrity, stability, and beauty of the biotic community. It is wrong when it tends otherwise."

Environmental Ethics in Action

One of the most famous clashes of environmental ethics involved the creation of a reservoir in the Hetch Hetchy Valley, a part of Yosemite National Park just north of the more famous Yosemite Valley (**Figure 1.27**). After John Muir first visited Hetch Hetchy in 1872, he wrote about it as a second Yosemite and campaigned relentlessly for the preservation of both it and the Yosemite Valley on the basis of an ecocentric environmental ethic. In 1890 Yosemite National Park, which included both valleys, was established. However, James Phelan, the mayor of San Francisco, proposed damming the Hetch Hetchy Valley to create a reservoir to supply the water needs of the city. He found an ally in Gifford Pinchot, the chief of the U.S. Forest Service, whose anthropocentric conservation ethic emphasized the wise use of natural resources for the benefit of the largest number of people.

Battle lines were drawn. Muir and the Sierra Club conducted a national campaign to defend the valley from the damming. Muir's ecocentric ethics represented the management philosophy of the National Park Service, which is dedicated to the goal of preserving intact wild lands. On the other side, Pinchot's position represented the U.S. National Forest, which is dedicated to managing the national forests for multiple uses, including economic gain. Pinchot won the battle, and in 1913 Congress authorized damming the valley. However, the struggle continues to this day as the Sierra Club hopes to remove the dam and restore the Hetch Hetchy Valley to its natural state.

Environmental Justice

Historically, environmental burdens, such as the location of waste dumps or highway systems, have been borne mainly by those most disadvantaged, because of socioeconomic status, race, or ethnicity. The **environmental justice** movement arose in the United States in response to these inequities.

The birth of the environmental justice movement is generally associated with a 1982 protest against a proposed chemical landfill in poor, mainly African American, Warren County, North Carolina. During the six weeks of protest against the landfill, 500 people were arrested (**Figure 1.28**). Although the toxic waste was eventually deposited at the proposed site, the protests sensitized the general public and policy makers, inspiring a movement

What does a person's opinion about whether or not Hetch Hetchy should be restored to its wild state reveal about his or her environmental ethics?

land ethic An ecocentric system of environmental ethics proposed by Aldo Leopold to promote the integrity, stability, and beauty of the biological community.

environmental justice The fair treatment and meaningful involvement of all people in the development, implementation, and enforcement of environmental laws, regulations, and policies.

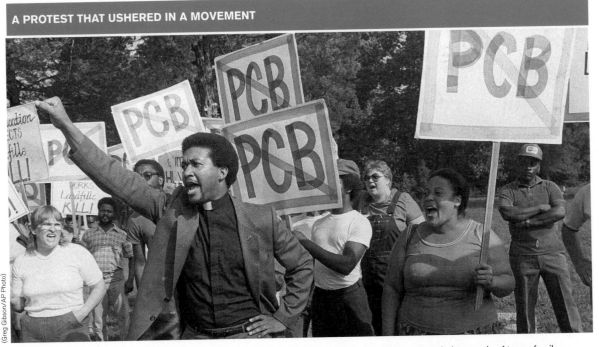

A PROTEST THAT USHERED IN A MOVEMENT

(Greg Gibson/AP Photo)

FIGURE 1.28 Weeks of protests in Warren County, North Carolina, in 1982 over plans to deposit thousands of tons of soil contaminated with toxic material in a landfill resulted in arrests of more than 500 protestors. The protest garnered national attention and inspired an international movement with the goal of working toward environmental justice for all people.

for fairness in environmental decisions. Pressures for environmental justice pushed the major environmental organizations, such as the Sierra Club, to place issues arising from the siting of landfills, toxic waste dumps, and other sources of pollution alongside their traditional focus on preservation of wilderness and other natural areas.

Why did it take nearly a century after the founding of Yosemite National Park for organizations like the Sierra Club to support the concept of environmental justice?

Religious and Cultural Perspectives

Your religious beliefs and cultural background will undoubtedly shape your own particular take on environmental issues. Nevertheless, a cursory review of belief systems around the world shows that people everywhere place value on protecting the environment (**Figure 1.29**). If cultures have converged on the conclusion that we should care for the environment,

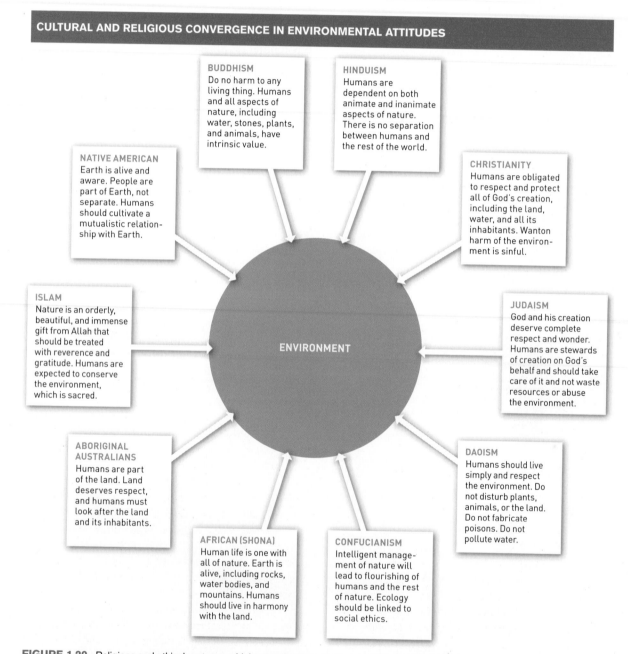

CULTURAL AND RELIGIOUS CONVERGENCE IN ENVIRONMENTAL ATTITUDES

BUDDHISM
Do no harm to any living thing. Humans and all aspects of nature, including water, stones, plants, and animals, have intrinsic value.

HINDUISM
Humans are dependent on both animate and inanimate aspects of nature. There is no separation between humans and the rest of the world.

NATIVE AMERICAN
Earth is alive and aware. People are part of Earth, not separate. Humans should cultivate a mutualistic relationship with Earth.

CHRISTIANITY
Humans are obligated to respect and protect all of God's creation, including the land, water, and all its inhabitants. Wanton harm of the environment is sinful.

ISLAM
Nature is an orderly, beautiful, and immense gift from Allah that should be treated with reverence and gratitude. Humans are expected to conserve the environment, which is sacred.

ENVIRONMENT

JUDAISM
God and his creation deserve complete respect and wonder. Humans are stewards of creation on God's behalf and should take care of it and not waste resources or abuse the environment.

ABORIGINAL AUSTRALIANS
Humans are part of the land. Land deserves respect, and humans must look after the land and its inhabitants.

DAOISM
Humans should live simply and respect the environment. Do not disturb plants, animals, or the land. Do not fabricate poisons. Do not pollute water.

AFRICAN (SHONA)
Human life is one with all of nature. Earth is alive, including rocks, water bodies, and mountains. Humans should live in harmony with the land.

CONFUCIANISM
Intelligent management of nature will lead to flourishing of humans and the rest of nature. Ecology should be linked to social ethics.

FIGURE 1.29 Religions and ethical systems, which arose independently around the world, developed similar perspectives on respect and care of the environment. (From Selin, 2003)

how do we explain the environmental damage that we can see around the world?

In some cases, harm to the environment may be the result of people, with few alternatives, trying to make a living or simply trying to survive. Environmental harm may also result from incomplete knowledge. For example, the manufacturers of the CFCs that have depleted the ozone layer did not know that their product would endanger this protective layer. In other cases, harm to the environment results from the practice of pursuing economic gain, while disregarding the environmental impacts of that activity. In addition, harm to one part of the environment must often be weighed against the benefits to another part of the environment or to human well-being. For instance, the potential harm that chemical insecticides may cause for wildlife must be weighed against the potential to increase food supplies or control populations of disease-carrying insects. Seldom are there simple solutions to the environmental problems that we face.

⚠ Think About It

1. What adjustments in moral responsibilities must be made in order to shift from an anthropocentric ethic to an ecocentric ethic? Which is better for human welfare? Does taking a short-term versus long-term frame of reference make a difference?

2. What is the relationship of the environmental justice movement and the various other environmental ethics perspectives?

3. What is the significance of values in addressing environmental issues? What are the limitations?

1.8 Sustainability as a pragmatic solution to environmental woes

As you learned earlier, scientists in the mid-1980s recognized that the release of CFCs had depleted the ozone layer so significantly that it was as if a hole had formed. The amount of ozone in the stratosphere is sustained by a balance between its formation, chemical reactions initiated by ultraviolet light from the Sun, and its destruction through a variety of processes. With the release of CFCs, the rate of ozone destruction exceeded its rate formation, resulting in the ozone hole over the Antarctic. Ozone depletion directly threatened human health, because an increase in ultraviolet rays reaching Earth's surface would increase the incidence of skin cancer.

When we use resources faster than they are replenished, our societies and economies are not sustainable. Because everything that humans need to survive comes from the natural environment, the principle of **sustainability** is about the wise use of resources to ensure our ability to endure and live healthy lives, without compromising the welfare of future generations. For instance, rather than relying on nonrenewable sources of energy, such as oil and gas, a sustainable strategy incorporates renewable energy such as solar and wind power. When we harvest from wild populations, as is the case in fisheries and forestry, we should do so cautiously in order to allow these populations to reproduce. When we use pesticides on farms, we need to ensure that they do not pollute the groundwater or disrupt the pest control services provided by other organisms. And when we expand cities and agriculture in dry regions, such as the desert Southwest, we will benefit from technologies and policies that conserve water. In short, sustainability is about reducing our ecological footprint in order to make sure that humans and nature can coexist.

On September 16, 1987, President Ronald Reagan and other world leaders made an important step toward creating sustainable policies to protect the ozone layer. Reagan agreed to the Montreal Protocol, which would reduce the production of CFCs and encourage the development of alternative refrigerants and propellants. Reagan, who himself had been treated for skin cancer, recognized the importance of sustainable policies in protecting human health and nature. In the beginning, the Protocol sought only to reduce the global production of the worst CFCs by 50%, but scientists soon recognized that no level of CFC production was sustainable; thus, CFCs were banned completely. Since that time, the Montreal Protocol has been adjusted five times to regulate almost 100 gases, and it is considered one of the most successful global environmental treaties in history, as well as a model for how countries might reduce greenhouse gas emissions. In fact, recent research has even suggested that the reduction in CFCs over the last 15 years has also slowed global warming, which we will later see threatens humans and nature in other ways.

Sustainability provides a pragmatic framework for solving environmental problems. It allows us to set aside differences in environmental philosophies and focus on the common goal of maintaining our health and prosperity, ensuring the survival of future generations.

sustainability The wise use of resources to ensure our ability to endure and live healthy lives, without compromising the welfare of future generations.

⚠ Think About It

1. Consider the damming of Hetch Hetchy. It flooded a valley in Yosemite, but provided freshwater to the people of San Francisco. Would you consider it an example of sustainable or unsustainable development?

2. Does the principle of sustainability fall under an ecocentric environmental ethic or an anthropocentric one?

Is any subject area making up the complex discipline of environmental science more significant than the others in addressing environmental issues?

environmentalism An ideological and social movement that advocates the protection of the environment from human harm through political action and education.

1.9 Environmental science provides a comprehensive framework for addressing environmental issues

Because of the complexity of the problems it addresses, environmental science is highly interdisciplinary (**Figure 1.30**). It draws from all the natural sciences, including chemistry, physics, biology, geology, meteorology, and climatology. And since it addresses practical problems, environmental science also draws heavily from the applied sciences, such as engineering, agricultural science, and toxicology. In addition, literature, art, and music provide the means to communicate the human significance of environmental issues. For instance, without Rachel Carson's highly developed writing skills, her influence on environmental policy would likely have been far less impactful.

How ozone depletion was addressed illustrates how multiple disciplines and interests contribute to solving complex environmental issues. Journalism, for instance, played a fundamental role. In 1974 Molina and Rowland published their first paper, describing how CFCs could contribute to the breakdown of the ozone layer. When their article in *Nature* drew little response, they issued a press release and held a news conference to warn the public and policy makers about the potential danger to the ozone layer.

These actions went beyond the boundaries of environmental science and into the realm of **environmentalism,** an ideological and social movement advocating protection of the environment from human harm through political action and education. Their press conference aroused intense public concern and more people wrote letters to the U.S. Congress than in response to any other issue in U.S. history up to that point except the Vietnam War. The public outcry created enough political pressure to move the federal government to action, passing legislation in 1977 to regulate any substances "reasonably anticipated to affect the stratosphere."

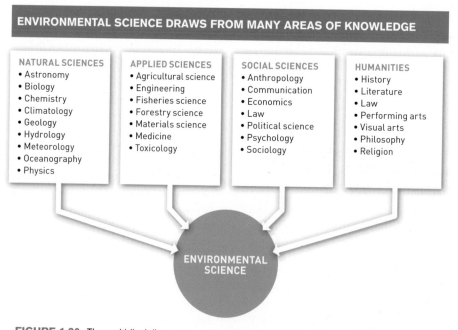

FIGURE 1.30 The multidisciplinary nature of environmental science is one of the factors contributing to the complexity of environmental science and the issues it addresses.

The business community generally plays a key role in addressing environmental issues, particularly where those issues collide with economic interests. Although chemical companies eventually accepted the results, they spent a decade trying to discredit Molina and Rowland's research. In the end, it was industrial chemists who developed a variety of substitutes for CFCs, which pose little or no threat to the ozone layer.

Where environmental impacts extend beyond national borders, diplomacy, negotiation and international treaties are essential. Discovery of the ozone hole in 1984 spurred worldwide concern. International regulation of CFC production eventually involved the United Nations and negotiations among countries around the world. The result was the Montreal Protocol.

In the coming chapters, we will discuss the most pressing contemporary environmental issues, build the science necessary to engage with those issues, and explore some of the solutions to them (see Figure 1.13). We will necessarily draw on many subject areas as we do so. As a result of drawing from so many areas of knowledge and skill, environmental science can provide a path to solving environmental issues, which threaten not just nature, but our health and livelihoods as well.

⚠ Think About It

1. The environmental science framework includes natural sciences, applied sciences, social sciences, and humanities. What roles do each of these areas of knowledge play in solving environmental problems?

2. Is environmental science in fact a "science"? If so, what type of science is it? If not, how would you classify the subject?

1.7–1.9 Solutions: Summary

Perspectives on nature and the environment range from very human-centric to large-scale, natural system–centric. Throughout history, these differing viewpoints have led to disputes about how to manage the natural environment. Though differing in details, religious and cultural traditions from all parts of the world advise protecting Earth and its diverse inhabitants. The environmental justice movement arose to combat historical inequities in the extent to which people are exposed to environmental hazards.

Sustainability includes developing policies, practices, and technologies to provide us with enough clean water, food, energy, and other resources crucial to human health over the long term. Environmental science can help us achieve this, by drawing on findings and ideas from the natural sciences, applied sciences, social sciences, and humanities.

Answer the following questions for each chapter section and then answer the Central Question.

Central Question: How do science and values help address environmental issues?

1.1–1.4 Science

- **What factors make up the environment?**

- **What is science and how is it useful?**

- **What is uncertainty and how can we use science to address it?**

- **Describe ethics as they relate to science and how can we uphold them?**

1.5–1.6 Issues

- **How have humans' impact and environmental awareness changed throughout history?**

- **In what ways is environmental impact by humans a global issue?**

Science, Environment, Values, and You

Chapter 1 explores what science is, how it works, and how science and values affect the way we address environmental issues. So that you can critically evaluate the science, issues, and solutions in the remainder of the text, here are some ways to dig deeper into what was covered in this chapter.

☐ *Practice the scientific process.*

Since the core of environmental science is scientific research, the more you get to know the process of science, the better you'll be able to evaluate evidence related to environmental issues. Ideally, you can involve yourself in scientific research at your college or university or with an independent research group. Your environmental science instructor may know of opportunities for such work, either as a volunteer or as a paid technician. Alternatively, your instructor may give you the opportunity to do a research project as part of your course.

☐ *Familiarize yourself with the scientific literature and practice thinking like a scientist.*

Whenever you encounter an environmental issue in the news, try to find the scientific research paper at the heart of the story. As you do so, keep track of the questions and hypotheses posed by the researchers. Review the research methods used, including the study design. Evaluate the results obtained and consider whether the conclusions reached by the authors are supported by the actual results. Propose additional research to learn more about the subject of the investigation.

☐ *Calculate your personal ecological footprint.*

Use the online footprint calculator at www.footprintnetwork.org to estimate the number of global hectares required to support your particular lifestyle. Using that website's ecological footprint calculator, you can also explore different living scenarios to see what you might personally do to reduce your own demands on the environment.

☐ *Determine your personal environmental ethic.*

Explore the world of environmental ethics and try to determine your own ethical perspective on the environment. If you are a member of an organized religion, learn more about its perspective on human moral responsibilities to other species and to the environment as a whole. Research other religious or ethical systems to learn of similarities and contrasts in values.

1.7–1.9 Solutions

- How do people's personal views and ethics affect how they value the environment?

- What is sustainability and how can it help us address environmental problems?

- What is environmental science and how does it address human impact?

Answer the Central Question:

_____

_____

_____

_____

_____

_____

Chapter 1
Review Questions

1. When was thinning of the ozone over the Antarctic first verified?
a. 2010
b. 1985
c. 1974
d. 1960

2. Which of the following is part of the environment experienced by humans?
a. Physical factors
b. Chemical factors
c. Cultural factors
d. All of the above

3. Which of the following is essential to any scientific investigation?
a. Field observations
b. Experiments
c. A hypothesis
d. Modeling

4. A theory, as used by scientists, is best described by which one of the following?
a. An idea that may have no support by research
b. An explanation for a natural phenomenon, widely supported by research results
c. A tentative answer to a research question
d. A proven idea with no uncertainty remaining

5. Which one of the following is not a form of research misconduct?
a. Not ordering sufficient research supplies for a particular experiment
b. Making up data to yield more interesting research
c. Copying the work of others and claiming it as your own
d. Altering the calibration of scientific instruments to produce false positive readings

6. When were some of the earliest observations of human environmental impact recorded in writing?
a. 10,000 years ago
b. Over 2,000 years ago
c. In the 1700s
d. In the 1960s

7. What eventually scuttled Benjamin Franklin's attempts to have the tanneries moved from the center of Philadelphia?
a. A lack of interest by the majority of Philadelphians living near the tanneries
b. The installation of pollution-control devices on the tannery discharge pipes
c. The belief that the odors of the tanneries were a source of good health
d. Entrenched economic and political interests supporting the tanneries

8. What did Rachel Carson advise regarding the use of chemical pesticides?
a. Total banning of chemical pesticides in the United States
b. Total banning of chemical pesticides around the world, including in the United States
c. Caution in the use of chemical pesticides
d. Greater use of chemical pesticides, particularly for control of significant agricultural pests

9. Which of the following is the most inclusive ethical perspective on the environment?
a. Anthropocentric ethics
b. Biocentric ethics
c. Ecocentric ethics
d. Research ethics

10. Which of the following best describes environmental science?
a. A scientific discipline concerned with controlling pollution
b. An interdisciplinary field that reaches beyond the natural sciences to address practical environmental problems
c. The study of the relationships between organisms and the environment
d. A largely political field concerned mainly with controlling human population growth

Critical Analysis

1. Scientists once thought they were on a quest to discover the absolute unchanging rules governing natural phenomena. Increasingly, scientists in every discipline have had to come to terms with the concept of uncertainty, a concept that even Albert Einstein had difficulty accepting. What are some of the implications of the inability of science to eliminate uncertainty entirely, particularly in regard to environmental issues, such as ozone thinning, involving great risks to harm human populations and with significant potential economic consequences following regulation?

2. Exploring nature and solving complex problems are often important motivations for scientists choosing a career path. For those successful in obtaining a position in scientific research, what responsibilities come with the privilege of working in this highly competitive field?

3. Awareness of human environmental impact was first recorded in ancient China and Greece, and that awareness has only increased over time, with the environment in the news nearly every day. What factors have contributed to the apparent increase in impact on the environment by humans? Do these contributory factors suggest potential ways to reduce environmental harm?

4. Figure 1.24 summarizes the components of the ecological footprint. If you were to propose an alternative measure of ecological footprint, what components would you choose? If you conclude that the existing ecological footprint index is the best one, without modification, justify its focus.

5. Why does environmental science include so many areas of thought and expertise in its domain? What are the relative roles played by the natural and applied sciences versus the social sciences and humanities in addressing environmental issues? Can we ignore the humanities and social sciences and just get on with the solving of environmental problems with the tools of the natural and applied sciences?

Find additional resources and links online at www.macmillanhighered.com/launchpad/molles1e.

Central Question: How can linking ecology and economics help reduce societies' environmental impacts?

Explain the nature and movement of matter and energy in ecosystems and economic systems.

SCIENCE

Ecosystems and Economic Systems

Analyze the environmental significance of energy demand, economic models, and the Tragedy of the Commons.

ISSUES

Discuss how linking environment to economics, property rights, and community-based management can affect environmental impacts.

SOLUTIONS

(Dieter Telemans/Panos)

The Maasai people, who are descended from cattle herders that migrated southward from the Nile River Valley many centuries back, long ago integrated with neighboring cultural groups in a traditional economic system based on bartering.

The Maasai People and the Economy of Nature

The Maasai way of life in the savannah ecosystem developed into a web of economic relations.

Life on the grassy plains of East Africa, considered the birthplace of humanity, changed dramatically when domesticated cattle arrived. In the 15th or 16th century, the Maasai people—who clothe themselves in red cloth and pierce their earlobes with thorns and elephant ivory—brought their cattle with them from the Nile River Valley in the north and settled in the savannahs of Kenya and Tanzania. Cattle provided the Maasai with nearly everything they needed: food in the form of milk, blood, and meat; dung for plastering the walls of huts; and hides for making shoes and other items. Cattle hold religious significance for the Maasai and are also a symbol of wealth

used as currency—to pay a family for their daughter's hand in marriage, for example.

Cattle, in other words, were prominent in the material culture of East Africa. For thousands of generations, humans in the region had survived by hunting wild antelope, and gathering fruits and nuts. But with cattle, the Maasai significantly modified life on the plains to better provide for their needs. This new stability allowed them to develop a complex society that included procedures for decision making by councils of experienced elders, division of labor, trade within their own and neighboring cultures, and a military system for defending their resources. For instance, the Maasai negotiated with neighboring people to allow movement of their herds in time of drought; Maasai women traded excess milk, meat, and skins for bananas, maize, and sweet potatoes grown by farming neightbors.

> "We must put the 'eco' back into economics and realise what the conditions and principles are for true sustainable living."
>
> David Suzuki, "The Challenge of the 21st Century: Setting the Real Bottom Line," 2008

As material culture developed, the Maasai joined with farmers and hunters in the area via an **economic system.** An economic system consists of a network of people, institutions, and commercial interests involved in the production, distribution, and consumption of goods and services. These principles apply to cattle trading and modern wealth economies alike.

economic system A network of people, institutions, and commercial interests involved in the production, distribution, and consumption of goods and services.

The Maasai livelihood continued to depend on climate and the condition of the soil, but it was also shaped by these social innovations. It turns out, there is a surprising similarity between how natural environments function and the modern economic systems that sprang from human civilization. Ernst Haeckel, who coined the term *ecology* in the 19th century, said, "By ecology we mean the body of knowledge concerning the economy of nature." We might also think of economics as the ecology of humans. Understanding the connection between economics and ecosystems is one of the key ways that environmental science allows us to find solutions to global challenges.

Though a modern economy may seem far different from that of the traditional Maasai, the differences result mainly from the level of technological development, the diversity of goods and services being traded, and the sheer amount of goods changing hands every day. Beyond those differences, however, our challenge, like that of the traditional Maasai and their neighbors, is the same: to sustain the healthy economies needed for thriving human societies while protecting the environment on which we depend for survival. This goal is generally referred to as **sustainable development,** development that meets the needs of the present generation without reducing the ability of future generations to meet their needs. This chapter's Central Question reflects the core of that challenge.

sustainable development A process of development that meets the needs of the present generation without reducing the ability of future generations to meet their needs. Development is sustainable when it does not, at a minimum, endanger Earth's natural life support system, including the atmosphere, waters, soils, and biological diversity.

Central Question

How can linking ecology and economics help reduce societies' environmental impacts?

(Jean Michel Labat/Ardea.com)

2.1–2.4 Science

In their traditional life on the grassy plains of Africa, the Maasai did not have major negative environmental impacts on the **ecosystem,** defined as all of the organisms living in a particular place and their interactions with each other and with the local climate and geology (**Figure 2.1**). The ecosystem perspective is not limited to natural environments. For instance, a farm or a city can be viewed as an ecosystem, and the study of urban ecosystems is very active and thus a growing area of ecological research. Historically, however, the study of how human-dominated systems, such as cities, function has been thought of as a topic for economics. **Economics** is the social science concerned with the production, distribution, and consumption of goods and services as well as the theory and management of economic systems.

2.1 Ecosystems and economic systems are built on matter

Whether we are talking about turtles in a pond, cattle dung, or solar panels, the entire material universe is composed of **matter.** Therefore, ecosystems have a material basis. In a forest ecosystem, for example, matter makes up the plants, animals, fungi, and all of the other organisms living in a forest, along with the soils in which they grow and the water contained in that soil. The same is true of economic systems. In an economic system, all the components, from people and buildings to machinery and food to money itself, are made of matter.

What's Matter Made Of?

Matter exists in three main physical states: as a solid, liquid, or gas. Water, which is central to life on Earth, has the rare characteristic of occurring in all three physical states—ice, water, and vapor—at temperatures commonly encountered on Earth (**Figure 2.2**).

The basic building block of matter is the **atom,** which we define as the smallest particle of a pure substance that still retains the chemical and physical properties of that substance. (For a detailed outline of basic chemistry, see

ecosystem The organisms living in a place and the biological, physical, and chemical aspects of the environment with which they interact. Ecosystem ecologists focus much of their research on the flux and transformation of matter and energy.

economics A social science that deals with the production, distribution, and consumption of goods and services, as well as the theory and management of economic systems.

matter Anything that occupies space and has mass; matter exists in three main physical states: as a solid, liquid, or gas.

atom The smallest particle of a substance that still retains the properties of the substance.

AN ECOSYSTEM

(© Stock Illustrations Ltd./Alamy)

FIGURE 2.1 Because it seems so self-contained, a pond may be one of the easiest natural systems to visualize as an ecosystem. A pond ecosystem consists of all the organisms, from bacteria and algae to turtles, inhabiting the pond and all of the biological, physical, and chemical factors with which they interact.

WATER IN ITS THREE PHYSICAL STATES

Ice skater on solid ice (Greg Epperson/Shutterstock)

SCUBA diver in liquid water (Rostislav Ageev/Shutterstock)

Water vapor coming off a natural hot spring (George Burba/Shutterstock)

FIGURE 2.2 Water exists in all three states of matter at temperatures commonly encountered in Earth's environment: as solid ice, as liquid water, and as gaseous water vapor.

Appendix A.) Substances composed of a single type of atom are called **elements.** Just six elements, carbon (C), hydrogen (H), nitrogen (N), oxygen (O), phosphorus (P), and sulfur (S), make up approximately 99% of the mass of all organisms, ranging from bacteria and plants to humans (Table 2.1).

element A substance composed of a single type of atom, such as hydrogen, helium, iron, or lead, that cannot be broken down into simpler substances via chemical or physical means.

TABLE 2.1 THE ATOMIC STRUCTURES AND BIOLOGICAL IMPORTANCE OF SIX ELEMENTS

The basic structure of atoms consists of a nucleus of positively charged protons plus uncharged neutrons, orbited by negatively charged electrons.

Symbol	Name	Structure	Biological Importance
H	Hydrogen		Hydrogen, present in all organic compounds, such as carbohydrates, fats, and proteins and part of water, is a major component of biological structure.
C	Carbon		Carbon is central to all organic compounds and forms the core of biological structure. Life on Earth is based on the chemistry of carbon.
N	Nitrogen		Nitrogen is an essential part of amino acids and therefore of proteins.
O	Oxygen		Oxygen is part of many organic molecules, is a component of water, and is critical for respiration in many organisms.
P	Phosphorus		Phosphorus is an essential part of the structure of RNA and DNA, is part of the energy-bearing molecule ATP, and is important to bone and tooth structure.
S	Sulfur		Sulfur is a key constituent of some amino acids, the building blocks of proteins, and helps determine the structure of enzymes and other proteins.

What are the implications of the fact that the elemental composition of all life forms is so similar?

TABLE 2.2 FOUR COMMON MOLECULES

The *electron shell model* shows the bonds that hold the atoms of a molecule together. Those bonds are formed by the sharing of electrons. The *covalent* bonds shown here involve sharing of one (e.g., carbon to hydrogen) or two (e.g., carbon to oxygen) pairs of electrons.

Type of Representation	Oxygen	Water	Carbon Dioxide	Methane
Molecular formula	O_2	H_2O	CO_2	CH_4
Electron shell model				

Molecules and Chemical Reactions

As atoms interact, they may bind together during chemical reactions to form substances called **molecules** (Table 2.2). For example, water is a molecule of two hydrogen atoms and an oxygen atom. Molecules (e.g., water) that are made up of atoms of two or more different elements are called **compounds.** Methane, which is made up of carbon and hydrogen, reacts with oxygen molecules to produce carbon dioxide and water (Figure 2.3). When we use natural gas, which is mostly methane, to heat our homes or to cook food, we are using the energy released during this reaction. Though atoms can combine in innumerable ways to form endless varieties of molecules, only a few types of molecules, such as proteins, fats, and carbohydrates, are the basis for the structure, energetics, and reproduction of all organisms.

⚠ Think About It

1. What must be true of a molecule for it to be considered a compound?

molecule Two or more atoms held together by chemical bonds; the constituent atoms may be of the same or different elements.

compound A substance composed of a fixed ratio of two or more elements (e.g., water, which consists of two hydrogen atoms and one oxygen atom: H_2O). Compounds can be broken down into the elements of which they are made via chemical or physical processes.

energy The capacity to do work. See *work*.

work A description of the transfer of energy; the work done on an object by a force is determined by the amount of force times the distance the object moves in the direction of the force. See *energy*.

USING MOLECULAR FORMULAS TO PICTURE A CHEMICAL REACTION

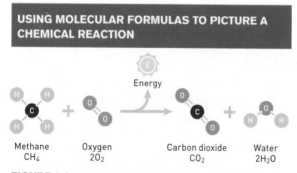

FIGURE 2.3 A molecule of methane, the main ingredient of natural gas, reacts with an oxygen molecule, producing one molecule of carbon dioxide and two molecules of water, and giving off energy in the process.

2. How would changes in global temperature potentially affect the amount of Earth's water in liquid, solid, and gas phases?

3. Are all molecules compounds? Explain using Figure 2.3.

2.2 Energy makes matter move

In a forest, a hawk pursues a fleeing sparrow through the sun-warmed canopy, both twisting and turning in flight as they avoid colliding with branches. In a nearby city, taxis rush headlong through streets to deliver passengers to their destinations. Stockbrokers walk hurriedly down the sidewalk on their way to air-conditioned buildings. All this activity, whether in a forest ecosystem or as part of a modern economy, is fueled by energy.

Energy and Work

Describing a person as having lots of energy means that he or she is vigorous or lively. Physicists use the term in a much more precise way, defining **energy** as the capacity to do **work.** Work is the product of the amount of force applied to an object and the distance the object is moved in the direction of the force. Certainly, a vigorous person moving his or her arms around will be performing a lot of work! Similarly, the amount of work required to put a book on a shelf depends on the weight of the book and the height of the bookshelf (Figure 2.4). Consequently, you burn more calories (i.e., energy) lifting a heavy book to a higher shelf than you would expend to lift a lighter book to a lower shelf. The same is true when it comes to a jet airliner filled with passengers and luggage ascending from sea level to an altitude of 10,000 meters (33,000 feet). That takes a huge amount of work and, consequently, requires substantial energy in the form of jet fuel.

WORK: A MATTER OF FORCE AND DISTANCE

(Jupiterimages/Getty Images)

(Taras Vyshnya/Shutterstock)

FIGURE 2.4 Work is the product of the amount of force applied to an object and the distance the object is moved in the direction of the force. Placing a book on a shelf is an example of a small amount of work, requiring the application of low force over a short distance. By contrast, a great deal of force must be applied to lift a loaded jet airliner from a runway to an altitude of 10,000 meters, which represents a great amount of work.

Use the concept of energy to explain how reading the words on this page and thinking about the concepts discussed qualify as work.

How are a compressed spring and a sugar molecule similar?

Forms of Energy

The energy to perform work comes in many different forms. Think about a battery in your desk drawer. It's not performing any work, but it has the *capacity* to do so, which we call **potential energy.** Likewise, the source of the energy we use to lift our arm and place a book on a shelf is stored as **chemical energy** in the bonds of molecules, such as sugars and fats. As the chemical energy fuels the contraction of your muscles, it is converted into **kinetic energy,** the energy of movement. When our book is placed on a shelf, the energy of placing it there is transferred to the book, as **gravitational potential energy,** which is a consequence of its position above ground. This potential energy would be released as kinetic energy (movement) if the book were to fall off the shelf and crash to the floor.

As fuel is processed, whether in the cells of your body while you are exercising or in a jet engine during takeoff, another form of kinetic energy is released: **Thermal** (or **heat**) **energy** results from the *motion* of molecules in a substance and therefore is a form of kinetic energy. The faster the molecules move, the hotter the substance becomes. When it's cold, our bodies use heat released through shivering to stay warm. Steam locomotives move trains down the track by harnessing the thermal energy in steam to move the pistons that turn their wheels.

Sunlight is made up of **radiant energy,** which is the energy of electromagnetic radiation, including visible light, infrared light, and ultraviolet light, as well as microwaves, radio waves, and X-rays. As the radiant energy from the Sun hits Earth, it drives many different energetic processes, melting snow, driving winds, and causing plants to grow (**Figure 2.5**).

Energy in all its forms is in action all around us. For example, solar energy drives the movement of atmospheric gases that we call wind, another example of kinetic energy. The heat radiating from a sidewalk on a hot day is an example of radiant energy being released by the sidewalk, which has been heated by sunlight. Some solar energy is converted to heat energy stored in the surface water of lakes and oceans, causing some water to evaporate, which transforms liquid water to gaseous water vapor. This water vapor may eventually condense to form clouds, one of the fundamental processes in the hydrologic cycle (see Figure 6.4, page 160). All the forms of energy operating at scales ranging from placing a book on a shelf to global atmospheric processes follow a set of strict physical laws.

The Laws of Thermodynamics

The **first law of thermodynamics** states that when one form of energy is transformed to other forms, the total amount of energy in a system and its surroundings remains the same. For instance, if we were to measure incoming solar energy and then add up the related energy on Earth and its surroundings, including the energy of wind or the energy absorbed by trees and stored in our own bodies, we would find that they are the same (**Figure 2.6**). We call this the *conservation* of energy.

However, if you were to actually measure the kinetic energy of the wind turning the blades of a wind turbine and then the electricity produced by the wind turbine, you would find that there is less electrical energy than wind energy (**Figure 2.7**). How can this be? The first law of thermodynamics states that the total amount of energy

potential energy The amount of energy an object has due to the configuration of its parts (e.g., a loaded spring), its chemical makeup, or its position in a force field (e.g., Earth's gravitational field).

chemical energy A form of potential energy; energy stored in the bonds of molecules, such as sugars, fats, or methane.

kinetic energy The energy of a moving object, which is equal to one-half the mass of the object times the square of its velocity.

gravitational potential energy The amount of potential energy an object contains due to its mass and height above a reference point, such as Earth's surface.

thermal (heat) energy A form of kinetic energy due to molecular motion in a mass of a substance, such as a mass of steam.

radiant energy The energy of electromagnetic radiation, including visible light, infrared light, ultraviolet light, microwaves, radio waves, or X-rays.

first law of thermodynamics A physical law concerned with the conservation of energy: Though one form of energy may be transformed to other forms, the total amount of energy in a system plus its surroundings is conserved; that is, the total amount of energy remains the same. See *second law of thermodynamics.*

SOME OF THE MANY FORMS IN WHICH WE ENCOUNTER ENERGY

KINETIC ENERGY	POTENTIAL ENERGY
Yellowstone Falls — Sunlight filtering through a forest canopy	Coal — A drawn bow

(Stewart Tomlinson/USGS) (Louise McLaughlin/NPS) (Jacek Fulawka/Shutterstock) (Coloroftime/Getty images)

FIGURE 2.5 From a raging hurricane to the opening of a flower on a spring morning, energy, in its many forms, does the work of the world.

in the system plus the surroundings must remain the same. Some of the difference can be accounted for by the production of heat by friction between the moving and stationary parts of the turbine mechanism. If you were to touch the wind turbine, you would find that it is warm.

This brings us to the **second law of thermodynamics,** which states that with each energy transformation, such as the conversion of wind energy to electrical energy by a wind turbine, the amount of energy available to do work decreases because of the production of unusable heat energy. The second law also states that over time, **entropy,** or the amount of disorder in a system, increases. And so maintaining order in a system—whether it's a wind turbine, the cells in your body, your room, or a motor vehicle—requires an input of energy (**Figure 2.8**). The loss of energy with each energy transformation and the need for energy input to keep machines in working

SOLAR-POWERED ATMOSPHERE, HYDROSPHERE, AND BIOSPHERE

second law of thermodynamics With each energy transformation, or transfer, the amount of energy in a system available to do work decreases. In other words, the quality of the energy declines with each energy transfer or transformation. See *first law of thermodynamics.*

entropy A measure of the amount of disorder in a system.

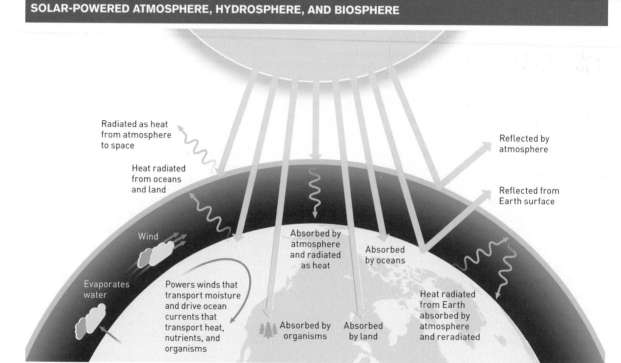

Radiated as heat from atmosphere to space

Heat radiated from oceans and land

Reflected by atmosphere

Reflected from Earth surface

Wind

Evaporates water

Powers winds that transport moisture and drive ocean currents that transport heat, nutrients, and organisms

Absorbed by atmosphere and radiated as heat

Absorbed by oceans

Absorbed by organisms

Absorbed by land

Heat radiated from Earth absorbed by atmosphere and reradiated

FIGURE 2.6 Solar energy interacting with the atmosphere, oceans, and biosphere does a wide variety of work and undergoes many transformations. But the total amount of energy originating as solar energy in the entire Earth system plus its surroundings remains the same.

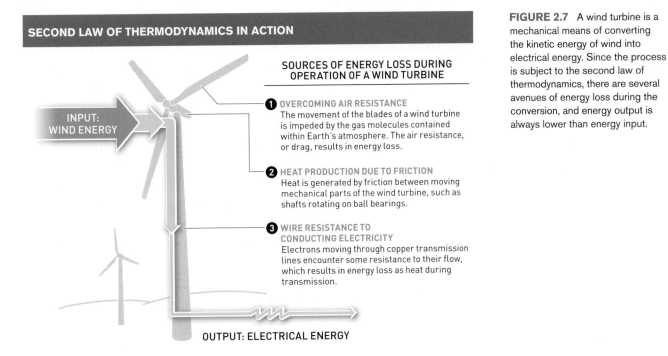

SECOND LAW OF THERMODYNAMICS IN ACTION

INPUT: WIND ENERGY

SOURCES OF ENERGY LOSS DURING OPERATION OF A WIND TURBINE

1 OVERCOMING AIR RESISTANCE
The movement of the blades of a wind turbine is impeded by the gas molecules contained within Earth's atmosphere. The air resistance, or drag, results in energy loss.

2 HEAT PRODUCTION DUE TO FRICTION
Heat is generated by friction between moving mechanical parts of the wind turbine, such as shafts rotating on ball bearings.

3 WIRE RESISTANCE TO CONDUCTING ELECTRICITY
Electrons moving through copper transmission lines encounter some resistance to their flow, which results in energy loss as heat during transmission.

OUTPUT: ELECTRICAL ENERGY

FIGURE 2.7 A wind turbine is a mechanical means of converting the kinetic energy of wind into electrical energy. Since the process is subject to the second law of thermodynamics, there are several avenues of energy loss during the conversion, and energy output is always lower than energy input.

order, as predicted by the second law of thermodynamics, are also true of natural ecosystems.

⚠ Think About It

1. As an engineer, what design features could you target to improve the energy efficiency of wind turbines?

2. Why is it impossible to build a mechanical device that is 100% efficient at converting wind energy into electrical energy?

3. Why is it impossible to engineer a mechanical system, such as an automobile, capable of being operated indefinitely with no physical maintenance?

2.3 Energy flows through ecosystems, while matter recycles

An influx of energy is required to maintain order in ecosystems and to sustain their physical and biological processes. For most ecosystems, the Sun is the primary

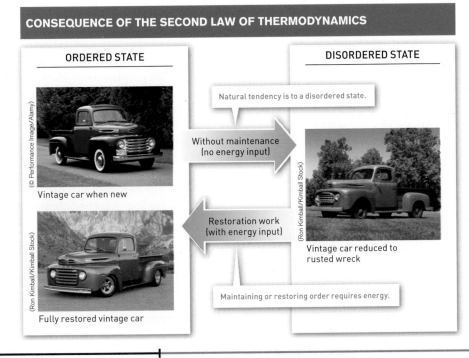

CONSEQUENCE OF THE SECOND LAW OF THERMODYNAMICS

ORDERED STATE — Vintage car when new

DISORDERED STATE

Natural tendency is to a disordered state.

Without maintenance (no energy input)

Vintage car reduced to rusted wreck

Restoration work (with energy input)

Maintaining or restoring order requires energy.

Fully restored vintage car

FIGURE 2.8 Without an input of energy, the level of disorder in a system, a property called entropy, will increase. Consequently, work is required to maintain or restore order in physical systems, which tend naturally to a state of disorder.

food web A set of feeding relationships among organisms indicating the flow of energy and materials in an ecosystem.

trophic level A step in the movement of materials or energy through an ecosystem or the position of a species in a food web.

primary producer (autotroph) An organism, generally a plant or alga, that converts the radiant energy in sunlight to the chemical energy in sugars through the process of photosynthesis.

photosynthesis A biochemical process employed by green plants, algae, and some bacteria that uses solar energy to convert water and carbon dioxide into the chemical energy in a simple sugar called glucose.

gross primary production The total amount of organic matter produced by the primary producers in an ecosystem over some period of time, for example, per year. See *net primary production.*

net primary production The net production of organic matter by the primary producers in an ecosystem, that is, gross primary production less the organic matter used by primary producers to meet their own energy needs. See *gross primary production.*

consumer An organism that meets its dietary needs by feeding on other organisms or on organic matter produced by other organisms. See *heterotroph.*

heterotroph An organism, incapable of producing its own food, that meets its energetic and nutritional needs by feeding on organic matter produced by plants and other primary producers or on other heterotrophs. See *consumer.*

SIMPLIFIED FOOD WEB FOR A FOREST EDGE-MEADOW ECOSYSTEM

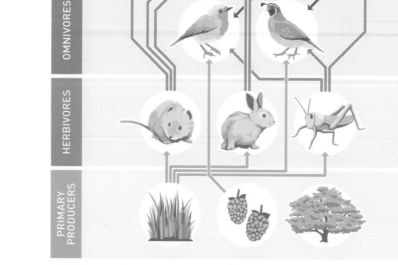

FIGURE 2.9 A food web is a diagram of feeding relationships within an ecosystem, with arrows showing the direction of energy flow between primary producers and consumers.

source of energy. Plants and algae convert this solar energy into chemical energy, which then moves from plant to animal to animal in the form of food. These movements of energy form a web of energy transfers called a **food web** (**Figure 2.9**). Food webs show who eats whom in an association of coexisting organisms. One of the basic elements in a food web is the **trophic level** of each organism, which identifies its position in the overall movement of materials or energy through an ecosystem.

The **primary producers,** or autotrophs, always occupy the first, or lowest, level in a food web. Primary producers, mostly plants on land and algae in aquatic ecosystems, produce the materials that form the basis of what all organisms need to stay alive. They do this by means of **photosynthesis,** a biochemical process that uses solar energy to convert water and carbon dioxide into the chemical energy in a simple sugar called *glucose* (**Figure 2.10**).

Ecologists refer to the total amount of organic matter produced by the primary producers in an ecosystem over some period of time as **gross primary production.** However, plants expend energy for all kinds of things besides growth: in respiration; in maintaining the structure of their tissues; and in defending themselves against attack by viruses, bacteria, fungi, insects, and other organisms. The amount of energy left over after subtracting these factors is called **net primary production.** Net primary production can also be looked at as the amount of food energy available to other organisms that feed on primary producers.

The organisms occupying trophic levels above the producers, which feed on organic matter produced by other organisms, are collectively referred to as **consumers,** or **heterotrophs.** Consumers meet their energy needs by extracting the energy from their food through **cellular respiration,** a process that occurs

PHOTOSYNTHESIS AND RESPIRATION

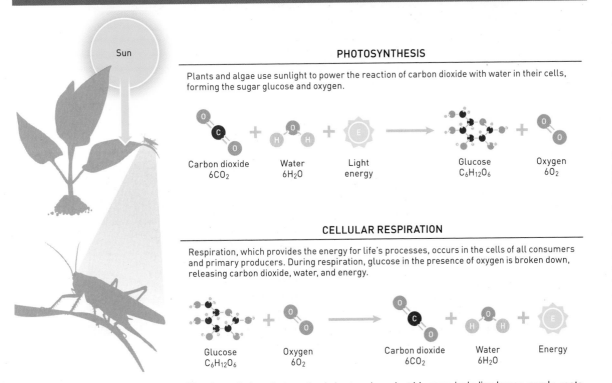

PHOTOSYNTHESIS

Plants and algae use sunlight to power the reaction of carbon dioxide with water in their cells, forming the sugar glucose and oxygen.

Carbon dioxide $6CO_2$ + Water $6H_2O$ + Light energy → Glucose $C_6H_{12}O_6$ + Oxygen $6O_2$

CELLULAR RESPIRATION

Respiration, which provides the energy for life's processes, occurs in the cells of all consumers and primary producers. During respiration, glucose in the presence of oxygen is broken down, releasing carbon dioxide, water, and energy.

Glucose $C_6H_{12}O_6$ + Oxygen $6O_2$ → Carbon dioxide $6CO_2$ + Water $6H_2O$ + Energy

FIGURE 2.10 Solar energy employed by plants during photosynthesis is stored as plant biomass, including leaves, seeds, roots, stems, flowers, and fruits; consumers, such as this grasshopper, release the chemical energy present in plant biomass through cellular respiration.

in the cells of all organisms (Figure 2.10). Consumers that feed on plants are called **herbivores** or primary consumers, while those feeding on other consumers are called **carnivores,** predators, or secondary consumers. Some consumers, called **omnivores,** eat both plants and animals. Finally, a critical group of consumers in any food web are the **detritivores,** or **decomposers.** These organisms, which feed on dead and decaying plant and animal material, are particularly important in the cycling of matter in ecosystems.

The net amount of consumer biomass, or energy, that goes into growth and reproduction, is referred to as **secondary production.**

As a result of energy losses at each trophic level (thanks to the second law of thermodynamics), a graph of the net production by all the organisms in each trophic level in an ecosystem takes the form of a pyramid, broader at the base, representing net primary production and narrowing progressively at higher trophic levels. **Figure 2.11** gives an example of an **energy pyramid** from the Silver Springs ecosystem in Florida. Because up to 90% of energy is lost between trophic levels, energy makes a one-way trip through an ecosystem.

ENERGY PYRAMID FOR THE SILVER SPRINGS, FLORIDA, ECOSYSTEM

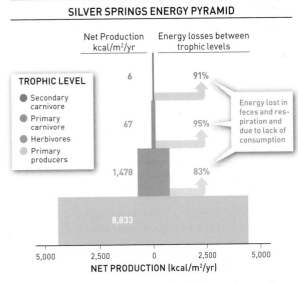

FIGURE 2.11 On average, approximately 10% of energy is passed between successive trophic levels, while 90% is lost. The result is a typical pyramid-shaped distribution of energy across trophic levels. (Data from Odum, 1957)

?

Does the conservation of matter mean that the amount of matter in the Earth system is not subject to change?

(U.S. Department of Agriculture)

FOREST FIRE: CONSERVATION OF MASS AT THE SCALE OF A FOREST ECOSYSTEM

FIGURE 2.12 While the trees and other plants, along with woody litter, may be destroyed in a forest fire, the matter that made up those plants and wood is not lost—it just changes form. Because the atoms that made up the forest before the fire are not destroyed, they can be recycled endlessly.

conservation of matter
A physical law describing how during chemical reactions, matter is neither created nor destroyed but conserved.

biogeochemical cycle
The cyclic path of an inorganic substance, such as phosphorus, nitrogen, or carbon, through the Earth system, including the atmosphere, Earth's crust, oceans, lakes, and rivers; key biological components are producers, consumers, detritivores, and decomposer bacteria and fungi.

carbon cycle The cycling of carbon through the Earth system; key biological processes in the carbon cycle include photosynthesis, respiration, and decomposition.

The effects of the second law of thermodynamics on food webs and trophic pyramids have some practical consequences. For instance, carnivores will always be less abundant than their prey species and, consequently, more likely to be in danger of extinction. Also, because of lower production at higher trophic levels, producing enough animal protein for everyone on Earth to have a diet comparable to that in the United States would not be sustainable.

Material Cycling

Matter, like energy, also flows through ecosystems. According to the principle of the **conservation of matter,** matter is neither created nor destroyed—it is conserved during chemical reactions, although it may change forms. Consider a forest fire that destroys millions of tons of wood (**Figure 2.12**). As the forest burns to the ground, the matter that made up the forest changes form—some is converted to carbon dioxide gas, some to smoke, and some to ash. However, if you could measure the process very precisely, you would find that there is no change in the mass of matter before and after the fire. Even in the hottest of forest fires, not a single atom is lost. This shows us that the movement of matter on Earth is cyclic—nothing is created or destroyed but can be recycled indefinitely.

You can demonstrate this fact for yourself by reviewing the chemical equations in Figure 2.10 (see page 41). Count the number of atoms on the two sides of the reactions in Figure 2.10, and you will find that, though the atoms have been rearranged, their numbers are the same.

Examples of substances that get cycled include water, nitrogen, carbon, phosphorus, sulfur, and iron. The cycles of these and other substances are called **biogeochemical cycles** because they involve biotic, or living components like plants and animals, and abiotic (nonliving) components, including Earth's water, minerals, and atmosphere. The biological components of biogeochemical cycles include the producers, herbivores, carnivores, detritivores, and decomposer bacteria and fungi represented in Figure 2.9 (see page 40). Decomposition of organic matter releases inorganic substances, such as nitrogen, phosphorus, or carbon, which are returned to soil, atmosphere, or water.

One molecule that has been making headlines for years is carbon dioxide, an important part of the global **carbon cycle.** Carbon plays a central role in the lives of all organisms as most of the molecules essential to the structure and functioning of living cells—including protein, DNA, and lipids—are built on a framework of carbon. Carbon is also a key player in two processes

THE CARBON CYCLE

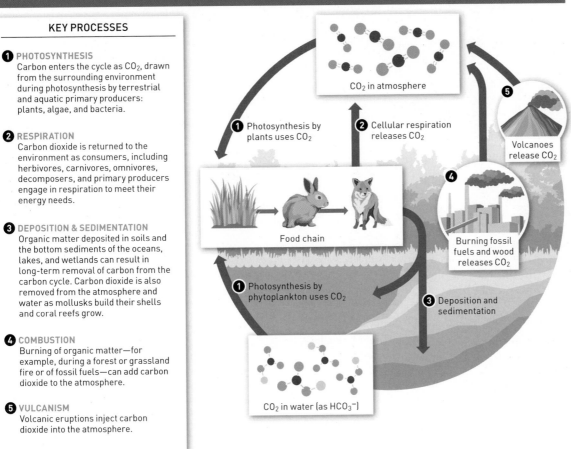

KEY PROCESSES

❶ PHOTOSYNTHESIS
Carbon enters the cycle as CO_2, drawn from the surrounding environment during photosynthesis by terrestrial and aquatic primary producers: plants, algae, and bacteria.

❷ RESPIRATION
Carbon dioxide is returned to the environment as consumers, including herbivores, carnivores, omnivores, decomposers, and primary producers engage in respiration to meet their energy needs.

❸ DEPOSITION & SEDIMENTATION
Organic matter deposited in soils and the bottom sediments of the oceans, lakes, and wetlands can result in long-term removal of carbon from the carbon cycle. Carbon dioxide is also removed from the atmosphere and water as mollusks build their shells and coral reefs grow.

❹ COMBUSTION
Burning of organic matter—for example, during a forest or grassland fire or of fossil fuels—can add carbon dioxide to the atmosphere.

❺ VULCANISM
Volcanic eruptions inject carbon dioxide into the atmosphere.

CO_2 in atmosphere

❶ Photosynthesis by plants uses CO_2

❷ Cellular respiration releases CO_2

❺ Volcanoes release CO_2

Food chain

❹ Burning fossil fuels and wood releases CO_2

❶ Photosynthesis by phytoplankton uses CO_2

❸ Deposition and sedimentation

CO_2 in water (as HCO_3^-)

FIGURE 2.13 The cycling of carbon is one of the most important biogeochemical cycles on Earth, since carbon atoms provide the basic structure for the organic molecules, including lipids, proteins, and DNA, that make up the cells of organisms.

we've already learned about—photosynthesis and respiration; carbon moves around on Earth, from the tissues of living creatures to sedimentary rocks, and from oceans to atmosphere; and carbon trapped in the bodies of once-living creatures is released millions of years later into the atmosphere by the burning of fossil fuels (**Figure 2.13**).

Because matter cannot be created or destroyed, biogeochemical cycles like the carbon cycle move essential elements around Earth continuously. For more on biogeochemical cycles, see Chapter 7 (nitrogen cycle), Chapter 8 (phosphorus cycle), and Chapter 13 (sulfur cycle). Another cycle, the hydrologic, or water, cycle plays a key role in the discussions of water-related issues in Chapter 6.

⚠ Think About It

1. Why is consumer production also known as secondary production?

2. How is the second law of thermodynamics related to the fact that net primary production is always less than gross primary production?

3. How do photosynthesis and respiration contribute to Earth's carbon cycle?

2.4 Economic systems and their currencies take several forms

Humans depend on natural systems and resources to support their lives and lifestyles. From ecosystems we draw resources that feed us, keep us warm, and give us livelihood. As we have seen, as the Maasai established systems for trading goods and services, they were building an economic system. Like natural ecosystems, human economies involve the movement of matter and energy. These flows are dictated not only by physical laws, such as

TWO ECONOMIC SYSTEMS

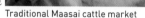

Traditional Maasai cattle market

Modern shopping mall

FIGURE 2.14 The traditional Maasai economic system involved production of cattle and dairy products, as well as limited trading with hunting and farming neighbors for wildlife products and crops. In contrast, a modern market economy is characterized by intense economic activity, involving the production and consumption of a great diversity of goods and services.

subsistence economy An economy in which individuals or groups produce or harvest enough resources to largely support themselves, with fewer resources gained through purchase or trade with other groups.

market economy An economy in which decisions about the production and consumption of goods and services are not centralized but made by businesses and individuals, generally acting in their own self-interest. See *centrally planned economy*.

supply and demand An economic model stating that the price of a good (or service) will reach equilibrium when the consumer demand for it at a certain price equals the quantity supplied by producers.

centrally planned economy An economy in which decisions about the production and consumption of goods and services are made by a central authority. See *market economy*.

black market The exchange of illegal goods and services.

the laws of thermodynamics, but also by social conventions and regulations. Economics adds a new layer of complexity atop our understanding of ecosystem principles.

Economic Systems

The traditional economy of the Maasai is an example of a **subsistence economy,** in which individuals or groups produce or harvest enough resources from the natural environment to largely support themselves, with fewer resources gained through purchase or trade with other groups.

In a modern **market economy,** such as that of the United States, the countries of the European Union, Canada, Japan, and other developed countries, decisions about production and consumption of goods and services are not made by some central governmental authority. Such decisions are made by businesses and individuals, generally acting in their own self-interest (**Figure 2.14**).

In a market economy, the price of goods and services is generally determined by their **supply and demand** (**Figure 2.15**). If the demand for oranges one year remains constant, but the supply increases, we end up with a surplus that causes their price to decline. The price will also drop if the demand declines but the supply stays the same. A shortage can occur if either the demand goes up or the supply goes down, leading to a spike in prices. Supply and demand affects the availability of a particular good. This is important to keep in mind, since most of the goods we consume are either natural resources or impact the environment in some way through their production or sourcing. For example, drilling for oil to meet the demand for fossil fuels directly impacts the drill site for that oil, as well as the atmosphere, when the fossil fuel is burned. For reasons like this, the economy and the environment are often portrayed as being in opposition to each other.

In a **centrally planned economy,** a central authority sets prices and makes decisions about production and consumption of goods and services. Because centrally planned economies are slower to react than market economies, they are particularly susceptible to developing shortages and surpluses, and residents often resort to illegal or **black markets** to obtain their goods. Contemporary examples of centrally planned economies are those of Cuba and North Korea.

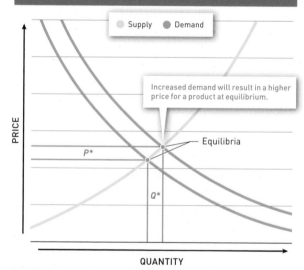

FIGURE 2.15 The law of supply and demand predicts that in a competitive market, the price of a product will eventually be determined by the relationship between the supply of that product and the consumer demand for it.

EVOLUTION OF MONEY

(Atelier_A/Shutterstock)
Wheat

(GreenTree/Shutterstock)
Cowrie shells

Ancient silver Greek coin

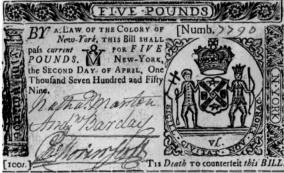
(© North Wind Picture Archives/Alamy)
New York colonial era money

FIGURE 2.16 Early in history, commodities including livestock and grain were used as currencies. Later, cowrie shells were used as a medium of exchange in China and Africa. Coins were first minted in China and then in Persia and Greece. The use of paper money began in China, then spread around the world.

Property

Each economic system places different limits on the ownership of property, which has consequences for environmental management. The term *property* is used to refer to land, manufactured goods, or resources such as freshwater, minerals, or fish stocks. Under a market economy, we can generally identify four property regimes: state property, private property, communal property, and open access. Government-owned property is considered **state property,** and private individuals may be able to make use of it as long as they follow the rules imposed by

the state, such as obtaining the proper permits. Examples of state property in the United States include national parks and national forests, wild game animals wherever they roam, and navigable rivers even if they pass through private property.

Private property, by contrast, implies that the owner has the full right to use and benefit from her property so long as she is following all applicable laws and not harming others. **Common property** is a form of private property, but rather than being owned by an individual, it is owned by a group. Members of the group control access to the property and exclude those who are not members. The last type of property, **open access,** implies that neither the government nor an individual has the right to exclude others. In other words, it's first come, first served.

Money

While traditional Maasai could get by using livestock as a form of currency, modern economies depend on money. **Money** is a medium of exchange for goods and services that has evolved over thousands of years (**Figure 2.16**). Originally, commodities such as cattle or grain, which carry inherent value in terms of both material (nutrients) and energy (calories), were used as modes of exchange. The bartering of commodities eventually gave way to the exchange of other currencies. One of the earliest currencies, cowrie shells, which can be gathered in shallow tropical waters, was used as money in China and some regions in Africa. Over time, the minting of metal coins and use of paper money rather than shells became widespread.

For a time, the value of paper money was tied to standard quantity of gold, a practice known as the *gold standard.* The gold standard was abandoned in the 20th century and has been replaced by complex international regulations governing the value of money.

While we may use the flow of money to measure economic activity, that flow is symbolic of the exchange of goods and services, which is tied to the flow and transformation of matter and energy. When you buy a tablet computer for a certain number of U.S. dollars, euros, or Chinese yuan, you are exchanging your money for the material, energy, engineering, labor, and proprietary information that went into the production, transport, and marketing of the computer. You are also paying for the service provided by the seller who makes it available for your purchase. Conversely, when you earn money, you are exchanging the energy, labor, and proprietary information that went into your work.

⚠ Think About It

1. How is money related to the movement of energy and matter in economic systems?

Europeans were shocked when they learned, in the 13th century, that paper currency was used in China. Why might this have seemed such a radical invention?

state property Property owned by federal, state, or local governments.

private property Property owned by individuals.

common property Property owned or controlled by a community, such as an indigenous tribe.

open access A property for which there are no restrictions about who may enter and exploit its resources.

money A medium of exchange using coins or paper bills.

2. Why does the law of supply and demand have more influence on the price of goods in a market economy than in a centrally planned economy?

3. What main factors likely influenced the evolution of money?

2.1–2.4 Science: Summary

The basic constituent of all matter, the atom, is the smallest particle of a substance that retains its properties. Substances composed of a single type of atom are called elements. Two or more atoms held together by chemical bonds form a molecule. All activity on Earth, whether in an ecosystem or as part of a modern economy, is fueled by energy. Energy, which comes in many forms of potential and kinetic energy, is defined as the capacity to do work. According to the first law of thermodynamics, one form of energy may be transformed into other forms, but the total amount of energy in a system plus its surroundings remains the same. A consequence of the second law of thermodynamics is that the overall quality of energy declines with each energy transformation, or transfer, reducing the energy available to do work. During chemical reactions, matter is neither created nor destroyed, but is conserved.

In ecosystems, matter is recycled through the biogeochemical cycles. The movement of energy and materials in an ecosystem is reflected in its food web and energy pyramid, which connect the organisms in an ecosystem through their feeding relationships. Matter and energy also flow through economic systems, which range from subsistence economies to market-based economies. Economies depend on various conceptions of public and private property and the use of money as a medium of exchange.

2.5–2.7 Issues

The economy in the land of the Maasai has continued to evolve over the last century. On July 7, 2000, the East African nations of Kenya, Tanzania, Uganda, Rwanda, and Burundi established an organization called the East African Community (EAC) to further economic growth. Between 2005 and 2010, the total value of goods and services produced in the EAC increased by 70%. And while agriculture remains the dominant economic sector of the region, there has been substantial growth in manufacturing and energy production. For example, between 2001 and 2010, the number of people employed in manufacturing in Kenya, the largest country in the EAC, grew by 24%, and consumption of electricity in the country increased by nearly 40%. This is all good news, but environmentalists are also concerned that a shift from low-impact subsistence lifestyles to high-impact consumer economies will lead to unfettered degradation of some of the last wild places in Africa.

When such concerns are raised about the environmental impact of economic growth, we often hear that the choice is between a healthy environment and a healthy economy. The alternative, some might say, is to leave the people of Africa living in poverty. How did we develop this view? In the traditional societies around Earth, the relationship between a healthy productive ecosystem and economic health was clear. When the Maasai people first brought their cattle to East Africa, they altered the existing ecosystem and established an economic system atop it. Their cattle, like the native wildlife, depended on the primary production of grasslands, which meant that their early economy was intricately linked to the health of the environment. Such clear connections are not always so obvious today, where people buying beef in shrink-wrapped packages from a city supermarket have little knowledge of how drought and environmental degradation are impacting grazing lands. The relationship between economics and ecology receded into the background with the development of modern industrial economies and of early economic theory.

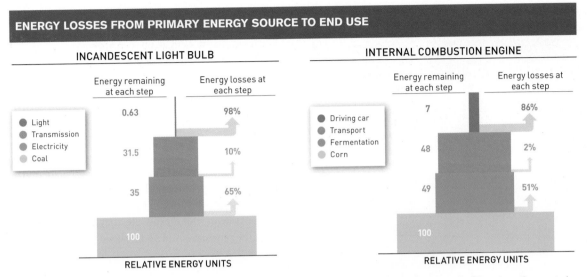

ENERGY LOSSES FROM PRIMARY ENERGY SOURCE TO END USE

INCANDESCENT LIGHT BULB

Energy remaining at each step / Energy losses at each step

- Light
- Transmission
- Electricity
- Coal

Energy remaining	Energy losses
0.63	98%
31.5	10%
35	65%
100	

RELATIVE ENERGY UNITS

INTERNAL COMBUSTION ENGINE

Energy remaining at each step / Energy losses at each step

- Driving car
- Transport
- Fermentation
- Corn

Energy remaining	Energy losses
7	86%
48	2%
49	51%
100	

RELATIVE ENERGY UNITS

FIGURE 2.17 Energy losses from a coal-fired generating station to a 60-watt incandescent light bulb. (Data from Graus et al., 2007; Leff, 1990; Agrawal et al., 1996) Energy losses from an ethanol-producing biofuels plant, which uses corn kernels as the feedstock, to a car driven by an internal combustion engine running on ethanol. (Data from Huang et al., 2011)

2.5 Energy fuels, and limits, the economy

Because economic systems are subject to the same physical laws governing ecosystems, they require inputs of energy and matter in order to keep functioning. The energy sources important to today's economic systems include the fossil fuels and renewable energy sources used to generate electricity, run transportation networks, and heat and cool homes and businesses. Energy and matter also enter the economy as the food consumed by people, pets, and livestock. Economists have historically considered the supply side of a market economy to be determined by human industry, such as manufacturing or resource extraction, but we now recognize that our resources are not unlimited and that our activities can have a wide range of impacts on our lives.

Let's take a look at how some of the physical laws we've explored impact the economy. As predicted by the second law of thermodynamics, the energy flowing through an economic system is reduced in quality with each transformation or with each bit of work done. Consider, for example, the path followed by the electrical energy powering a 60-watt incandescent light bulb in a desk lamp. An average electrical power station using coal as a fuel source has an efficiency of about 35%, which means that of the energy contained within 100 units of coal burned, 65% is lost as heat. As the 35 units of electrical energy are transmitted along power lines to the desk lamp, approximately 10% will be lost as heat and sound energy. At its final destination, about 2% of the energy delivered to the 60-watt incandescent light bulb will be converted to light energy, while the remaining 98% will

be lost as heat. As a consequence of energy losses from the point of production to use, less than 1% of the energy burned in the coal-fired power station is used to light the desktop (**Figure 2.17**).

Although less drastic, similar energy losses occur along the energy pyramid describing the flow of energy from the production of corn-based ethanol fuel to driving an internal combustion vehicle powered by that fuel. In this case, approximately 7% of the energy present in the corn kernels entering this flow path is used for driving the motor vehicle (Figure 2.17).

You can observe that the overall pattern of energy distribution along these flow paths takes the shape of a pyramid, much like that observed by ecologists in ecosystems. Human economies sit at the top of the energy pyramid and grow by tapping into natural ecosystems and Earth's resources. Because of energy losses resulting from the second law of thermodynamics, both ecosystems and economic systems require continued inputs of energy to sustain their functions. This need to keep energy flowing through modern economies to keep them productive has been a source of environmental disruption and loss of human lives (**Figure 2.18**).

⚠ Think About It

1. How is the flow of energy through natural ecosystems similar to energy flow through economic systems? How are they different?

2. Why is the functioning of modern economies so intimately tied to the production and marketing of energy sources such as fossil fuels and electricity?

ENVIRONMENTAL RISK ASSOCIATED WITH FUELING ECONOMIC PRODUCTION

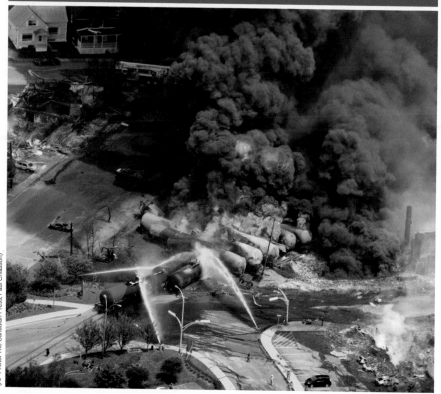

(AP Photo/The Canadian Press, Paul Chiasson).

FIGURE 2.18 Energy development associated with the energy-intensive modern economy comes with environmental impacts, including air, water, and soil pollution. The direct human costs can also be serious. On July 6, 2013, a train carrying crude oil from North Dakota derailed in Lac-Mégantic, Quebec, about 10 miles from Maine. The resulting explosions and fires killed 47 people.

2.6 How we represent economic systems can have environmental consequences

When the human population was small and our technology simple, our environmental impact was negligible and there was little need to consider how we might be affecting Earth at anything but the most local scale. But with an ever-expanding human population, environmental impact by the human species has become significant. In many ways, how we define our economic systems influences our relationship to the environment.

Traditional Closed Economic Model

positive feedback A stimulus in which an increase in some factor in a system, such as an economic system or ecosystem, produces additional increases in that factor within the system or in which a decrease in a factor causes additional decreases.

Western economists historically modeled an economy as a closed system. In such a model, processing and manufacturing industries produce goods that are marketed and distributed to consumers. When consumer spending on goods and services (demand) rises, this stimulates the processing and manufacturing industries to meet the new demand by making more goods. The industry may have to hire new employees to help produce the goods and meet the demand. People who are employed and make money

spend money in the economy, so as industry grows to meet demand, employment rates go up, spending goes up, and demand continues to go up. The system is spurred on continuously. This is an example of **positive feedback** in a system, when an increase in some factor produces additional increases in that factor within the system (**Figure 2.19**).

A closed economic system leaves out several important elements: It ignores anything that, in real life, comes from outside the system itself—like raw materials and exchanges of energy, such as fossil fuels or solar energy. To be sure, leaving out these elements simplifies the models and makes it easier to think about complex economic systems. But meanwhile, they disregard the limits of the physical world and its laws.

Open Economic Model

In the real world, economic systems are not closed but open, since they cannot function without inputs of materials and energy. In a model that considers the

MODEL OF AN ECONOMIC SYSTEM AS A CLOSED SYSTEM

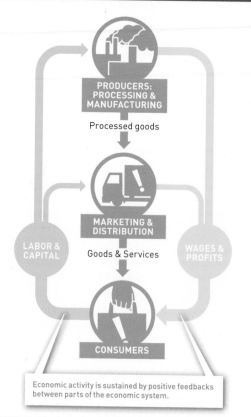

FIGURE 2.19 A closed model leaves out exchanges of material and energy between the economic system and the environment. Economic activity in the model is sustained by positive feedbacks between parts of the economic system.

environment, raw materials such as oil, metal ores, and agricultural crops must come from somewhere. These materials are then processed by the economic system to produce goods or services that are distributed and marketed to consumers using energy sources such as fossil fuels. Along the way, each of these economic processes will result in waste energy and waste materials that impact the environment (**Figure 2.20**).

In this type of open system model, the environment is considered to be outside or "external" to the economic system. The environmental impacts that may occur as a consequence of economic activity, such as the dumping of waste into a river or the spewing of gases into the atmosphere, are expected to have little impact on prices of products. That's because historically, companies did not have to pay for causing such damage; those costs were instead borne by society at large. An **economic externality** is therefore a cost or benefit to the environment or to society that is not included in the market price of the product. As we shall see in the next section, economic externalities can contribute to a **market failure,** a situation in which society is harmed because of the inefficiency in how goods and services are distributed.

⚠ Think About It

1. How might a closed model of an economic system be useful despite its not being realistic?

2. Under what circumstances would it be reasonable to not include the impact of economic activity on the environment in calculating the cost of goods and services produced?

2.7 Unregulated use of resources can lead to a "Tragedy of the Commons"

Economists have argued that businesspeople pursuing their own self-interest would do what was best for society and the environment. But, in his classic 1968 environmental essay "Tragedy of the Commons," ecologist Garrett Hardin disputed this view. He argued that unregulated use of a common resource would lead inevitably to its ruin.

Imagine a pasture available for community use—in other words, an open-access property on which everyone in a community is free to graze cattle and cannot exclude anyone else. They all seek to maximize the number of cattle they graze on the commons and by doing so maximize their own profits; for each head of cattle an individual farmer adds to the commons, he or she reaps all the profit of eventually selling the animal. So each farmer will graze as many cattle as possible on the property to maximize his or her own profits.

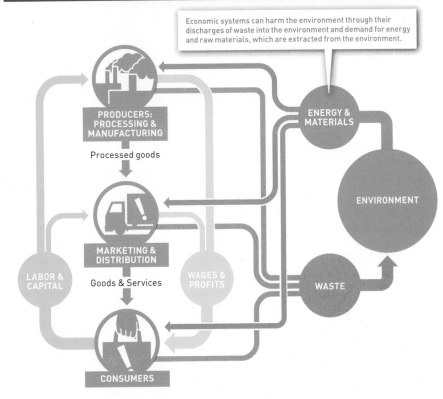

MODEL OF AN ECONOMIC SYSTEM AS AN OPEN SYSTEM

Economic systems can harm the environment through their discharges of waste into the environment and demand for energy and raw materials, which are extracted from the environment.

PRODUCERS: PROCESSING & MANUFACTURING

Processed goods

MARKETING & DISTRIBUTION

Goods & Services

LABOR & CAPITAL

WAGES & PROFITS

CONSUMERS

ENERGY & MATERIALS

ENVIRONMENT

WASTE

FIGURE 2.20 An open economic model includes exchanges of materials and energy with the environment, but the environment is treated as if it is external to the economic system. Therefore, in this model, the effects of the economic system on the environment are considered "externalities."

Adding cattle will damage the commons by reducing plant cover, increasing soil erosion, and decreasing future productivity of the grazing land. What are the costs to the individual farmer of damaging the land? Since all the farmers in the community share the costs of overgrazing, the cost to each one of them is only a fraction of the total cost. Therefore, in this situation, profit exceeds costs, so it is profitable for each individual farmer to add to his or her herd. Eventually, the productivity of the pasture is severely reduced and no one benefits (**Figure 2.21**). Hardin thus concluded in his essay that "freedom in a commons brings ruin to all."

Hardin extended the logic of his analysis to the use of many other so-called **common-pool resource**s, including rangelands in the western United States (Chapter 7, marine fisheries (Chapter 8), and water and air into which industries discharge pollutants (Chapter 13). In each of these situations, unrestricted use of a community resource leads to a market failure causing environmental harm. These represent some of the greatest challenges to environmental policy in the 21st century.

What are some potential environmental consequences of viewing the environment as outside or "external" to the economy?

economic externality
A cost or benefit to the environment or to society resulting from the production and use of a product that is not included in the market price of the product.

market failure A situation in which free markets do not allocate goods and services efficiently, such as when the price of a product does not include its environmental impact.

common-pool resource
A resource owned and utilized in common by a community (e.g., a community forest or grazing land).

GARRETT HARDIN'S SCENARIO LEADING TO A TRAGEDY OF THE COMMONS

(Peter Essick/Aurora)

FIGURE 2.21 Hardin proposed that a common-pool resource, such as grazing lands open to all, would be ruined ultimately as individual users attempt to maximize their profits. The results of such unregulated use might appear like the scene in this photo, showing a severely overgrazed pasture.

⚠ Think About It

1. What are some assumptions of Hardin's Tragedy of the Commons argument?

2. What are some possible scenarios that would avert a Tragedy of the Commons without governmental intervention?

2.5–2.7 Issues: Summary

Human economies sit at the top of the energy pyramid and grow by tapping into natural ecosystems and Earth's resources. Because economic systems are subject to the same physical laws governing ecosystems, they require inputs of energy and matter in order to keep functioning. Traditional closed economic models include no interactions between the economic system and the environment. The environment is included in open economic models, in which the economic system exchanges energy and materials with the environment. However, in these open models, the environment has been historically represented as "external" to the economic system.

Keeping potential environmental harm external to the economic system fails to account fully for the potential costs of economic activity. In his "Tragedy of the Commons" article, Garrett Hardin argued that unregulated use of a common resource would lead inevitably to its ruin.

2.8–2.10 Solutions

Economic development does not have to lead to environmental degradation. Certainly, some limited resources will inevitably be depleted over time, but history shows us that improvements in resource use efficiency coupled with smarter regulations can reduce the impact of growing economies. Consider the story of sulfur-dioxide pollution. Sulfur dioxide is produced from the burning of fossil fuels, including gasoline and coal. This invisible gas can cause respiratory problems and aggravate heart conditions. It can also cause acid rain, which damages buildings and statues along with forests and aquatic ecosystems (see Chapter 13).

As nations' economies grew in the 1960s, they put more vehicles on the road and built more factories, which led to greater amounts of sulfur-dioxide pollution. But in the late 1970s and early 1980s, many of these countries reached an economic turning point as the average income of their citizens increased. As these citizens became better off, they demanded more regulations to protect their health and their environment. Their governments were also in a better position to enforce their laws. Since then, sulfur-dioxide pollution has decreased substantially, and acid rain has been substantially curtailed in many countries. Today, economic growth has created many similar challenges for society, and finding solutions for them hinges on applying our knowledge of ecosystems and economic systems to address pressing environmental issues.

2.8 Economics should include environmental costs and benefits

Every time you drive to work or school in your car, you add to the traffic on the road and cause a little damage to the pavement. As thousands of others in your hometown do the same every day of the year, the damage accumulates and the road will inevitably need to be repaired and, perhaps, widened. In the previous section, we brought up the concept of *economic externalities,* which are the costs or benefits to society or the environment not included in the price of a product or activity. In this case, road damage is an externality associated with driving motor vehicles on roads maintained at public expense. Government regulations exist to prevent such market failures and take several forms.

Command-and-Control Regulations

The most obvious solution to preventing harm to society is through **command-and-control regulation,** which is direct regulation of an industry or activity, indicating what is and is not allowed, along with a schedule of penalties for breaking the rules. One way to prevent damage to the road is to legislate what type of tires are allowed on vehicles. For instance, towns in cold areas frequently prohibit snow chains within their limits because they tear up the road. In the city of São Paulo, Brazil, planners reduce traffic by regulating which cars can be on the road on any given day based on their license plate number.

In Chapter 1, we introduced the discovery that CFCs (chlorofluorocarbons) were causing an ozone hole over the Antarctic. Companies that produced CFCs were profiting as they harmed the environment—and they had no interest in voluntarily ending production. In fact, they spent millions on public relations campaigns and lobbying efforts to *resist* attempts to limit the production and use of CFCs; but once alternative chemicals were developed, they became supportive. In 1978 the United States instituted the first ban on CFCs, eliminating their use as propellants in aerosol spray cans. The Montreal Protocol, signed in 1987, would eventually lead to the phasing out of CFCs in all products.

The success of such command-and-control regulations depends on enforcement of the rules, which includes the government's ability to catch violators and the level of fine or other punishment required to discourage rule-breaking. If a penalty is too low or the chance of getting caught is unlikely, then businesses will consider breaking the rules. Penalties for illegally importing CFCs into the United States today can include prison time.

Pigovian Taxes

Command-and-control instruments are not suited to preventing all environmental damage, just as they aren't ideal for preventing routine damage to a highly trafficked road. After all, the economy depends on people using the roads. Nevertheless, this raises the question of who should foot the bill for such repairs. Should everyone in town contribute an equal amount? Should car companies pay for the roads? What about truck drivers whose vehicles cause even more damage? And what about people who choose to walk or bike and have less of an impact on

the roadways? Should they really be paying the same amount?

In 1920 an economist named Arthur C. Pigou became the first to propose taxing economic externalities, such as the damage caused by driving. His ideas caught on, and today we pay for roads in the United States through vehicle registration fees and taxes on fuel, which generally correlate with the number of miles a person drives. Since 2003 the city of London has charged a 10-pound "congestion charge" for every vehicle entering the city center on weekdays. So-called Pigovian taxes, or "sin taxes," are also placed on cigarettes and alcohol to discourage their use due to the harm they cause to society.

These taxes also contribute to the field of study called **environmental economics,** which includes the environment in its models. In its analyses, environmental economics draws mainly from the field of economics in its assessments and management of costs and benefits of economic impacts on the environment. For example, the way that environmental economists control damage to the environment by economic externalities, such as pollution, is to put a tax on the externality proportional to its damage to the environment. For instance, Pigovian taxes have been proposed for the emission of carbon dioxide and other pollutants in the atmosphere, where an outright ban is impractical. Pigou also thought that economic activities that benefit the environment should be encouraged through government subsidies. As with command-and-control regulations, one of the challenges with this strategy is determining the right level of Pigovian tax or subsidy.

Ecological Economics

For economists, money is the fundamental metric of value. But even when they have tried to incorporate externalities into their economic models, some ecologists believe that they still undervalue **natural capital,** which is all of Earth's natural assets, including plants and animals, minerals and soils, and air and water. In contrast to environmental economics, the field of **ecological economics** draws from many disciplines, including economics, in its studies of the relationships between economic activity and its impact on natural capital. Whereas conventional economics has seen people and their institutions as nearly alone in the world, and traditional ecology has focused most of its studies on ecosystems occupied by species other than humans, ecological economics attempts to build a conceptual bridge between humans and human institutions and the rest of nature (**Figure 2.22** on the next page).

Rather than seeking only to maximize financial capital through short-term economic productivity, ecological economists are focused on sustaining natural capital, which provides humans with a flow of goods and services. A healthy fish stock will replenish itself each

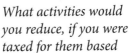

What activities would you reduce, if you were taxed for them based on their environmental impacts?

command-and-control regulations Laws and regulations that control activities and industries through the use of subsidies and penalties prescribed by the government.

environmental economics A branch of economics that draws mainly from the field of economics as it assesses and manages the costs and benefits of economic impacts on the environment.

natural capital The value of the world's natural assets (e.g., minerals, air, water, and living organisms).

ecological economics A branch of economics that draws on many disciplines in studies of the influence of economic activity on the environment in an attempt to build a conceptual bridge between humans and human institutions and the rest of nature.

FIGURE 2.22 Ecological economics attempts to bridge the traditional approach of conceptually separating economic systems and ecosystems. By quantifying the exchanges of materials and energy between natural and human-dominated systems, ecological economics identifies ecosystem services, such as water purification and climate moderation, to which it assigns monetary value, placing natural ecosystems squarely within the realm of economics.

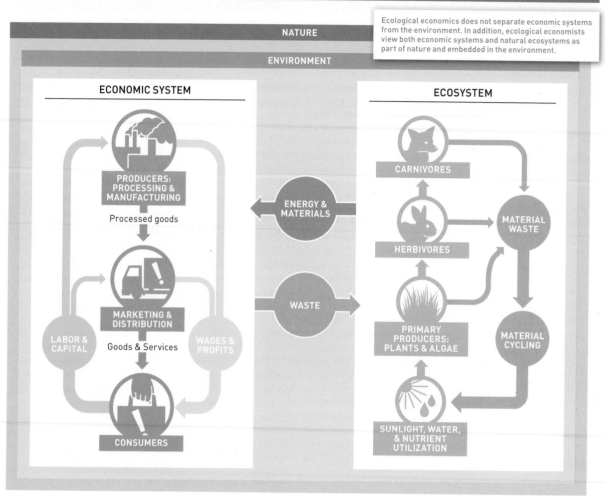

ECOLOGICAL ECONOMICS PERSPECTIVE ON THE RELATIONSHIP AMONG ECONOMIC SYSTEMS, NATURAL ECOSYSTEMS, AND THE ENVIRONMENT

Ecological economics does not separate economic systems from the environment. In addition, ecological economists view both economic systems and natural ecosystems as part of nature and embedded in the environment.

year, providing food to humans indefinitely. A forested valley will provide clean water. And a pristine beach may simply be pleasant to walk on, which has social or cultural value. All these things have value and one of the central challenges of ecological economics is determining how much they are worth to society now and in the future. Knowing these values can help us make smart, sustainable economic decisions. Ecological economists who have placed a monetary value on such natural goods and services, called **ecosystem services**, have found that they far exceed the value of all goods and services produced by the world's economic systems. We will discuss ecosystem services in detail in Chapter 4.

ecosystem services
The benefits that humans receive from natural ecosystems such as food, water purification, pollination of crops, carbon storage, and medicines.

⚠️ Think About It

1. Why are the analyses of ecological economists necessarily more complex than those of environmental economists?

2. What aspect of CFC depletion of the ozone layer made that issue better addressed by a command-and-control approach than did using Pigovian taxes?

3. What are possible reasons why some have resisted putting a monetary value on nature's ecosystem services?

2.9 Property rights can lead to environmental preservation

The establishment and protection of privately owned property are considered the defining roles of government and the foundation of modern economies. The 17th-century political philosopher John Locke wrote that the fundamental purpose of government was to elevate people from their natural state and ensure their right to life, liberty, and property. Indeed, one solution to Hardin's

Tragedy of the Commons is to distribute property to individuals, who will therefore have an economic incentive to maintain the integrity of that resource.

The Vicuña

The vicuña (*Vicugna vicugna*) is the wild cousin of the llama and alpaca. It lives in the highlands of South America and is prized for its wool. After the arrival of Europeans in the region, vicuña were treated as an open-access resource and their populations were decimated by hunting. In the late 1960s and early 1970s, regulations were put in place to protect the animals by banning all national and international trade in vicuña products. Although vicuña populations began recovering as a result of the reduction in hunting, their former grazing lands were now overrun by domesticated llamas and alpacas. Locals could own domesticated llamas and alpacas and therefore benefit from them.

Vicuña, on the other hand, provided no economic benefit. In 1979 the Vicuña Convention authorized a tightly regulated trade in vicuña wool, which allowed locals to harvest wool and sell the wool without killing the animals. In Argentina and Chile, vicuña were considered private property and raised in captivity by individual families. In Peru and Bolivia, local communities engaged in periodic round-ups, where they sheared and released the vicuña, and shared in the profits. It turned out that giving ownership of the vicuña, or their wool, to the people fostered the survival of the species: Between 1969 and 2001, the vicuña population increased from 14,500 to 227,500.

Water Use Rights

Few environmental issues are as emotional as freshwater use rights in the western United States. Economic growth is limited by freshwater availability, but freshwater is a limited resource that depends on climate, geology, and ecosystems. Allocation of water in western rivers is governed by treaties and compacts that date back more than a hundred years, when water was far more plentiful than it is today. Unlike goods on the free market, the price that farmers and city-dwellers pay for water does not increase as the availability of water declines. It is set by local utilities and water management agencies. Moreover, people who dig their own wells and install their own pumps often have to pay nothing at all, even though they may be tapping into an aquifer shared by their neighbors. Overexploitation of freshwater resources is another example of a Tragedy of the Commons.

Cities facing water shortages often resort to command-and-control regulations, such as strict regulations about watering lawns and penalties for individuals who exceed water use quotas. Facing a drought in May 2014, the City of Santa Cruz, California, enacted such a measure for water use. Single-family homes were allowed to use 249 gallons per day with hefty penalties for those exceeding the limits.

Many economists believe that such command-and-control approaches would be unnecessary and water use more sustainable if farmers and residents were able to buy and sell their rights to water on a free market. Using such a **market-based approach,** water allocations could be traded among residents and the price of water would fluctuate with supply and demand. Because water would suddenly become more valuable to farmers, they might be more willing to invest in technologies such as drip irrigation to reduce their water use and sell their excess to municipalities. Meanwhile, residents who are willing to pay more for a green lawn in summertime can do so by paying the market price for the water they use.

One of the advantages of market-based strategies over command-and-control regulations or Pigovian taxes is that policy makers only have to set the overall goal for water use and do not need to identify a penalty or tax schedule. The market will establish the appropriate price for water. Taking a market-based approach to managing water may in many cases require provisions to protect the public interest in, for example, maintaining wetlands and fish habitat. Such an approach has never been implemented for water, but tradable quotas have proven successful in reducing sulfur-dioxide pollution (Chapter 13)—which caused acid rain—and reducing overfishing (Chapter 8).

⚠ Think About It

1. Since the Republic of South Africa began allowing private ownership and management of wildlife, big-game populations have increased dramatically. Explain.

2. What role would the law of supply and demand play in setting price if water management were market-based?

3. What are some dangers of making water management entirely market-based?

2.10 Alternative paths to sustainability: Tragedy of the Commons revisited

Although clearly defined property rights are often the key to sustainability, the Western conception of private ownership is not the only solution. When Kenya came under British colonial rule at the end of the 19th century, the colonial government forced the migratory Maasai to settle on specific plots of land. The stated goals of the colonial government were to improve the condition of the land and to increase cattle production. However, the colonial government's policies achieved the opposite.

Compared with Great Britain, the amount of rainfall across the lands historically occupied by the Maasai fluctuates greatly. In contrast to the settled lifestyle of British cattle raisers, the Maasai adapted to the swings between dry and wet periods by adopting a semi-nomadic

market-based approach
An alternative to command-and-control regulation that seeks to encourage adherence to social or environmental goals using the principles of supply and demand.

OVERSTOCKED AND OVERGRAZED MAASAI LANDS

(Fred Hoogervorst/Hollandse Hoogte/Redux)

FIGURE 2.23 Confining Maasai livestock to relatively small areas and prohibiting long-distance movement of cattle to more productive lands during droughts have led to overgrazing and soil erosion.

How might differences in the climate and culture of Britain versus Kenya have contributed to the damage done by colonial administrators to the Maasai economy?

How does Elinor Ostrom's analysis affect Hardin's Tragedy of the Commons model?

lifestyle. In years with abundant rain, the low-lying, driest parts of Maasai country were valuable for sustaining cattle during the wet season. During the dry season and during extended droughts, the Maasai moved their cattle to wetter environments at higher elevations, where usually there would be enough food for the herds.

Seeing some traditional Maasai lands vacant during certain periods, the colonial government gave them to colonial farmers. A total of 20,000 km² of the most productive land (approximately equal in area to the state of New Hampshire or half the area of Switzerland) were converted to private colonial farms and ranches, and the Maasai were restricted to reservations. Although the Maasai retained 100,000 km², 20% of their reservation lands were arid or semiarid and another 10% were infested with vectors of serious diseases of humans or cattle. Most seriously, however, the Maasai and their cattle were cut off from the lands that they had depended on for grazing during the dry season and during extended droughts (**Figure 2.23**).

Colonial restrictions also prevented the Maasai from moving their cattle north to trade cattle with their traditional trading partners. Rather than fostering sustainability and economic improvements, the establishment of strictly demarcated property rights was devastating.

Today, the Maasai are a people in transition. While they still rely mainly on livestock for their livelihood, contemporary Maasai also participate in the market economy. The Maasai Foundation, an organization dedicated to preserving Maasai culture as it fosters integration of the Maasai into the modern world, promotes "education, health, environmental conservation, and economic development." Partly due to such efforts, increasing numbers of Maasai children are receiving a formal education. How these future leaders will influence the cultural and economic development of the Maasai remains to be seen.

Elinor Ostrom and the Commons

In the 1990s, the late political economist Elinor Ostrom studied the management of Maasai grazing lands to learn why Kenyan institutions have failed to foster sustainability of that system (**Figure 2.24**). Her interest was stimulated by the observation that, historically, the Maasai had actually developed a robust system for managing their lands before colonial governments disrupted it.

Ostrom had studied similar systems around the world and identified a number of principles that have allowed commons to be managed sustainably. First, those who have a right to share in the resource must be clearly identified, and the boundaries of the resource pool must be clearly defined. Next, the share of the resource to which an individual or household has a right must be proportional to its contribution to the costs of sustaining the resource system—for example, the costs of sustaining a community-managed irrigation system. In addition, the state of the resource needs to be monitored and the use of resources must match local conditions. Those who do the monitoring and set the resource use levels must

ELINOR OSTROM, NOBEL LAUREATE

(© Steve C. Mitchel/epa/Corbis)

FIGURE 2.24 In her Nobel Prize–winning research, Elinor Ostrom successfully challenged the idea that governments are always better at protecting natural resources than are organized communities of resource users.

SUSTAINABLE MANAGEMENT OF COMMON-POOL RESOURCES BY LOCAL COMMUNITIES

A community-managed irrigation system in Nepal

Community grazing lands in the Basque Country

FIGURE 2.25 Communities around the world have owned and sustainably managed common-pool resources for centuries without outside intervention. For example, irrigation systems in Nepal (left), which have been built and successfully managed for centuries by communities of local farmers, supply water to over 70% of the irrigated lands in the country. Community management has also sustained the productivity and health of grazing lands (right) in the Basque Country of northern Spain for centuries.

be accountable to the users of the resource or consist of the resource users themselves, which will ensure accountability for poor management decisions.

Ostrom, who won a Nobel Prize for her work on common-pool resources in 2009, also pointed out that it is important for the individuals affected by any rules governing resource use to participate in decision making. The system must include a tactic for penalizing those who break the rules and low-cost, rapid mechanisms for conflict resolution. In addition, the users of the common-pool resources, or people subject to the users, must assign the penalties for violating resource use rules. Ostrom's work suggests that these penalties should be graduated, that is, smaller for first or minor offenses and larger for repeated or major offenses. Finally, the rights of the local users of a common pool to organize and manage the resource must be recognized by external authorities.

There are many examples of communities that have managed common-pool resources sustainably for centuries, such as those pictured in **Figure 2.25**. It was precisely the existence of such communities that nurtured Ostrom's challenge to the proposed inevitability of Hardin's Tragedy of the Commons. Her findings offer hope as we search for a sustainable relationship with Earth's environment and its myriad common-pool resources.

⚠ Think About It

1. How might British colonial administration of Maasai lands have been improved by including the traditional ecological knowledge of the Maasai in their management schemes?

2. Why is monitoring of the resources by users or people responsible to the users critical in Ostrom's model for successful local management of common-pool resources?

2.8–2.10 Solutions: Summary

Linking environment with economics has the potential to contribute significantly to sustainable development. Although most economists view the environment as outside the economic system, ecological economists think of the economy as embedded in the environment. A major effort in ecological economics has been its attempts to assign economic value to nature. One solution to Hardin's Tragedy of the Commons is to assign and enforce private property rights. In the example of the wild vicuña, allowing local people to harvest and sell the wool contributed to the recovery of that species. Other resources, such as water, are overexploited because their price does not vary with supply and demand. Allocating tradable use rights can lead to more sustainable use of such resources. Nobel Laureate Elinor Ostrom has shown that, under specific conditions, communal management is a viable alternative to private ownership or regulation by a central government authority.

Answer the following questions for each chapter section and then answer the Central Question.

Central Question: How can linking ecology and economics help reduce societies' environmental impacts?

2.1–2.4 Science

- What makes up economies and ecosystems?
- What are the forms of energy and what are the rules that govern it?
- How do matter and energy move through ecosystems?
- How do matter and energy move through economic systems?

2.5–2.7 Issues

- What factors fuel and limit economic growth?
- How can economic system models have environmental consequences?
- What is the Tragedy of the Commons and how does it impact the environment?

Ecosystems, Economic Systems, and You

A sustainable economy starts with you! We can all play a role in working toward a society that uses resources wisely and also benefits from economic growth. Get to know the economic foundations of your community and investigate the environmental impacts of the industries in your community.

☐ *Study the local and regional economics and environment.*

What is the economic base for your community? Is the economy dominated by one or a few industries—for example, agriculture, manufacturing, or mining—or is it a mixed economy? How is employment divided among the main sectors of the regional or local economy? What developmental initiatives are currently being pursued in your community or region? What are the areas of growth or decline?

Familiarize yourself with the local and regional ecology and the main environmental issues in your community. Learn how economic development may be affecting the local or regional ecosystems. How is the local economy impacting air or water quality? What challenges does the local economy create for wildlife in your area? What steps are being taken to reduce the environmental impacts of water pollution, air pollution, or land use changes? Are there particular parts of your community being impacted or neglected?

☐ *Promote economic and environmental sustainability.*

Does your community have a long-term plan for sustainability, which attempts to promote economic development with reduced environmental impact? For example, does your community have a program to promote energy efficiency and reduce waste through recycling programs? If so, try to become involved in activities and planning related to the sustainability program. If not, join with other interested individuals to encourage such a program by writing opinion pieces for the local media and letters to local or regional leaders. It may be most efficient to focus on a particular issue, such as recycling or water or energy efficiency, of special significance in your area.

☐ *Make environmentally constructive choices.*

How each of us affects the surrounding environment begins with our personal choices. As a consumer, every time you purchase a product, you send a message to the manufacturer that you support the values behind its product. Use this purchasing power to support businesses that work to reduce their negative impacts on the environment or that have a positive environmental influence. Join friends or a Meetup group with like-minded economic values and communicate your choices on a blog or Facebook page dedicated to your topic of interest.

2.8–2.10 Solutions

- **What economic strategies and policies support wise resource use?**

- **How can property rights and ownership lead to environmental preservation?**

- **What are some alternative economic approaches to sustainability?**

Answer the Central Question:

_____

_____

_____

_____

_____

_____

_____

Chapter 2

Review Questions

1. Which of the following statements about matter is false?
a. All matter is composed of atoms.
b. An element consists of a single kind of atom.
c. Molecules consist of two or more atoms.
d. Two bound oxygen atoms are a compound.

2. Which of the following is a form of potential energy?
a. Kinetic energy
b. Heat energy
c. Chemical energy
d. Radiant energy

3. Which of these statements is true of energy flow through an ecosystem?
a. Primary production exceeds production at higher trophic levels.
b. Most energy flows through carnivores.
c. Most energy flows through detritivores.
d. Herbivore production equals primary production.

4. Which of the following is a consequence of the second law of thermodynamics?
a. Organisms expend energy for maintenance.
b. Energy in a system is conserved.
c. Energy in a system and surroundings decreases.
d. Energy for work increases over time.

5. What is an externality in an open economic model?
a. Effects of manufacturing on consumer activity
b. Environmental damage due to economic activity
c. Environmental damage to an economic system
d. Effects of consumers on an economic system

6. How does supply and demand influence the price of a product?
a. Price is influenced entirely by demand.
b. Increased supply will lead to increased price.
c. Increased supply will lead to decreased price.
d. Decreased supply will lead to decreased price.

7. Which of the following does not fit the Tragedy of the Commons model?
a. A family grazing cattle on their own land
b. Farmers grazing sheep on community lands
c. Independent boat owners fishing the open sea
d. Several utility companies emitting air pollutants

8. Why was access to large areas with varying local climates critical to sustaining the traditional cattle-based economy of the Maasai?
a. It allowed raising unlimited numbers of cattle.
b. It avoided cooperation with neighboring groups.
c. It included grazing lands that remained productive during droughts.
d. It provided a means to avoid predators of cattle, such as lions.

9. What role did the business community play in the campaign to reduce production of CFCs?
a. They first resisted regulation, but became supportive once alternatives were developed.
b. They resisted control efforts throughout the campaign and still do.
c. They were initially supportive of CFC regulation but later resisted.
d. They remained neutral throughout.

10. What key observation inspired the economist Elinor Ostrom to challenge Hardin's Tragedy of the Commons model?

a. Her ideas were not inspired by observations; they were entirely theoretical.
b. She observed that many local communities have managed common-pool resources for centuries without outside regulation.
c. She observed that common-pool resources are almost never depleted in the absence of regulation.
d. She observed that traditional communities do not require rules for sustaining resources.

Critical Analysis

1. How are ecosystems and economic systems similar? How are they different?

2. What are the average energy losses in each of the pyramids in Figures 2.16 and 2.17? What are some implications of the differences in energy loss among these energy pyramids?

3. Examine Figure 2.9. Describe how eliminating all decomposers and detritivores would affect the cycling of carbon in the global ecosystem.

4. Explain why the energy flow through carnivores in an ecosystem cannot exceed primary production in the ecosystem.

5. Discuss how Pigovian taxes and subsidies can be used to protect the environment (e.g., to protect a nation's forests from deforestation).

Find additional resources and links online at www.macmillanhighered.com/launchpad/molles1e.

Central Question: How can we protect species in an increasingly human-dominated world?

Explain the ecology of populations and interactions in communities.

SCIENCE

Conservation of Endangered Species

Analyze the threats to survival of species.

Discuss the legal, social, and economic factors that help conserve and restore threatened and endangered species.

ISSUES

SOLUTIONS

(Joel Sartore/National Geographic Stock)

(Joel Sartore/National Geographic Stock)

(Joel Sartore/National Geographic Stock)

Multiple threats endanger various species.

Threatened and Endangered Species

The Mexican spotted owl is threatened by logging of its old growth forest habitat. Koalas face a variety of threats to their existence, including infectious disease and wildfires. The highly restricted relict darter has become endangered as a result of habitat destruction.

The morning of March 24, 1995, Yellowstone National Park was still blanketed in snow when six gray wolves—five males and one female—stepped out of their pen in the Lamar Valley. As they broke an infrared beam that spanned the opening of their cage, a radio signal alerted anxious park employees stationed at a distance. More than 24 hours later, Doug Smith, a field biologist with a bushy mustache, spotted the wolves for the first time,

frolicking and exploring the unfamiliar landscape. "They were cavorting, playing and checking things out," he told the *New York Times*. It was, he said, a celebration of their "recent liberation."

This wolf pack was the first to set foot in one of the world's most iconic national parks in more than 50 years. The canine predators had been eradicated from Yellowstone and the surrounding area by ranchers protecting their livestock and by the federal government, which had laid out poison carcasses to kill them, dynamited their dens, and offered bounties to hunters who brought in their heads and skins. Their 1995 reintroduction was hugely controversial among ranchers and others, who unsuccessfully sued to stop it. As that battle played out, the wolves were live-trapped in Canada, outfitted with radio collars, and housed at this pen where they were fed elk, deer, moose, and bison, while becoming accustomed to the sights, sounds, and smells of their new home.

"We reached the old wolf in time to watch a fierce green fire dying in her eyes. I realized then, and have known ever since, that there was something new to me in those eyes—something known only to her and to the mountain."

Aldo Leopold, from *Thinking Like a Mountain,* 1949

By 2003, 31 reintroduced wolves had multiplied and Yellowstone boasted 174 wolves in a dozen packs. But the wolves ventured farther and farther outside the park boundaries, once again creating conflicts with ranchers. Over the years, Doug Smith became known as the "Wolf Man" and continued to follow the wolves' movements and activities with radio collars. On October 3, 2009, he learned that an alpha female he had been following—wolf 527F—had been legally shot by a hunter in Montana. Smith was devastated. The cycle of exploitation and conservation had come full circle. "I have a deep-seated, fierce love of nature, and I'm afraid that slowly, piece by piece, we're losing it all," he later told the *Christian Science Monitor.*

In this chapter and the next, we focus on the challenge of restoring and conserving **biodiversity** in the modern world. Biodiversity refers to biological variety, from genes and species to diversity at the scale of ecosystems and the globe. Some of the reasons to conserve biodiversity are very practical. The extinction of species results in the irreplaceable loss of potential sources of food, medicinal drugs, industrial chemicals, and other materials and services potentially useful to humans. In addition, as we will discuss in Chapter 4, some species may play key roles in sustaining the health of the ecosystems on which all populations depend. Beyond these practical reasons, biocentric ethics, such as those expressed by Doug Smith, commonly counsel stewardship of nature and doing no harm to other species.

biodiversity Biological variety from genes and species to diversity at the scale of ecosystems and the globe.

Central Question

How can we protect species in an increasingly human-dominated world?

(Jim Peaco, Yellowstone National Park, NPS)

3.1–3.5 Science

species A group of interbreeding, or potentially interbreeding, populations, reproductively isolated from other populations.

populations All the individuals of a species that inhabit a particular place at the same time.

endangered species A species whose populations have become so small that they may become extinct in the near future.

population ecology Branch of ecology that is concerned with the factors influencing the structure and dynamics of populations, including population size, distribution, and growth.

genetic diversity The sum of the different genes and gene combinations found within a single population of a species and across populations of the same species.

genes Stretches of DNA that direct the growth, development, and functioning of organisms.

To understand the goals of conservation, we need to take a step back and understand what it is we are conserving. What is it that makes a wolf a wolf? Biologists define a **species** as a group of interbreeding, or potentially interbreeding, populations that are reproductively isolated from other populations. For example, the gray wolf, *Canis lupus,* is a widely distributed species with populations across the Northern Hemisphere from western Europe through eastern Canada. The species is made up of many **populations,** which are defined as all the individuals of a species that inhabit a particular place at the same time. Those 31 new wolves in Yellowstone became the founding members of a new population. Any attempt to restore or conserve an **endangered species,** one whose populations have become so small that they may become extinct in the near future, draws on the branch of ecology called **population ecology.** Population ecology is concerned with the factors influencing the structure and dynamics of populations, including population size, distribution, and growth.

3.1 Genetic diversity is essential to the evolution and survival of populations

The genetic diversity of populations is critical to conservation. Mexican gray wolves living in the mountains of southern Arizona and New Mexico represent a regional population that is genetically distinct from the wolves in Yellowstone. These genetic differences contribute to the overall genetic diversity of gray wolves worldwide. We define **genetic diversity** as

the sum of the different genes and gene combinations found within a single population of a species, as well as across populations of the same species. **Genes** are stretches of DNA that direct the growth, development, and functioning of organisms (**Figure 3.1**). For example,

DNA, GENES, AND PROTEINS

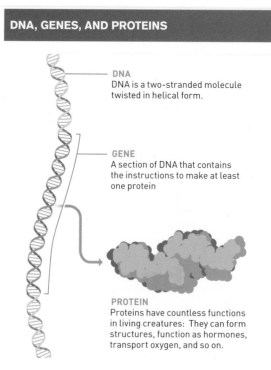

DNA
DNA is a two-stranded molecule twisted in helical form.

GENE
A section of DNA that contains the instructions to make at least one protein

PROTEIN
Proteins have countless functions in living creatures: They can form structures, function as hormones, transport oxygen, and so on.

FIGURE 3.1 DNA is the hereditary molecule common to all life on Earth. Within its code are the instructions to build a living organism, including how it will grow and develop, as well as its unique physical features.

VARIATION IN FACIAL MARKINGS

(Wendy Shattil and Bob Rozinski/Getty Images)

(Kennan Ward/Corbis)

(Aaron Ferster/Getty Images)

FIGURE 3.2 Distinctive facial markings and hair color are visible expressions of genetic variation among these Mexican gray wolves, *Canis lupus baileyi*.

some genes code for proteins that function as structural molecules, making up tendons or ligaments, whereas proteins, called enzymes, facilitate specific chemical reactions. For instance, the enzyme lactase speeds up the digestion of the lactose sugar in milk.

We can see abundant evidence of genetic diversity in facial differences among individual people. These differences allow us to instantly recognize friends and acquaintances in a crowd. Such facial variation can also be seen in the facial markings of Mexican gray wolves (**Figure 3.2**). However, most genetic variation—such as the differences in DNA that cause disease resistance, or tolerance of high or low temperatures—is invisible to casual inspection and is best studied using DNA sequencing.

Conservation biologists have demonstrated that genetic variation, or genetic diversity, increases the probability that populations will survive in the face of environmental challenges. Consider, for example, genetic variation in the maximum temperature a fish can withstand without dying, called *lethal maximum temperature*. As **Figure 3.3** shows, a population with low genetic variation may not include any individuals capable of surviving temperatures reached during an unusual heat wave. As a result, the population dies out entirely. Meanwhile, some individuals in another population with higher genetic variation do survive. Genetic diversity not only helps ensure survival in the face of environmental challenges, but it also enables

a population to change and adapt to new environmental circumstances—an important factor if species are to survive on a planet with a human population changing the environment as it grows.

Genetic Diversity and Selective Breeding

Genetic diversity has been crucial to **domestication,** the deliberate change of a wild animal or plant species to better meet the needs of humans. Humans have produced the many existing varieties of food crops and beautiful flowering plants and animal breeds, such as dairy cows and racehorses, by selectively breeding certain individuals within an ancestral wild population. The individuals chosen for breeding during domestication were those that possessed certain desired characteristics, such as the ability to produce large amounts of milk or the ability to run fast.

Domestic dogs descended from wolves, and their breeds have been refined for particular abilities, such as herding livestock, hunting game, or guarding property. The Border collie, for instance, is a specialist at herding sheep (**Figure 3.4**). Over generations, shepherds have kept and bred only those Border collies with characteristics helpful in controlling flocks of sheep without injuring them, including some of the hunting behaviors of its wolf ancestors like stalking, chasing, and staring down a prey animal. Meanwhile, a Border collie

domestication The deliberate change of a wild animal or plant population through selective breeding to better meet the needs of humans.

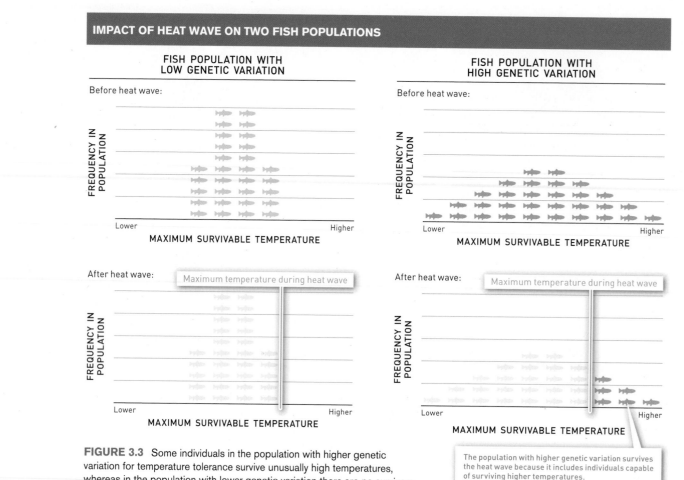

IMPACT OF HEAT WAVE ON TWO FISH POPULATIONS

FISH POPULATION WITH LOW GENETIC VARIATION

FISH POPULATION WITH HIGH GENETIC VARIATION

FIGURE 3.3 Some individuals in the population with higher genetic variation for temperature tolerance survive unusually high temperatures, whereas in the population with lower genetic variation there are no survivors.

The population with higher genetic variation survives the heat wave because it includes individuals capable of surviving higher temperatures.

artificial selection A process in which humans "select" which individuals in a population mate to produce descendants with desired characteristics.

that shows behaviors associated with immobilizing and killing prey—behaviors essential to wild wolves—is not kept as a working dog, nor is it bred (**Figure 3.5**). This kind of selective breeding is called **artificial selection**—wherein humans "select" which individuals will reproduce to form future generations.

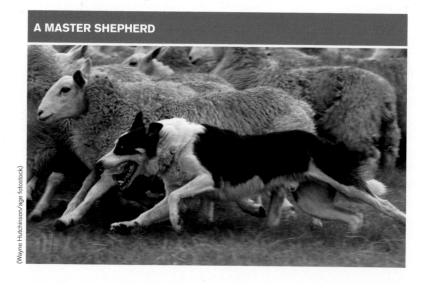

A MASTER SHEPHERD

(Wayne Hutchinson/age fotostock)

FIGURE 3.4 The Border collie, a herding dog that originated on the border between Scotland and England, employs several behaviors present in its wolf ancestors to control the movement of sheep.

In the plant world, sunflowers are grown worldwide as a source of oil and seeds. The wild ancestor of the domestic sunflower, *Helianthus annuus,* is native to North America, where it was domesticated and grown as a crop by Native Americans more than 4,000 years ago. Native Americans, and later Russian plant breeders, selectively bred the sunflowers that had the largest seeds and *inflorescences* (a tight cluster of small flowers) with other sunflowers that had large inflorescences and seeds. Over time, this resulted in a plant with a giant flowering head, bearing large seeds (**Figure 3.6**).

Genetic Diversity and Natural Selection

Charles Darwin, the famed 19th-century naturalist, recognized a parallel between changes in populations produced by plant and animal breeders and changes in natural populations. He proposed that as individuals are born into a

ARTIFICIAL SELECTION FOR BEHAVIOR

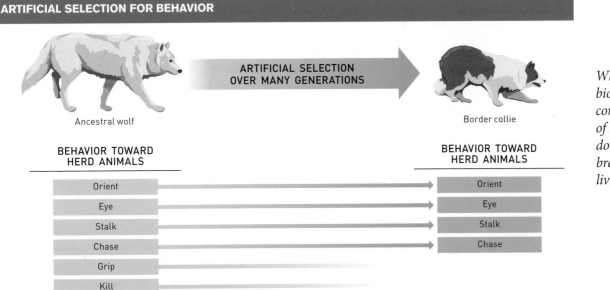

Ancestral wolf

ARTIFICIAL SELECTION OVER MANY GENERATIONS

Border collie

BEHAVIOR TOWARD HERD ANIMALS

Orient
Eye
Stalk
Chase
Grip
Kill

BEHAVIOR TOWARD HERD ANIMALS

Orient
Eye
Stalk
Chase

Why are conservation biologists working to conserve populations of rare varieties of domestic plants and breeds of domestic livestock?

FIGURE 3.5 The herding behavior in modern Border collies is the result of a long process of artificial selection by shepherds for a dog capable of controlling the movement of sheep without injuring them.

population, the characteristics of some individuals better fit the requirements of the environment than others. Those individuals with the more favorable traits survive and reproduce at a higher rate. Consequently, the traits more closely matching environmental requirements increase in frequency in the population—more individuals in the next generation possess the favorable traits compared with the number that possessed the traits in the parental generation. Darwin called this process **natural selection.**

A vivid example of how natural selection works can be found in the American chestnut trees that once grew in abundance in the forests of eastern North America. Individual chestnut trees grew to heights of over 30 meters (100 feet) with trunk diameters of 1 to 2 meters (3 to 6 feet) (Figure 3.7). Then around 1900, the blight fungus (*Cryphonectria parasitica*) arrived in North America, carried by imported Asian chestnut trees, which were resistant to the fungus.

natural selection A process of interaction between organisms and their environment that results in different rates of reproduction by individuals in the population with different physical, behavioral, or physiological characteristics; can change the relative frequencies of particular genes in the population—that is, in evolution.

ARTIFICIAL SELECTION FOR BEHAVIOR

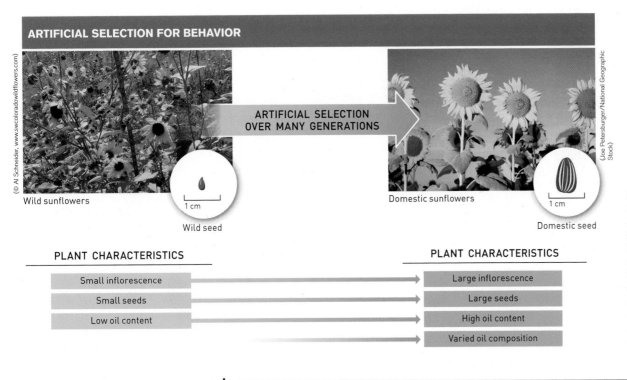

Wild sunflowers — Wild seed — 1 cm

ARTIFICIAL SELECTION OVER MANY GENERATIONS

Domestic sunflowers — Domestic seed — 1 cm

PLANT CHARACTERISTICS

Small inflorescence
Small seeds
Low oil content

PLANT CHARACTERISTICS

Large inflorescence
Large seeds
High oil content
Varied oil composition

FIGURE 3.6 Artificial selection in sunflowers, *Helianthus annuus,* has emphasized larger inflorescences (a cluster of small flowers called *florets*), larger seeds, higher oil content, and variation in the composition of oils to serve diverse purposes—from stability when frying to a delicate taste as a salad oil.

FIGURE 3.7 The American chestnut, *Castanea dentata,* was once one of the giant trees of the temperate forests of eastern North America, where it produced vast amounts of food for wildlife and was a source of income for local people. Today, the American chestnut survives mainly as shrubby sprouts from the stumps of trees infected by the chestnut blight fungus, *Cryphonectria parasitica.*

Was the genetic diversity in the American chestnut population likely to be higher or lower after the chestnut blight fungus swept through the population? Explain.

evolution A change in the genetic makeup of a population as a consequence of one of several different processes, including natural selection and selective breeding.

LOSS OF NATURAL WEALTH

(Eben Lehman/Forest History Society)

In less than 50 years, chestnut blight swept across the American chestnut's entire natural range, killing nearly 4 billion trees. In most circumstances, only the root systems survived, but here and there large American chestnut trees survived because they either grew in soil and climate conditions unfavorable to the fungus or they possessed genes that gave them some resistance to the blight fungus. In the small population of surviving trees, many have blight-resistant genes. In other words, the genes that give resistance to the blight fungus have increased in frequency in the population.

The increase in frequency of resistance genes relative to susceptible genes in the population is a good example of **evolution** by natural selection (**Figure 3.8**). In general, we define evolution as a change in the genetic makeup of a population as a consequence of one of several different processes, including natural selection and selective breeding.

⚠ Think About It

1. What is the most likely reason that Asian chestnut trees were resistant to the chestnut blight fungus when it was introduced to North America, whereas the American chestnut tree was not?

2. How does genetic variation help a population survive in the face of changing, uncertain environments?

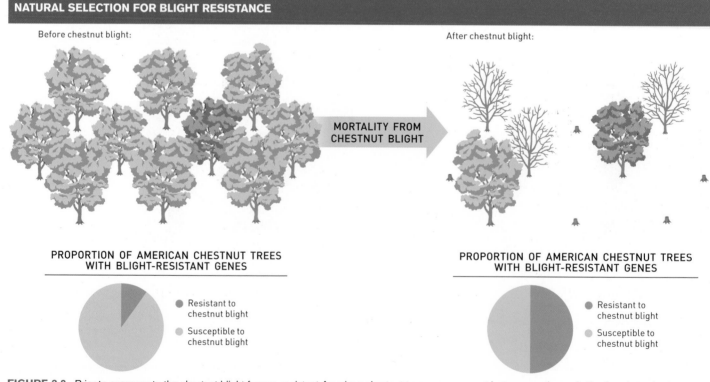

NATURAL SELECTION FOR BLIGHT RESISTANCE

Before chestnut blight:

After chestnut blight:

MORTALITY FROM CHESTNUT BLIGHT

PROPORTION OF AMERICAN CHESTNUT TREES WITH BLIGHT-RESISTANT GENES

● Resistant to chestnut blight
● Susceptible to chestnut blight

PROPORTION OF AMERICAN CHESTNUT TREES WITH BLIGHT-RESISTANT GENES

● Resistant to chestnut blight
● Susceptible to chestnut blight

FIGURE 3.8 Prior to exposure to the chestnut blight fungus, resistant American chestnut trees were present but apparently rare in the American chestnut population. However, following massive blight-induced mortality, resistant trees represented a much higher proportion of the population of surviving large trees.

3.2 Distribution and abundance are key indicators of population security

It is no coincidence that endangered populations, such as the Venus fly trap of North and South Carolina, have small populations and often live in relatively small geographic areas. While thousands of geographically restricted species face the prospect of imminent **extinction,** the loss of all members of a species, other species are abundant and have wide **distributions.** For example, the familiar house sparrow, *Passer domesticus,* now lives on all continents except Antarctica (**Figure 3.9a**).

In contrast, the endangered mountain gorilla, *Gorilla beringei beringei,* lives in just two small areas in East Africa (**Figure 3.9b**). The total distribution of the mountain gorilla is approximately 700 square kilometers (270 square miles), about one-fourth the area of Rhode Island, the smallest U.S. state, or about one-fortieth the area of the tiny African country of Rwanda.

Some invertebrate species survive in even smaller areas. The Bay checkerspot, an endangered butterfly, survives in about 75 square kilometers (29 square miles), a small remnant of its former range in the central coastal region of California, approximately one-tenth the area occupied by mountain gorillas (**Figure 3.9c**).

Why might a species be abundant in one place and absent in another? Climate turns out to be a major

How might the fact that house sparrows are associated with towns and cities explain its wide geographic distribution?

extinction The loss of all members of a species.

distribution The geographic range of a species.

RANGE OF GEOGRAPHIC DISTRIBUTIONS

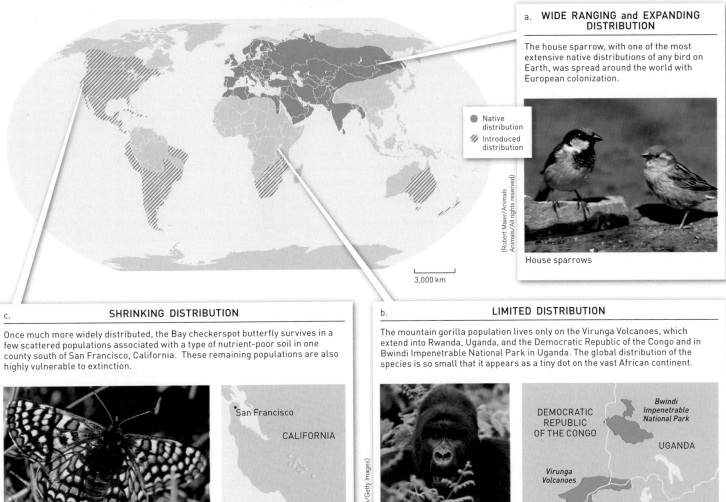

a. **WIDE RANGING and EXPANDING DISTRIBUTION**

The house sparrow, with one of the most extensive native distributions of any bird on Earth, was spread around the world with European colonization.

● Native distribution

▨ Introduced distribution

House sparrows

3,000 km

c. **SHRINKING DISTRIBUTION**

Once much more widely distributed, the Bay checkerspot butterfly survives in a few scattered populations associated with a type of nutrient-poor soil in one county south of San Francisco, California. These remaining populations are also highly vulnerable to extinction.

Bay checkerspot butterfly

San Francisco

CALIFORNIA

50 km

b. **LIMITED DISTRIBUTION**

The mountain gorilla population lives only on the Virunga Volcanoes, which extend into Rwanda, Uganda, and the Democratic Republic of the Congo and in Bwindi Impenetrable National Park in Uganda. The global distribution of the species is so small that it appears as a tiny dot on the vast African continent.

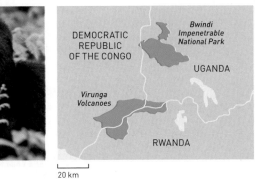

Mountain gorilla

DEMOCRATIC REPUBLIC OF THE CONGO

Bwindi Impenetrable National Park

UGANDA

Virunga Volcanoes

RWANDA

20 km

FIGURE 3.9 The wide-ranging distribution of the house sparrow, *Passer domesticus,* contrasts sharply with the distributions of endangered species such as the mountain gorilla, *Gorilla beringei beringei,* and the Bay checkerspot butterfly, *Euphydryas editha bayensis.*

reason why all species, except for humans, have restricted geographic ranges. For example, the **adaptations**—traits favored by natural selection for surviving and reproducing in a particular environment—that enable an organism to withstand the cold of Arctic winters do not prepare it for the heat of a desert summer. Other reasons might be that one region lacks suitable physical habitat, such as caves, which are critical to cave-dwelling species like bats, or that a region is home to a fatal disease to which a species is susceptible.

Rare to Abundant

When population ecologists refer to the **abundance** of a species, they mean the number of individuals in a population, also called *population size*. We can represent population size in two ways: as total population, the total number of individuals in the population, or as **population density,** the number of individuals inhabiting some defined area, such as numbers per square kilometer (km²) or per square mile (mi²). For example, the total house sparrow population has been estimated to number half a billion. In contrast, the global population of mountain gorillas totals approximately 700. Ecologists monitoring one of the last remaining populations of Bay checkerspot butterflies estimated that its population ranged from 12,000 during poor years to over 500,000 during a favorable year.

Population densities also vary widely among species. Larger species generally live at lower densities than smaller species and have lower total population sizes. For instance, the overall population density of mountain gorillas is approximately one per square kilometer. In contrast, the population density of the caterpillars of the Bay checkerspot butterfly can vary from 10,000 to more than 1 million per square kilometer.

⚠ **Think About It**

1. Why might larger species be generally more vulnerable to extinction compared with smaller species?

2. Why might herbivores be at less risk of extinction compared with carnivores in the same ecosystem? (Consider Figure 2.11, page 41.)

3.3 **Populations change**

An animal population is not static, but rather changes over time. The geographic distribution of a population

(David Parsons/Getty Images)

FIGURE 3.10 Female plains cottonwood trees, *Populus deltoides*, produce millions of tiny wind- and water-dispersed seeds each year. A few seeds among these millions will land on sites where they can germinate and grow to a mature tree.

can expand or contract, and the number of organisms can increase or decrease with each passing year. Understanding the factors that contribute to population size is critical to protecting and rehabilitating populations of endangered species.

Rates of Reproduction

Reproductive rates vary widely among species. For example, a plains cottonwood tree produces approximately 25 million seeds per year (**Figure 3.10**). In contrast, a Bay checkerspot butterfly lays an average of about 730 eggs during its one season of life. Female mountain gorillas generally give birth to one offspring once every four years. Expressed in terms of births per 1,000 females per year, mountain gorillas have a reproductive rate of approximately 250 births per 1,000 females per year, whereas the annual rate of reproduction for the Bay checkerspot butterfly population is approximately 730,000 eggs per 1,000 females per year. The great contrast between species in reproductive rates translates into great differences in potential for population growth.

Population Growth

Populations can grow in one of two stereotypical ways. The first pattern ecologists recognize is **J-shaped population growth,** which is also called **exponential population growth.** During exponential growth, a population grows by a fixed rate, called r, the rate of population increase per capita (per individual); this is calculated as the birthrate minus the death rate. The population growth rate can also be expressed as a percentage of population size, for example, 4% per year.

Why do you think expanding distribution and increasing abundance are generally included in endangered species management plans?

adaptations Traits favored by natural selection for surviving and reproducing in a particular environment.

abundance (population size) The number of individuals in a population.

population density The number of individuals inhabiting some defined area.

J-shaped (exponential) population growth Population growth that occurs at a constant, or fixed, rate per capita and that produces a characteristic J-shaped pattern of increase in population size over time.

J- AND S-SHAPED POPULATION GROWTH

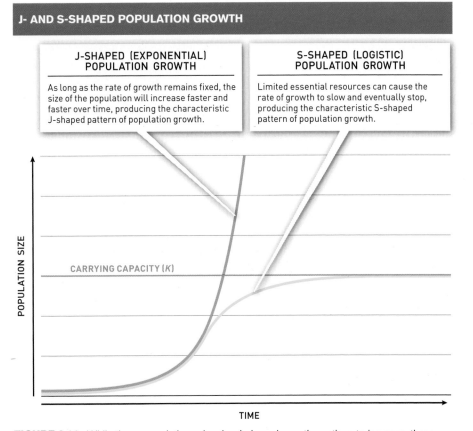

J-SHAPED (EXPONENTIAL) POPULATION GROWTH

As long as the rate of growth remains fixed, the size of the population will increase faster and faster over time, producing the characteristic J-shaped pattern of population growth.

S-SHAPED (LOGISTIC) POPULATION GROWTH

Limited essential resources can cause the rate of growth to slow and eventually stop, producing the characteristic S-shaped pattern of population growth.

POPULATION SIZE

CARRYING CAPACITY (K)

TIME

FIGURE 3.11 While those populations showing J-shaped growth continue to increase, those undergoing S-shaped growth stop growing when something limits the continued growth of the population—predators, limited food supplies, finite habitat space, and so on. The maximum size of a population able to survive over the long term in a particular area is called the carrying capacity.

As long as the rate of growth remains fixed, the size of the population will increase faster and faster over time, because each year the population is larger, producing the characteristic J-shaped pattern of population growth (see the orange line in **Figure 3.11**).

We often see exponential growth when a population is living at a low density and environmental conditions, including climate and food resources, are favorable. For example, Rocky Mountain bighorn sheep reintroduced to the Wheeler Peak Wilderness in northern New Mexico grew at an average rate of approximately 28% annually for seven years. At this constant rate, the population grew from 32 adults to 180. Because a fixed 28% is added to the population annually, the *number* of individuals added, in this case each year, increases. Continued exponential population growth would increase the size of this population to more than 4,000 in just 20 years, producing a J-shaped pattern of growth (**Figure 3.12**).

J-shaped growth cannot continue indefinitely, since the population would eventually exhaust essential resources such as energy, space, and nutrients. As such

resources are depleted, the per capita rate of growth decreases, resulting in a plateau of population size. This is what ecologists call **S-shaped,** or **logistic, growth** (see the yellow line in Figure 3.11). If the bighorn sheep population shown in Figure 3.12 continues to grow, the sheep would strip their mountain home of its plant cover. Dwindling food supplies might slow population growth and eventually stop it, unless some other factor such as disease or predators did so first.

The population size at which S-shaped population growth stops is called the **carrying capacity,** generally abbreviated with the letter **K.** Populations approaching carrying capacity may not necessarily level off as they approach *K,* but may "overshoot" it before declining again.

For an animal population, the carrying capacity may be determined by food supply, the number of nesting sites or breeding territories, or access to shelter from predators or severe weather. For a plant population, the factors determining carrying capacity include the availability of nutrients, water supply, or access to light. Carrying capacity differs from one species to another. For example,

If the per capita rate of increase, r, is held constant, why does a population grow faster and faster over time during J-shaped (exponential) population growth?

S-shaped (logistic) growth Population growth in which the per capita rate of growth decreases with increasing population size as a result of predation or reduced availability of food, space, or other resources; eventually levels off at carrying capacity.

carrying capacity (K) The number of individuals in a population that an environment can support over the long term.

FIGURE 3.12   The population grew exponentially at a fixed rate of 28% per year, from 32 to 180 individuals in seven years (numbers rounded to nearest individual); the population growth shown from 180 to 4,460 sheep is potential growth, assuming a continued constant rate of exponential growth. Growing exponentially, this population would increase by nearly 140 times in 20 years.

POPULATION OF ROCKY MOUNTAIN BIGHORN SHEEP IN THE WHEELER PEAK WILDERNESS, NEW MEXICO

Rocky Mountain bighorn sheep

YEAR	NUMBER ADDED TO THE POPULATION	POPULATION SIZE
0	–	32
1	9	41
2	11	52
3	15	67
4	19	86
5	24	110
6	31	141
7	39	180
8	51	231
9	64	295
10	82	377
11	107	484
12	135	619
13	173	792
14	222	1,014
15	$0.28 \times 1,014 = $ 284	$+ 1,014 = $ 1,298
16	363	1,661
17	466	2,127
18	550	2,722
19	762	3,484
20	976	4,460

The number added to the population is generated by multiplying 0.28 (the growth rate) by the previous year's population size.

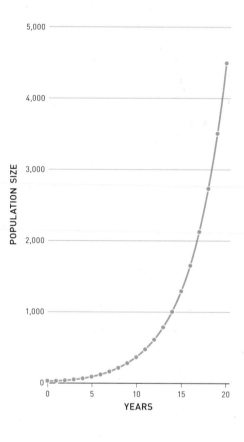

population biologists estimate that the carrying capacity of bighorn sheep in the Wheeler Peak Wilderness is approximately 180, or 3.5 individuals per km² (9 per mi²). By contrast, the Wheeler Peak Wilderness might support a single mountain lion, one of the main predators of bighorn sheep.

Because carrying capacity is determined by environmental conditions, it may vary over time. During periods of abundant rain, deserts may support high levels of plant growth, which, in turn, supports large populations of small herbivorous mammals, mainly rodents. However, droughts in these same deserts may last for a decade or more and reduce the carrying capacity of both plant and herbivore populations.

Controls on Population Size

Most populations tend to fluctuate around some average number, which is typically at or below the carrying capacity. For instance, during the last half of the 20th century, the mountain gorilla population on the Virunga Volcanoes ranged between approximately 250 and 450 individuals (**Figure 3.13**). Meanwhile, a population of the Bay checkerspot butterfly on Jasper Ridge on the Stanford University campus ranged from fewer than 100 to approximately 5,000 individuals.

Such fluctuations are the result of interactions between factors that promote growth and reproduction and factors that suppress them. Soil moisture promotes growth in plants, whereas abundant food might signal to animals that it is time to reproduce. By contrast, infectious disease or a hard frost might limit growth.

Mechanisms of population control that change with the density of a population are called **density-dependent factors.** Infectious disease is considered a density-dependent factor because it can more easily spread through a population living at high densities than one living at low densities. Similarly, predation pressure also increases with population density as predators focus on areas with lots of prey.

Density-independent factors are controls on populations *not* affected by population density (**Table 3.1**). Generally, they include physical aspects of the environment such as drought, floods, and extreme temperatures.

All populations are subject to both density-dependent and density-independent factors, but the relative importance of each varies substantially from one

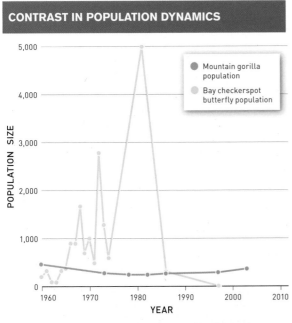

CONTRAST IN POPULATION DYNAMICS

- ● Mountain gorilla population
- ● Bay checkerspot butterfly population

FIGURE 3.13 The population of mountain gorillas on the Virunga Volcanoes has varied by 200, from a low of 250 to a high of 450 individuals. Meanwhile, the number of Bay checkerspot butterflies on Jasper Ridge varied by nearly 5,000, from 80 to 5,000 between 1960 and 1997, when it became locally extinct.

TABLE 3.1 LIST OF COMMON DENSITY-INDEPENDENT AND DENSITY-DEPENDENT POPULATION-REGULATING FACTORS	
Density-Independent Factors	**Density-Dependent Factors**
Floods	Intraspecific competition (competition among members of the same species)
Salinity	Interspecific competition (competition among members of different species)
Drought	Predation (one organism preys on another)
Extreme Temperatures	Parasitism (parasites harm a host organism)
Pollution	Disease
Fire	Fire (for a population of fire-prone plant)

population to another. For instance, the chief causes of death recorded among mountain gorillas are density dependent, including disease, parasitic infections, and occasional predation by leopards. By comparison, Bay checkerspot butterfly populations are more subject to density-independent factors, particularly weather conditions.

⚠ **Think About It**

1. In Section 3.3, we have emphasized the influence of birth and death rates on population growth. How might migration affect population growth?

2. How can fire act as a density-dependent control factor for some species and not for others?

3.4 **The life history of a species influences its capacity to recover from disturbance**

Some species' numbers grow slowly and steadily and have stable populations over long periods of time, whereas others have populations that fluctuate wildly in response to changes in environmental conditions. Such differences are due to variation in **life history** among species, which includes such variables as the age at which individuals begin reproducing, the number of offspring they produce, and the rate at which the young survive.

Whales, bears, wolves, and gorillas all have populations that can stabilize close to their carrying capacity, which is why we call them *K*-selected species. Ecologists propose that living near the carrying capacity favors individuals that excel at competing for limited resources in a crowded environment. These species have a long life span, tend to reproduce later in life, and have a small number of large

How could mountain lions preying on bighorn sheep act as a density-dependent population-regulating factor? What other environmental factors might be important in regulating populations of bighorn sheep?

density-dependent factors Mechanisms of population control that change with the density of a population (e.g., infectious disease, predation).

density-independent factors Controls on populations that are not affected by population density (i.e., physical aspects of the environment such as drought, floods, and extreme temperatures).

life history Characteristics of a species, such as the age at which individuals begin reproducing, the number of offspring they produce, and the rate at which the young survive.

***K*-selected species** Organisms with populations that generally stabilize close to their carrying capacity and are often regulated by density-dependent factors.

TABLE 3.2

CHARACTERISTICS OF *K*-SELECTED AND *R*-SELECTED SPECIES

Life History Trait	*K*-Selected Species	*r*-Selected Species
Maturity	Late	Early
Reproductive rate	Low	High
Survivorship	High	Low
Length of life	Long	Short
Population stability	High	Low

Information from Pianka, 1994.

Can we classify all organisms as either r- or K-selected, or are these endpoints in a continuum? Explain.

As humans alter the environment for our own benefit, what sorts of life histories among wild species are we favoring?

offspring that receive intensive parental care. In general, density-dependent regulating factors, such as disease and competition, exert significant controls on populations of *K*-selected species.

Mice, rabbits, dandelions, and cockroaches are examples of **r-selected species,** which grow rapidly when the environment is favorable. You might sum up their philosophy as "live fast, die young." Unlike large, enduring animals such as whales, *r*-selected species are small and subject to catastrophic mortality from harsh weather, fires, and other density-independent factors. They have undergone strong natural selection to reproduce prolifically when the time is right. Table 3.2 contrasts the major characteristics of *K*-selected and *r*-selected life species.

Let's look closely at some animal and plant examples and see how the life history concept applies. Mountain gorillas, which are large and have low reproductive rates, are a perfect example of a *K*-selected species. Female mountain gorillas may weigh up to 100 kilograms (220 pounds) and give birth every fourth year. In comparison, pygmy marmosets, which weigh only

0.1 kilogram (3.5 ounces) and may have two sets of twins per year, appear to be *r*-selected (**Figure 3.14**). As a result of their higher reproductive rates, a pygmy marmoset can produce 16 offspring in the time that a female gorilla produces one. Not surprisingly, while mountain gorillas live at population densities of about one per km² (2.6 per mi²), pygmy marmoset population densities may be as high as 200 per km² (518 per mi²).

The concepts of *r*- and *K*-selection apply to plants as well. For example, the American chestnut, which produces large seed-bearing fruits that number in the thousands, contrasts sharply with the plains cottonwood, which produces millions of tiny wind-dispersed seeds each spring (**Figure 3.15**).

The life history of an organism influences its capacity to recover from environmental disturbances. Ecologists define **disturbance** as a discrete event, such as a fire, earthquake, or flood, that disrupts a population, ecosystem, or other natural system by changing the resources, such as food, nutrients, or space, available to organisms or by altering the physical environment. In general, populations of *r*-selected species are faster to recover from disturbance than are *K*-selected species. For instance, a population of mountain gorillas decimated by massive mortality during a volcanic eruption would take a longer time to recover compared with a population of pygmy marmosets subjected to a similar disturbance. It should not be surprising, then, that endangered species often have *K*-selected life histories.

⚠ Think About It

1. In terms of life histories, why are populations of K-selected species generally slower to recover from disturbance compared with r-selected species?

r-selected species Organisms with populations that generally fluctuate widely in size; subject to catastrophic mortality from harsh weather, fires, and other density-independent factors.

disturbance A discrete event (e.g., a fire, earthquake, or flood) that disrupts a population, ecosystem, or other natural system by changing the resources available or by altering the physical environment.

K-SELECTED VERSUS *r*-SELECTED PRIMATE SPECIES

Mountain gorillas

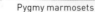
Pygmy marmosets

FIGURE 3.14 On average, a female mountain gorilla produces a single young every four years. In contrast, the tiny pygmy marmoset commonly produces two sets of twins each year.

K-SELECTED VERSUS r-SELECTED TREE SPECIES

American chestnut seeds

Plains cottonwood seeds

Seeds

(Courtesy The American Chestnut Foundation)

(© Iryna Rasko/Alamy)

FIGURE 3.15 An American chestnut tree produces thousands of large seeds annually. In contrast, female plains cottonwood trees produce millions of tiny seeds.

2. When studying life histories, why is it generally more informative to compare closely related organisms, for example, two primates, than to compare two very different organisms, for example, a primate and a tree or a primate and a butterfly?

3.5 Species interactions define biological communities

An **ecological community** consists of all the species—plants, animals, fungi, and microbes—that exist and interact in a given location. The interactions among the community members, such as those between a predator and its prey, shape ecological patterns and influence the evolution of populations (**Figure 3.16**). The web of interactions can be baffling. For example, hummingbirds dip their bills into deep flowers to drink nectar; at the same time, they inadvertently pollinate these plants in a mutually beneficial relationship, called a **mutualism.** Meanwhile, small insects, such as bees and flies, might visit these same flowers and compete with hummingbirds for nectar. If these bees are poor pollinators, then they impose a cost on both the flowers and the hummingbirds. Finally, it's a little-known fact that hummingbirds sometimes eat insects for protein, so they may well snatch one of these flying insects in their flexible bills!

Nature is filled with elaborate mutualisms that go beyond pollination. For instance, Acacia ants live inside the thorns of some Acacia trees and protect them from herbivores; in exchange, they feed on a nutrient-rich liquid the plants secrete. Mycorrhizal fungi growing around plant roots help the plants acquire soil nutrients in exchange for energy-rich carbohydrates.

Environmental Requirements and the Niche

The ecological **niche** is a description of the physical and biological requirements of a species. A species' niche includes where an organism lives—its **habitat**—and its trophic position in the ecosystem, for example, whether it's a primary producer, a predator, or an herbivore (see Chapter 2 for more on trophic relations). However, the niche may also include such factors as the temperature range the species lives in, its water requirements, as well as when and where it feeds.

For instance, mountain gorilla habitat is mountain tropical rain forest, whereas that of the Bay checkerspot butterfly is temperate grassland. Both are herbivores; however, their approaches to herbivory are quite different. Mountain gorillas feed on more than 140 species of plants, which makes them a "generalist herbivore." In contrast, Bay checkerspot caterpillars are "specialist herbivores" that feed mainly on just two plant species (**Figure 3.17**).

The Competitive Exclusion Principle

Competition involves interactions between individuals that depend on the same resources, such as food or space, for their survival and reproduction. Where the resources involved are in limited supply, competition for those resources may impact one or both of the competitors by reducing growth rates, reproductive rates, or probability of survival. Competition is most likely to occur between organisms with similar niches, since by definition they have similar resource needs. Because individuals of the same species share the same niche, competition

Why is a species' habitat not the same as its niche?

ecological community All the species—plants, animals, fungi, and microbes—that exist and interact in a given location.

mutualism Mutually beneficial relationship between organisms.

niche A description of the physical and biological requirements of a species.

habitat Where an organism usually lives (e.g., a forest, coral reef, or marsh).

competition Interactions among individuals that depend on the same resources; generally results in reduced growth, reproduction, or survival of one or both competitors.

INTERACTIONS BETWEEN SPECIES IN COMMUNITIES

(Shutterstock)
Predation

(© Image Source/Alamy)
Herbivory

(© Carlos Ordonez/age fotostock)
Pollination

(© Moodboard/Alamy)
Protection mutualism

FIGURE 3.16 Ecological communities are characterized by a complex web of interactions among species. (a) A black-backed jackal hunting sand grouse in Africa typifies predation, the killing and eating of another organism for food. (b) An elephant stripping foliage from a tree shows an herbivore harvesting living plant biomass. (c) Dusted with yellow pollen, a bee is photographed in the act of transporting pollen from one plant to another as it gathers nectar from the flowers it visits. (d) In another mutualism, a tropical anemonefish finds shelter among the tentacles of a stinging anemone for which the anemonefish provides chemical nutrients from its wastes and food scraps.

How are competing businesses, for example, restaurants or car manufacturers, similar to species living in a natural ecosystem? How are they different?

intraspecific competition
Competition among individuals of the same species.

interspecific competition
Competition among individuals of different species.

competitive exclusion principle If two species with identical niches compete for a limited resource (e.g., nectar), one or the other will be a better competitor and will eventually eliminate the other species.

resource partitioning
Coexisting species use different resources, such as food, nesting sites, and feeding areas.

is often strongest among them. This competition is called **intraspecific competition,** one of the density-dependent factors that regulate populations. Competition among individuals of different species, **interspecific competition,** is most likely to occur where the organisms have similar niches (**Figure 3.18**).

Interspecific competition may influence the number and kinds of species that live together in a community. For instance, one species may exclude another from an ecological community and so reduce the total number of species living together. According to the **competitive exclusion principle,** if two species with identical niches compete for a limited resource (e.g., nectar), one or the other will be a better competitor and will eventually eliminate the other species. It follows from the competitive exclusion principle that when species live together, they generally have slightly different niches.

Studies of animals as diverse as insects, birds, fish, rodents, and lizards have shown that coexisting species use different resources, such as food, nesting sites, and feeding areas. Called **resource partitioning,** one of the

CONTRASTING APPROACHES TO DIET

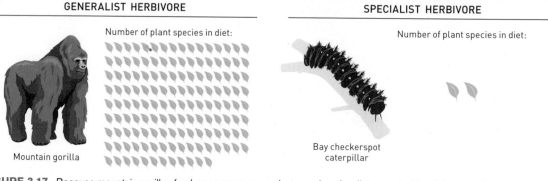

GENERALIST HERBIVORE

Number of plant species in diet:

Mountain gorilla

SPECIALIST HERBIVORE

Number of plant species in diet:

Bay checkerspot caterpillar

FIGURE 3.17 Because mountain gorillas feed on so many more plant species, the dietary part of its niche is much broader compared with that of the Bay checkerspot butterfly, which has a narrow feeding niche.

COMPETITION WITHIN AND BETWEEN SPECIES

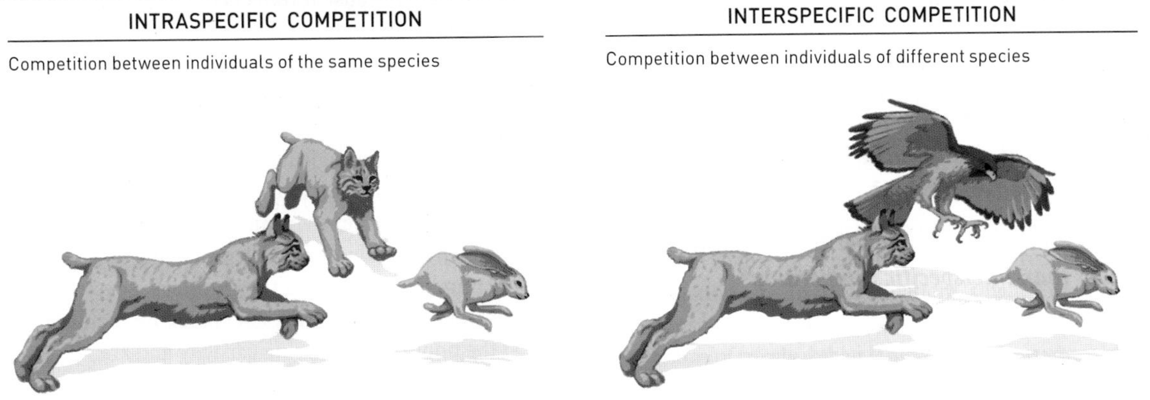

INTRASPECIFIC COMPETITION	INTERSPECIFIC COMPETITION
Competition between individuals of the same species	Competition between individuals of different species

FIGURE 3.18 Both intraspecific (within species) competition and interspecific (between species) competition are important to the ecology of species. The intensity of interspecific competition depends on the amount of overlap in niches between two species.

best-known examples of the phenomenon occurs among coexisting species of woodland warblers. Up to five species of woodland warbler may live together in the forests of northeastern North America despite the fact that all species are about the same size and all eat insects. Since each species forages for insects in a different part of the trees, however, the spatial aspects of their feeding niches are different (**Figure 3.19**). Similarly, many

PARTITIONING OF RESOURCES

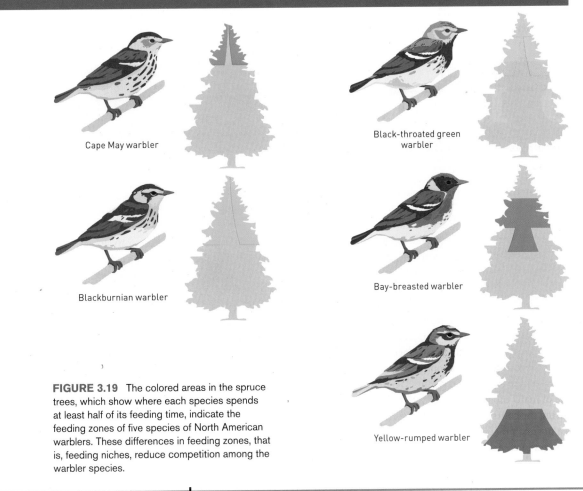

Cape May warbler

Black-throated green warbler

Blackburnian warbler

Bay-breasted warbler

Yellow-rumped warbler

FIGURE 3.19 The colored areas in the spruce trees, which show where each species spends at least half of its feeding time, indicate the feeding zones of five species of North American warblers. These differences in feeding zones, that is, feeding niches, reduce competition among the warbler species.

A SYMBOL OF OLD GROWTH TEMPERATE RAIN FOREST

(Daisy Gilardini/Getty Images)

FIGURE 3.21 The temperate rain forests of southeastern Alaska and northwest Canada are rich in biodiversity, including the Kermode, or spirit, bear, *Ursus americanus kermodei*, a white subspecies of the black bear found only in the central and northern coastal regions of British Columbia.

The typical focus of conservation efforts is on large, impressive species, such as Siberian tigers or Kermode bears. How does this help or hinder the conservation of ecosystems?

Will the relative rate of infection of invasive non-native species by parasites and pathogens remain constant in their new environment over the long term? Why or why not?

invasive species An introduced species that poses a serious threat to native populations.

Temperate habitats are also under threat. More than 50% of historic grasslands and more than 30% of desert habitat have been lost. The Kermode (or spirit) bear, a white subspecies of the black bear, is one temperate species that is seriously threatened by habitat destruction. The bear lives in the temperate rain forests of northwest Canada (**Figure 3.21**). The bear's habitat is shrinking due to the logging of old growth forests in the region and will shrink even further as a result of a recently approved project to build an oil pipeline on Princess Royal Island in British Columbia.

Marine and freshwater species are also increasingly impacted by human activity. According to the IUCN, freshwater species are going extinct faster than either terrestrial or marine species. In 2014 the IUCN estimated that human pressures on freshwater ecosystems threatened one-third of freshwater species with extinction. The impact of humans on freshwater wetlands and other aquatic ecosystems is considered in Chapter 6, which focuses on managing freshwater resources and associated environmental issues. Pollution and warming of the oceans have placed 33% of reef-building corals in danger of extinction. In addition, overfishing has severely depleted marine fish populations around the world. Ways to reverse these impacts on fish stocks are discussed in Chapter 8.

⚠ Think About It

1. What factors likely contribute to the greater vulnerability of freshwater species to extinction, compared with the vulnerability of terrestrial and marine species?

2. Does a growing human population inevitably produce habitat destruction? What other factors besides our population growth contribute to habitat destruction?

3.7 Invasive species threaten native species

After the brown tree snake arrived on the island of Guam following World War II, 10 out of the island's 12 native forest bird species went extinct—all fallen prey to a predator they neither recognized nor had learned to fear. The brown tree snake is what we call an **invasive species,** one that, when introduced to a new environment, poses a serious threat to native populations. Increasing globalization, including the rise in airline travel and transportation of goods on giant cargo ships, is increasing the movement of species across Earth.

Invasive species may be predators, as in the case of the brown tree snake, or they may be pathogens or competitors. Introduced species often compete fiercely with native species because they are not hindered by their own predators, parasites, and pathogens (see Table 3.1, page 71). As a result, non-native species can become overwhelmingly abundant in their new environment and competitively displace native species (**Figure 3.22**).

The cane toad is an invasive species that has hurt biodiversity in Australia in an unexpected way. Here's what happened: Sugarcane farmers introduced the softball-sized cane toad, hoping the toads would eat beetles that infested the sugarcane crop. The toads proved to be a weak beetle-controlling mechanism. But the bigger problem was their effect on animals besides beetles. The cane toad secretes a toxin from glands on the back of its head. In South America, where the toad is native, predators had either evolved resistance to the toxin or knew to stay away from the cane toad. In Australia, however, monitor lizards and endangered carnivorous mammals known as quolls had no resistance to the toad's toxin and simply saw it as an easy meal. So far, it has caused serious declines in species' populations. It remains to be seen how well Australian animals will recover and adapt to the toads' presence in the future.

Introduced species can have indirect effects as well. Invasive plants have pushed the Bay checkerspot butterfly to the brink of extinction largely by altering the makeup of the native grasslands of California. Similar stories are repeated around the globe, which is why the World Resources Institute, an environmental think tank

INVASIVE COMPETITORS

Kudzu vines can cover entire landscapes. In the setting shown here, it has invaded the southern United States.

In the arid regions of the United States, *Tamarix*, or saltcedar, has choked riverside habitats, simplifying the habitat and increasing the frequency and intensity of fires.

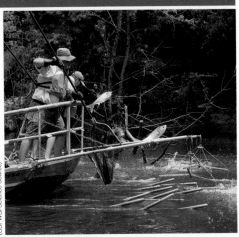

Several species of Asian carp have increased to large population sizes in the upper Mississippi River system, threatening native aquatic populations.

FIGURE 3.22 Released from predators, parasites, and pathogens present in their native environments, populations of invasive species can grow explosively in a new environment, overwhelming native species in the process.

dedicated to finding ways to protect the environment and improve people's lives, cites invasive species as the second greatest threat (after habitat destruction) to endangered species.

⚠ Think About It

1. How may similarity in niches affect competitive interactions between native and invasive species?

2. What sorts of evolutionary pressures do invasive predators and competitors exert on the native species with which they interact?

3.8 Plant and wildlife trafficking are growing dangers to species

With rhinoceros horns selling for more than $100,000 per kilogram in Asia as a purported cancer treatment, it's no surprise that rhinoceros poaching in South Africa increased 30-fold between 2007 and 2011. The black rhino in western Africa has already been declared extinct, and the northern subspecies of white rhino is on the brink of extinction. Illegal trading in wildlife and plants is one of the top three threats to endangered species worldwide (**Figure 3.23**). Demand for exotic pets, flashy hunting trophies, bush meat, and traditional medicines drive wildlife trafficking. Some of the primary targets include rare parrots, for the pet trade; tigers, for skins and traditional medicines; elephants, for ivory; bears, for

trophies and traditional medicine; and exotic tropical woods, for furniture. The Congressional Research Service has estimated that the black market value of

CARVED IVORY CONFISCATED BY CUSTOMS AGENTS

FIGURE 3.23 Wildlife trafficking involves the movement of endangered live animals and plants and wildlife products, such as ivory and skins. This illicit trade is valued in billions of U.S. dollars annually.

illegally traded wildlife is worth between $5 billion and $20 billion each year. The trade is not just a problem for African countries or other distant places: In 2012 two Manhattan jewelry store owners pled guilty to selling more than $2 million worth of elephant ivory.

In many cases, the impacts of wildlife trafficking extend beyond the wildlife killed or captured. For example, the use of cyanide to stun coral reef fish sold in the aquarium trade has done massive damage to coral reef ecosystems, which are one of the most productive and diverse ecosystems on Earth. The wildlife trade also holds the potential to spread disease to humans and other wildlife species. For instance, prior to the 1970s, the African clawed frog (*Xenopus laevis*) was widely traded as a pregnancy test—frogs lay eggs when exposed to the urine of a pregnant woman—and, today, it remains a model laboratory animal for a variety of experiments. Scientists believe trade in this frog may be at least partly to blame for the global spread of the chytrid fungus, a pathogen that has led to amphibian die-offs around the world.

How might the poachers who kill animals for the illegal wildlife trade potentially play a useful role in the protection of the wildlife they hunt?

Unfortunately, criminal networks that focus on wildlife have increased in recent years because of globalization and increasing wealth in countries such as China, where there is high demand for traditional medicine derived from animals, such as snake blood wine and foods like shark fin soup.

⚠ Think About It

1. How are drug trafficking and plant and wildlife trafficking similar? How are they different?

2. There would be no plant and wildlife trafficking without demand. What tools could be used to substantially reduce demand for illegal plant and wildlife products?

3.9 Pest and predator control have pushed species to the brink of extinction

When Europeans arrived in North America, gray wolves (*Canis lupus*) and red wolves (*Canis rufus*) occupied most of the continent. However, wolf numbers were quickly depleted as state and federal agencies in the United States organized campaigns to exterminate wolves, which were seen as a threat to livestock and game animals. The systematic hunting and poisoning that followed were so effective that, by the early 20th century, nearly all wolves in the United States south of Canada and north of Mexico had been eliminated (**Figure 3.24**). The last wolf in Yellowstone National Park was killed in 1926. Today, the United States Department of Agriculture's Wildlife Services continues to intentionally trap, shoot, and poison tens of thousands of animals every year, including predators ranging from gopher snakes to coyotes.

Predators aren't the only animals that have been targeted for elimination by humans. Pest species such as rodents and insects, which threaten agriculture, and migratory birds, which can interfere with airplane traffic, are also targeted for killing; but their higher numbers and reproductive rates make extinction unlikely. Nevertheless, control aimed at "pest" species can have unintended impacts on other species, such as the peregrine falcon, *Falco peregrinus*.

Historically, the peregrine falcon was found across much of the Northern Hemisphere. The fastest of all birds, the peregrine can dive on prey at speeds of 320 kilometers per hour (200 miles per hour). From the 1940s to 1970s, the number of peregrine falcon nesting pairs in North America declined precipitously from an estimated historical number of approximately 3,875 to just 324 pairs. The decline was traced to DDE

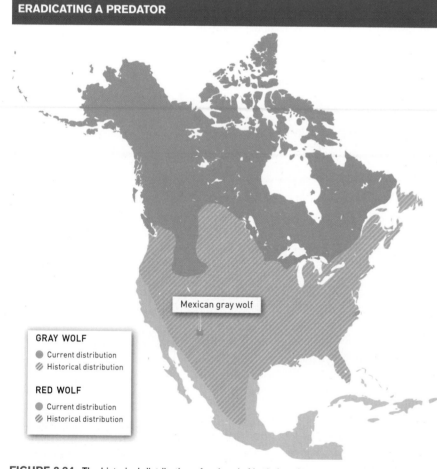

ERADICATING A PREDATOR

Mexican gray wolf

GRAY WOLF
● Current distribution
▨ Historical distribution

RED WOLF
● Current distribution
▨ Historical distribution

FIGURE 3.24 The historical distribution of wolves in North America once encompassed nearly the entire continent. However, by the early 20th century, wolves had been almost entirely eliminated south of the United States–Canada border.

(Geoff Kuchera/iStock/360/Getty Images)

FIGURE 3.25 The peregrine falcon, *Falco peregrinus,* is the fastest-flying of all living bird species. Its recovery from near extinction in North America is one of the great successes in the history of endangered species restoration.

(dichlorodiphenyldichloroethylene), a highly persistent chemical produced during the breakdown of the insecticide DDT (dichlorodiphenyltrichloroethane), which caused their eggshells to thin and reduced reproductive success. DDT was ingested by plant-eating insects, and it moved up the food chain as those insects were eaten by small birds. With each transfer in a food chain, the DDT by-products became more concentrated in the tissues of animals, a process called **biomagnification,** which we will consider in detail in Chapter 11 (see page 340). As a consequence, top predators such as the peregrine falcon ingested a very high dose of the chemical. DDT was banned in the United States in 1972, and the numbers of peregrine falcon began to bounce back in the decades that followed (**Figure 3.25**).

⚠ **Think About It**

1. Why was DDT more of a threat to peregrine falcon populations than to the populations of the birds it fed on?

2. What factors should be considered in a cost-benefit analysis of pest and predator control efforts?

3.6–3.9 Issues: Summary

Species are threatened by a variety of human activities. Habitat destruction and alteration are considered the most serious threat to species, affecting nearly 90% of endangered amphibians, mammals, and birds. Invasive species can harm native species through unchecked predation or competition. Their impact is particularly devastating on islands, where they have caused numerous species to go extinct. Illegal plant and wildlife trafficking for traditional medicine, the pet trade, hunting trophies, and bush meat causes significant harm to threatened species. Globalization and rising wealth in countries such as China have led to an increase in criminal gangs that focus on wildlife trafficking. Finally, predator control programs extirpated wolf populations, and pest control has caused collateral damage to nontarget populations. The 20th-century decline in the North American population of peregrine falcons was traced to the insecticide DDT.

biomagnification The process of certain chemicals becoming more concentrated in the tissues of animals with each transfer in a food chain.

3.10–3.14 Solutions

Can laws, such as those protecting endangered species, be effective if they do not reflect the prevailing values of society as a whole?

How can we possibly reverse the sixth mass extinction? In 2013 *National Geographic* magazine ran a controversial cover story titled "Reviving Extinct Species." It was not a work of science fiction, but an exploration of genetic technologies that are edging closer to reconstructing life from bits and pieces of DNA found in bones, hair, and other pieces of fossilized or preserved material. Imagine woolly mammoths from the Ice Age walking Earth again! Or how about reviving the passenger pigeon, which once numbered in the billions and was wiped out by hunting?

While genetic resurrection may be useful in bringing back a handful of the most spectacular and important animals on the planet, it is hardly a viable solution to the biodiversity crisis. Even if all the species that human activity has destroyed thus far were effectively brought back, the threats that hindered their survival in the first place still pose great danger. In the foreseeable future, it is likely to be far cheaper and easier to prevent an extinction than to try to reverse one. The first step involves establishing a legal framework to protect species.

3.10 National laws and international treaties protect endangered species

While laws regulating the hunting, fishing, and trapping of wildlife have a long history, it was not until rather recently that fears of species extinctions resulted in legal protection. The **Endangered Species Act of 1973,** or **ESA,** provides legal protection in the United States to both domestic and foreign endangered species and declares plants and *all* invertebrate animals eligible for protection.

In addition, the ESA *requires* federal agencies to develop programs for the conservation of two categories of species: endangered species, which face the highest danger of extinction, and threatened species, which face a lower danger of extinction. Agencies are also prohibited from authorizing, funding, or engaging in any action that would jeopardize an endangered or threatened species or that might destroy or alter so-called **critical habitat,** areas essential for the survival of a listed species. The ESA

Endangered Species Act of 1973 (ESA) Legal protection in the United States for both domestic and foreign endangered species that declared plants and *all* invertebrate animals eligible for protection.

critical habitat Areas that are essential for the survival of a listed endangered or threatened species.

TABLE 3.3

EVOLVING SPECIES PROTECTION U.S. federal endangered species law has developed over time through a series of legislation and amendments

Year	Title	Main Effect
1900	Lacey Act	First federal law prohibiting the trade of illegal fish and wildlife, aimed at preserving hunting game and wild birds.
1966	Endangered Species Preservation Act (ESPA)	Authorized identifying endangered native animal species and giving them limited protection.
1969	Endangered Species Conservation Act (ESCA)	Amended 1966 ESPA to prohibit importation and sale of species threatened with worldwide extinction and changed name of act to ESCA.
1972	Marine Mammal Protection Act	Prohibited hunting, killing, capturing, or harassing of marine mammals, including sea otters, polar bears, whales, and dolphins.
1973	Endangered Species Act (ESA)	Replaced ESCA, distinguished between endangered and threatened species, extended protection to endangered plants and all invertebrate animals, required federal agencies to conserve listed species, implemented CITES protection in the United States.
1978	ESA Amendment	Required the designation of critical habitat at the time of species listing, required consideration of economic and other impacts of designating critical habitat.
1982	ESA Amendment	Restricted the determination of endangered status to be based solely on biological information, excluded economic and other effects of listing.
1988	ESA Amendment	Required population monitoring of species proposed as candidates for listing and for species that have been recovered from their endangered status.
2004	National Defense Authorization Act of 2004	Exempted the Department of Defense from designating critical habitat, required a management plan approved by the Secretary of the Interior.
2008	Lacey Act Amendment	Expanded the act to cover plants and plant products, making it possible to prosecute importers of illegal timber.

CONSERVATION SUCCESS

Grizzly bears (Yellowstone population)   Bald eagles (lower 48 states)   Whooping cranes

FIGURE 3.26 Species listed for protection under the Endangered Species Act of 1973 (ESA) include the Yellowstone population of grizzly bears, the bald eagle in the lower 48 states, and the whooping crane. Marking the success of the recovery programs supported by the ESA, the Yellowstone grizzly bear and bald eagle populations have recovered sufficiently to be delisted. Meanwhile, the whooping crane population is growing exponentially.

was followed by 30 years of legislative acts that generally broadened the range of protected species and tied their legal protection more closely with science (Table 3.3).

Protection of endangered species and their critical habitats often clashes with economic activities. An endangered fish called the snail darter prevented a dam from being built in Tennessee for many years, while the Northern spotted owl has blocked the cutting of old growth forests in the Pacific Northwest. As a consequence, a variety of economic interests have tried repeatedly to weaken the Endangered Species Act, in part by limiting the funding that gets appropriated to it each year.

Nevertheless, the law has been remarkably successful in conserving and restoring a wide variety of species. Approximately 1,400 species in the United States and 2,000 globally have been listed and placed under the protection of the ESA. Some of the most well-known species listed under the ESA include the Florida panther, blue whale, California condor, and whooping crane. It is a sign of the success of the ESA that several formerly endangered species have recovered sufficiently to be removed from the endangered species list, a process called *delisting*. Many species have now been delisted, including the Yellowstone grizzly bear population and the bald eagle (Figure 3.26).

The CITES Treaty

The major treaty regulating international trade in wildlife is the Convention on International Trade in Endangered Species of Wild Fauna and Flora, or CITES (pronounced "sight-ees"), which came into force in 1975 and has been signed by nearly 180 nations. CITES protects

approximately 5,000 animal species and 29,000 plant species, which are listed in three appendices. Appendix I of the Convention lists species threatened with extinction, which can be traded only under exceptional circumstances. Appendix II includes species that are not presently threatened by extinction, but for which trade is regulated by import and export quotas to avoid threats to their survival. Appendix III consists of species protected by at least one country that has asked for assistance from other signatories of CITES for help in regulating trade of the species.

The nations that have signed CITES and many prominent nongovernmental organizations, such as the World Wildlife Fund or World Wide Fund for Nature (WWF), Conservation International, and the Wildlife Conservation Society, are cooperating to reduce wildlife trafficking and the threat it presents to endangered species. In the United States, one of the critical tools to enforcing CITES is the **Lacey Act.** First passed in 1900 and amended in 2008, the Lacey Act forbids the trading of illegally harvested plants and animals. For example, if an endangered ebony tree is illegally harvested in Madagascar, outside of U.S. jurisdiction, one can still be prosecuted for importing it to or trading it in the United States. In 2012, for instance, Gibson Guitar pled guilty to importing ebony for its guitars in violation of the law.

⚠ Think About It

1. How might including the potential for negative economic impact as a criterion for the listing of endangered species affect how the ESA is applied?

How can communication and education complement the laws and treaties protecting endangered species?

Lacey Act First passed in 1900 and amended in 2008, this law forbids trade in illegally harvested plants and animals.

2. Why is it critical to have both national laws and international cooperative treaties regarding trading in endangered species?

3.11 Banning of a toxin and captive breeding brought peregrine falcons back from the brink of extinction

Using what you learned about population growth earlier in the chapter, predict how the pattern of growth shown in Figure 3.28 will or will not change over the long term. (Hint: Consider Figure 3.11.)

How would reduced genetic diversity among North American peregrine falcons between 1985 and 2007 have made the population less able to adapt to future environmental challenges?

After the peregrine falcon was declared an endangered species in the United States in 1970, the greatest threat to its survival needed to be eliminated. In 1972 the United States banned the insecticide DDT, which had caused eggshell thinning. As a result of decreased DDT use in the United States, Canada, and Latin America, the amount of DDE, the breakdown product of DDT associated with eggshell thinning, has gradually decreased in the tissues of peregrine falcons (**Figure 3.27**). However, that was only the first step toward a solution.

To restore falcon populations, the U.S. Fish and Wildlife Service entered into a partnership with state natural resource agencies and nongovernmental organizations to establish captive rearing programs. From 1974 to 1997, these programs released more than 6,000 peregrine falcons to their historic range in 34 states. In addition, critical habitat for the peregrine falcon was identified and protected.

The initial goal of the captive rearing and release program was to build the population back up to 631 nesting pairs in the United States. By the mid-1990s, the population had grown well beyond this target (**Figure 3.28**), and the U.S. Fish and Wildlife Service

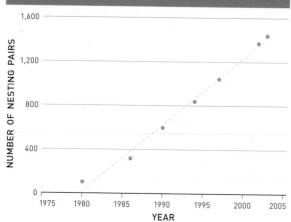

FIGURE 3.28 The number of pairs of nesting peregrine falcons in the lower 48 (contiguous) United States grew rapidly following the banning of DDT, as did the release of falcons produced by the captive breeding program. (Data from U.S. Fish and Wildlife Service 2003, 2006)

announced its intention to remove the American peregrine falcon from the endangered species list.

The American peregrine falcon was removed from the endangered species list on August 25, 1999, and the population continued to grow. By 2003 the American peregrine falcon population in the United States, Canada, and Mexico reached an estimated 3,005 breeding pairs. Follow-up genetic studies of migratory peregrine falcons have shown no reductions in genetic diversity from 1985 to 2007, indicating that the restored peregrine falcon populations of North America are not in decline.

⚠ Think About It

1. Why was a captive rearing program necessary? Why was a ban on DDT alone not sufficient for the peregrine falcon to recover on its own?

2. What were some advantages of having a cooperative breeding program for the peregrine falcon, involving federal and state scientists along with private foundations and citizens?

3.12 Population ecology provides a conceptual foundation for wolf restoration

Population ecology provides a general framework for guiding restoration of endangered species and for predicting how their populations will respond during restoration. Gray wolf restoration provides a clear

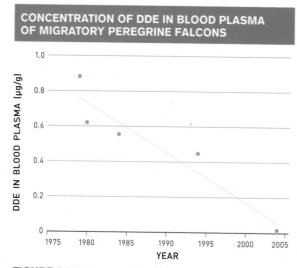

FIGURE 3.27 The concentration of DDE, a breakdown product of DDT, in the blood plasma of peregrine falcons captured at Padre Island, Texas, during spring migration declined rapidly between 1979 and 2004. (Data from Henny et al., 2009)

example of the utility of population ecology for managing endangered species.

Promoting Genetic Diversity

Genetic analysis gives environmental scientists essential tools for restoring endangered species populations. For instance, restoration of wolf populations in Yellowstone, Idaho, and along the Arizona–New Mexico border would require that founding populations have diverse genetics. That's why, following the initial introduction of wolves into Yellowstone in 1995, an additional 17 Canadian wolves were brought in the subsequent year.

Compared with the wolves of the Northern Rocky Mountains, there is much less genetic variation among the small population of surviving Mexican gray wolves (see Figure 3.24, page 80), which are the descendants of three unrelated but highly inbred lineages. However, biologists discovered sufficient remaining diversity in these three surviving lineages to improve the chances of success by the population. By crossbreeding descendants of the three lineages, they have been able to more than double litter sizes from an average of 3.5 or 3.6 pups to 7.5 pups per litter (Figure 3.29). In addition, crossbred Mexican wolf pups also have an 18% to 21% higher chance of surviving to 6 months of age. Such positive effects of increasing genetic diversity are called *genetic rescue*.

Logistic Population Growth in a *K*-selected Species

The wolf population of the Northern Rocky Mountains Recovery Area grew quickly following their reintroduction to Yellowstone and Idaho. Conservation biologists set a restoration goal of 30 breeding pairs, defined as a male and female that successfully rear at least two pups which survive to the end of the calendar year. That goal was met in 2000 and was quickly exceeded (Figure 3.30a). By 2007 there were over 100 breeding pairs across the Northern Rocky Mountain Recovery Area, which includes parts of Idaho, Montana, and Wyoming, and the number of breeding pairs appeared to be stabilizing, a pattern suggesting S-shaped, or logistic, population growth (see Figure 3.11, page 69). Since generally only the dominant male and female in a wolf pack breed, the total number of wolves in the recovery area is far higher than the number of breeding pairs.

In fact, the population grew to over 1,700 wolves by 2011 (Figure 3.30b), then dropped slightly to 1,600 individuals in 2013. Looking at just one part of the range, the Yellowstone wolf population peaked in 2003 at 172, then declined to fewer than 100 by 2009, where their numbers have approximately stabilized since. Again, the growth of the restored wolf population appears to be following an S-shaped pattern at both regional and local scales, as we would expect from a highly territorial, *K*-selected species like the gray wolf.

Wolves remain a charged subject in Rocky Mountain states, where ranchers—and their supporters in state government—oppose their preservation. As a consequence of their dramatic recovery and pressure from state officials and interest groups, the U.S. Fish and Wildlife Service classified the gray wolf population of the Northern Rocky Mountain states as recovered and took them off the U.S. endangered species list in 2012. With delisting, responsibility for management of gray wolf populations was returned to the states of Idaho, Montana, and Wyoming, opening a controversial hunting season on the wolves. In September 2014, however, the Federal District Court for the District of Columbia restored wolves in Wyoming to endangered species status.

⚠ **Think About It**

1. How would population recovery for an r-selected species, such as the Bay checkerspot butterfly (see page 67), likely compare to that of gray wolves?

2. What aspects of gray wolf life history make it unlikely that any restored population will grow exponentially in the future?

3.13 Restoration of North American gray wolves has required working through conflict

Conservation success can sometimes introduce its own problems. In many cases, a species' revitalization is inhibited by an environmental threat the peregrine falcon

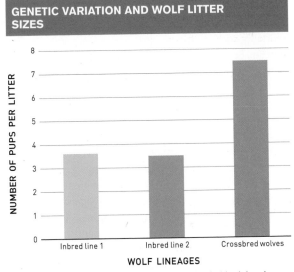

GENETIC VARIATION AND WOLF LITTER SIZES

FIGURE 3.29 By combining the genes carried by inbred lines of Mexican gray wolves through crossbreeding lineages, biologists have been able to more than double the number of pups in litters. (Data from Fredrickson et al., 2007)

Should restored gray wolf populations be subject to hunting and other forms of management control, or should they continue to receive protection as an endangered species?

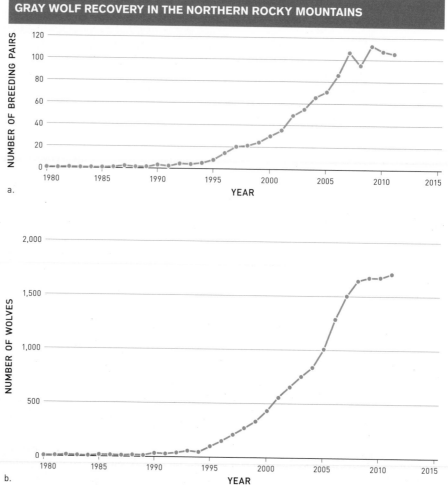

FIGURE 3.30 (a) The number of gray wolf breeding pairs in the Northern Rocky Mountains states of Idaho, Montana, and Wyoming grew rapidly beginning in the 1980s and then showed signs of leveling off after 2007. (b) The total gray wolf population across the region showed a similar pattern of growth and apparent stabilization. (Data from U.S. Fish and Wildlife Service, 2013)

did not face: conflict with entrenched economic and cultural interests. Such is the case with the reintroduction of the wolf to Yellowstone and the surrounding area. The greatest opposition comes from livestock ranchers whose cattle may be attacked by wolves. One poll in the southern Rocky Mountains found that 53% of ranchers opposed restoration of wolves, while only 28% of nonranchers were opposed.

There is no doubt that wolves prey on livestock, especially cattle and sheep that are left unattended for long periods of time, and that ranchers could potentially suffer the economic consequences from this predation. In the 19-year span between 1987 and 2011, 1,669 cattle and 3,261 sheep were confirmed to have been killed by wolves in the Northern Rocky Mountain Recovery Area (**Figure 3.31**). But to put these numbers in context, we need to consider all sources of livestock loss. First, fewer sheep and cattle are lost to predators of all kinds than to other sources of

mortality, such as disease, extreme weather, injuries, and so forth. Second, although wolves have killed substantial numbers of livestock in the Northern Rocky Mountain states, other predators, such as coyotes and mountain lions, take many more (**Figure 3.32**).

What can be done to reduce the economic consequences of wolf restoration? The main mitigation strategy is to compensate ranchers for livestock losses. The Defenders of Wildlife, a nongovernmental organization, took the lead role in compensating ranchers for livestock lost to wolves. The position of the Defenders of Wildlife is that the loss of even a single animal to one rancher can be significant, and it established a fund of private donations for economic compensation. That fund, the Bailey Wildlife Foundation Wolf Compensation Trust, paid nearly $1.4 million in compensation to ranchers between 1987 and 2010, when compensation programs were

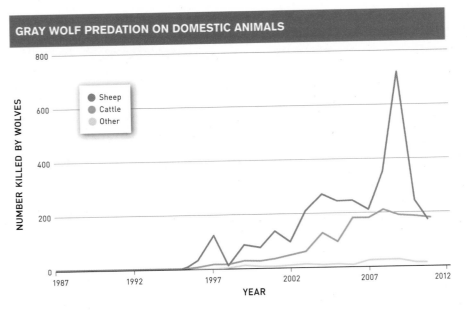

GRAY WOLF PREDATION ON DOMESTIC ANIMALS

FIGURE 3.31 From 1987 to 2011, gray wolves were confirmed to have killed approximately 5,000 domestic animals in the Northern Rocky Mountain states of Idaho, Montana, and Wyoming. Most of the animals killed were cattle and sheep; others included dogs, llamas, goats, and horses. (Data from U.S. Fish and Wildlife Service, 2013)

transitioned to the states with wolf populations. Now that the states have taken responsibility for compensating livestock producers for losses to wolves, Defenders of Wildlife has begun to invest its resources into developing nonlethal ways to promote coexistence between livestock production and wolves (**Figure 3.33**).

⚠ Think About It

1. Many ranchers graze livestock on public lands. Should public funds be used to eliminate predators on public lands to protect private livestock?

2. A published study of Mexican gray wolves living in Arizona and New Mexico found that 96% of their diet was based on wild prey, mainly elk and deer, and just 4% made up of cattle. How should we factor this information into an assessment of the threats that these wolves pose to cattle producers?

3. How should the economic burden of wolf restoration be spread across society?

Livestock production coexisted with wolf populations for thousands of years before modern traps, poisons, and firearms provided the means to exterminate wolves. What does this fact indicate about the possibility of coexistence in today's world?

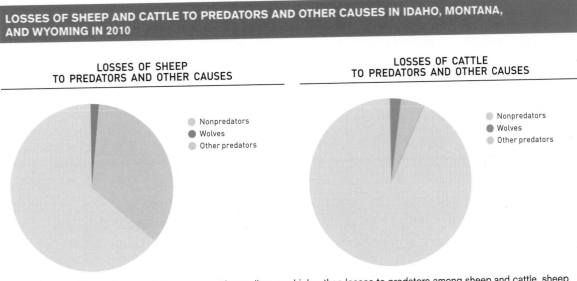

LOSSES OF SHEEP AND CATTLE TO PREDATORS AND OTHER CAUSES IN IDAHO, MONTANA, AND WYOMING IN 2010

FIGURE 3.32 Although nonpredatory sources of mortality were higher than losses to predators among sheep and cattle, sheep suffered far more losses to predators, especially to coyotes. Wolves were the cause of a higher percentage of predatory losses in cattle than in sheep. (Data from USDA, 2013; National Agricultural Statistics Service, www.nass.usda.gov/)

SCIENCE   ISSUES   SOLUTIONS

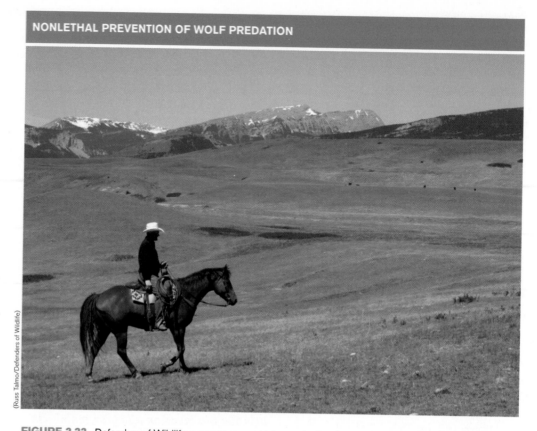

NONLETHAL PREVENTION OF WOLF PREDATION

(Russ Talmo/Defenders of Wildlife)

FIGURE 3.33 Defenders of Wildlife, a nongovernmental organization that took the lead in compensating livestock raisers for livestock predation by wolves, is now contributing to the application of nonlethal methods for promoting the coexistence of livestock production and wolves, such as hiring range riders to oversee livestock and discourage wolves.

3.14 Wild populations are sources of significant economic benefits

Sometimes saving an individual species can also be good for the economy. Restoration and conservation of banner species, such as mountain gorillas, eagles, and wolves, provide viewing opportunities for wildlife enthusiasts. Travel to natural areas to observe animals is just one facet of **ecotourism,** recreational travel that helps conserve the environment and improves the well-being of local people. The Ugandan government charges tourists $500 to spend just one hour watching mountain gorillas in the Bwindi Impenetrable Forest Reserve—and that doesn't include all the money spent on lodging, food, and travel. A University of Montana study estimated that the presence of wolves in Yellowstone has brought in $35 million in tourist dollars. Wildlife watchers in the United States spend nearly $40 billion annually. Globally, ecotourism is valued at hundreds of billions of U.S. dollars.

Hunting for trophies and meat can also play a positive role in biodiversity, so long as it is properly regulated. In Namibia, wildlife conservancies managed by local

ecotourism Recreational travel that helps conserve the environment and improves the well-being of local people.

indigenous communities with the help of the World Wildlife Fund allow limited hunting of big game species, which would otherwise be extirpated from the landscape. Money spent on hunting fees goes toward conservation initiatives and is funneled back into the communities, creating a sustainable income stream. Studies have shown that individual trophy hunters spend more money and time visiting more remote local communities than do ecotourists.

One 2007 study in the journal *Biological Conservation* found that in 23 countries, 1.4 million square kilometers of land was set aside for trophy hunting, an area that exceeded all other protected areas in the region. Since the country of Kenya banned wildlife trophy hunting in 1977, much of the land has been converted to livestock and agriculture, resulting in losses of 60% to 70% of large mammals. By contrast, wildlife is booming in southern African countries that allow hunting, and the importance of the industry means that these countries have more funds to fight illegal poaching. In addition, the wildlife-associated industries provide many local residents with jobs, which has contributed to a decrease in poaching.

Species can provide other economic benefits to humans as well. A 2005 survey found that about half of the drugs used to treat human disease were first isolated as natural chemicals produced by plants, fungi, bacteria, or animals. In 1997, 118 of the top 150 prescription drugs used in the United States were derived from natural sources. Of these, 74% were based on compounds isolated from plants, 18% from fungi, 5% from bacteria, and 3% from a species of snake. For example, the rosy periwinkle, *Cantharanthus roseus,* a plant native to the tropical dry forests of Madagascar, is the source of two important drugs. One drug, used to treat childhood leukemia, increased the odds of surviving the disease from 20% to 99%. The other is used to treat Hodgkin's disease. According to the World Resources Institute, plant-derived medicines are worth over $40 billion worldwide each year, and there is still great potential for discovery of new medicines within the great diversity of plants.

It's not always possible to predict how a specific species will benefit humanity. In 1966 Thomas Brock of Indiana University isolated a bacterial species he named *Thermus aquaticus* from several springs in Yellowstone National Park (see Figure 4.2), one of Earth's most unique and harsh ecosystems. This thermophilic, or heat-loving, bacterium thrives in hot temperatures ranging from 50°C to 80°C (122°F to 176°F). He deposited cultures of *T. aquaticus* in the American Type Culture Collection, where the newly discovered bacterium would be available to other researchers.

Twenty years later, scientists from Cetus Company obtained *T. aquaticus* samples from the culture from which they extracted an enzyme called **Taq polymerase.** This enzyme could be used to amplify small quantities of DNA, a process requiring an enzyme that is stable at high temperatures, and it is widely used in modern medical diagnosis and forensics. Had Yellowstone not been protected 100 years earlier, that enzyme might never have been discovered. Undoubtedly, intact natural ecosystems harbor countless other species, the value of which cannot be calculated or anticipated at this time. In Chapter 4, we will explore how ecosystems as a whole provide valuable services to humanity and how conserving large areas can protect these services. Marine-protected areas are discussed in Chapter 8.

⚠ Think About It

1. Should we use only economic criteria for species conservation and restoration? Elaborate on your answer.

2. Can aesthetic and economic reasons for conserving endangered species coexist, or are they mutually exclusive? Explain.

3.10–3.14 Solutions: Summary

Finding solutions to the biodiversity crisis involves developing both a legal and social framework. The U.S. Endangered Species Act of 1973 provides legal protection to both domestic and foreign endangered animal and plant species. The CITES treaty, signed by nearly 180 nations, regulates international trade in wildlife. Saving a species typically involves eliminating the factors that led to its decline and putting it back on the path to recovery. In the case of the peregrine falcon, legally protecting the species and banning DDT were the first steps. Once that foundation had been established, a cooperative captive rearing and reintroduction program could begin.

By comparison, the restoration of wolves reveals how complicated it can be to navigate a web of economic and political forces opposed to conservation. Nongovernmental and governmental programs have paid ranchers for livestock losses partly to soften their opposition. On the other hand, wild species can provide economic benefits. Restoration and conservation of large, "charismatic" species, such as mountain gorillas and wolves, can boost local economies through ecotourism. Trophy hunting can provide a sustainable income stream for rural communities and encourages conservation programs. Wild species, particularly plants, are also a valuable source of medicinal drugs, worth billions of U.S. dollars annually.

What are the relative merits of utilitarian (anthropocentric) versus intrinsic (biocentric) value of species as justifications for saving endangered species?

Taq polymerase An enzyme isolated from a bacterium discovered living in hot springs in Yellowstone National Park; used to amplify small quantities of DNA.

Answer the following questions for each chapter section and then answer the Central Question.

Central Question: How can we protect species in an increasingly human-dominated world?

3.1–3.5 Science

- **What is the significance of genetic diversity in populations?**
- **How do distribution and abundance vary among species?**
- **How do populations grow and how are they regulated?**
- **What are the typical life histories of species**
- **How does species interaction influence communities?**

3.6–3.9 Issues

- **How have humans affected extinction rates?**
- **What are the three main factors that threaten species populations?**
- **How have predator and pest control impacted vulnerable species?**

Endangered Species and You

Protecting endangered species involves work across many areas, including science, politics, law, and economics. It is impossible for any individual to singlehandedly save and protect an endangered species. However, the solutions to endangered species issues can be addressed in our daily lives in many significant ways that collectively make a very big difference.

☐ *Get involved with endangered species and wildlife conservation.*

There are many opportunities to work as a volunteer or intern in endangered species and wildlife conservation programs that will give valuable experience in preparation for future studies or a career in a conservation-related field. For example, the U.S. Fish and Wildlife Service sponsors large numbers of volunteers to work on endangered species management, habitat restoration, and many other conservation-related projects. Most state fish and game agencies or departments of natural resources support and value volunteers and interns who work in similar areas. Most modern zoos are also focused on endangered species conservation and the majority have volunteer programs. A summary of government-related volunteer programs across the United States may be found at www.volunteer.gov.

☐ *Support predator-friendly products and organizations.*

Livestock producers around the United States are becoming certified as producers of "predator-friendly" products, including wool, lamb, and beef. These are ranchers and farmers who use nonlethal means of reducing losses to predators, including shepherds, protective fencing, and livestock guard dogs. If you are a consumer of such products, buying from these certified producers will help sustain their efforts. You may also consider becoming active in organizations, such as Defenders of Wildlife, that promote the coexistence of livestock and predators.

☐ *Help control invasive species.*

The massive problem of invasive species has created virtually unlimited opportunities for involvement in control and prevention programs. You can begin at home. If you are a gardener, consider growing native plant species to help reduce the flow of invasive species into the environment. If you keep exotic pets, do not release them into the wild. Pets released into the wild generally die in a short time. Also, wildlife refuges, nature centers, national or state forests, and others often enlist volunteers to help with invasive species control and habitat improvement.

☐ *Do not encourage wildlife trafficking.*

Taking care with your buying habits can reduce the environmental impact of wildlife trafficking. The pet trade is a major source of damage to habitats, particularly coral reefs, and wild populations. Millions of birds, primates, amphibians, and reptiles are captured and sold every year. If you keep pets, you can reduce these impacts by purchasing only captive-reared birds, aquarium fish, and reptiles and by strictly avoiding any purchase of wild-caught animals. To reduce the destruction of reef habitats around the world, do not buy coral or live reef rocks. Avoid all products made of ivory or turtle shells.

3.10–3.14 Solutions

- How do laws and international treaties protect endangered species?

- How was the peregrine falcon saved from probable extinction?

- What contributions has population ecology made to restoring wolf populations?

- What approaches have been used to reduce conflict resulting from wolf recovery?

- What economic benefits do wild populations offer?

Answer the Central Question:

_____
_____
_____
_____
_____
_____
_____
_____

Chapter 3

Review Questions

1. Genetic diversity in populations is important to which of the following?
a. Domestication of plants and animals
b. Survival of species faced with climate change
c. Survival of species challenged by pathogens
d. All of the above

2. Which of the following species would most likely be in danger of extinction?
a. A widespread, abundant, genetically diverse species
b. Drought-caused deaths in a plant population
c. A rare, geographically restricted species with low genetic diversity
d. A rare, geographically restricted species with high genetic diversity

3. Which of the following describes an example of density-dependent population of regulation?
a. High temperatures reducing insect numbers
b. Drought-caused deaths in a plant population
c. The release of toxic pollutants into a lake
d. A disease that spreads easily through a population

4. _K_-selected species are least likely to have which of the following qualities?
a. Large size
b. Short average life span
c. Long average life span
d. Late maturity

5. Competitive exclusion is most likely to occur under which of the following circumstances?
a. Competition between two species of carnivores
b. Competition between two species of herbivores
c. Competition between two species with nearly identical niches
d. Competition between two species of plants

6. Which of the following is considered to be the greatest threat to the existence of species?
a. Habitat destruction
b. Wildlife trafficking
c. Invasive species
d. Predator control programs

7. Does saving endangered species have any potential positive economic consequences?
a. No, saving endangered species just costs money.
b. Yes, saving endangered species can provide a focus for commercial nature films.
c. Yes, but only saving endangered plant species from which medicines can be made.
d. Yes, since plants, animals, and insects make a broad range of contributions to the economy.

8. What species are protected by the Endangered Species Act of 1973?
a. Endangered species native to North America
b. Endangered mammals such as the gray wolf
c. Endangered animals, including invertebrate animals, and plants from anywhere in the world
d. Only endangered species with no potential economic value

9. Why was the banning of DDT critical to saving North American peregrine falcons from extinction?
a. DDT was killing adult peregrine falcons.
b. A breakdown product of DDT was causing reproductive failure in peregrine falcons.
c. DDT was killing the species peregrines prey on.
d. DDE, a breakdown product of DDT, was causing blindness in adult peregrine falcons.

10. What do the patterns of loss of sheep and cattle indicated by Figure 3.32 suggest about these two types of livestock?
a. Sheep are much less subject to predation than cattle.
b. Compared with sheep, a higher percentage of cattle are lost to predators, especially to wolves.
c. A higher percentage of sheep are lost to predators, but a higher proportion of cattle losses to predators are the result of wolf predation.
d. Losses of sheep and cattle to predators, as a percentage, are approximately equal.

Critical Analysis

1. In places where there are fewer species of warblers in the community, compared with those pictured in Figure 3.19, the feeding zones of the remaining warblers expand. What does this suggest about the feeding zones of these warblers?

2. What steps might be taken to protect species from wildlife trafficking? Consider all aspects of the activity.

3. If DDT were the only insecticide available for controlling malaria-carrying mosquitoes, should the use of the insecticide have been continued, even at the cost of the extinction of peregrine falcons, bald eagles, and other birds of prey? Justify your answer.

4. Do changing societal attitudes toward wolves over the past century reflect changes in the relative influences of anthropocentric versus biocentric or ecocentric ethics (see Chapter 1, page 22)?

5. How can stakeholders reach across cultural and economic divides to establish cooperative, mutually beneficial approaches to wolf restoration?

Find additional resources and links online at www.macmillanhighered.com/launchpad/molles1e.

Central Question: How can we protect Earth's diverse ecosystems?

Explain influences on patterns of species, ecosystem, and geographic diversity.

SCIENCE

Species and Ecosystem Diversity

Analyze human impacts that threaten
species and ecosystem diversity.

Discuss the keys to sustaining
species and ecosystem diversity.

ISSUES

SOLUTIONS

(Jim Peaco/Yellowstone National Park)

Its vast size and careful protection have made Yellowstone National Park a haven for wildlife. The park has succeeded in sustaining all the large wildlife species that have occupied this landscape for centuries and may be a model for the design and management of protected areas.

Protecting Yellowstone's Diverse Ecosystems

Yellowstone National Park, established to protect unique geologic features, became a global model for protecting large areas with high ecological and **conservation** value.

When the first American explorers visited the Yellowstone region in the early 1800s, they were astounded. Their accounts of boiling pools and spouting geysers in a landscape bathed in sulfurous fumes were thought to be the tall tales of frontiersmen. It was only after official government expeditions confirmed these early reports that the descriptions were believed.

In 1872 the U.S. Congress designated Yellowstone as the first national park in the world. Covering nearly 9,000 square kilometers (3,475 square miles) of mountainous landscape, Yellowstone National Park exceeds the combined areas of the states of Rhode Island and Delaware. Today,

conservation The protection, management, or restoration of species, ecosystems, or natural resources.

Yellowstone is the largest area of ecologically intact north temperate landscape on Earth. Its ecosystem is home to the largest herds of elk in North America, and it is one of the few remaining areas in the lower 48 states where the magnificent grizzly bear still roams in significant numbers. Greater Yellowstone serves as wintering ground for the rare trumpeter swan, and is home to the largest free-ranging herd of bison in the lower 48 states. Mountain lions and wolverine still roam its mountains, bighorn sheep scramble among its cliffs, moose browse its willows, and eagles grace the open sky.

"Thank you but we do not print fiction."

Editor in response to early descriptions of the Yellowstone landscape, *Lippincott's Magazine* (1869)

As we learned in Chapter 3, restoring the wolf population to the landscape was just one part of the task required to maintain Yellowstone's functioning ecosystems. Because Yellowstone is surrounded by landscapes highly modified by human actions and occupation, there are a number of threats to its ecological health, and the national park requires active management to maintain its ecological integrity. Park managers must prevent bison populations from expanding too rapidly, which would harm the landscape and cause conflicts with ranchers outside the park. They also allow moderately sized fires to burn regularly in the forests to prevent catastrophic ones and to maintain meadow ecosystems. In addition, they monitor and control some 212 species of exotic plants. In other words, conserving Yellowstone's biodiversity requires strategies that are more comprehensive than the species-by-species model we looked at in Chapter 3. Instead, we now move up a level from individual organisms and their populations to communities of organisms and their ecosystems.

One reason to preserve ecosystems is that we enjoy the experience of hiking through a beautiful forest, for example, and spotting wildflowers or animals. Another reason is that maintaining biodiversity provides practical benefits, ranging from clean water to climate control.

Central Question

How can we protect Earth's diverse ecosystems?

(Cheryl Jaworowski/USGS)

4.1–4.5 Science

Why does pollution often reduce species richness and evenness in terrestrial and aquatic ecosystems?

species diversity A measure of diversity that combines the number of species in a community and their relative abundances.

species richness The number of species in a community or living in a local area or region; higher species richness increases species diversity.

species evenness How evenly individuals are apportioned among the species inhabiting a community; higher evenness increases species diversity.

ecosystem diversity A measure of the variety and extent of ecosystems in an area.

Biologists have described and named more than 1.75 million species of organisms, but the true number of species on Earth likely ranges between 3 and 100 million. In fact, we do not even know the full diversity of any single ecosystem on the planet. Nevertheless, biologists have developed tools to help us understand and study diversity and its importance to ecosystem functioning.

4.1 Species and ecosystem diversity are key elements of biodiversity

When you think about biodiversity, you probably imagine a lush, green tropical rain forest like Madidi National Park in Bolivia, which has more than 600 tree species. By comparison, an oak-hickory forest in the Missouri Ozarks contains just 46 tree species. One of the most basic elements of biodiversity is **species diversity,** which is determined by two components: the number of species in a community and their relative abundances. The species diversity of a community increases with the number of species, which is called **species richness.** However, species richness alone does not give a complete picture of species diversity. We must also consider **species evenness,** a measure of the relative abundances of species in a community. This is important because even though a community has high species richness, it may be dominated by a single species, making it, effectively, less diverse.

Consider butterflies living in three hypothetical meadow ecosystems (**Figure 4.1**). With six butterfly

species, ecosystem (b) clearly supports a higher diversity of butterflies compared with ecosystem (a), which supports only a single species of butterfly.

In contrast, ecosystems (b) and (c) both have six butterfly species. Still, the butterflies in ecosystem (c) give a visual impression of higher butterfly species diversity. The visual contrast between ecosystems (b) and (c) is created by higher species evenness in (c). In other words, butterfly species diversity is higher in ecosystem (c), as a consequence of higher species evenness.

Diversity has consequences for the functioning of ecosystems. More diverse communities are usually more stable over time and recover more quickly after disturbances, such as a treefall or a forest fire. In addition, ecosystems with higher species richness generally support higher levels of primary production, compared with less diverse ecosystems (see Chapter 7, page 190).

Ecosystem Diversity

Zooming out from the species level, we see that the landscape is a patchwork of different ecosystems ranging from forests to wetlands. **Ecosystem diversity** is a measure of the variety and extent of ecosystems in an area. Because of differences in the kinds of species in different ecosystems and differences in their patterns of energy flow and nutrient cycling, each ecosystem in a landscape adds to overall biodiversity. For instance, the butterflies pictured in Figure 4.1 live in meadow ecosystems.

A nearby woodland ecosystem would support a different assemblage of butterfly species, and an

CONTRIBUTIONS OF SPECIES RICHNESS AND SPECIES EVENNESS TO SPECIES DIVERSITY

a.   LOWEST (ZERO) SPECIES DIVERSITY

Community includes just one butterfly species.

b.   INTERMEDIATE SPECIES DIVERSITY

Species richness is high (6 species) but species evenness is low.

c.   HIGHEST SPECIES DIVERSITY

High species richness and high species evenness (all species equally abundant).

FIGURE 4.1 (a) A population of a single butterfly species has zero species diversity. (b) A community of six butterfly species that is dominated numerically by one of the six species has low species evenness. (c) A community of six species is equal in species richness to community (b) but, with all species equally abundant (highest possible species evenness), is higher in species diversity compared with community (b).

agricultural field would likely support yet another collection of butterflies. Furthermore, the butterflies that live in these three different ecosystems would be pollinating different plants, which, in turn, are eaten by different insects.

Yellowstone National Park's ecosystems include forests, meadows, large rivers, ponds, marshes, and lakes, along with the geothermal ecosystems for which the park is best known. In fact, Yellowstone may contain a higher diversity of ecosystems than anywhere else in North America (**Figure 4.2** on the next page).

⚠ Think About It

1. Should any particular component of biodiversity (species, ecosystem, etc.) receive a higher priority for conservation (see Chapter 1) than others?

2. How do conservation for species diversity and ecosystem diversity complement each other?

3. Does a natural ecosystem need to be rich in species to merit conservation attention?

4.2 Geographic patterns and processes influence biodiversity

Now that we have reviewed species and ecosystem diversity, which reflect biodiversity within communities and across landscapes, let's consider patterns of biodiversity at larger regional and global scales. In

order to make conservation decisions, we need to know something about the biological differences among Earth's various regions, such as islands and continents, and climatic zones, such as those on cold mountaintops or in the tropics. Such knowledge will help us choose the best places for biological reserves and will inform management of these areas.

The Terrestrial Biomes

The **biomes** are large geographic areas recognized by their distinctive biological structure—associations of plants, animals, and other organisms—and characteristic vegetation or plant growth forms, such as trees, shrubs, grasses, or vines. For example, the plant and animal species that live in tropical forests and tundra are entirely different, and both support species different from those living in the Mediterranean scrub biome. However, because of their large scale, biomes are made up of many interacting communities and ecosystems. Variation in climate, soils, and other physical factors determine the type of biome covering a geographical region (**Figure 4.3** on page 99). For example, the cold climates of the far north are dominated by the low herbs and dwarf trees of the tundra biome, as well as the vast expanses of coniferous trees characteristic of the taiga or boreal forest biome. Meanwhile, tropical forests are found in regions where it is warm year-round and very rainy. Deserts develop where there are dry climates.

As we will see in Chapter 7 (pages 193 and 195), the soils of the terrestrial biomes are also distinctive

What relationship do you expect between ecosystem diversity and species diversity in a landscape?

How might the map in Figure 4.3 change with global warming?

biome Associations of plants, animals, and other organisms that occur over large areas and that are characterized by distinctive biological structure, especially by characteristic growth forms (e.g., trees, shrubs, or grasses on land; corals, kelp, or mangrove trees in aquatic environments).

A LANDSCAPE OF DIVERSE ECOSYSTEMS

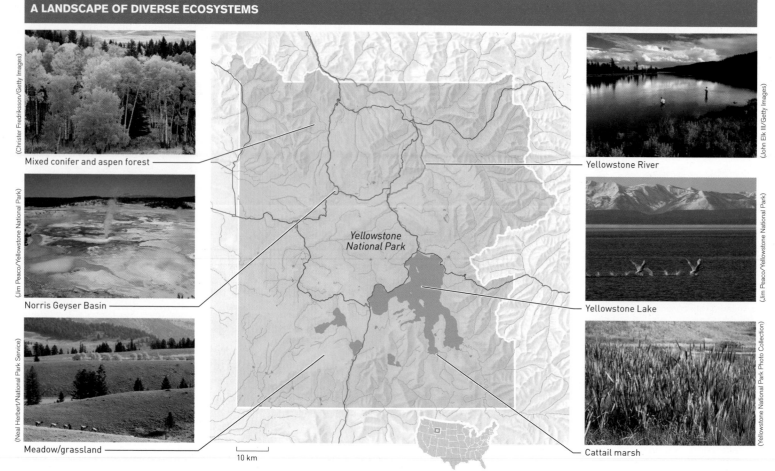

Mixed conifer and aspen forest
(Christer Fredriksson/Getty Images)

Norris Geyser Basin
(Jim Peaco/Yellowstone National Park)

Meadow/grassland
(Neal Herbert/National Park Service)

Yellowstone River
(John Elk III/Getty Images)

Yellowstone Lake
(Jim Peaco/Yellowstone National Park)

Cattail marsh
(Yellowstone National Park Photo Collection)

Yellowstone National Park

10 km

AREA ENLARGED

FIGURE 4.2 Yellowstone National Park includes an exceptionally high level of ecosystem diversity, which consists of the variety and relative extents of ecosystems in the landscape. The presence of diverse and unique ecosystems, a natural wonderland that still impresses the modern visitor, was a likely contributor to Yellowstone being named as the world's first national park.

and support very different types of economic activities: Temperate grasslands support extensive growing of grains, such as wheat and maize, whereas the boreal forests support vast wood harvesting for lumber and wood pulp for paper manufacturing.

The Aquatic Biomes

The major marine biomes include the open ocean, ocean floor, coral reefs, and kelp forests (**Figure 4.4** on page 100). While coral reefs are limited to warm tropical oceans, kelp forests are found in cool-temperate coastal waters through cold regions, such as the coasts of southern California to Alaska. The diversity of species living in and on intact coral reefs rivals that of tropical forests, and the organisms making up a coral reef community are very different from those inhabiting a freshwater marsh. Aquatic biomes that occur at the

transition between marine and terrestrial environments include salt marshes and mangrove forests. Meanwhile, freshwater wetlands occupy the transition between freshwater and terrestrial environments. Because of highly variable physical conditions and greater nutrient availability, these transitional biomes have low biodiversity but exceptionally high production.

Latitudinal Diversity Gradients

Consider that the number of bird species living in tropical Suriname is over 7 times higher than in Newfoundland, although they are roughly the same size (163,820 km² versus 111,390 km²). Other groups of organisms such as mammals, trees, reptiles, fish, and insects also increase in species richness from the poles to the tropics. Similar increases in species richness occur among freshwater and marine organisms, such as fish,

PORTRAITS OF EARTH'S TERRESTRIAL BIOMES

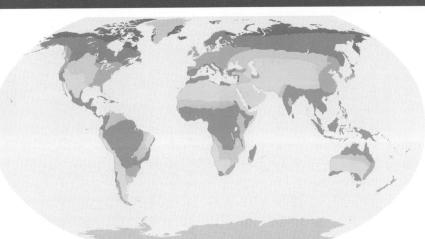

● **TROPICAL FOREST**

CLIMATE: Wet year-round or wet and dry seasons

VEGETATION: Broad-leaved evergreen or deciduous trees, vines, and epiphytes such as orchids and ferns

● **TEMPERATE FOREST**

CLIMATE: Moderate winters, warm summers, medium to high precipitation

VEGETATION: Deciduous or coniferous trees, shrubs, and herbaceous plants in understory

● **TEMPERATE GRASSLAND**

CLIMATE: Cold winters; wet and hot summers

VEGETATION: Grasses and other herbaceous plants, varying in height with annual precipitation

● **DESERT**

CLIMATE: Hot summers and mild or cold winters, dry

VEGETATION: Drought-tolerant succulents, deep-rooted shrubs and trees, fast-growing ephemeral plants

● **TROPICAL SAVANNA**

CLIMATE: Warm year-round; distinct wet, dry seasons

VEGETATION: Dry season fires help maintain grassland with scattered trees

● **MEDITERRANEAN SCRUB**

CLIMATE: Mild and wet winters; hot, dry summers

VEGETATION: Tough-leaved, fire-tolerant shrubs and trees, spring flowering grasses and herbaceous plants

● **TAIGA OR BOREAL FOREST**

CLIMATE: Cold winters; short, mild summers; medium precipitation

VEGETATION: Vast expanses of coniferous trees, especially spruce, fir, and larch; some deciduous aspen, birch, and willow

● **TUNDRA**

CLIMATE: Cold winters; brief, cool summers; low precipitation

VEGETATION: Lichens, low herbs, and dwarf trees

FIGURE 4.3 The plant growth forms and animals characteristic of Earth's diverse biomes reflect evolutionary adaptations to the planet's geographic variation in climate.

mollusks, and algae. As a consequence, regions near the equator contribute disproportionately to global species richness (**Figure 4.5** on page 101).

Why are the tropics so diverse? There are a few possible reasons for this phenomenon. The simplest is that there is simply more room in the tropics. The continents are centered around the equator, and the amount of land area there exceeds that in temperate regions. Other factors may include the amount of sunlight that the equator receives year-round, which contributes to greater levels of plant growth. The tropics also tend to have relatively stable climates, compared with the temperate zone, where seasonal changes limit growth in cold months. However, we do not yet have an entirely

Where would species unknown to science most likely be found?

PORTRAITS OF EARTH'S AQUATIC BIOMES

STREAMS & RIVERS

PHYSICAL ASPECTS: Turbulent, downstream flow of well-mixed water; high oxygen

KEY ORGANISMS: Algae and riverside plants source of food for aquatic invertebrates, amphibians, and fish

LAKES & PONDS

PHYSICAL ASPECTS: Conditions vary from shore to open water, variable salinity and nutrients

KEY ORGANISMS: Plants, algae, fish, and invertebrates vary from shore to deeper water

OPEN OCEAN

PHYSICAL ASPECTS: Extensive open water, light limited to upper layers, well mixed

KEY ORGANISMS: Drifting algae and grazing animals fed on by larger fish, seabirds, and whales

OCEAN FLOOR

PHYSICAL ASPECTS: Mainly sand or mud, light gradient from shallows to ocean depths

KEY ORGANISMS: Invertebrates such as shrimp and fish abundant in shallows and at hydrothermal vents

CORAL REEFS & KELP FORESTS

PHYSICAL ASPECTS: *Coral Reefs:* Warm, well-lighted, shallow water with stable salinity and temperature; *Kelp Forests:* Seasonally variable cool to cold, well-lighted, shallow water

KEY ORGANISMS: Reef-building corals, or macro-algae supporting highly diverse fish and invertebrate populations

MANGROVE FORESTS

PHYSICAL ASPECTS: Margins of warm tropical oceans, gradient from salt to freshwater

KEY ORGANISMS: Mangrove trees; high diversity of birds, fish, insects, and other invertebrates

SALT MARSHES

PHYSICAL ASPECTS: Generally twice daily tides; highly variable salinity, oxygen, and temperature

KEY ORGANISMS: Herbaceous plants tolerate saline, oxygen-poor soils; abundant invertebrates, birds, and fish

FRESHWATER WETLANDS

PHYSICAL ASPECTS: Flooding saturates soils with water for varying lengths of time

KEY ORGANISMS: Plants tolerate oxygen-poor, water-saturated soils; abundant, diverse animals

FIGURE 4.4 The biological communities of aquatic biomes are determined mainly by physical and chemical aspects of the environment, especially temperature, current and wave energy, salinity, and oxygen content.

biodiversity hotspot A region that supports at least 1,500 endemic plant species, approximately 0.5% of the world total, and that has been reduced in area by at least 70%.

endemic species Local or regional species of organisms found nowhere else on Earth.

equilibrium model of island biogeography The hypothesis that the number of species on an island is determined by a balance between rates of immigration of new species and rates of species extinction on the island, where rates of species immigration and extinction are determined by island size and isolation from sources of immigrants.

satisfactory explanation for this long-known pattern, and ecologists continue their research.

Biodiversity Hotspots

Although the tropics are generally the most diverse parts of the globe, particular areas seem to be "biodiversity hotspots," with spectacularly high numbers of species. Conservation International, a nongovernmental conservation organization, has identified 34 **biodiversity hotspots** on Earth, which they define as regions that have been reduced from their historic area by at least 70% and that support at least 1,500 **endemic species** of plants—that is, local or regional species found nowhere else on Earth. These biodiversity hotspots represent only 2.3% of Earth's land area yet support half the world's plant species (150,000 species) and nearly half the terrestrial vertebrate species. Many of these hotspots are located

in the temperate zone, including areas around the Mediterranean Sea and the California region (**Figure 4.6**).

Species Richness on Islands: Effects of Area and Isolation

Islands are natural laboratories for the study of biodiversity. For instance, biologists Robert MacArthur and E. O. Wilson proposed that the number of species on an island remains relatively constant over time but that the composition of species on an island changes. According to their theory, known as the **equilibrium model of island biogeography,** the number of species on an island is the result of a balance between the rate at which new species immigrate (arrive from somewhere else) there and the rate at which species already on the island become locally extinct (although they may persist elsewhere).

SPECIES RICHNESS VARIES WITH LATITUDE

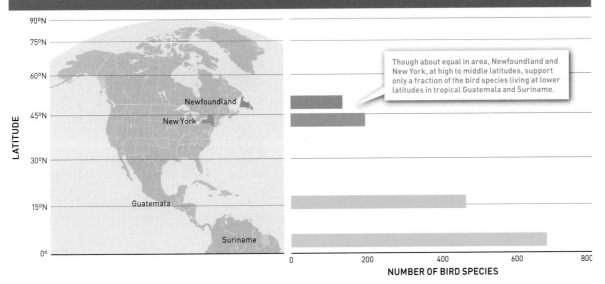

FIGURE 4.5 Bird species richness increases from higher to lower latitudes, a common pattern among most groups of organisms, including plants, mammals, insects, and fish. (Data from various sources)

Let's look at an example. The birds on Delos, a small Greek island of only 6 km^2, were surveyed during the middle of the 20th century and then again 35 years later (**Figure 4.7** on the next page). Between surveys, three bird species became extinct on the island, while two new species immigrated. The result was an insignificant

BIODIVERSITY HOTSPOTS MAY MERIT CONSERVATION PRIORITY

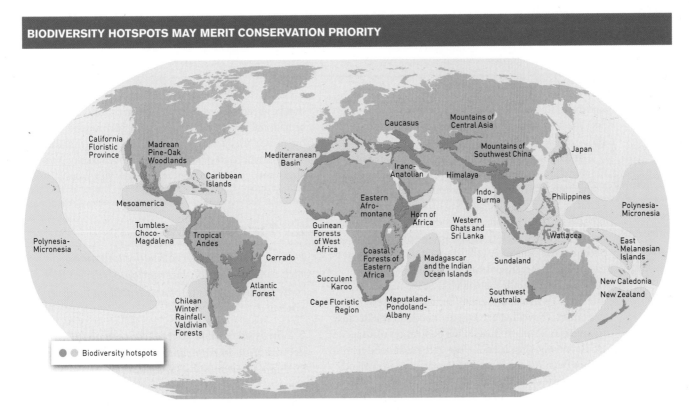

FIGURE 4.6 A relatively small number of biodiversity hotspots are home to a very large portion of Earth's species. Because biodiversity hotspots, shown by red shading, support half of Earth's plant species and a large fraction of animal species, many conservation scientists suggest that they be a focus for biodiversity conservation. (After Mittermeier et al., 2005)

ON DELOS ISLAND, GREECE, SPECIES EXTINCTION AND IMMIGRATION ARE NEARLY EQUAL

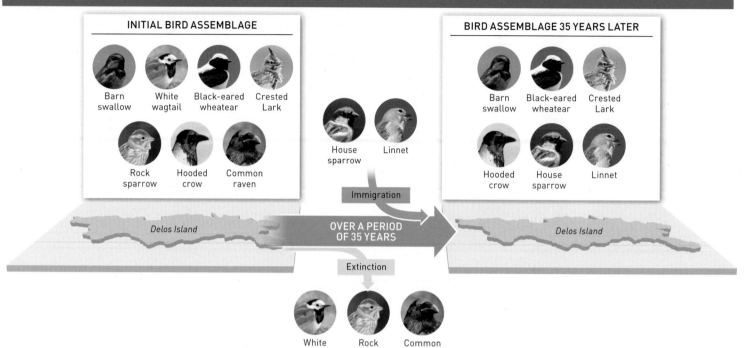

FIGURE 4.7 Species richness on islands, such as Delos, has been shown to result from a dynamic balance between local extinction and immigration. Despite substantial extinction and immigration of species, the number of bird species on Delos Island changed little in 35 years. (Data from Foufopoulos & Mayer, 2007) (Left panel: U.S. Fish and Wildlife Service, Dennis Jacobsen/Shutterstock, skapuka/Shutterstock, Bildagentur Zoonar GmbH/Shutterstock, John Navajo/Shutterstock, Vishnevskiy Vasily/Shutterstock, Edwin Butter/Shutterstock; Right panel: U.S. Fish and Wildlife Service, skapuka/Shutterstock, Bildagentur Zoonar GmbH/Shutterstock, Vishnevskiy Vasily/Shutterstock, Andrew Williams/Shutterstock, Florian Andronache/Shutterstock; Center top: Andrew Williams/Shutterstock, Florian Andronache/Shutterstock; Center bottom: Dennis Jacobsen/Shutterstock, John Navajo/Shutterstock, Edwin Butter/Shutterstock)

If you were to study the bird community on Delos Island 50 years from now, what would you expect to find?

decline in the number of bird species from seven to six. The bird community on Delos supports the theory proposed by MacArthur and Wilson.

MacArthur and Wilson proposed that the area of an island and its distance from the mainland source population determine rates of immigration and extinction. Larger islands tend to have higher species numbers for two reasons. First, they provide a larger target area and therefore receive more immigrants. Second, they have lower extinction rates because they can maintain larger populations of resident species. Now consider islands that are close to the mainland. They receive more immigrants because of close proximity. They also have lower extinction rates because immigration can help sustain the population of a species. Consequently, large islands close to the mainland tend to harbor the most species, whereas small, isolated islands harbor the fewest (**Figure 4.8**).

The equilibrium model of island biogeography can be broadly applicable to any habitat that occurs as isolated patches, such as remnant patches of woods in a landscape

that has been cleared to establish agricultural fields or national parks or wildlife refuges, surrounded by intensively used forest or grazing lands. The implication of the equilibrium model of island biogeography is that larger and less isolated natural habitats will support higher species richness.

⚠ Think About It

1. Why do some conservation organizations, such as Conservation International, suggest that conservation efforts be concentrated on biodiversity hotspots?

2. Why is there more general concern about deforestation in tropical forests, such as those in Suriname, compared with boreal forests, such as those of Newfoundland?

3. Are there unique species and ecosystems that will be lost if we focus conservation entirely on

AREA AND ISOLATION INFLUENCE SPECIES RICHNESS ON ISLANDS

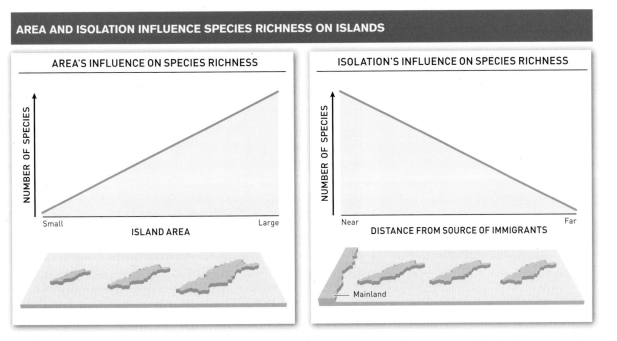

FIGURE 4.8 On average, larger islands support more species (higher species richness), whereas islands farther from sources of immigrants (e.g., a continent) support fewer species. As a result, small islands far from sources of immigrants support the fewest species, and large islands near sources of immigrants support the most species.

biodiversity hotspots and tropical latitudes? (Consider the polar bear.)

4.3 Some species influence biodiversity much more than others

The species in an ecosystem can be thought of loosely as a sports team, where each species has a different role to play. While the loss of any player might hurt the team's prospects, the loss of certain ones, such as the quarterback in football, would be absolutely devastating. In nature, it's also the case that some species appear to be more vital to the integrity of communities and ecosystems than others.

Keystone Species and Ecosystem Engineers

Species that exert significant effects on biodiversity despite low abundance within the community are called **keystone species.** Losing one of these species is like the collapse of a stone archway following the removal of its keystone. Keystone species are usually top-level predators like sea stars, sea otters, and jaguars that influence the structure of communities through their feeding activities (**Figure 4.9**). When restored to Yellowstone National Park, the gray wolf revealed its role as a keystone species. Before the restoration of wolves, woody vegetation along the Lamar River was limited to a scattering of mature cottonwood and willow trees, with very few young trees

in the populations. Elk had been feeding on the young cottonwood and willow trees so that few grew to maturity. However, soon after the reintroduction of wolves, elk

WHAT IS THE SOURCE OF THE KEYSTONE SPECIES HYPOTHESIS?

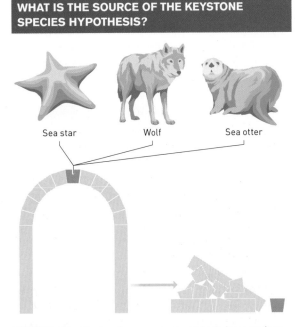

FIGURE 4.9 The keystone species hypothesis is an analogy drawn from architecture. The stone at the top of a stone arch is called the "keystone." If the keystone is removed, the arch will collapse upon itself. The removal of a "keystone species" can have an analogous effect on an ecological community.

keystone species A species with substantial influence on community structure, despite its low biomass or numbers relative to other species; the influence of keystone species is often exerted through feeding activities.

WOLVES AND RECOVERY OF RIPARIAN TREES IN YELLOWSTONE

(R. L. Beschta)

FIGURE 4.10 Wolf reintroduction is fostering cottonwood and willow reproduction and growth in Yellowstone National Park. Shown here are the results of successful establishment and growth of young cottonwoods (the smaller, densely spaced saplings under the mature cottonwood trees) in a riverside area along the Lamar River, where the local topography and the presence of wolves have reduced feeding by elk and bison.

BEAVERS ENGINEER STREAM ECOSYSTEMS

(Robert Cicchetti/Shutterstock)

FIGURE 4.11 By building dams, beavers introduce ponds and marshes along streams, which diversifies the ecosystem by adding new environments along areas that would otherwise be limited to the alternating pools and riffles of a typical stream.

Why is starvation no longer a major source of mortality among elk wintering in Yellowstone National Park?

avoided riverside areas, where they were vulnerable to wolf attacks. With the elk gone, young trees survived at a higher rate (**Figure 4.10**).

These effects trickled down through the ecosystem. As willows increased in size and abundance, there was an increase in the feeding and breeding habitat available to birds. Beaver populations grew, too, and physically altered the landscape as they became more abundant (**Figure 4.11**). Species, like the beaver, that manipulate the physical environment and influence ecosystem structure and processes are called **ecosystem engineers.** By preying on elk throughout the year, wolves also left

behind carrion that other species, including grizzly bears, coyotes, foxes, ravens, and eagles, could feast on.

Foundation Species

Foundation species create a physical framework for the community with their own bodies. Think, for example, of a forest, where the trees form a canopy and where shrubs and herbaceous vegetation on the forest floor create a habitat for other species. In the ocean, corals or kelp create a complex structural framework for fish and other animals to live in and feed in (**Figure 4.12**).

FOUNDATION SPECIES OCCUR IN MOST ECOSYSTEMS

ecosystem engineer
A species, like the beaver, that influences ecosystem structure and processes by altering the physical environment.

foundation species
A species that strongly influences community structure by creating environments suitable for other species by virtue of its large size or biomass.

(Borisoff/Shutterstock)

Coral reef

(Dennis Burns/Getty Images)

Grove of aspen trees

FIGURE 4.12 Corals create foundations for species-rich ecosystems in shallow, tropical marine environments. Aspen trees are an important foundation species in Yellowstone National Park.

INFLUENCE OF WOLVES RIPPLES THROUGH THE YELLOWSTONE ECOSYSTEM

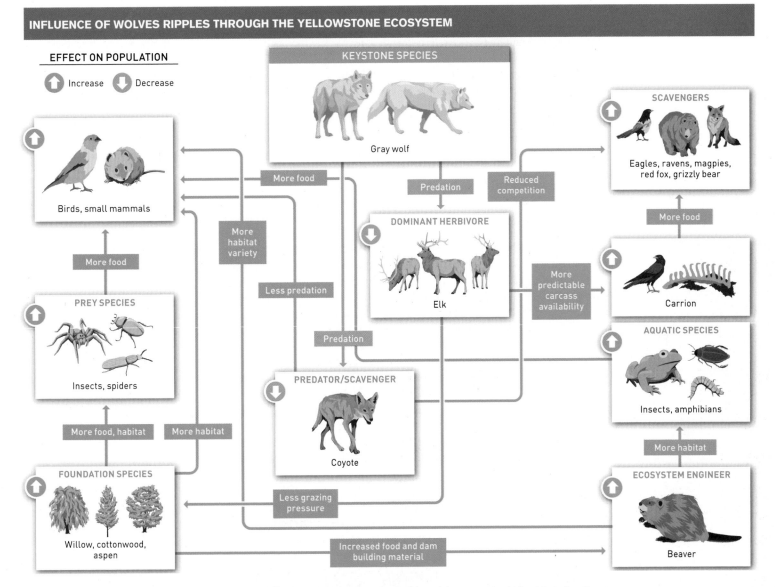

FIGURE 4.13 Predation by wolves directly suppresses the populations of some species, mainly coyotes and elk, while indirectly stimulating growth in many other populations, including ecosystem engineers and foundation species. The far-reaching effects of the gray wolf in the Yellowstone National Park ecosystem, which appear to qualify it as a keystone species, are the subject of ongoing long-term scientific studies.

Willows and cottonwoods, since they appear to have been released from control by elk grazing, may resume the status of foundation species in the Yellowstone National Park ecosystem. In addition to providing food for beavers, willows are a nesting habitat for a diversity of birds, including the endangered Western willow flycatcher. They provide shelter and food to a great diversity of insects, as well as cover for small mammals. By shading the ground, willows change the temperature and moisture conditions experienced by soil organisms, including worms, insects, and fungi. Again, wolves, by chasing elk out of streamside and riverside areas, seem to be releasing willows from herbivore control, further underscoring the keystone role of wolves (**Figure 4.13**). These far-reaching effects of

wolf reintroduction show how the presence, or absence, of a single species in an ecosystem can make a profound difference to biodiversity.

Indicator, Umbrella, and Flagship Species

There are many other species that can play important practical roles as we work to conserve ecosystems and landscapes. Viable populations of some species can demonstrate the health of the ecosystems on which they depend. Consequently, monitoring these **indicator species,** such as the California gnatcatcher, which is associated with coastal sage scrub (see Chapter 3, page 77), can provide information about the state of the ecosystem in which they live. Management designed to

How might the views of an elk hunter versus a trapper of beavers for the fur trade differ in regard to wolf restoration in the Northern Rocky Mountains?

indicator species A species that provides information about the state of the ecosystem in which it lives.

sustain populations of indicator species, as well as others such as the grizzly bears of Yellowstone or the spirit bear of the temperate rain forests of western Canada (see Figure 3.21, page 78), is another entry into ecosystem conservation. Indicator species can also act as **umbrella species,** since protecting them provides protection for the entire ecosystem on which they depend. Yet other so-called **flagship species** can attract and sustain human interest in protecting ecosystems. Examples of such charismatic species, which have become symbols for ecosystem conservation, include redwood trees, giant pandas, and the great whales.

⚠ **Think About It**

1. Can an organism be a keystone species and an ecosystem engineer? How about a keystone species and a foundation species?

2. Why do conservation biologists take threats to keystone species, ecosystem engineers, and foundation species very seriously?

3. How might too much attention to known keystone and foundation species harm conservation efforts?

4.4 **Ecological succession affects community composition and diversity**

After a disturbance, such as an intense forest fire, disrupts an established biological community, or when a new substrate, such as ash from a volcano, is laid down, microorganisms, plants, and animals begin to colonize the affected area. As these organisms establish themselves, they modify the environment, starting a process of ecological change called **succession,** which is the gradual change in a community over time following a disturbance.

Succession on a bare geologic surface, such as a recent lava flow, is called **primary succession.** It is generally a slow process, requiring hundreds or thousands of years, because it begins in the absence of well-developed soil. One example of this is the succession that takes place on the stones and gravel exposed by a retreating glacier (**Figure 4.14**). The earliest colonizers must be tolerant of low nutrient conditions. Many of these "pioneer" species, including plants and lichens (an association between a fungus and cyanobacteria), do not depend entirely on soil nutrients because they are capable of getting nitrogen from the air (for more on the nitrogen cycle, see Chapter 7, page 192).

The earliest community to develop during succession, called a **pioneer community,** is made up mainly of species that are tolerant of exposure to full sun, capable of living under conditions that are stressful to many other species. These pioneer species also generally have high reproductive rates and good dispersal abilities. Gradually, as these early organisms grow and shed detritus, which is consumed by fungi, worms, and other detritivores, nutrient-rich soil forms. As soil builds, conditions become suitable for the establishment of grasses and herbs, then shrubs and trees. In time, a forest could develop on the site.

Succession following a disturbance of an established community that doesn't destroy all living creatures or the soil is called **secondary succession,** because the process occurs on a landscape already much modified

umbrella species A species whose protection provides protection for the entire ecosystem on which that species depends.

flagship species A species that attracts and sustains human interest in protecting ecosystems.

succession The gradual change in a community over time following a disturbance.

primary succession Succession on a bare geologic surface, such as a recent lava flow.

pioneer community The earliest community to develop during succession.

secondary succession Succession following disturbance of an established community that doesn't destroy all living creatures or the soil.

PRIMARY SUCCESSION FOLLOWS GLACIER RETREAT

(NPS/Kay White)

FIGURE 4.14 As glaciers retreat, they expose mineral substrates, including stones, gravel, and sand. Over time, this surface is gradually colonized, initiating a successional process that may take many centuries.

EXAMPLE OF SECONDARY SUCCESSION: OLD-FIELD SUCCESSION

(Natural Area Teaching Laboratory, University of Florida, Gainesville, FL, US)

(Ronald van der Beek/Shutterstock)

15 JUNE, 2007 Cleared field with some herbaceous plant cover

(Natural Area Teaching Laboratory, University of Florida, Gainesville, FL, US)

(© Orokiet/Alamy)

21 JULY, 2008 Field dominated by tall ragweed

(Natural Area Teaching Laboratory, University of Florida, Gainesville, FL, US)

14 JULY, 2010 A more diverse mix of herbaceous plant cover on field

(Natural Area Teaching Laboratory, University of Florida, Gainesville, FL, US)

15 JULY, 2013 Woody shrubs and young sun-tolerant trees have colonized field

FIGURE 4.15 During old-field succession in eastern North America, the succession of communities will generally pass through several predictable stages. The initial stage consists of bare soil immediately following abandonment of the field. Second, an early successional community consists mainly of annual plants capable of colonizing bare soil. Third, a later stage of mixed perennial and annual herbaceous develops. Fourth, an early forest stage consists mainly of young, fast-growing, sun-tolerant trees and shrubs. Finally (not shown here), there is an ultimate successional stage of trees that can germinate and grow in the shade of mature, sun-tolerant trees.

by organisms. For example, after an agricultural field is abandoned, it may gradually become forested again. Early in succession, annual plants—those that germinate from seeds, then grow, set seeds, and die within one year—with high reproductive rates and rapid dispersal ability will dominate the plant community. Later, perennial herbs, grasses, and shrubs, with more restricted environmental requirements, will establish themselves, displacing many of the earliest, pioneer plant species. Next, we might see fast-growing pine trees. Eventually, the pines will be overgrown and replaced by slower-growing deciduous trees, such as oak and hickory. These will become the foundation species of the **climax community,** which will remain stable until the next disturbance (**Figure 4.15**). The species making up the climax community will, of course, depend on the biome in which succession occurs.

As succession unfolds, the sequence of species that arrive have different life histories (see Chapter 3, page 71). The communities during the early stages of succession are dominated by the *r*-selected species (many offspring, wide dispersal) characteristic of disturbed environments, while the later stages of succession favor *K*-selected species (few offspring, slow growth). Between these two endpoints, mid-successional communities are formed from a mixture of *r*- and *K*-selected species. As a result, the middle stages of succession are generally richer in species than either early or late successional stages (**Figure 4.16** on the next page).

⚠ Think About It

1. Considering that disturbance is a natural part of the environment, why might human-caused disturbance often have a more negative impact on biodiversity?

2. What would the life histories of species be like in a world with frequent and intense disturbances everywhere?

3. If pioneer species are so good at colonizing and dominating space during the early stages of succession, why does their dominance not extend through the later stages of succession?

Should the goal of conservation be to eliminate all forms of disturbance? Explain.

climax community The community at the end of a successional sequence that persists until a disturbance disrupts it sufficiently to restart succession.

4.1–4.5 Science: Summary

Species diversity is a function of the number of species—species richness—and their relative abundances—species evenness. Ecosystem diversity consists of the number and kinds of ecosystems in an area. Species richness generally increases from the poles to the equator. Biodiversity hotspots, inside and outside the tropics, support especially high numbers of species.

Diversity is shaped by a number of factors. Species richness on islands, determined by species immigration and extinction, is usually lower compared with similar-sized areas on continents. Keystone species, foundation species, and ecosystem engineers disproportionately influence the biodiversity of their communities. Diversity at any given point

in time depends on a region's history of a disturbance. Following severe disturbance, organisms will begin to colonize the area, setting in motion an ecological process called primary succession. Succession after less severe disturbances is called secondary succession. Species richness generally peaks during the middle stages of a successional process, in which the community commonly consists of a mixture of early (*r*-selected) and late (*K*-selected) successional species.

Species diversity is also the product of evolution. During allopatric speciation, a population is divided into two geographically separate populations. The two separated populations then accumulate differences over time, becoming separate species when they no longer interbreed. During sympatric speciation, new species form without geographic isolation.

4.6–4.8 Issues

It's fair to say that no place on Earth has been left untouched by human activity. In places like Manhattan's Greenwich Village, there's almost no trace of Minetta Brook, which once flowed into a marsh where Washington Square Park now sits. The wildlands of the western United States are crisscrossed with roads cut by ranchers, loggers, and oilfield workers. Even deep within the Amazon, evidence of ancient metropolises can be seen in soil charcoal patterns and irrigation canals. And even in ecosystems where humans have never set foot, our carbon-dioxide emissions are altering the ecosystems' function. This is the global reach of human impact on biodiversity.

habitat fragmentation
A subdivision of a formerly continuous habitat into isolated habitat patches as a result of activities such as deforestation, road building, and dam construction on rivers.

4.6 Habitat fragmentation reduces biodiversity

Every time we clear land for roads or farms, build subdivisions, or excavate a mine, we break up natural ecosystems. **Habitat fragmentation** transforms continuous ecosystems into smaller, increasingly isolated

patches of habitat, or habitat fragments. A patch of forest surrounded by a large area of land cleared for cattle pasture is a typical example. Habitat fragmentation affects the entire community of organisms and can reduce species richness because the populations of species dependent on the remaining patches of native ecosystem are generally reduced and their long-term survival threatened. Like islands, the smaller the fragment, the fewer species they can support.

Tropical Forest Fragmentation

In addition to the 90,000 km² of tropical forest that is cleared annually, some 20,000 km² are not entirely destroyed each year but rather damaged by logging, burning, and road building, resulting in habitat fragmentation (**Figure 4.19**). For the last three decades, the Biological Dynamics of Forest Fragments Project, a cooperative effort by Brazilian scientists and others from all over the world, has been investigating this phenomenon. Within a 1,000-km² study area, researchers

SATELLITE IMAGES REVEAL THE MASSIVE SCALE OF TROPICAL DEFORESTATION

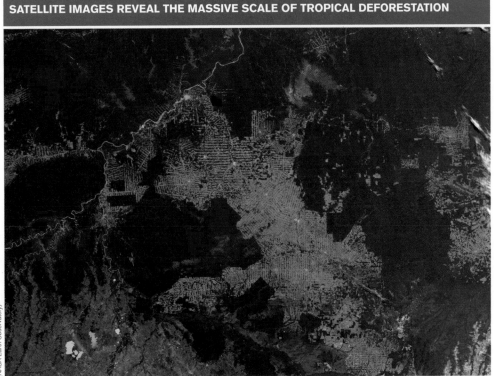

(NASA Earth Observatory)

FIGURE 4.19 Current estimates of deforestation in the tropics total tens of thousands of square kilometers annually.

How would ecological succession (see page 106) generally affect forest fragmentation?

established forest fragments of various sizes to observe and study. Each fragment was surrounded by land that had been cleared of forest to establish cattle pasture (**Figure 4.20**). The controls for this research are similar-sized forest plots completely surrounded by and connected to large expanses of forest (more than 200 kilometers in extent).

The results of the study, which have been reported in hundreds of scientific papers, are clear: Compared with larger fragments, smaller forest fragments support fewer

ECOLOGISTS STUDY FRAGMENTATION EXPERIMENTALLY

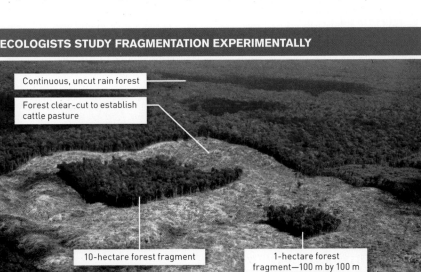

Continuous, uncut rain forest

Forest clear-cut to establish cattle pasture

10-hectare forest fragment

1-hectare forest fragment—100 m by 100 m

(Dr. Richard O. Bierregaard)

FIGURE 4.20 The Biological Dynamics of Forest Fragments Project is one of the largest and longest-running ecological experiments ever. Shown here are small and large forest fragments separated from continuous rain forest by cleared land.

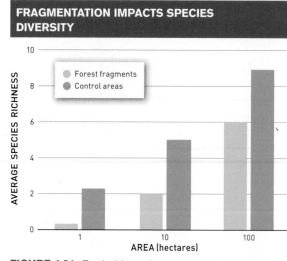

FRAGMENTATION IMPACTS SPECIES DIVERSITY

Legend:
- Forest fragments
- Control areas

Y-axis: AVERAGE SPECIES RICHNESS (0 to 10)
X-axis: AREA (hectares) — 1, 10, 100

FIGURE 4.21 Tropical forest fragmentation reduces the number of forest-dwelling species in forest fragments, including insect-feeding birds. Compared with control areas, which are equal-size areas of unfragmented forest, forest fragments support fewer species of insect-feeding birds. (Data from Laurance et al., 2002)

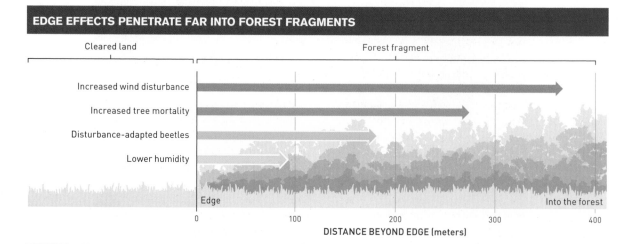

EDGE EFFECTS PENETRATE FAR INTO FOREST FRAGMENTS

Cleared land | Forest fragment

- Increased wind disturbance
- Increased tree mortality
- Disturbance-adapted beetles
- Lower humidity

Edge | Into the forest

X-axis: DISTANCE BEYOND EDGE (meters) — 0, 100, 200, 300, 400

FIGURE 4.22 Here are a few of the dozens of edge effects, which extend anywhere from 10 to 400 meters from the edge of a forest patch, that have been documented in studies of isolated forest patches in the Amazon River Basin. (Data from Laurance et al., 2002)

?

Was there ever a time when tropical forest interior species may have been more common than tropical forest edge species?

edge effects Environmental conditions occurring near the edge of an ecosystem (e.g., near the edge of an isolated forest fragment); conditions at edges differ from those deep in the ecosystem interior.

species of large mammals, primates, understory birds, dung beetles, ants, bees, termites, and butterflies. Most significant, forest fragments of all sizes support fewer species of forest animals compared with similar-sized areas embedded in continuous forest (**Figure 4.21**). This study and many others around the world have demonstrated repeatedly that habitat fragmentation results in the loss of biodiversity.

Edge Effects

When considering the size of an isolated fragment of forest, it's important to remember that the newly created edge of that patch is different from the interior. Edges may be windier, drier, and hotter and might be subject to more insect pests. These **edge effects** often extend deep into the patch of forest. In the Amazon River Basin, 30 different edge effects have been identified, extending as far as 400 meters (1,300 feet) into forest fragments (**Figure 4.22**). While some species thrive where fragmentation produces edge effects, those that require the physical conditions of the interior of an ecosystem (e.g., higher humidity or lower sun exposure) do not. Consequently, as edge conditions proliferate with increasing fragmentation, these "interior species" decline in abundance and at some point disappear from the landscape.

⚠ Think About It

1. What are some ways to counteract the effects of isolation that result from habitat fragmentation?

2. How are habitat patches—for example, forest patches surrounded by agricultural fields—and oceanic islands similar? How are they different?

4.7 Valuable services of ecosystems are threatened

Beyond losing the aesthetic benefits of nature, what are the consequences of losing an ecosystem? In the late 1990s, a panel of 11 ecologists, led by Gretchen Daily of Stanford University, identified a long list of goods and services that humans derive from natural ecosystems. The goods can be readily assigned economic value because they are regularly bought and sold—seafood, timber, feed for livestock, and biomass fuels.

Many of those goods are already being threatened by unsustainable harvesting (see Chapter 8). According to the United Nations Food and Agriculture Organization, one-quarter of the world's 600 monitored fish stocks are overexploited or depleted. Another 52% are being

TABLE 4.1
ECOSYSTEM SERVICES

1	Purification of air and water
2	Mitigation of droughts and floods
3	Generation and preservation of soils
4	Waste detoxification and decomposition
5	Cycling of nutrients
6	Pollination of crops and wild plants
7	Seed dispersal
8	Agricultural pest control
9	Maintenance of biodiversity
10	Reduction of coastal erosion
11	Uptake and storage of greenhouse gases
12	Moderation of climate
13	Protection from harmful ultraviolet light
14	Aesthetic, cultural, and intellectual values

Information from Daily et al., 1997.

exploited at maximum harvest rates. One of the principal problems with current attempts to regulate fisheries is that species are often viewed in isolation when, in fact, they are components of interconnected ecosystems. For instance, we are currently overharvesting Atlantic menhaden for fish oil, which harms other species that feed on them, including tuna, mackerel, and cod, along with birds such as ospreys.

In addition to goods, ecosystems provide many services to the local and global environment: Plant roots and soil filter water, trees filter the air, mangrove and coastal forests provide flood control. These **ecosystem services** have economic, social, or cultural value to human populations (**Table 4.1**). For example, wild native insects provide ecological services such as pollination, control of native herbivore pests, and even recreation. A 2006 estimate put the annual value of ecological services provided by insects at $57 billion in the United States alone. However, these services are threatened by factors such as the overuse of pesticides and herbicides. For instance, colony collapse disorder, a mysterious condition that has been wiping out approximately 10% of honeybee colonies each year, has been linked to neonicotinoid pesticides (insecticides chemically similar to nicotine), although other factors may be involved.

Other animals may control pest insect populations: A colony of 30,000 southeastern bats in Florida eats more than 15 tons of mosquitoes each year. Historically, bats have been persecuted by humans who have dynamited, poisoned, and shot them. Today, the biggest threat to bats' survival comes from a fungal disease called white nose syndrome, which was likely spread by recreational cavers and has killed more than 5.7 million bats in eastern North America.

Functioning ecosystems can even save lives by providing clean water, reducing the impact of forest fires,

or damping storm surges. For instance, the spidery roots of mangrove forests reduce the power of waves crashing against the coast during violent storms. An economist with India's Institute of Economic Growth has estimated that mangrove forests prevented 20,000 deaths during the 1999 Orissa Cyclone in the Indian Ocean. But mangroves are one of the world's most endangered ecosystems. They are being chopped down for firewood and building material and cleared for shrimp farms. In the last 50 years, their distribution has been reduced by 60%. Destroying mangroves rarely makes economic sense. One study in Mexico has shown that each acre of mangrove forest brings in $15,000 per year in seafood, more than 200 times the market value of their wood.

Of course, the economic value of the full gamut of ecosystem services is not always so easy to quantify.

The challenge was taken up by teams of researchers led by Robert Costanza of the Australian National University in Canberra. In a study published in 2014, Costanza and his colleagues estimated that the total value of the goods and services provided by natural ecosystems in 2011 was $125 trillion. Their estimate was almost twice the total global annual gross product at the time. However, many of the goods and services provided by ecosystems, particularly extensive ones such as climate control by the world's oceans, are simply irreplaceable.

⚠ Think About It

1. Are ethical and aesthetic values versus economic values mutually exclusive?

2. If some ecosystems, such as the Amazon rain forest or the boreal forests of Canada and Russia, provide ecosystem services that are global in significance, should the international community compensate the countries that sustain them?

3. What sorts of ecosystem services are simply irreplaceable and therefore invaluable?

4.8 Many invasive species harm ecosystems

We already learned in Chapter 3 that invasive species can endanger populations of native species. However, the impacts of invasive species can extend to entire ecosystems. The Fynbos vegetation of South Africa is one of Earth's biodiversity hotspots, where rolling green meadows border the southern Indian Ocean. Unfortunately, invasive pine and acacia species in the Fynbos have sucked so much water out of the ground that they have reduced water flow in nearby river basins by nearly one-third. These invasive plants have also impaired a wide range of other valuable services: They

Some have criticized attempts to put an economic value on ecosystem services. What are some of the sources of such criticisms?

How would you go about assigning cash value to a cooling sea breeze on a summer evening?

ecosystem services
Aspects of ecosystem structure and function that have economic, social, or cultural value to human populations (e.g., flood control by wetlands or water purification by forest and river ecosystems).

INVASIVE GRASS SPECIES CHANGES A FIRE REGIME

(istockphoto/Getty Images)

FIGURE 4.23 One of the most serious threats to native ecosystems is the result of changed fire regimes fueled by flammable invasive species, such as cheat grass, *Bromus tectorum*, shown here fueling a fire across a rangeland in the western United States.

Is it sometimes appropriate to consider invasive species problems as threats to public safety and security in addition to associated ecological issues?

displace plants used for traditional medicines and tea, and invade coastal dunes and riparian areas, which restricts access for tourism and fishing.

Woody invasive plants have also increased the intensity of fires by increasing the amount of flammable plant biomass in the ecosystem. In the southwest United States, a thirsty tree called saltcedar, *Tamarix* species, has dried up streams and changed the **fire regime** by increasing the frequency and intensity of fires in these ecosystems. Similarly, cheat grass, *Bromus tectorum,* which has

invaded vast areas in western North America, not only increases fire frequencies but also is a poor source of forage for livestock or native animals. Prior to the cheat grass invasion, wildfires occurred every 60 to 110 years; now they happen every 3 to 5 years (**Figure 4.23**).

Invasive Species and Aquatic Ecosystems

The impact of invasive species is not limited to terrestrial environments. Invasive aquatic species have been spread around the world with devastating consequences to ecosystem goods and services. Countries across Africa spend millions of dollars each year to control invasive aquatic weeds that choke waterways important for fishing and commerce. The introduction of the Nile perch into Lake Victoria, one of the great lakes of Africa, has resulted in hundreds of native fish species going extinct (**Figure 4.24**), but it has also had consequences for the ecosystem as a whole. For instance, local people once caught smaller native fish, which they could sun-dry on racks. The larger Nile perch must be smoked, so people need to cut down trees for fire. This has increased erosion and led to increases in pollution by agricultural runoff.

In Asia, the golden apple snail has ravaged rooted aquatic plants, reducing their biomass and releasing nutrients into the ecosystem. As a result, previously clear waters are now turbid and choked with algae. In the United States and Canada, zebra mussels have become a nuisance by attaching to nearly any submerged structure, including the intakes to water supply systems.

Economic Impact of Species Invasion

The executive director of the United Nations Environmental Programme estimated that the costs

How does one weigh the loss of hundreds of fish species found nowhere else on Earth against establishing the highly profitable Nile perch fishery in Lake Victoria, the largest lake fishery in the world?

fire regime The frequency and intensity of fires that typically occur in a particular ecosystem.

INVASIVE AQUATIC SPECIES REDUCE ECOLOGICAL SERVICES IN AFRICA

(AP Photo/Danny Wilcox Frazier)

Water hyacinth in Lake Victoria

(Woodfall Wild Images/Photoshot)

Nile perch caught in Lake Victoria

FIGURE 4.24 Water hyacinth in Lake Victoria interferes with fishing and transport. Another invasive species, the introduced Nile perch, coupled with nutrient pollution has led to the extinction of hundreds of native fish of Lake Victoria, which had been the traditional source of protein for millions of people living around the lake.

of invasive species to the global economy exceeded $1.4 trillion in 2010. The economic costs of controlling just the saltcedar invasion of western North America have been estimated at $127 to $291 million annually for control and estimated water losses. Meanwhile, zebra mussel control associated with water systems around the Great Lakes approaches $70 million each year. However, some invasive species exact even greater costs. For instance, the water losses resulting from invasive species in the Fynbos region are valued at $1.4 billion, and the golden apple snail, introduced from South America, causes over a $1 billion in annual losses in rice cultivation in the Philippines. David Pimentel and colleagues from the College of Agriculture and Life Sciences at Cornell University estimated that as of 2005, invasive species have caused over $120 billion in damages annually in the United States alone.

⚠ Think About It

1. During control efforts, do we have an obligation to treat invasive species, especially vertebrate animals, ethically?

2. Does the use of militaristic terminology (e.g., battle, eradicate, overrun, and kill) to frame relationships with invasive species foster an adversarial relationship with nature rather than a sustainable one?

4.6–4.8 Issues: Summary

Habitat fragmentation, which transforms continuous ecosystems into smaller, isolated habitat patches, results in significantly reduced biodiversity. Increased wind or reduced moisture on the edges of ecosystem patches gives them a smaller effective size than they appear on a map. Ecological economists estimate that the total value of goods and services provided by natural ecosystems, including maintenance of biodiversity, may be $125 trillion, or nearly twice the total global annual gross product. Many ecosystem services are threatened by habitat destruction and the spread of infectious diseases, such as white nose syndrome in bats. These services are not only valuable, but they can also save human lives by reducing coastal storm surges, reducing forest fires, and providing clean water. Invasive species can also harm ecosystems by altering hydrologic and fire regimes. The economic impacts of invasive species, which amount to hundreds of billions of dollars annually, result from control costs and reduced ecological services.

4.9–4.12 Solutions

The two major focal points for biodiversity conservation and restoration are on *species* (emphasized in Chapter 3) or on *ecosystems*. Often, these approaches are presented as polar opposites. However, it is apparent that conserving endangered species requires healthy, functioning ecosystems. Wild species do not live apart from the ecosystems on which they depend. Similarly, wherever we conserve or restore natural ecosystems, we sustain many species populations. Some of the most extensive programs to protect whole ecosystems have involved setting aside protected areas.

Concern that the world's ecosystems were under threat led 168 countries of the world to sign the **Convention on Biological Diversity** following the "Earth Summit" meeting in Rio de Janeiro, Brazil, in 1992. It was a defining moment in the history of the relationship between humans and the biosphere, as it came with the recognition that protecting ecosystems is fundamental to economic development. Even the site of the meeting was significant—it was held near the Amazon River basin, perhaps the last greatest reservoir of biological diversity on Earth, which provides immeasurable services to the people of the Amazon basin and beyond. The Convention's specific objectives were "the conservation of biological diversity, the sustainable use of its components, and the fair and equitable sharing of the benefits arising out of the utilization of genetic resources."

The Convention has charted a clear path toward finding a solution to the biodiversity crisis, requiring countries to meet a variety of milestones, including setting aside 26% of their land area and 17% of their territorial seas as protected areas. In addition, the nations need to ensure the effectiveness of their national parks and other protected areas by developing management plans for them and identifying the impacts of climate change. As of 2015, 196 nations had ratified or accepted

Convention on Biological Diversity An international agreement negotiated under the sponsorship of the United Nations Environmental Programme to promote the conservation of biological diversity, the sustainable use of its components, and the fair and equitable sharing of the benefits arising out of the utilization of genetic resources.

the Rio Convention (although U.S. president Bill Clinton signed the Rio Convention, the Senate has not ratified it). To date, most countries have failed to meet their goals, but there have been significant conservation successes over the last 20 years.

4.9 The number of protected areas has grown rapidly

One of the principal means of conserving an ecosystem is to protect it. The Convention on Biological Diversity defines a **protected area** as a geographically defined area that is designated or regulated and managed to achieve particular conservation objectives. The various types of protected areas range from strict nature reserves, set aside mainly for scientific research and in which tourism is prohibited, to areas protecting biodiversity but managed for sustainable harvest of natural resources such as timber and game animals. Generally, local laws regulate the use of protected areas.

The world's oldest protected areas were sacred sites valued as the homes of the gods or as resting places for the dead. Protection of the resources produced by an ecosystem came later. In Japan, for instance, forests providing timber for the construction of Shinto temples have been protected for over 2,000 years. Hunting preserves were established in northern India around the same time. The first large natural area protected for purely aesthetic reasons was Yellowstone National Park.

The founding of Yellowstone inspired the establishment of national parks and other types of protected areas the world over. Between 1872, the date of Yellowstone's founding, and 1962, the number of protected areas grew to over 10,000. Today, there are more than 100,000 protected areas across the globe, encompassing nearly 19 million km², more than the combined areas of the United States and Brazil. In total, about 12% of Earth's land surface is afforded some degree of conservation protection (**Figure 4.25**).

Size and Connections in Protected Area Design

When it comes to nature reserves, bigger is better. Unfortunately, that's not always practical. One key to sustaining functioning ecosystems in smaller reserves is to ensure that they have connections, or **habitat corridors,** to other reserves. Habitat corridors increase the movement of organisms among protected areas; increase gene exchange between separated populations, which helps maintain genetic diversity in local populations; and reduce the chance of extinctions in habitat fragments. In southern China, for instance, environmental scientists have identified potential habitat corridors between existing reserves for the giant panda in the Minshan Mountains (**Figure 4.26**). If adopted by the government, this conservation plan would reduce habitat fragmentation and multiply the size of the panda protected area, making the future of the species more secure.

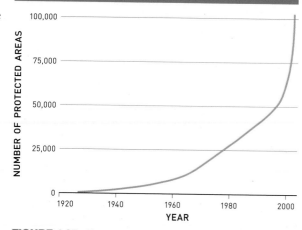

NUMBER OF PROTECTED AREAS HAS GROWN EXPONENTIALLY

FIGURE 4.25 The number of protected areas worldwide has increased rapidly from the few designated at the beginning of the 20th century to over 100,000 by the end of the century. (Data from Mulongoy & Chape, 2004)

Marine Protected Areas

Protected areas in coastal areas and the oceans, called **marine protected areas,** or **MPAs,** help conserve ecosystems critical for biodiversity, such as coral reefs and salt marshes, and sustain populations that supply us with fish and other marine resources. For many years, protecting marine habitats was ignored, compared with protecting terrestrial habitats, but that's slowly changing. In 2006 President George W. Bush declared 36 million hectares of marine habitat in the Pacific Ocean the Papahanaumokuakea Marine National Monument, which became one of the largest MPAs in history. Studies have shown that properly managed MPAs provide a refuge for fish to breed and replenish exploited fisheries. In fact, social scientists have found links between MPAs and a reduction in poverty among small fishing communities in places such as Papua New Guinea and Indonesia—making it a win–win for conservation and society (see also Chapter 8, page 251).

⚠ Think About It

1. The United Nations Environmental Programme notes in a recent report that most of the world's largest protected areas are in remote, low-diversity regions, including ice caps and deserts. What does this bias reveal about the actual commitment to biodiversity protection?

2. What political, economic, and social conditions could lead to the creation of large protected areas in high-diversity areas where human population densities are high?

---

protected area A geographically defined area designated or regulated and managed to achieve particular conservation objectives (e.g., national parks, national forests, wildlife refuges).

habitat corridor A strip of suitable habitat linking protected areas intended to increase the movement of wildlife between protected areas to sustain genetic variation and reduce the likelihood of extinction of protected populations.

marine protected areas (MPAs) Protected areas in coastal regions and the oceans that help conserve ecosystems critical for biodiversity (e.g., coral reefs and salt marshes) and sustain populations that supply fish and other marine resources.

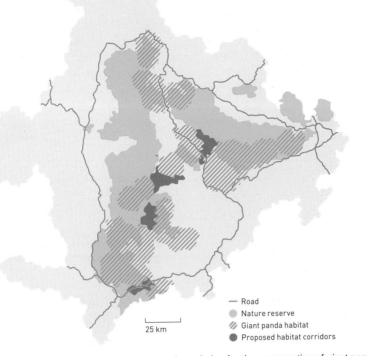

- Road
- Nature reserve
- Giant panda habitat
- Proposed habitat corridors

25 km

FIGURE 4.26 Conservation planners have developed a landscape-based plan for the conservation of giant pandas in the Minshan Mountains, in southern China. The key element in their plan is to link existing panda reserves with a network of habitat corridors to reduce fragmentation. (After Shen et al., 2008)

4.10 Nongovernmental conservation complements governmental programs

During the past three decades, nongovernmental organizations (NGOs) and private individuals have taken the lead in protecting ecosystems around the world. One of the main contributions to conservation that they make stems from their greater flexibility, compared with that of governments. When private individuals are involved, setting up a protected area can be decided by a personal commitment. For example, entrepreneur Douglas Tompkins, who founded the clothing companies North Face and Patagonia, and media mogul Ted Turner, who launched CNN, have created protected areas around the world totaling approximately 16,000 km² (6,240 mi²)—about twice the combined areas of the U.S. states of Rhode Island and Delaware.

In the case of a nongovernmental conservation organization, a decision to allocate funds to protect an area with potential conservation value may only require a positive vote from a board of directors. There are hundreds of international and national nongovernmental conservation organizations actively working to sustain biodiversity, and they have preserved

areas totaling over 2.5 million km². Some of the best known organizations include the World Wildlife Fund, Conservation International, and The Nature Conservancy.

The Nature Conservancy

The Nature Conservancy is a nonprofit, nongovernmental conservation organization established in 1951 to protect ecologically important areas around the world. Over the course of its more than half-century of operation, the organization has been active in the conservation of terrestrial, freshwater, and marine ecosystems, protecting nearly 500,000 km² of terrestrial ecosystems and over 8,000 kilometers of rivers. It also manages more than 100 conservation projects in the marine environment.

The Nature Conservancy originally focused on conservation in the United States and protecting critical habitat for rare and threatened species or conserving very small examples of regionally distinctive environments. Today, the organization has increasingly taken a larger-scale, more inclusive approach, emphasizing the conservation of whole ecosystems and landscapes. Wherever The Nature Conservancy operates, it strives to integrate its conservation programs with the broad economic interests of the local community, working

THE NATURE CONSERVANCY IS ACTIVE IN MORE THAN 30 COUNTRIES AROUND THE WORLD

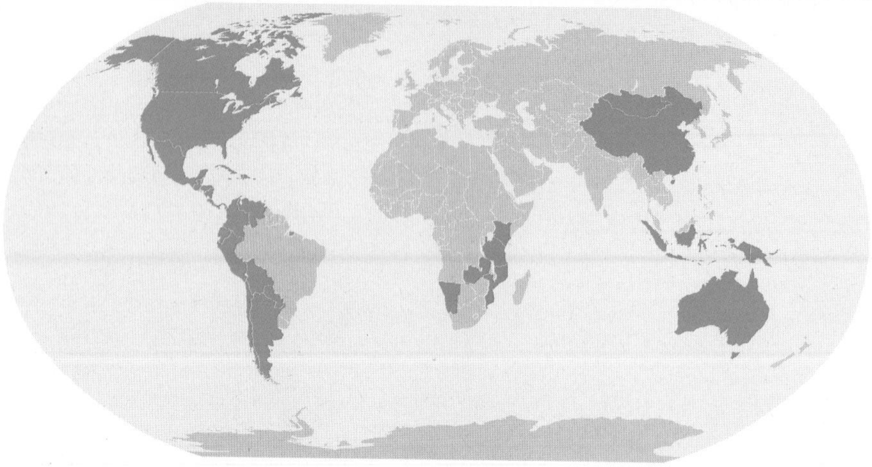

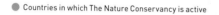

● Countries in which The Nature Conservancy is active

FIGURE 4.27 Beginning as a nongovernmental organization dedicated to conserving biodiversity in the United States, The Nature Conservancy has expanded its mission and now works around the globe. (Information from www.nature.org)

Does the involvement of business and financial interests in The Nature Conservancy's operations help or hinder its conservation mission?

The Nature Conservancy approach has emphasized working mostly out of the public eye to acquire and protect areas important to biodiversity conservation. What are some complementary approaches to this work adopted by other NGOs?

debt-for-nature swap
A transaction wherein a developed nation forgives the debt of a developing nation in exchange for conservation pledges.

with stakeholders ranging from academic scientists and conservationists to industrial and financial leaders. It has more than 1 million members and is active in more than 30 countries (**Figure 4.27**).

Forever Costa Rica

The Forever Costa Rica project provides an example of how The Nature Conservancy helps protect natural areas through partnerships with governments and private individuals. Costa Rica is famous for being one of the most biologically rich spots in the world—a tiny country that is home to 5% of all species. By the early 2000s, it had set aside 25% of its land in preserves, which meant that it was closing in on one of the most important metrics under the Convention on Biodiversity.

Unfortunately, its terrestrial parks were struggling with financial survival, and the country was protecting very little of its marine diversity. In 2010 The Nature Conservancy brokered one of the largest **debt-for-nature swaps** with the U.S. government. A debt-for-nature swap occurs when a developed nation forgives the debt of a developing nation in exchange for conservation pledges. Under the 1998 Tropical Forest Act, for instance, the U.S. Treasury can forgive up to $20 million worth of a country's debt each year if that country can make a case that it will use the funds to preserve a tropical forest in perpetuity. Like many developing countries, Costa Rica was indebted to the United States for loans from the U.S. Agency for International Development (USAID). The Nature Conservancy used funds freed up from a debt-for-nature swap, combined with funds from conservation partners, to create a $56 million trust

fund that could support Costa Rica's national parks in perpetuity.

⚠ Think About It

1. In many regions of the world, the business community resists the establishment of new protected areas, arguing that such areas depress the economy. How might the conservation community address such concerns?

2. How might governments, conservation-related NGOs, and academic institutions partner to protect ecosystems and landscapes? (What each does best: Governments administer and protect; NGOs use flexibility to pursue emerging conservation opportunities and raise funds from private sources; academic institutions teach and research.)

4.11 Sustaining biodiversity and ecosystem services requires active management

In most situations, you cannot simply declare an area "protected" and expect that the species and ecosystems included in the area will sustain themselves. Managing protected areas to sustain their conservation value requires attention to the many environmental factors known to influence biodiversity, such as keystone species and disturbance.

Sustaining Crucial Species

One of the most important parts of managing a natural area is to ensure that important species are present and their populations are healthy. These crucial species generally fall into the categories we mentioned earlier as *foundation species, ecosystem engineers,* and *keystone species*. Reintroducing wolves to Yellowstone National Park was a conscious decision, in part, to help restore an ecosystem degraded by high populations of elk. After the wolves' reintroduction, it has been crucial for field biologists like Doug Smith to monitor the wolf population and track their movements inside and outside the park. Although the wolf has been relatively successful at multiplying on its own, other species, such as the peregrine falcon, required years of captive breeding to help recover their numbers.

Controlling Invasive Species

In ecosystems seriously damaged by invasive species, many land managers have instituted programs to control them and restore native species. For example, several such programs are aimed at controlling large-scale infestations of saltcedar across the American West. Historically, managers have used a wide range of tools for control and restoration, such as mechanical removal combined with spraying herbicides and managing flooding to favor reproduction of native cottonwood and willows. More recently, beetles that feed on saltcedar have been imported from saltcedar's native range in Asia and released in heavily invaded areas in North America. Using a combination of these tools, managers have succeeded in controlling saltcedar and restoring native vegetation (**Figure 4.28**).

Fire, Succession, and Management

Fire is a critical factor for maintaining a wide variety of ecosystems, including grasslands. In protected areas within the temperate grassland biome, periodic fires are necessary to keep succession from converting grasslands to shrublands or forests (**Figure 4.29**). Lightning causes fires in grasslands naturally. However, humans began setting fire to grasslands long ago to encourage fresh growth by grasses that would attract game. Today, managers of grasslands in places like the Konza Prairie Preserve in eastern Kansas and the Tallgrass Prairie National Preserve use fire to prevent encroachment by trees and shrubs. Without this active fire management, the grasslands these preserves were created to protect would disappear and be replaced by forests. In other conservation areas, managers may suppress fires to achieve other management goals such as restoring previously cleared forest.

⚠ Think About It

1. Assuming that grazing and fire are sources of disturbance to grasslands, how would you manage grazing intensity and fire frequency to maximize species richness in a grassland preserve?

2. Why is simply putting a fence around an area and declaring it protected no guarantee that the biodiversity in the area is secure?

What safeguards need to be in place before introducing another species to control an invasive species?

What does the need for active management of protected areas suggest about the stability of ecosystems?

CONTROLLING INVASIVE SALTCEDAR, *TAMARIX* SPECIES, IN THE WESTERN UNITED STATES

Mechanical control of saltcedar

(Courtesy of Anna Sher Simon)

Tamarix beetle feeding on saltcedar

(USDA photo by Bob Richard)

Restored cottonwood-willow riparian forest

(International Boundary and Water Commission)

FIGURE 4.28 Managers have used mechanical removal to control invasive saltcedar in many protected areas. Some beetles that primarily feed on saltcedar, in its native range, have been imported and released in western North America to help control the invasive species. Thousands of hectares of native cottonwood and willow forest have been restored following saltcedar control.

FIRE IS IMPORTANT IN TERRESTRIAL ECOSYSTEM MANAGEMENT

(U.S. Forest Service)

FIGURE 4.29 Managers of grasslands commonly use controlled burns, such as the one shown here, to discourage replacement of grassland vegetation by shrubland or forest.

What economic activities should be allowed in each of the zones pictured in Figure 4.31? Which should be prohibited?

buffer zone A zone around a nature reserve or protected area in which limited economic activity is allowed.

4.12 Integrating conservation with local communities can help sustain protected areas

One of the most serious threats to protected areas is encroaching development in surrounding areas. As the environment surrounding a protected area becomes increasingly degraded and fragmented, the protected area itself becomes an increasingly isolated habitat (**Figure 4.30**). In addition, many protected areas are further compromised when local people illegally remove plants and animals from the protected area itself. However, such problems have been reduced substantially where managers and residents of surrounding communities work cooperatively. Such cooperative relationships arise where local communities derive clear economic benefits from the protected area and its surroundings.

One method of ensuring these direct benefits is to establish **buffer zones** around the core of a protected area in which the local community can pursue a number of economic activities, such as harvesting of wood, fishing, and some agriculture, while highly restricting activity in the core of the protected area (**Figure 4.31**). Involving the local community in the management of protected areas can yield additional benefits.

Community Integration Provides Protection

The Guanacaste Conservation Area in Costa Rica provides a model for the integration of a protected area with the local community. Situated in the dry tropical forest on the Pacific coast, Guanacaste has more than 100 full-time employees, 80% of whom are from the surrounding communities. In addition, the conservation area offers dozens of part-time and seasonal positions, and over the years local people have been heavily involved in ongoing biodiversity studies. The scientific knowledge resulting from studies in the reserve is also being used to help farmers and ranchers in the surrounding agricultural lands.

The Guanacaste Conservation Area has also taken creative approaches to protecting the preserve by employing known poachers as game wardens, as well as known fire-starters for fire protection. Finally, Guanacaste has developed a variety of educational materials, emphasizing environmental education, and has become an educational center, where students from the surrounding area take several field trips per year (**Figure 4.32**). Through these multiple efforts, the Guanacaste Conservation Area has become a source of economic gain, educational opportunity, and pride for the people of northwest Costa Rica.

Looking Beyond the Boundaries

Protected areas can be threatened by events taking place far away. For example, toxic chemicals dumped in the water can decimate freshwater and marine populations in distant protected areas. Toxic gases and pollutants can drift from industrial and urban areas into parks. Carbon dioxide emitted in one country can alter global temperature and precipitation patterns all over the world, threatening all protected ecosystems on Earth.

Protecting natural ecosystems and the biodiversity they sustain is a key to sustaining Earth's biodiversity. However, protected areas cannot be sustained as islands of exceptional biodiversity in a matrix of degraded ecosystems. The long-term security of protected areas

ENCROACHMENT BY DEVELOPMENT CAN ISOLATE AND THREATEN PROTECTED AREAS

(Jeffrey Greenberg/Science Source)

FIGURE 4.30 Where intensive development encroaches, a protected area becomes increasingly island-like and more vulnerable to invasion by invasive species. Such conditions eventually lead to reduced biodiversity within the protected area.

A BUFFER ZONE CREATES A GRADIENT OF ECONOMIC ACTIVITY

PROTECTED AREA CORE
No economic activity

BUFFER ZONE
Low economic activity

BUFFER ZONE
Moderate economic activity

FIGURE 4.31 Buffer zones allow some level of economic activity in regions around protected areas, which decrease in intensity toward the core of a protected area. By providing economic benefits, such buffer zones can help build and sustain support for protected areas within nearby communities.

depends on developing a comprehensive and integrated approach to sustaining healthy regional and global environments.

⚠ **Think About It**

1. What dangers to biodiversity conservation may occur if human communities are allowed unrestricted access and use of the protected areas?

2. What are the dangers if protected areas are completely isolated from human communities?

3. What would be a viable middle ground regarding protection versus use of biodiversity reserves?

What are the implications of global climate change to the functions of protected areas over the long term?

EDUCATIONAL OUTREACH CONNECTS PROTECTED AREAS WITH COMMUNITIES

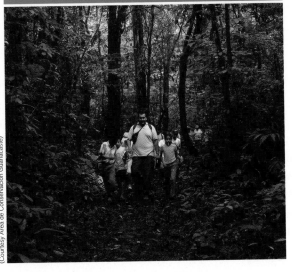

(Courtesy Área de Conservación Guanacaste)

FIGURE 4.32 The Guanacaste Conservation Area acts as a center for environmental education for the surrounding communities. Local students, in grades 4 to 8, take several guided field trips, such as this one, each year to educational centers located in the conservation area.

4.9–4.12 Solutions: Summary

The Convention for Biological Diversity set the stage for efforts to protect ecosystems and their biodiversity. Protected areas are a key part of this solution, and today there are more than 100,000 of them across the globe. A second key to effectively conserving Earth's biodiversity is involving a broad range of stakeholders, including local communities and NGOs such as The Nature Conservancy. Sustaining biodiversity requires not only the establishment of protected areas, but also management through maintenance of keystone species, regulation of the fire regime, and control of invasive species. In many situations, sustaining the biodiversity values of protected areas may be enhanced by integration with human communities beyond their boundaries.

Answer the following questions for each chapter section and then answer the Central Question.

Central Question: How can we protect Earth's diverse ecosystems?

4.1–4.5 Science

- How do species and ecosystem diversity differ?

- What are the major characteristics of terrestrial and aquatic biomes?

- How does species diversity vary by latitude, in biodiversity hotspots and on islands?

- What types of species have especially great influences on biodiversity?

- How does biodiversity change during ecological succession?

- What are the similarities and differences between allopatric and sympatric speciation?

4.6–4.8 Issues

- How does habitat fragmentation affect biodiversity?

- What are valuable ecosystem services and what factors threaten them?

- How do invasive species change terrestrial and aquatic ecosystems?

Species and Ecosystem Diversity and You

It is the rare individual who can establish a large protected area on his or her own. However, there are many opportunities to contribute to the protection of natural ecosystems through organizations dedicated to nature protection or through work on established protected areas. The first step in getting involved is to learn about the opportunities.

☐ *Learn about conservation opportunities in your region.*

Familiarize yourself through the Internet with the size, location, and conservation mission of nearby protected areas. Once you have explored neighboring nature reserves online, visit those that interest you most, if possible.

☐ *Get involved with protected area management.*

Most conservation areas operate on limited budgets and need volunteers or interns to help with the management of protected areas. There are often opportunities to work on habitat restoration, invasive species control, and many other conservation-related projects. Check out local state fish and game agencies or departments of natural resources who need volunteers. A summary of government-related volunteer programs across the United States may be found at www.volunteer.gov.

☐ *Join and become active in conservation organizations.*

If you don't have enough time to volunteer, consider becoming an active member of a conservation organization such as Sierra Club, Audubon Society, The Nature Conservancy, or the Natural Resources Defense Council. As a member, your voice will be amplified through an organization with far more influence than any single individual. Such membership also provides many opportunities to learn more about conservation programs and challenges across your region and around the world.

☐ *Support wildlife-friendly products.*

Whenever you can, put your buying power to work by purchasing wildlife-friendly products. Again, the average person cannot make a large financial difference, but as such products become more popular, they become more competitive. There are many initiatives around the world to help sustain wildlife and their habitats through wildlife-friendly farming, ranching, and forestry practices. You can easily learn more about such efforts online. Check out www.goodguide.com to see how many household products "score" on an environmental level.

- **What are the stated goals of the Convention on Biological Diversity?**

- **How has the number of protected areas grown over time?**

- **What unique roles can NGOs play in conserving species and ecosystem diversity?**

- **What factors make active management necessary for sustaining protected areas?**

- **How does integrating protected areas with local communities help sustain biodiversity?**

Answer the Central Question:

_____
_____
_____
_____
_____
_____
_____
_____
_____

Chapter 4

Review Questions

1. Which represents the highest species diversity?
a. 10 species with 1 abundant and 9 rare species
b. 10 species, each equally abundant
c. 5 species, each equally abundant
d. 9 species but no estimates of relative abundance

2. What produced biodiversity hotspots?
a. Higher speciation rates
b. Lower extinction rates
c. Higher species immigration and speciation rates and lower extinction rates
d. Higher species immigration rates

3. How does an ecosystem engineer affect ecosystems?
a. By altering the physical environment
b. Through its predatory behavior
c. Through its large size and abundance
d. By competing with keystone species

4. What life history characteristics make _r_-selected species good pioneer species?
a. Large size   c. High competitive abilities
b. High dispersal rates   d. Late maturity

5. What is the approximate global value of ecosystem services?
a. Equal to the annual gross product of China
b. Equal to the annual gross product of the E.U.

c. Equal to the annual gross product of the U.S.
d. Nearly twice the total gross product of the globe

6. Which of the following will likely support the fewest forest interior species?
a. A small forest fragment near a large forest
b. A large forest fragment far from a large forest
c. A small forest fragment far from a large forest
d. A large forest fragment near a large forest

7. Invasive species can have which of the following impacts?
a. Reduced river flows
b. More frequent wildfires
c. Extinctions of native species
d. All of the above

8. What is the main way that habitat corridors improve the functioning of protected areas?
a. Provide convenient water sources during drought
b. Promote the movement of individuals through protected areas
c. Provide habitat for keystone species
d. Increase rates of extinction among pests

9. How can ecological succession affect management of protected areas?
a. Once areas are protected, ecological succession does not influence protected areas.
b. Management of protected areas is aimed at reducing disturbance and promoting succession.
c. The effects of succession on management depend on the purpose of the protected area.
d. In general, managers of protected areas attempt to suppress succession.

10. Which policies will most likely secure a protected area over the long term?
a. Strict isolation of the protected area from the surrounding community
b. Restricting access to the protected area to professional scientists and students
c. Allowing local communities unlimited access to the protected area
d. Integrating management with the local community and providing direct economic benefits

Critical Analysis

1. Are there any biomes (see Figures 4.3 and 4.6) underrepresented among biodiversity hotspots? Are there any that are overrepresented?

2. What criteria should be used to decide where to establish protected areas? How large should protected areas be?

3. How might the discoveries of Elinor Ostrom (see Chapter 2, page 54) be used to forge a sustainable relationship between protected areas and surrounding communities?

4. How can a country balance forest conservation and economic development?

5. Using Figure 4.13 as a guide, explain how the Yellowstone ecosystem is affected by the removal or addition of wolves.

Find additional resources and links online at www.macmillanhighered.com/launchpad/molles1e.

(AP Photo/Pavel Rahman)

Reflecting a population so large that services are overwhelmed, a crowd of Bangladeshis occupy the roof of an overcrowded train, seeking transportation to their homes to celebrate an annual festival, while thousands more wait for their chance.

Bangladesh's Lessons for Environmental Scientists

The challenges faced by the people of Bangladesh dramatically demonstrate issues arising from a population growing beyond the capacity of its environment to support them.

Tink. Tink. Tink. This is the sound of monsoon rains pinging against the tin roof of a shack in Bangladesh. Imagine, for a moment, that this is where you grew up: Each night you sleep on the dirt floor, squeezed into the same tiny room with four or five other members of your family. Your entire home is about 100 square feet in size, 25 times smaller than a typical American home. When your mother cooks on the wood-burning stove, choking black smoke fills the house. You make do with just 10 liters of water per day for both your washing and drinking water. By comparison, the average American uses 380 liters

each day. One of your siblings has diarrhea, another hunger pangs.

Life in Bangladesh, one the world's most densely populated countries, is not easy. Nearly 150 million people live on this flood-prone river delta between India and Burma, a region about the size of Iowa. Today, the country is notorious for having some of the most poorly paid workers, who labor under harsh conditions at garment factories that produce clothing for international companies such as Benetton and Walmart. In April 2013 an eight-story factory collapsed, killing 1,129 workers and injuring more than 2,500. For people who live in dire poverty and need to feed their families, such jobs are the only option they have.

> "The key problem facing humanity in the coming [21st] century is how to bring a better quality of life—for 8 billion or more people—without wrecking the environment entirely in the attempt."
>
> Edward O. Wilson, distinguished ecologist

Sadly, their options will only get worse as Bangladesh's population continues to expand. Researchers estimate that population will reach 218 million by the middle of this century, with most residents living in slums, lacking proper sewage disposal and electricity. The country hasn't produced enough food for its people since the 1950s, and agricultural land has only declined since, replaced by living spaces. By 2060 there may be none left at all.

Bangladesh's struggles provide a cautionary tale for the wider world. Prior to the Industrial Revolution, the human population on Earth numbered less than 1 billion. It has soared over the last 200 years. Halloween day of 2011 was a particularly frightening day for environmental scientists interested in **demography,** the study of the statistics of human populations. The United Nations christened it the "Day of Seven Billion," when the world's population crossed that threshold. "Our world is one of terrible contradictions," Ban Ki-moon, the UN secretary-general, said at a news conference. "Plenty of food, but one billion people go hungry. Lavish lifestyles for a few, but poverty for too many others."

Rich or poor, these 7 billion bodies require food, water, and other natural resources to build their homes and feed their families. Not only does this place a heavy toll on the functioning of natural ecosystems and our global economy, but many environmental scientists believe that the resources we depend on are running out. The richest may see their quality of life decline, while the poorest among us have little hope. The World Wildlife Fund's Living Planet Report predicts that, if present trends continue, by 2050 the productive capacity of nearly three Earths will be needed to meet the needs of the world's 8 to 11 billion residents (see Chapter 1, page 20). It's a problem that ecologist Paul Ehrlich famously called the "population bomb," and population growth represents one of the most fundamental challenges when it comes to balancing human rights and environmental sustainability.

demography The statistical study of populations, generally human populations, including their density, growth, age structure, birthrates, and death rates.

Central Question

How can we achieve sustainable human populations?

(Jörg Hackemann/Shutterstock)

5.1–5.3 Science

What might be the impacts on the environment of concentrating the population in densely populated cities versus spreading the population across the landscape?

population density The number of individuals in a population per unit area.

immigration The movement of individuals into an area, or country, to which they are not native.

emigration The movement of individuals out of one area, or country, to another.

Human population density lies at the root of environmental damage. In order to minimize the environmental impact of human societies, we need to know where people live today and where they will be concentrated in the future.

5.1 Human population density varies significantly across Earth

In Bangladesh, people are crowded together with an average density of nearly 1,000 people on every square kilometer of land (or 0.39 square miles). Australia, by comparison, has an average of just 3 people in the same-size area. Such regional contrasts provide a graphic demonstration of how human **population density** varies. Within countries, the highest population densities are generally found along coasts and river valleys. The lowest population densities occur mainly in extreme environments such as deserts and arctic tundra. The Ganges river delta in Bangladesh was historically a very advantageous place to settle because it could support a relatively high population density with its fertile soil and ample water, whereas Australia's outback, in its recent history, has been better suited to kangaroos and lizards than to people (**Figure 5.1**).

We know that people don't always stay in the place where they were born. Population density results from a combination of the rates of birth, death, **immigration** (movement of individuals into a population from the outside), and **emigration** (movement of individuals away from a population to another area) (**Figure 5.2**). On a regional scale, in places where rates of birth and immigration exceed rates of emigration and death, population density will increase. In regions where birth and immigration rates are lower, population density declines.

Today, populations around the world are becoming increasingly concentrated in cities as people migrate from rural to urban areas. By 2014 over half the world's population was living in cities, with that proportion expected to rise to two-thirds by 2050. However, in some countries the proportion of the population living in cities is already much higher. For example, in 2014, over 80% of the populations of the United States and Canada were urban, while in Japan and Belgium that proportion was over 90%. One of the consequences of this migration has been the formation of mega-cities with populations of more than 10 million (**Table 5.1**). The urban population of Earth will continue to grow, as rural to urban migration remains high through the middle of the century.

TABLE 5.1
THE 10 LARGEST MEGA-CITIES

In 2014 there were 28 mega-cities scattered around the world. Shown here are the 10 with the largest populations—six in Asia, two in Latin America, one in Africa, and one in the United States.

City	Population in Millions
Tokyo, Japan	38
Delhi, India	25
Shanghai, China	23
Mexico City, Mexico	21
Mumbai, India	21
São Paulo, Brazil	21
Osaka, Japan	>20
Beijing, China	<20
Cairo, Egypt	18.5
New York–Newark, USA	18.5

Data from United Nations, Department of Economic and Social Affairs, Population Division, 2014.

HUMAN POPULATION DENSITIES VARY GREATLY ACROSS THE GLOBE

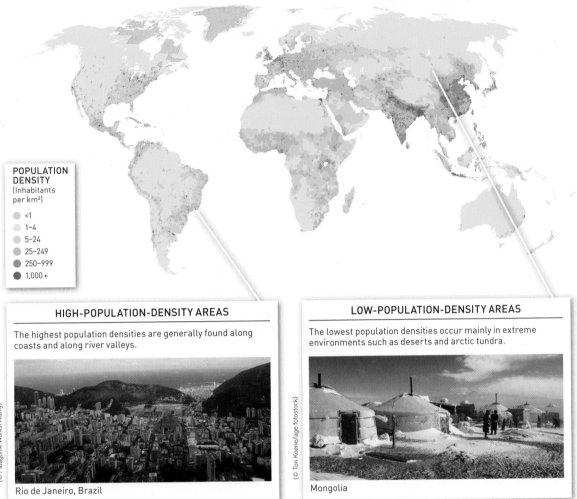

POPULATION DENSITY
(Inhabitants per km²)

- <1
- 1–4
- 5–24
- 25–249
- 250–999
- 1,000+

HIGH-POPULATION-DENSITY AREAS

The highest population densities are generally found along coasts and along river valleys.

Rio de Janeiro, Brazil

(© Patagonik Works/Alamy)

LOW-POPULATION-DENSITY AREAS

The lowest population densities occur mainly in extreme environments such as deserts and arctic tundra.

Mongolia

(© Ton Koene/age fotostock)

FIGURE 5.1 Some environments typically support higher population densities compared with others, often depending on the local climate and natural resources like water, soil quality, and raw materials.

POPULATION DENSITY: A DYNAMIC BALANCE BETWEEN OPPOSING PROCESSES

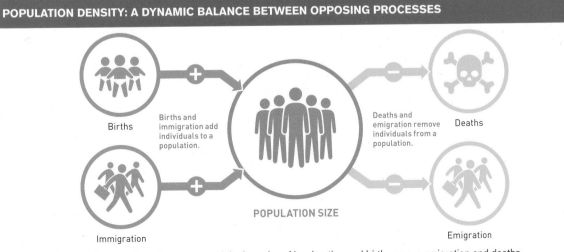

Births

Births and immigration add individuals to a population.

Immigration

POPULATION SIZE

Deaths and emigration remove individuals from a population.

Deaths

Emigration

FIGURE 5.2 Population density is the outcome of the interplay of immigration and births versus emigration and deaths.

⚠ Think About It

1. How do you explain the higher population densities in coastal areas around the world?

2. Europe, Canada, and the United States are within about 10% of each other in terms of land area. However, their population densities differ markedly (Europe: 70 per km²; United States: 31 per km²; Canada: 3.2 per km²). What factors likely account for these differences?

3. What factors are attracting people from rural areas to cities?

5.2 The global population will grow into the middle of this century

Some 60,000 years ago, a few thousand people in East Africa began a historic migration into the Middle East, Europe, Asia, and beyond. While there is some debate among archaeologists and geneticists about the exact timing and process of human dispersal from Africa, we know that, in the early years, the global human population grew very slowly. Then, with the shift from hunting and gathering to agriculture, approximately 10,000 years ago, populations began to swell—but numbers remained in the few hundred million. About 500 years ago, the global population began to grow more rapidly, climbing to 1 billion by 1804. The rate of population growth increased with the Industrial Revolution in the mid-1800s, because of improved sanitation, health care, and increased agricultural production. Consequently, the global population grew to 2 billion by 1930 and to 3 billion just 29 years later, in 1959.

One way to think about the quickening pace of human population growth is to calculate the **population doubling time,** the amount of time required for a population, growing at a particular rate, to double its size. As shown in **Table 5.2,** the doubling time for the global population has been declining—that is, the amount of time it takes for the

At its rate of growth in 2015, the U.S. population would double in 91 years. Given this growth rate, what sorts of changes in housing, population density, and infrastructure would you expect to see in your lifetime?

population doubling time The amount of time required for a population, growing at a particular rate, to double its size.

population to grow to twice its size is growing shorter. For example, it took all of human history for the population to reach 3.2 billion in 1963. The population grew by another 3.2 billion people by 2004, requiring only 41 years. As of 2015, the rate of global population growth had slowed to the point where the doubling time for the global population had increased to 66 years. However, it is unlikely that the population will double again, since the trend is for continued declines in population growth rates. As a result, the best current population models predict a leveling off of the global population at a size far lower than double its size in 2015.

The global population has followed a J-shaped curve over much of the past five centuries (**Figure 5.3a**). Although the global population (total population number) continues to increase, it is not increasing as quickly as it formerly did (**Figure 5.3b**). In other words, the growth *rate* is slowing down. The population growth rate peaked in the late 1960s and early 1970s and has been decreasing steadily since then, as birthrates have fallen worldwide. The current theoretical doubling time of the population is 66 years. However, population biologists predict that the growth rate will keep slowing into the next half-century (**Figure 5.3c**), which will increase doubling time as population growth slows. A slowing global population growth rate will resemble an S-shaped curve (for more on J- and S-shaped curves, see Chapter 3).

Does leveling off of the human population sometime this century mean that we will be at carrying capacity? In Chapter 3, we defined carrying capacity as the average number of individuals in a population that the environment is capable of supporting *over the long term*. Recall that if present trends in population growth and resource consumption continue, by 2050 we will need the productive capacity of nearly three Earths to support the expected human population over the long term. Some scientists have estimated that Earth could supply enough food to support a population of 9 to 10 billion humans—but only if we were all vegetarians. However, human carrying capacity is influenced by a host of environmental factors besides food, including supplies of freshwater, availability of critical energy supplies, and capacity of built and natural systems for waste disposal. What is clear is that the human carrying capacity will be determined by a combination of limits set by the natural environment and by the choices we make as individuals and societies.

⚠ Think About It

1. What environmental factors produced the very rapid growth of human populations following the Industrial Revolution?

2. Why is it impossible for the global population to sustain J-shaped population growth indefinitely? (Hint: Check out Chapter 3 for more on J-shaped growth curves.)

3. What environmental factors may ultimately limit global population growth?

TABLE 5.2

DOUBLING TIMES FOR THE GLOBAL POPULATION

The doubling time for the global population has decreased markedly during historic times.

Years	Population Change	Doubling Time in Years
500 BCE to 600 CE	100 to 200 million	1,100
600 to 1200	200 to 400 million	600
1200 to 1750	400 to 800 million	550
1750 to 1900	800 million to 1.6 billion	150
1900 to 1963	1.6 billion to 3.2 billion	63
1963 to 2004	3.2 to 6.4 billion	41

Data from U.S. Census Bureau, International Data Base (IDB), 2010.

THREE VIEWS OF GLOBAL POPULATION
GROWTH

HUMAN POPULATION GROWTH

a. During the past 12,000 years, the global population has followed a J-shaped trajectory.

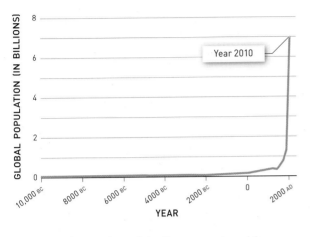

b. However, the annual rate of growth as a percentage of the population has been in decline since the late 1960s.

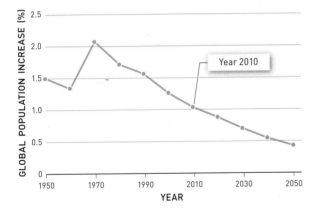

c. The recent and projected slowing of global population growth can be seen in a growth pattern that is beginning to show signs of leveling off.

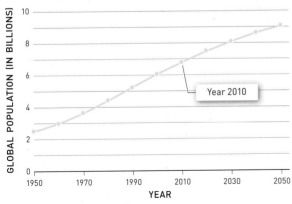

FIGURE 5.3 The perspective you develop on population growth is significantly affected by how you examine it. For example, although total global population is still increasing, the pace at which the population is increasing has slowed over the past 40 or so years. (Data from U.S. Census Bureau, International Data Base, 2010)

5.3 The age structure of a population gives clues to its growth or decline

You may have heard that the U.S. Social Security system, which provides support to retirees, is under threat. How is it that every working member of the population is required to give part of his or her salary to Social Security, but the fund is still running dry? The answer comes back to demography. Because people are living longer and having fewer children, the number of workers supporting retirees has been decreasing over time. In 1960 there were approximately five workers for each person receiving Social Security benefits. By 2005 the ratio of workers to beneficiaries had fallen from 3.3 to 1, and demographers predict that the ratio of workers to beneficiaries will fall to 2:1 by 2060. Today, about 12% of our population is over the age of 65, but by 2080 that number will almost double. The only way to pay for these retirees is to increase taxes on the working, decrease benefits for retirees, or reduce the number of retirees by raising the retirement age.

As you can see, demography is critical to planning for the needs of any population. In an aging population, that could mean Social Security and elder-care facilities. In a young, growing population, that might mean building more schools. To predict future trends in population growth, demographers study the **age structure** of a population. Age structure is a tally of how many males and females there are at different ages in a particular population, region, or country. The distribution of ages of a population reflects whether a population is growing, stable, or declining, and we can visualize it with an *age structure diagram* (**Figure 5.4**).

Contrasting Population Trends

One way to predict growth trends is to compare the relative number of children in a population with the relative number of reproductive-age adults. Consider the triangular-shaped age structure of the population in Yemen (Figure 5.4), a country on the southwestern tip of the Arabian Peninsula. The age structure diagram reveals that young children far outnumber adults. As these children grow up, they will reproduce and contribute to a future population boom. In contrast, the age structure in Ukraine, a country southwest of Russia bordering the north shore of the Black Sea, is constricted at its base, indicating a relatively small proportion of children—too few to replace the adults in the population. That population is in decline. Finally, the straight-sided age structure of Iceland's population indicates that the number of children is just sufficient to replace the adults in the population—this population is nearly stable.

The contrasting age structures of Yemen, Iceland, and Ukraine result mainly from differences in **total fertility rate,** which is an estimate of the average number

Imagine visiting a city in Iceland, Yemen, or Ukraine on a festival or market day, when a cross section of people is on the streets. Given their contrasting age structures, what would be your likely impression of each population?

age structure The proportions of individuals of various ages in a population; the relative proportions of individuals of reproductive and pre-reproductive age indicate whether a population is growing, stable, or declining.

total fertility rate An estimate of the average number of children that a woman in a population gives birth to during her lifetime.

CONTRASTING POPULATION AGE STRUCTURES

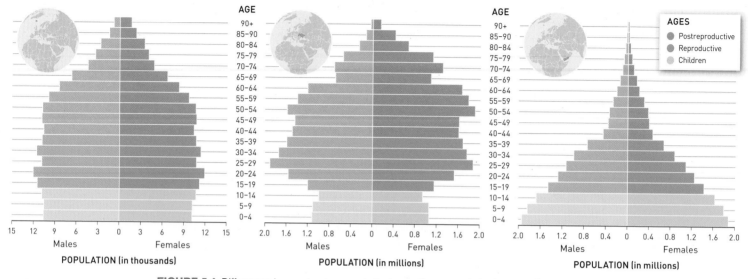

ICELAND: Approaching stability

Iceland's population has a relatively even distribution of individuals between reproductive-aged adults and children.

POPULATION (in thousands)

UKRAINE: Declining

Ukraine's population includes relatively few children.

POPULATION (in millions)

YEMEN: Growing rapidly

In contrast, a relatively large proportion of Yemen's population is made up of children.

POPULATION (in millions)

AGES
- Postreproductive
- Reproductive
- Children

FIGURE 5.4 Differences in age structure can indicate whether a population is stable (or approaching stability), in decline, or growing. (Data from U.S. Census Bureau, International Data Base, 2013)

replacement-level fertility
The total fertility rate required to sustain a population at its current size, which varies from approximately 2.1 births per woman in the more developed countries to 2.5 or higher in the least developed countries, where mortality rates are higher.

of children that a woman in a population gives birth to during her lifetime. A fertility rate of fewer than 2 births per woman tells us that a population is not at **replacement-level fertility.** The total fertility rate in Yemen (4.5) far exceeds that of Iceland (1.9) and Ukraine (1.3), meaning that a woman in Ukraine is not producing the two babies necessary to replace each pair of adults in the population. In fact, true replacement-level fertility tends to be slightly higher than 2. That's

because some children die before reaching reproductive age. In developed countries, such as Iceland and Ukraine, replacement-level fertility is approximately 2.1 births per woman. In developing countries, replacement-level fertility is 2.5 births per woman or higher. These differences in total fertility rate translate to the rapidly growing population of Yemen, the stabilizing population of Iceland, and the declining population of Ukraine (**Figure 5.5**).

THREE PATTERNS OF POPULATION CHANGE

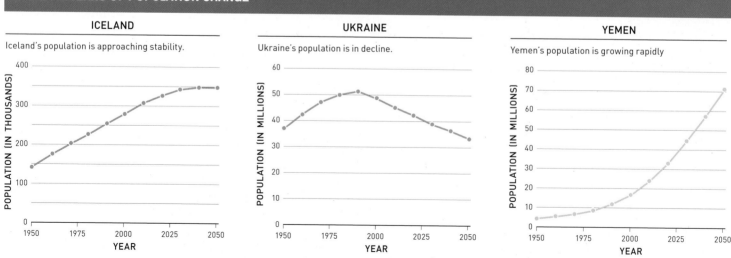

ICELAND

Iceland's population is approaching stability.

UKRAINE

Ukraine's population is in decline.

YEMEN

Yemen's population is growing rapidly

FIGURE 5.5 The patterns of population change in Iceland, Ukraine, and Yemen nearly span the range of population change among countries in the early 21st century. (Data from U.S. Census Bureau, International Data Base, 2006)

U.S. AGE STRUCTURE DOES NOT PREDICT CURRENT POPULATION TRENDS

UNITED STATES

a. Despite an age structure that suggests a stable population or one approaching stability. . .

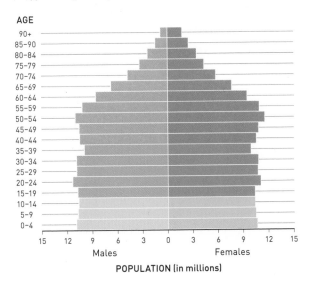

b. . . . the population of the United States is projected to continue growing beyond midcentury.

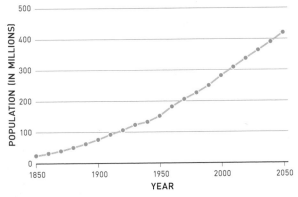

FIGURE 5.6 (a) The age structure of the U.S. population suggests a stable population. (b) Analysis by the U.S. Census Bureau predicts that the U.S. population will continue to grow through 2050. (Data from U.S. Census Bureau, International Data Base, 2013)

Population Trends in the United States

With a replacement-level fertility rate of 2.1, the United States seems like a perfect example of a stable population (**Figure 5.6a**). However, the U.S. population is still growing at a rate of 0.9% annually, and demographers predict it will reach nearly 400 million people in 2050 (**Figure 5.6b**). What's going on?

The tendency of a population to grow even though fertility has fallen to replacement levels or lower is called **population momentum.** A lag period occurs because a large proportion of individuals remain in their childbearing years, most of whom will still be living long after their children are added to the population. In 2015 the annual birthrate per 1,000 individuals in the United States was 12; the death rate was 8 per 1,000. Population momentum in the United States will eventually stop once the birthrate equals the death rate, but it will be especially significant in a nation such as Yemen (see Figure 5.4), which has a large proportion of children in the population. The same is true of almost all the countries of sub-Saharan Africa.

Reproduction at replacement rate should eventually cause the U.S. population to stabilize—but this assumes no migration into or out of a population. We know that the U.S. population is not closed and has long received large numbers of immigrants. Immigration, which was responsible for approximately 48% of the growth in the U.S. population in 2015, will continue to add significantly to U.S. population growth well into midcentury.

⚠ Think About It

1. Immigration accounts for some of the projected population growth in the United States shown in Figure 5.6. What contribution does immigration make to the projected growth for the world population shown in Figure 5.3?

2. Could a population such as that of Ukraine (see Figures 5.4 and 5.5) show population momentum if total fertility rates were to increase to replacement levels? What would be the nature of this momentum?

5.1–5.3 Science: Summary

After thousands of years of relatively slow growth, the human population began to grow rapidly during the past 500 years, mainly in response to increased food supplies and improved sanitation and health care. Population density, which varies around the world, results from a dynamic interplay between rates of birth, death, immigration, and emigration. Demographers predict that the global population will begin to level off sometime during the mid-21st century, eventually approximating an S-shaped pattern of population growth. The distribution of people of different ages can indicate population trends. Population growth occurs in young populations, such as in Yemen, whereas population declines occur in aging populations, such as the Ukraine. Even though the U.S. population has a stable population structure, it continues to grow due to population momentum and immigration.

population momentum
Population growth as a consequence of a large number of women reaching childbearing age.

5.4–5.7 Issues

The human population and the demands it makes on the environment lie at the core of environmental science. The food we eat, including the variety, its level of processing, and the amounts, is one the most fundamental indicators of consumption by a population. **Figure 5.7** contrasts the foods consumed by three families during a typical week. This stark contrast shows clearly that the pressure a population puts on Earth's resources is a product of not just the size of a population, but also the rate at which individuals in a population consume resources. Typically, the level of resource use in a population increases with economic development.

5.4 Fertility ranges greatly among countries and regions

In 2007 developed countries around the world had total fertility rates at or below the replacement level of fertility of 2.1 (**Figure 5.8**). Consequently, populations in these countries, such as those of Iceland and Ukraine, are now generally stable, approaching stability, or decreasing in size. Meanwhile, high levels of fertility and rapid population growth, as in Yemen, continue in many nations of sub-Saharan Africa, parts of the Middle East, and southern Asia.

Each of these population trends presents a unique challenge. Populations with very low levels of fertility have a shrinking workforce of younger people who face the prospect of supporting a larger number of aged individuals in the population. On the other hand, high levels of fertility mean a large number of young in need of education, proper nutrition, and health care.

As a consequence of differences in fertility and population momentum, population trends differ a great deal among regions (**Figure 5.9**). Studies by the United Nations and others indicate that Europe's population will decline over the next half-century, whereas the populations of Latin America, Asia, and North America will grow slowly. The same studies also predict moderate population growth in North Africa and Oceania. In contrast, rapid growth will continue in sub-Saharan Africa, where projections indicate that the population will more than double in the 40 years between 2010 and 2050.

Is population planning more of a global or regional problem? Perhaps both? Explain your answer.

If you were planning for social services, what changes would you need to anticipate over the next 40 years in sub-Saharan Africa? In Europe?

FOOD FOR A WEEK CONTRASTS LEVELS OF CONSUMPTION IN DIFFERENT POPULATIONS

(Ben Lister/Daily Mail/Solo Syndication)

Family in the United Kingdom

(Abir Abdullah/Oxfam)

Family in Sri Lanka

(Tom Pietrasik/Oxfam)

Family in Ethiopia

FIGURE 5.7 These families, who posed with all the food they will eat in a week, give a visual demonstration of the great differences in consumption across societies.

TOTAL FERTILITY RATE DIFFERS GREATLY AROUND THE WORLD

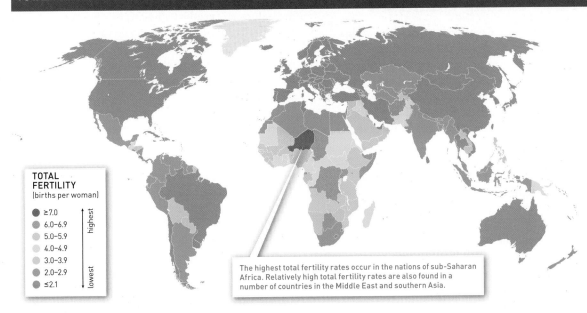

TOTAL FERTILITY (births per woman)

- ≥7.0
- 6.0–6.9
- 5.0–5.9
- 4.0–4.9
- 3.0–3.9
- 2.0–2.9
- ≤2.1

highest ↕ lowest

The highest total fertility rates occur in the nations of sub-Saharan Africa. Relatively high total fertility rates are also found in a number of countries in the Middle East and southern Asia.

FIGURE 5.8 Many populations around the world have total fertility rates at or below the replacement level of 2.1 live births per woman. Many other populations are very close to replacement-level fertility. However, regional population trends in places like sub-Saharan Africa should not distract us from the great variation in trends among individual nations, each with its own specific qualities and needs. This is particularly significant in Africa, where, for example, variation in total fertility rate is higher than in any other major region. (Data from United Nations Human Development Report, 2009)

⚠ Think About It

1. What types of age structure do you think you'd find in sub-Saharan Africa? In Europe? (Hint: See Figure 5.4.)

2. Which regions have the highest total fertility rates? Which regions have the greatest variation in total fertility rates among countries?

VARIATION IN REGIONAL POPULATION GROWTH

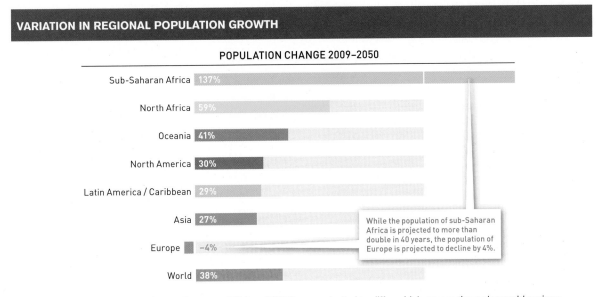

POPULATION CHANGE 2009–2050

- Sub-Saharan Africa 137%
- North Africa 59%
- Oceania 41%
- North America 30%
- Latin America / Caribbean 29%
- Asia 27%
- Europe –4%
- World 38%

While the population of sub-Saharan Africa is projected to more than double in 40 years, the population of Europe is projected to decline by 4%.

FIGURE 5.9 Population growth rates between 2009 and 2050 are projected to differ widely across the major world regions. (Data from U.S. Census Bureau, International Data Base, 2010)

5.5 Development varies widely among countries

The king of Bhutan proposed that noneconomic aspects should be given greater emphasis in evaluating development and proposed an index called the "Gross National Happiness." How would you go about assessing happiness?

As discussed earlier, Bangladesh is one of the world's poorest and least developed countries, which obviously has a wide range of consequences for the population in terms of its health, educational level, and quality of life. Those consequences are also reflected in the population's impact on the environment. Compared with the average American, a Bangladeshi citizen uses far fewer resources. He eats less meat, requires less electricity, and is less likely to throw out an old cell phone in order to buy the latest model. Still, because of Bangladesh's high population density, the population's local impact on its surroundings has already been significant: polluted water supplies, dirty air, and vanishing farmland.

Development, therefore, is a double-edged sword with the potential to raise the standard of living while also threatening future prosperity through the depletion of resources. To some extent, this is happening in Bangladesh. As poverty has declined from 57% of the population in 1992 to 32% in 2010, the average Bangladeshi has had a larger impact on the environment. This means that understanding the impact of a population on the environment requires an accurate measurement of that population's level of development.

Measuring the Health and Well-Being of Populations

When it comes to development, the basic currency is money: how much people make and how much they spend. After all, the more money they have available,

the more choices they have over the lives they lead. Economists typically compare countries based on their **gross domestic product,** or **GDP,** which is the total market value of all the goods and services produced within the borders of a nation during a given year. To compare economic conditions among countries, we divide GDP by population size, resulting in per capita GDP. As shown in **Figure 5.10,** per capita GDP is over 100 times higher in countries with the highest levels of development compared to those with the lowest development. Bangladesh has a per capita GDP of $1,883, which puts it above quite a few African countries, but far below the $49,965 **per capita GDP** of the United States. Countries with a low GDP generally have a lower cost of living. However, importing manufactured goods and food from richer countries can be a challenge.

To take a closer look at what money really means for development, it's often useful to examine population statistics. For instance, **life expectancy at birth,** the predicted average life span of individuals born during a particular year, reflects the state of health in different regions and countries. In highly developed countries, life expectancy at birth averages 80 years. Compare that with many countries in sub-Saharan Africa and Afghanistan, where people are expected to die by their early to mid-50s. In Africa, this reflects the ravages of AIDS in many populations, while in Afghanistan, it might be due to poor sanitation, minimal access to medical care, and decades of war.

Another measure of health is **child mortality rate,** the number of children per 1,000 live births that die before reaching 5 years of age. The level of child mortality can be as much as 20 times higher in the poorest countries than in the richest ones (**Figure 5.11**).

The poorest countries often have the worst educational systems, and many children never learn to read. Lack of educational opportunities may be one of the key impediments to improving the lives of people in such countries, while high levels of literacy and commitment to education are essential to sustaining high levels of development in rich countries.

As you might imagine, although a country's GDP is the engine that drives development, a country's history and politics also have significant influences. To gauge levels of development, the United Nations (UN) combines health, economic development, and education into the **Human Development Index,** or **HDI.** The HDI ranges from 0 to 1.0, a theoretical situation in which the health, education, and economic potential of individuals are all maximized. The HDI uses life expectancy at birth as an indicator of health in a population, average number of years of schooling as an indicator of educational opportunities, and per capita income as a measure of economic development. When the UN evaluated 187 countries in 2011, their

gross domestic product (GDP) The total market value of all the goods (e.g., manufactured articles or agricultural crops) and services (e.g., transportation and banking services) produced within the borders of a nation during some period of time. See *per capita GDP.*

per capita GDP The market value of the goods (e.g., manufactured articles or agricultural crops) and services (e.g., transportation and banking services) produced within the borders of a nation per individual in its population.

life expectancy at birth The predicted average life span of individuals born during a particular year.

child mortality rate The number of infants per 1,000 live births who die before reaching 5 years of age.

Human Development Index (HDI) An index of national development that includes life expectancy at birth, educational opportunities, and economic productivity.

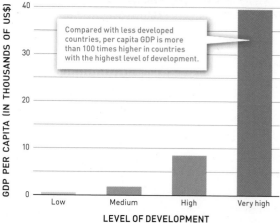

GROSS DOMESTIC PRODUCT (GDP) VARIES DRAMATICALLY WITH LEVEL OF DEVELOPMENT

Compared with less developed countries, per capita GDP is more than 100 times higher in countries with the highest level of development.

GDP PER CAPITA (IN THOUSANDS OF US$)

LEVEL OF DEVELOPMENT

FIGURE 5.10 A measure of material wealth, per capita GDP differs more among countries with different development status than many of the other major indicators of development. (Data from United Nations Human Development Report, 2009)

RELATIONSHIP BETWEEN DEVELOPMENT AND CHILD MORTALITY

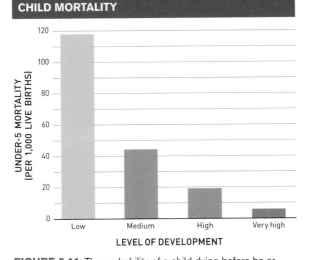

FIGURE 5.11 The probability of a child dying before he or she reaches age 5 decreases dramatically with development. In 2009 the proportion of children dying before 5 years of age in countries with the lowest development is nearly 20 times higher than in countries that have attained very high development. (Data from United Nations Human Development Report, 2011)

HDI values ranged from 0.286 for the Democratic Republic of the Congo, which ranked 187th, to 0.943 for Norway, which ranked 1st (**Figure 5.12**). Bangladesh ranked 146th.

⚠ Think About It

1. What factors are used to calculate the Human Development Index, or HDI? What does each indicate about a population?

2. Are there any aspects of human development left out of the HDI? What other factors would you add to the index, if you were to make your own independent assessment of human development around the world?

5.6 Population growth and development generally increase environmental impact

In 1971 ecologists Paul Ehrlich and John Holdren made one of the first attempts to quantify the impact of human populations on the environment. They recognized that a population's environmental impact is not simply a matter of the number of people in the population. It's also important to consider how affluent the population is and measure that affluence in terms of per capita resource consumption and waste production. Generally, the average level of resource consumption increases with rising levels of affluence.

In addition, one needs to factor in technology used in production of, for instance, consumer goods, because some

The Human Development Index (HDI) is calculated on the basis of three different factors (education, life expectancy, and per capita GDP). Are these elements independent or is there, in fact, a causal interrelationship among them?

HUMAN DEVELOPMENT INDEX (HDI) SCORES

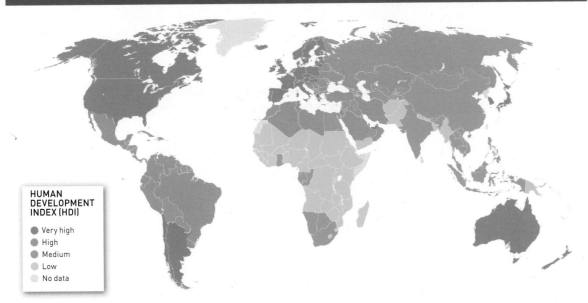

FIGURE 5.12 Geographical variation in human development shows regional clustering of higher and lower levels. HDI scores are highest in North America, Western Europe, Japan, and South Korea in the Northern Hemisphere and in Australia, New Zealand, Chile, and Argentina in the Southern Hemisphere. Meanwhile, the lowest HDI scores are concentrated in sub-Saharan Africa and southern Asia. (Data from United Nations Human Development Report, 2011)

How might technological developments reduce the environmental impacts of populations?

technologies consume more energy and produce more waste than others. Ehrlich and Holdren recognized that some technologies, such as more fuel-efficient vehicles, can lower the environmental impact of a population. In their article in the journal *Science,* Ehrlich and Holdren proposed what they called the IPAT equation:

$$I = P \times A \times T, \text{ where:}$$

I = **i**mpact on environment (loss of resources, degradation of ecosystems)
P = **p**opulation (total number in population)
A = **a**ffluence (a shorthand for per capita resource consumption)
T = **t**echnology (processes and products that require energy and resources)

One of the main conclusions we can draw from the IPAT equation is that individuals in less affluent countries, such as those of sub-Saharan Africa and southern Asia,

have less environmental impact than do individuals in more affluent countries, such as the United States.

The Ecological Footprint and Development

The IPAT equation was, in many ways, a precursor to the ecological footprint, which we introduced in Chapter 1; it is an estimate of the area of land and water required for a human population to provide the natural resources it uses (see page 20). You may notice that as a country's development level increases, its per capita ecological footprint also tends to increase (Figure 5.12 and **Figure 5.13**). That's not always true, and while development in many countries has been bought at a very high environmental cost, some rich countries have achieved high levels of development with lower impacts.

The total impact of a population on the environment can be estimated by multiplying the number of people by their per capita ecological footprint. **Figure 5.14** shows the relative population sizes in the world's major regions, along

GLOBAL VARIATION IN PER CAPITA ECOLOGICAL FOOTPRINT

FIGURE 5.13a The largest per capita ecological footprints in global hectares (gha) are concentrated in North America, Europe, northern Asia, Australia, and New Zealand. The lowest per capita ecological footprints occur in Africa, southern Asia, and Latin America. (Data from WWF [World Wildlife Fund], 2006; WWF, 2012)

PER CAPITA ECOLOGICAL FOOTPRINT (gha)

○ Very large >6.0
○ Large 4.0–6.0
○ Medium 2.0–3.9
○ Small <2.0
○ No data

PER CAPITA ECOLOGICAL FOOTPRINT INCREASES WITH DEVELOPMENT

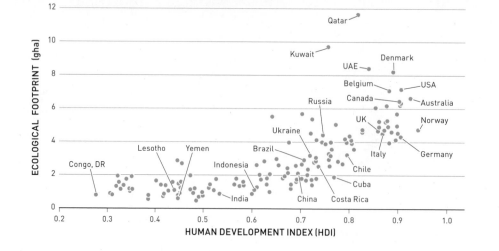

FIGURE 5.13b Human development requires investment in health care, education, and economic infrastructure. Resource consumption increases as a consequence. However, *how* much resource consumption increases with development varies widely. Many countries have attained high levels of development with much smaller ecological footprints than others. (Data from United Nations Human Development Report, 2011; WWF [World Wildlife Fund], 2012)

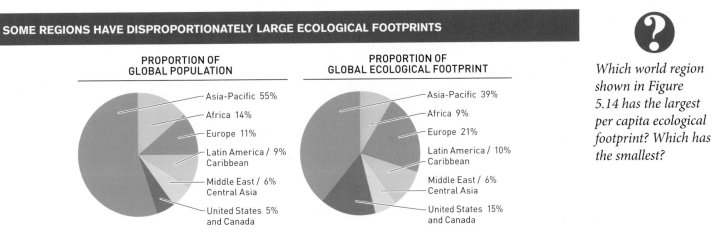

SOME REGIONS HAVE DISPROPORTIONATELY LARGE ECOLOGICAL FOOTPRINTS

PROPORTION OF GLOBAL POPULATION

- Asia-Pacific 55%
- Africa 14%
- Europe 11%
- Latin America / Caribbean 9%
- Middle East / Central Asia 6%
- United States and Canada 5%

PROPORTION OF GLOBAL ECOLOGICAL FOOTPRINT

- Asia-Pacific 39%
- Africa 9%
- Europe 21%
- Latin America / Caribbean 10%
- Middle East / Central Asia 6%
- United States and Canada 15%

FIGURE 5.14 The United States, Canada, and Europe have much larger total ecological footprints relative to their population sizes, compared with other regions. With a combined population of just over 1 billion, these regions have a combined ecological footprint nearly as large as the Asia-Pacific region, with a population of nearly 4 billion. Africa, by contrast, has a proportionately smaller footprint relative to population size. (Data from WWF [World Wildlife Fund], 2012]

Which world region shown in Figure 5.14 has the largest per capita ecological footprint? Which has the smallest?

with their total ecological footprints. One of the clearest conclusions we can draw from this comparison is that, relative to their sizes, the populations of North America—mainly the United States—and Europe have a much higher total impact on the environment than do the populations of other regions. Consider that if everyone on Earth lived like Americans did in the year 2012, we would need four Earths to sustain the current population.

⚠ Think About It

1. Using Figure 5.13, compare the relative ecological footprints of an individual in the United States and an individual in China.

2. Using the estimates you determined above, compare the total ecological footprints of the two countries. (Note: In 2012 the population of China was 1,343,000,000 and that of the United States was 314,000,000.)

3. How does your own ecological footprint compare with that of other individuals living around the world? (See the ecological footprint calculator at www.footprintnetwork.org.)

5.7 Developmental differences between populations create migration pressures

The Statue of Liberty in New York Harbor is one of the most well-known monuments to immigration. Indeed, the Americas have received large numbers of immigrants from all over the world. For instance, in the wake of the

Irish potato famine over 150 years ago, 1 million Irish came to the United States. However dramatic, this event involved a small fraction of the millions of people from all over Earth who now seek better prospects: Sri Lankan boat people seeking asylum in Australia, Nicaraguans entering richer Costa Rica, Eastern Europeans moving into Western Europe, young Africans stumbling onto the beaches of southern Spain, or Mexicans immigrating to the United States (**Figure 5.15**).

DEVELOPED COUNTRIES FACE AN INCREASING FLOW OF ILLEGAL IMMIGRANTS

(© Franco Cufari European Press Agency/Newscom)

FIGURE 5.15 Here, Italian immigration police detain a boatload of people attempting to enter Italy illegally.

How does one reconcile the aspirations of immigrants with the concerns of many residents in countries attracting high flows of immigrants?

These encounters among people, often with different cultures and languages, have commonly involved competition for land, jobs, and other resources. Consequently, immigration is one of the most sensitive and complicated issues related to human population dynamics, and history is filled with attempts to stop or control the flow of immigrants (**Figure 5.16**).

Migration and Population Dynamics

Today, migration augments population growth in some regions and contributes to population declines in other regions. The main sources of immigrants are poor countries in Asia, Latin America, the Caribbean, and Africa, whose principal destinations are rich countries in North America, Europe, Australia, and New Zealand (**Figure 5.17**). In the richest countries, the number of immigrant residents totaled over 111 million in 2013, or about 11% of the overall population. About 2.5 million people migrate from one country to another every year. Of this total, approximately 1 million, or 40%, legally immigrate to the United States. Another several hundred thousand persons per year illegally enter the United States. Most of the others go to Europe, Canada, or Australia.

High rates of immigration have been associated with some incidents of political and social conflict. In South Africa, for example, controversy over immigration from surrounding countries, both legal and illegal, exploded into widespread rioting and violence in 2008 and 2015. In the United States, a number of local governments and states frustrated by the number of illegal migrants have made life harder for immigrants by preventing them from getting driver's licenses or arresting them when they cannot show proof of legal status. These legal moves have drawn opposition from the U.S. federal government and have become flashpoints for debate among political parties, citizen groups, and the U.S. Supreme Court (**Figure 5.18**).

Immigration policy in the United States is governed by the Immigration and Naturalization Act, which limits permanent immigrants to 675,000 annually. U.S. immigration policy has historically given priority to reuniting families, admitting immigrants with skills valuable to the U.S. economy and providing shelter to refugees. Many naturalized citizens, human rights groups, and business owners worried about their labor force have called for immigration policy reform.

⚠ Think About It

1. In setting immigration policy, what criteria can be used to screen immigrants? Are some criteria unethical? Explain your answers.

2. Do developed countries have an obligation to accept immigrants from developing countries with large populations? Do poor countries have a right to expect that rich countries will accept their immigrants?

BARRIERS BUILT TO CONTROL HUMAN MOVEMENTS HAVE A LONG HISTORY

(Blasco de Avellaneda/AFP/Getty Images)

Border fence between Spain and Morocco

(Frederic Brown/AFP/Getty Images)

Border fence between the United States and Mexico

FIGURE 5.16 The border fence between the Spanish city of Melilla and Morocco is intended to reduce illegal migration into Spain from Africa. The border fence between the United States and Mexico is intended to reduce illegal migration into the United States along its southern border.

IMMIGRANTS USUALLY GO TO AREAS OF BETTER ECONOMIC OPPORTUNITY

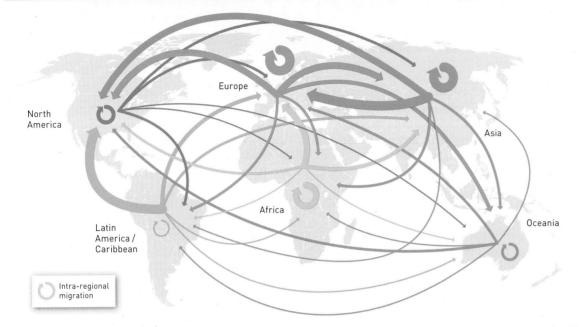

Intra-regional migration

FIGURE 5.17 The major flow paths of international immigration mainly connect Latin America, the Caribbean, Africa, and Asia with North America, Western Europe, Australia, and New Zealand. However, significant flows of immigration also occur between countries within regions, where there are differences in economic opportunity—for example, to South Africa from nearby less developed countries. (Data from United Nations Human Development Report, 2009)

IMMIGRATION: A SIGNIFICANT ISSUE IN MANY COUNTRIES

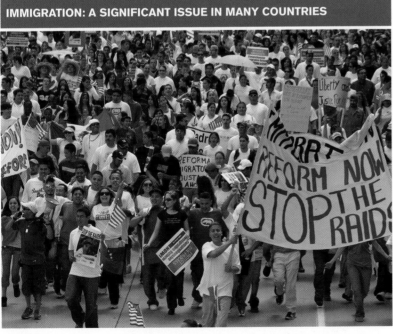

(Michael Rieger/Zuma Press/Newscom)

FIGURE 5.18 In the United States, intense debate swirls around the issues of how to reform immigration policy and to deal with the millions of illegal immigrants in the country. Here, some demonstrators protest a controversial immigration law passed by a state legislature, while others show support for it.

5.4–5.7 Issues: Summary

Understanding the impact of a population on the environment requires an accurate measurement of its level of development. The United Nations incorporates life expectancy, wealth, and access to education into a single index of development called the Human Development Index, or HDI. The two main factors influencing its impact are the number of individuals in the population and the per capita consumption of natural resources. These effects can be summarized by the IPAT equation: I (environmental impact) = P (population) × A (affluence or resource consumption) × T (technology). Ecological footprints, which are on average higher in more developed countries, also quantify the impact of populations on the environment. Migration has had a significant influence on human populations throughout history. Today, migration augments the population growth of many countries, while contributing to population declines in others. Immigration can be a significant contributor to population growth trends and, in some situations, may lead to social conflict.

5.8–5.10 Solutions

How might cultural and ethical considerations help or hinder efforts to control population growth? Give specific examples.

A conversation about unsustainable population growth represents an opportunity to preserve the environment and improve people's lives. Consider that in southern Somalia, young girls get married in their teenage years and start having children almost immediately. Across Africa, women have an average of six children. The lives of these women are almost entirely consumed by childbearing, cooking, cleaning, and water collection. That means there's no time for education or entrepreneurship. There's little chance to advance in their lives or to give their children more options than they themselves had.

Having a large family is, of course, a personal choice, which is one reason why population growth remains one of the most daunting and controversial environmental challenges of our day. Nevertheless, when countries design ethical policies to encourage population stability, everyone has a chance to benefit.

5.8 Most nations have national policies aimed at managing population growth

The problem of high fertility rates is not limited to sub-Saharan Africa. Women in Asia and other parts of the world are also bearing children at an unsustainable rate. Consequently, some of these countries have adopted national policies to lower fertility. By contrast, Japan and some European nations with the lowest fertility levels and a declining workforce are promoting higher fertility. Still other nations, particularly in the Americas, pursue policies of nonintervention in fertility decisions (**Figure 5.19**). Before considering how individual countries attempt to manage fertility, let's step back and consider the factors that have traditionally led families to produce many children.

The Historical Norm: High Fertility Rates

The lack of effective contraceptives in the past is one of the major reasons why family size has historically been so large. But there are also good reasons why people have wanted to have large families. Poor nutrition, sanitation, and health care meant that many children would die shortly after birth or at a very young age. Large families provide a hedge against loss of children to disease and accidents. As you probably know from growing up, children are a free source of labor to their parents. They can do chores such as tending crops and livestock, and they can help sell goods at the market. As

NATIONAL POLICIES FOR MANAGING FERTILITY RATES VARY WIDELY

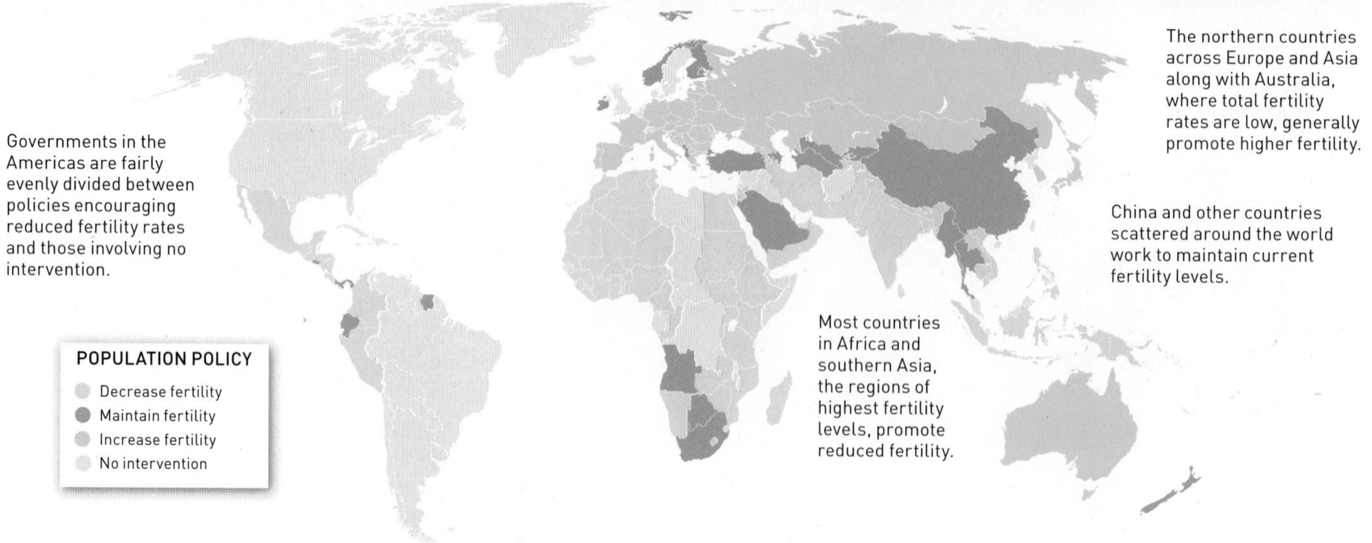

The northern countries across Europe and Asia along with Australia, where total fertility rates are low, generally promote higher fertility.

Governments in the Americas are fairly evenly divided between policies encouraging reduced fertility rates and those involving no intervention.

China and other countries scattered around the world work to maintain current fertility levels.

Most countries in Africa and southern Asia, the regions of highest fertility levels, promote reduced fertility.

POPULATION POLICY
- Decrease fertility
- Maintain fertility
- Increase fertility
- No intervention

FIGURE 5.19 The population policies of countries range from those promoting lower fertility, increased fertility, or maintaining current fertility levels to those involving no efforts to influence fertility decisions. (Data from United Nations, World Population Policies, 2007)

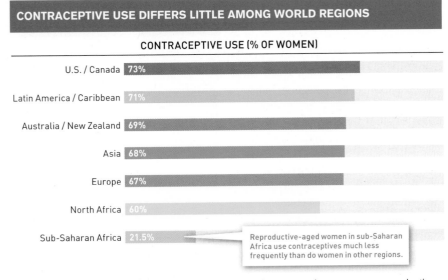

CONTRACEPTIVE USE DIFFERS LITTLE AMONG WORLD REGIONS

CONTRACEPTIVE USE (% OF WOMEN)

U.S. / Canada	73%
Latin America / Caribbean	71%
Australia / New Zealand	69%
Asia	68%
Europe	67%
North Africa	60%
Sub-Saharan Africa	21.5%

Reproductive-aged women in sub-Saharan Africa use contraceptives much less frequently than do women in other regions.

FIGURE 5.20 The major exception to the prevalence of contraceptive use among reproductive-aged women, defined by the United Nations as women between the ages of 15 to 44, occurs in sub-Saharan Africa. (Data from United Nations, World Contraceptive Use, 2007)

parents and grandparents age, children can help take care of them.

Access to Contraceptives and Birth Rates

Today, birth control pills, condoms, sterilization, and other forms of contraception are a key element in most national programs for reducing population growth. The United Nations encourages the use of contraceptives to reduce birthrates where population growth is rapid. Abortion of unwanted and dangerous pregnancies, though controversial, also plays a role in some programs. The UN has stated, "All couples and individuals have the basic right to decide freely and responsibly the number and spacing of their children and to have the information, education, and means to do so." The United Nations recognizes the importance of controlling population growth in a manner that respects local laws and customs and encourages all countries to provide "universal access to a full range of safe and reliable family-planning methods and related health services which *are not against the law* [emphasis added]."

In a 2007 study, the United Nations found that contraceptive use varies little among most world regions (**Figure 5.20**). Worldwide, the range of contraceptive use by reproductive-age women among most regions varied only from 60% to 73%. The major exception was sub-Saharan Africa, where the rate of contraceptive use averaged 21.5%. However, contraceptive use varies significantly *within* all regions, including sub-Saharan Africa. The pattern shown in **Figure 5.21** indicates a clear correlation between access to contraceptives and reduced birthrates. This relationship suggests that providing access to family-planning information and contraceptives

can reduce fertility rates substantially. For instance, in African countries where contraceptive use is prevalent, birthrates are nearly as low as in the most developed countries.

Let's examine the history of two countries that have invested heavily in population planning. Two of the most prominent examples of national population policies are

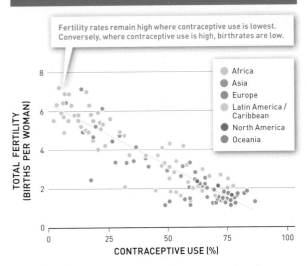

AS CONTRACEPTIVE USE INCREASES, TOTAL FERTILITY RATE DECLINES

Fertility rates remain high where contraceptive use is lowest. Conversely, where contraceptive use is high, birthrates are low.

Legend:
- Africa
- Asia
- Europe
- Latin America / Caribbean
- North America
- Oceania

Y-axis: TOTAL FERTILITY (BIRTHS PER WOMAN), 0, 2, 4, 6, 8
X-axis: CONTRACEPTIVE USE (%), 0, 25, 50, 75, 100

FIGURE 5.21 There is a clear relationship between contraceptive use, as a percentage of adult population, and total fertility rate. A significant decline in the fertility rate in countries with increased contraceptive use is found in all world regions. (United Nations, 1999; U.S. Census Bureau, International Data Base, 2000)

those of India and China, home to over 35% of the global population.

India: A Pioneer in Population Policy

Concerned that India's economic development would be disrupted by the rapid growth of its population, India developed the world's first national policy on population in 1952. Under India's National Population Policy, the government provides all citizens with the information and means to make informed choices regarding childbearing. However, participation in any family-planning service is entirely voluntary (**Figure 5.22**). A key element of India's population program is collaboration between national planners and local communities to improve prenatal and postnatal health care and to provide access to contraceptives. The program also offers economic incentives to couples that wait until the mother is 21 to have their first child and to stop reproducing after their second child. Public-health campaigns reinforce the advantages of small families over large ones.

Though available funds have been inadequate to entirely meet the goals of its population policy, India has made remarkable progress over the past half-century. After 1951 India's total fertility rate decreased from 6 children per woman to 2.6 children in 2012 (**Figure 5.23**). During the same period, the infant mortality rate declined from 146 deaths to 32 deaths per 1,000 live births, while life expectancy at birth increased from 37 years to 67 years. However, some regions of India have achieved exceptional results. For instance, the Indian state of Kerala adopted a population-control plan centered around three "e's":

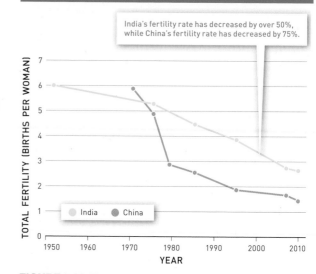

TOTAL FERTILITY HAS DECREASED DRAMATICALLY IN INDIA AND CHINA

India's fertility rate has decreased by over 50%, while China's fertility rate has decreased by 75%.

FIGURE 5.23 The population policies of the two most populous countries on Earth have been successful in decreasing total fertility to a fraction of their former level in a period of decades. (Data from United Nations, World Population Policies, 2007; U.S. Census Bureau, International Database, 2010)

education, employment, equality. Kerala's school system boasts a 90% literacy rate, identical for boys and girls. Educated women join the workforce before having children, and 63% (compared with 48% in India as a whole) use contraceptives. This has stabilized the state's birthrate at 2.0. Despite this progress, India's population will continue growing, though at slower and slower rates, through 2050.

China's One-Child Family Policy

In 1970 the Chinese government introduced a voluntary program that simply encouraged smaller families. However, its population continued to swell, and by 1979 China was home to nearly one-quarter of the world's population, living on 7% of Earth's land area. That's when the government made the radical decision to restrict the number of children a family could have, punishing those who broke the rules.

The one-child family policy restricted most families to a single child, a limit that is strictly enforced for government employees and urban dwellers. In rural areas, couples are generally allowed a second child if their first child is a girl. However, the policy stipulates a span of five years between the first and second child. The policy also allows ethnic minorities, such as Mongols, Tibetans, and Uyghurs, as well as other people living in remote areas with low population densities, to have a third child. The one-child policy is supported by a complex system of campaigns

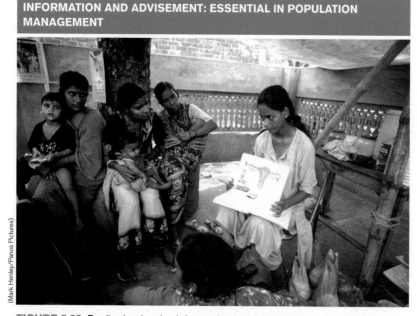

INFORMATION AND ADVISEMENT: ESSENTIAL IN POPULATION MANAGEMENT

(Mark Henley/Panos Pictures)

FIGURE 5.22 Family planning that brings trained advisors together with individual women and couples is an essential part of India's National Population Policy.

promoting small families, reproductive education, universal access to contraceptives, and economic rewards. These measures are backed up by an array of punishments for noncompliance, including substantial fines and confiscation of property.

The program had dramatic impacts on China's population dynamics (see Figure 5.23). The total fertility rate in China, which averaged 5.9 children per woman in 1970, fell to 2.9 during the period of voluntary family planning. However, following the establishment of the strict one-child family policy, the total fertility rate fell further, reaching 1.5 children per woman by 2010, well below replacement-level fertility.

Demographers predict that China's population will peak sometime after 2030 and begin declining thereafter. In contrast, India's population is projected to continue growing past mid-century to become the most populous nation on earth (**Figure 5.24**). While effective, China's one-child policy is considered unethical by some, particularly in Western nations, since it restricts the basic human right to reproduction and has been accompanied by reports of forced sterilizations and abortions. Such concerns have led to some reforms in the policy.

China has been gradually loosening its infamous policy. In 2013 it began allowing a second child when one of the parents was an only child. Previously, both parents had to be an only child in order to qualify. Then, in late 2015, it adopted a blanket two-child policy amid growing concerns about the economic consequences of an aging population. A second concern has been over an imbalance in the number of men and women in China's population—because the culture has traditionally shown a preference for male children. This phenomenon is not limited to China, however.

The Missing Daughters

The natural **sex ratio at birth** of human populations, defined as the ratio of male to female newborns, is about 103 to 107 male births for every 100 female births. In other words, there are 3% to 7% more male births than female births. This imbalance evens out with age to a ratio of approximately 100 males to 100 females in the population. That's because males at all ages are more likely to die from disease and accidents. Recently, however, sex ratios at birth have become *highly* biased toward males in a number of Asian populations.

The prejudice among parents in North Africa, the Middle East, and Asia is that sons are more valuable than daughters. People in these regions believe that sons are better able to do agricultural work and are more capable of supporting aged parents. They are also, by custom or law, the inheritors of family property and can continue the family line. Today, even as the historical justifications

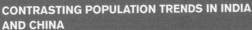

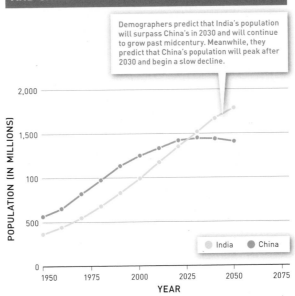

FIGURE 5.24 The populations of the world's two most populous countries are on different trajectories as both, according to demographers, approach population stability. (Data from U.S. Census Bureau, International Database, 2008)

for this bias have faded, families continue to prefer sons, a quest that has been assisted through sex-screening technology, such as ultrasound. Consequently, many of these populations have unbalanced sex ratios.

An imbalance in the proportions of males and females may be one of the unforeseen outcomes of China's one-child family policy, where the sex ratio at birth was 117 boys to 100 girls in 2001. This difference translates into approximately 1 million excess male births in China each year. Highly male-biased sex ratios at birth also occur in other Asian populations, such as those in India and South Korea, countries that do not have restrictive family-size policies in place. In fact, demographers have documented sex ratios at birth as high as 126:100 in some regions of India. This suggests that prenatal sex screening and sex-selective abortions may be the most important factor distorting the sex ratio.

Countries across Asia have become alarmed about unbalanced sex ratios. Young men in China, concerned by their limited marriage prospects, sometimes take organized tours to neighboring countries, such as Vietnam, in search of a mate. Policy makers fear that populations with millions of single males with no prospects for marriage and family may lead to more violence and social instability. In response, several Asian countries, including India and China, have outlawed prenatal sex screening and sex-selective abortion. India has also increased educational and job opportunities for girls. The government of China has passed laws making

How would you react if your government attempted to restrict your reproductive rights?

sex ratio at birth The ratio of male to female newborns.

MASS COMMUNICATION IS IMPORTANT IN CHINA'S POPULATION PROGRAM

FIGURE 5.25 The Chinese government actively campaigns to promote the advantages of small families and the value of both daughters and sons.

How might male-biased sex ratios lead to increased international tensions?

it easier for daughters to inherit property and to provide financial benefits, such as waived school fees, to daughter-only families.

In addition, China has produced public awareness campaigns aimed at warning of the social dangers of an imbalanced sex ratio and extolling the inherent value of a child of either sex (**Figure 5.25**). Public attitudes appear to be changing in China, where recent surveys have

indicated that 37% of women expressed no preference for a son or daughter, and equal numbers, about 6%, preferred either a single daughter or a single son. The remainder of those surveyed said that their ideal family would include one son and one daughter. In India, similar campaigns have begun to reduce sex-selective abortions. In 2008, for the first time in many decades, female births slightly exceeded male births in Delhi.

Global Trends in Fertility

As in India and China, total fertility rates are falling rapidly the world over due to a "reproductive revolution" caused by national policies, education, and contraception. As shown in **Figure 5.26**, between 1990 and 2010, total fertility rates fell significantly in countries at virtually all levels of development. The richest countries showed no change, in part, because fertility rates were already at 1.7 births per woman, well below replacement levels. At the global level, these declines translate into a decrease in fertility from 3.1 to 2.6 births in just two decades. Demographers predict that total fertility for the world will decline to the replacement level of 2.1 sometime before 2050.

⚠ Think About It

1. When the least developed nations wish to provide family-planning services and contraceptives for their people but lack the funds to do so, should the most developed countries provide the necessary funding?

What are some of the serious social and economic costs of having a population with fewer young than older members? What are some ways to reduce those costs without creating other problems?

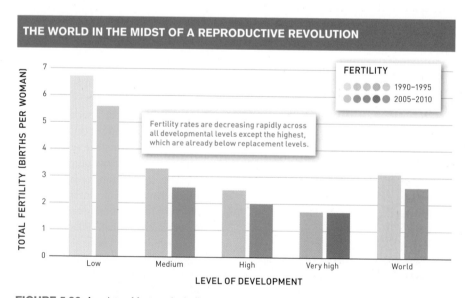

FIGURE 5.26 A variety of factors, including government population policies that provide access to family-planning information and contraceptives, have combined to produce significant decreases in fertility worldwide. In 2009 populations with the highest development were below replacement levels of fertility, whereas medium-development populations were rapidly approaching replacement levels. (Data from United Nations Human Development Report, 2009)

THE FOUR STAGES OF DEMOGRAPHIC TRANSITION

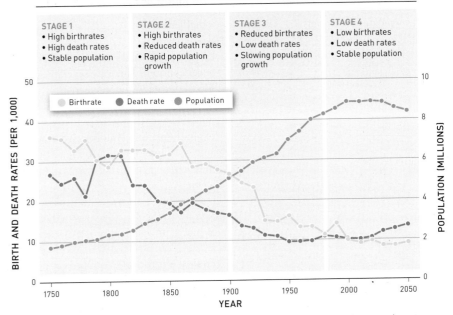

FIGURE 5.27 Demographic transition is a theoretical process proposed by demographers to explain changes in rates of fertility and death rates observed during the history of mainly European populations. How well the demographic transition model predicts demographic change in populations outside of the developed countries where it has been observed has been a subject of debate. (Data from Statistics Sweden; U.S. Census Bureau, International Data Base, 2006)

2. At what point do international campaigns for population control interfere with the rights of nations to manage their own internal affairs?

5.9 Human development is associated with lower fertility and reduced emigration

One question that development experts ponder is when Somalia and other sub-Saharan African countries will shift from having high death rates and birthrates to low death rates and birthrates. Such a **demographic transition,** which occurs in response to improved living conditions, has begun to take place in some African cities; it can be seen clearly through the history of a rich European nation such as Sweden in response to improved living conditions. We can break down the demographic transition into four stages. During stage 1, birthrates and death rates are high. Because birthrates and death rates are approximately equal, a population in stage 1 is stable or grows very slowly. As death rates decline, due to more dependable food supplies, improved sanitation, and drinking water supplies, populations enter stage 2.

Because birthrates remain high, stage 2 populations grow rapidly. As you can see in **Figure 5.27,** Sweden entered stage 2 during the late 18th and early 19th centuries. Many developing countries, particularly in sub-Saharan Africa, have fertility rates (see Figure 5.8, page 135) and death rates that would place them in this second stage of demographic transition.

During stage 3, death rates keep dropping—but so do birthrates, which translates to slower population growth. Declining birthrates may result from improved economic conditions, higher life expectancy, higher literacy particularly among women, and improved access to contraceptives and family-planning information. Today's developed countries moved into stage 3 sometime during the 20th century. Most developing countries in Asia, Latin America, and the Caribbean are currently in stage 3.

Populations stabilize at stage 4 when birthrates approximately balance death rates. The developing countries currently in stage 3 of demographic transition are expected to reach stage 4 sometime during the 21st century. For example, a Central American country rapidly approaching stage 4 is Costa Rica (**Figure 5.28**). As Figure 5.28a shows, Costa Rica's population is expected to stabilize by the year 2050. The quality of life in Costa Rica has improved dramatically as it has passed from stage 2 to 3 and approaches stage 4. By 2014 life expectancy at

demographic transition
A theory proposing that, with improved living conditions, human populations will undergo a gradual change from an earlier state of high death rates and birthrates to a state of low death rates and birthrates, with improved living conditions. The demographic transition model fits the history of today's developed countries well.

COSTA RICA IS IN THE MIDST OF RAPID DEMOGRAPHIC CHANGE

DEMOGRAPHIC TRANSITION: COSTA RICA

a. Costa Rica appears to be undergoing demographic transition. Birth and death rates are expected to be equal by 2050, when the country's population is expected to stabilize.

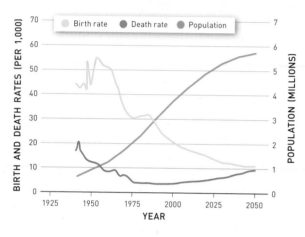

b. Life expectancy at birth has increased from levels far lower than any country on Earth today to equal or exceed those in many highly developed countries. Infant mortality rates have shown similar degrees of improvement.

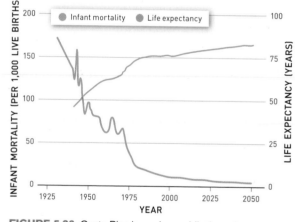

FIGURE 5.28 Costa Rica is moving rapidly through demographic transition, which is reflected in greatly reduced infant mortality and increased life expectancy at birth. (Data from Instituto Nacional de Estadistica y Censos de Costa Rica, 2006; U.S. Census Bureau, International Data Base, 2006)

Costa Rica was one of the first countries in the world to institute free and obligatory education for all (starting in 1869). Today, Costa Rica has one of the highest literacy rates in Latin America— over 95%. How might education have helped accelerate the country through demographic transition?

What factors may sustain high rates of fertility in some populations that have attained a high level of development?

development appears to help stabilize the populations of such countries (**Figure 5.29** and **Figure 5.30**).

Educating and Empowering Women

Whether in Asia, Africa, the Americas, or Europe, women with more education and who are sufficiently empowered to make critical choices in their lives bear fewer children. There are several reasons for this pattern. First, educated women marry later and they wait longer to have their first child. In addition, educated women are more likely to use contraceptives and are more receptive to family-planning information. Educated women also tend to spend more time on their careers and have less time to focus on raising many children. Finally, educated women have greater influence on those around them, and they become formal and informal educators themselves, passing on their perspectives to those around them and to succeeding generations.

Easing Migration Pressures

As we learned earlier, population density depends on not only population growth within a country's borders, but also immigration from other countries. One way to reduce migration pressure is to lessen economic disparities across borders. Such investment occurred in Western Europe following the expansion of the European Union (EU) in 1986 to include some of the then poorer countries of Western Europe, including Greece, Ireland, Portugal, and Spain. The EU invested heavily in their

birth in Costa Rica increased to 78, two years lower than in the United States, and its total fertility rate had fallen to 1.9 births, lower than in the United States in 2014 and equal to that in Iceland (Figure 5.28b).

While demographic transition theory explains the history of today's highly developed countries, demographers have suggested that some nations may be trapped in a lower state of development by poverty, overpopulation, and low literacy. That's why investing in

TOTAL FERTILITY RATE DECLINES WITH INCREASED DEVELOPMENT

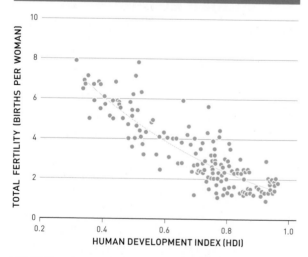

FIGURE 5.29 On average, countries with higher Human Development Index scores have significantly lower total fertility rates. (Data from United Nations Human Development Report, 2006; U.S. Census Bureau, International Data Base, 2006)

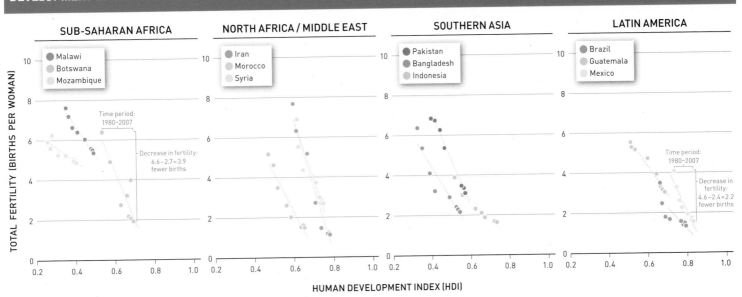

FIGURE 5.30 Improved Human Development Index scores from 1980 to 2007 were accompanied by reduced total fertility in populations of developing countries in geographic regions around the world. Interpretations provided for the plots of Botswana and Mexico as examples of how all country plots can be read. (Data from United Nations Human Development Report, 2009; U.S. Census Bureau, International Data Base, 2010)

economic infrastructure; as a result of such investments and local entrepreneurship, these countries entered the 21st century much better off economically. As a consequence of improved opportunities at home, fewer people emigrated from these nations; instead, they became a destination for immigrants. However, the economic downturn of 2008, which contributed to an economic crisis across Europe, again sent many residents of these countries abroad, seeking better opportunities elsewhere.

At the forefront of these developments was Ireland, which quickly developed into one of the strongest economies anywhere. In 2008 Ireland ranked 5th among the nations of the world on its Human Development Index score. At least partly influenced by improved economic prospects at home, the status of Ireland changed from being a major source of immigrants to being a destination for immigrants from around the world. However, the economic recession near the end of 2008 greatly reduced immigration to Ireland and by 2009 more people were again leaving Ireland than entering. However, with economic recovery, immigration to Ireland began increasing again after 2010; by 2015 one-third of Ireland's population growth was the result of immigration. These patterns suggest that investment in development in poorer countries will have multiple benefits, including reduced population growth and decreased immigration pressures.

⚠ Think About It

1. What do you think living conditions are like during the four stages of demographic transition?

2. How could you speed up the pace of demographic transition? Be specific.

3. Can you imagine a circumstance in which a country underwent significant development but did not evolve into a society in which both death rates and birthrates were low? Explain your response.

5.10 The challenge: Achieve high development and sustainable resource use

We often forget that at some point in their histories, the populations of all of today's developed countries went through a transformation from very low development to high development. Consider Iceland, which has an infant mortality rate of less than 3 deaths per 1,000 live births, one of the most favorable in the world. However, conditions were quite different as late as the 19th century. From 1840 to 1890, Iceland's infant mortality rate was 250 to 300 deaths per 1,000 live births, and in one disastrous year, 1846, Iceland's infant mortality rate rose

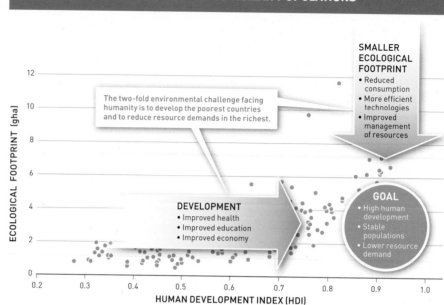

TWO MAJOR CHALLENGES RELATED TO HUMAN POPULATIONS

FIGURE 5.31 The first challenge is development of the least developed countries, including improved health care, education, and economic infrastructure. The second challenge is reducing the ecological footprints of the more developed countries through lower consumption, more efficient technologies, and better management of natural resources.

to 600 deaths per 1,000 live births. This is more than 3 times the infant mortality rate in any country on Earth today!

According to an analysis by Icelandic economist Thorvaldur Gylfason, Iceland's per capita economic output in 1901 was about $2,600, approximately that of Bangladesh in 2011. Since then, Iceland has grown its economic sector by about 2.6% annually, mainly through careful management of productive fishing grounds and renewable energy sources, rising to a per capita GDP of about $42,000 in 2013, according to the World Bank. However, the most important contributor to this growth may come as a surprise: education.

Literacy in Iceland stands at 99.9% of the population. Education is compulsory through age 16 and attendance through that age is estimated at 99%. According to Statistics Iceland, close to 27,000 Icelanders (nearly 10% of the entire population) received university degrees between 1999 and 2008. Of this total, nearly twice as many degrees went to women as men.

Iceland has a democratically elected parliamentary system of government in which all citizens 18 years or older are entitled to vote for president and members of parliament. Empowerment of women extends to the highest levels of involvement in Iceland. Women made up 30% of the Icelandic parliament in 2004, compared with 20% of the U.S. Congress. As a final testament to women's empowerment, Jóhanna Sigurdardóttir was selected as

Iceland's first woman prime minister, the head of the government, in 2010.

Improvements in health, along with high levels of education and literacy, a rich natural resource base, and universal suffrage and empowerment of women, contribute to Iceland's high level of development. They are also key factors in its movement toward a stabilized population. While we might consider Iceland exceptional in many ways, other countries have also made rapid developmental progress in the recent past. Costa Rica, a tropical Latin American country, is making rapid progress toward population stability. Like Iceland, Costa Rica's success owes much to heavy investment in education at all levels, universal suffrage, and empowerment of women. In 2014 the percentage of women in Costa Rica's Legislative Assembly was 33%, compared with 18.5% in the 113th U.S. Congress. In addition, like Iceland, Costa Rica elected its first woman head of government in 2010: President Laura Chinchilla Miranda.

The Challenge

The challenge for humanity is to invest in human development while reducing demand on Earth's resources (**Figure 5.31**). In this chapter, we have focused primarily on the need to reduce human population growth, which may be achieved by providing access to family-planning information and contraceptives and promoting the development of the least developed countries. The second task at hand is to reduce resource demands, which depends on reducing per capita consumption by developing more efficient technologies and management methods. These subjects are the main focus of Chapters 6 through 14.

The slowing of world population growth that we've already reviewed (Figures 5.3b and 5.3c, page 131) is a good sign of progress. Progress toward curbing explosive population growth is, of course, being played out within individual populations. Bangladesh, the country discussed at the beginning of this chapter, has made dramatic strides toward that goal, with family-planning programs contributing to a reduction in its total fertility rate from 6.6 children in 1981 to 2.4 children in 2015. In addition, concern continues to grow in both developed and developing countries about the environment, including overconsumption of resources and waste. Perhaps most important, there is widespread concern among the people of developed countries about the increasing gap between rich and poor worldwide and strong public support for giving aid to developing countries. Achieving a sustainable relationship with the biosphere is perhaps the most daunting challenge ever faced by our species, and it will take all the intelligence, wisdom, and sensitivity that we can muster.

⚠ Think About It

1. On average, what is the relationship between a nation's Human Development Index and its per capita ecological footprint?

2. What is the environmental and sociological significance of differences in ecological footprints among countries with similar Human Development Index scores?

3. How do the countries of Iceland and Costa Rica differ? How are they similar?

5.8–5.10 Solutions: Summary

Most countries have national population policies that promote decreasing, increasing, or maintaining fertility levels, depending on the population trend in individual countries. Other nations, particularly in the Americas, have no official policies regarding fertility levels. Educating and empowering women, which includes encouraging contraceptive use, can reduce fertility rates in rapidly growing populations. The national population policies of China and India have succeeded in reducing their rates of population growth. Prenatal sex screenings, coupled with sex-selective abortions, have produced an excess number of males in several countries throughout Asia, an unintended result that is being addressed forcefully.

Historical improvements in living conditions in today's developed countries led to demographic transition, a shift from high to low death rates and birthrates. Most developing countries are now in the middle to late stages of demographic transition. Worldwide investment in development may accelerate the passage of developing countries through demographic transition and slow the rate of global population growth. By reducing economic disparities across borders, investment in development in less developed countries may reduce incentives to immigrate.

Development generally comes at an environmental cost; on average, countries with higher developmental scores have larger ecological footprints. However, some countries have achieved high levels of development at a relatively lower cost to the environment. The two-fold challenge before humanity is to invest in human development, while reducing demand on Earth's resources.

If you had the authority and the means to do so, how would you meet the twin global challenges identified in this section?

How might development among the least developed countries be achieved without making them dependent on aid from more developed nations?

Answer the following questions for each chapter section and then answer the Central Question.

Central Question: How can we achieve sustainable human populations?

5.1–5.3 Science

- What environmental factors influence variation in human population density?

- How has the global population varied during human history?

- What can you learn from age distributions?

5.4–5.7 Issues

- How do fertility rates vary among countries and regions?

- What are the variables used to calculate the Human Development Index?

- What is the general relationship between development and environmental impact?

- How do development differences affect immigration rates and destinations?

The Human Population and You

The issue of human population is a personal one, related to our own reproduction and how we conduct our lives. Now there is more urgency than ever, because it appears that we have already exceeded the long-term capacity of Earth to sustain us. In the face of such a challenge, what can you do to make a difference?

☐ *Keep informed about the rapid pace of change.*

Population issues around the world are in a period of dynamic change. Likewise, perspectives and ideas about these critical issues quickly become dated and irrelevant. Stay informed on human population and consumption issues by reading world news on population and sustainability—also check out the population data sheet (produced by the Population Reference Bureau) and the CIA World Factbook.

☐ *Support international educational programs.*

Education is key to solving any problem, including problems related to the human population. Not only does education support the democratic process, but there is also a direct relationship between education and reduced fertility rates. You can support education by becoming a teacher yourself or by supporting the educational mission of the schools in your community. There are also opportunities to contribute to international education initiatives, for example, the Central Asia Institute, which has built more than 130 schools in Pakistan and Afghanistan that emphasize educating girls.

☐ *Support developmental initiatives.*

Development encourages reduced fertility rates and stabilized populations. Developmental efforts include initiatives to provide health care, improve economic infrastructure, or increase food production and availability. In the United States, there are many opportunities to help directly with development through programs such as the Peace Corps, a government-run program, or Los Amigos de las Americas, a nongovernmental, nonprofit program promoting development and understanding in the Americas.

☐ *Make commitments in your personal life.*

One of the most direct contributions that any of us can make is to adjust our personal lives to make them consistent with the goals of stabilized populations and sustainable resource use. If you have children or plan to, you may consider reproducing at a replacement rate of two children. Many tactics to reduce our personal ecological footprints are discussed in the chapters to come.

5.8–5.10 Solutions

- How have immigration policies affected populations around the world?

- What is the relationship between human development and population characteristics?

- How do we transition to sustainable human populations?

Answer the Central Question:

_____

_____

_____

_____

_____

_____

Chapter 5

Review Questions

1. Which of the following is the closest estimate of the global population in 2012?
a. 6 billion   c. 8 billion
b. 7 billion   d. 9 billion

2. For the past several centuries, the global population has shown a J-shaped pattern of increase. Does this type of increase continue in the global population?
a. Yes, J-shaped population growth continues.
b. No, the global population is now stable.
c. No, the rate of population growth (%) is declining.
d. No, the size of the global population is decreasing.

3. The age structure of the United States suggests that its population is stable, yet it continues growing. Which of the following best explains continued population growth in the country?
a. A relatively young population combined with immigration
b. High immigration rates
c. Population momentum due to the relative youth of the population
d. A total fertility rate above replacement level

4. Which of the following regions is growing the most rapidly?
a. North Africa   c. North America
b. Latin America   d. Sub-Saharan Africa

5. Which one of the following factors is not included in the United Nations' Human Development Index, or HDI?
a. Health   c. Education
b. Emotional well-being   d. Economic productivity

6. Which of the following potentially contribute to the environmental impact of a population?
a. Population size
b. The types of technologies used in the population
c. The level of resource consumption
d. All of the above

7. How, in general, does economic opportunity influence immigration patterns?
a. Levels of economic opportunity are unrelated to immigration patterns.
b. Immigrants generally move from areas of lower economic opportunity to areas of higher economic opportunity.
c. Economic opportunity influenced immigration historically but no longer does.
d. Immigrants generally move from areas of higher economic development to avoid higher taxes.

8. What is the present global trend in total fertility rate?
a. The global total fertility rate has already decreased below replacement level.
b. The global total fertility rate has stabilized.
c. The global total fertility rate continues to increase.
d. The global total fertility rate is decreasing.

9. How are death rates and birthrates related during demographic transition??
a. Declines in birthrates follow declines in death rates.
b. Declines in death rates follow declines in birthrates.
c. Increased death rates stimulate decreased birthrates.
d. Death rates and birthrates are unrelated to demographic transition.

10. How does higher human development appear to affect fertility rates in populations?
a. Higher human development appears to encourage higher fertility.
b. Higher human development does not appear to affect fertility rates.
c. Higher human development levels are associated with lower fertility rates.
d. Higher development levels are associated with lower fertility rates on some continents but not others.

Critical Analysis

1. The doubling time of a population can be estimated by dividing 70 years by a population's annual growth rate percentage. What are the estimated doubling times for the populations of Lesotho, with an annual growth rate of 0.3% in 2010; the United States, with an annual growth rate of 1%; and Yemen, with an annual growth rate of 2.7%?

2. The per capita ecological footprints of individuals in the United States and Canada are very similar. However, the total ecological footprint of Canadians has not yet exceeded Canada's capacity to produce resources, whereas the total of the U.S. population has exceeded the country's productive capacity. How does this difference affect the way you view the impacts of the two populations?

3. How much variation is there in ecological footprints among countries with Human Development Index scores of 0.8 (the UN's approximate threshold for "very high" development) or higher (see Figure 5.13, page 138)? What are some of the implications of this variation?

4. Immigration, both legal and illegal, is a significant issue in countries around the world, particularly those that receive large numbers of immigrants. What policies on immigration would you institute, if you had the power to do so? Explain how your policies would benefit the countries receiving immigrants, as well as the home countries of migrants.

5. The Icelandic economist Thorvaldur Gylfason has suggested that investing in education has been key to Iceland's achieving a high level of human development. Can you imagine a population living in the midst of natural resources as rich as those available in Iceland, but with a much lower level of human development? Elaborate on your answer.

Find additional resources and links online at www. macmillanhighered.com/launchpad/molles1e.

Central Question: How can we meet human needs for freshwater, while avoiding or reducing environmental impact?

Discuss the hydrologic cycle and how climate can affect it.

(NASA)

Sustaining Water Supplies

Analyze the global demand for water, as well as the factors and industries that affect its availability.

ISSUES

Discuss the individual, industrial, and societal tactics for sustaining water supplies.

SOLUTIONS

By 2015, a historic three-year drought turned much of California's San Joaquin Valley, one of the richest agricultural regions in the world, into a virtual desert.

(Photo by Cynthia Mendoza, U.S. Department of Agriculture)

Fresh Water Supplies Shrink as Demand Increases

The worst drought in 1,000 years destroys crops in California's San Joaquin Valley and triggers water rationing, a reminder that freshwater is essential to human welfare.

Inch by inch, California's Central Valley is sinking. Three unrelenting years of drought have wreaked economic, agricultural, and environmental havoc in a region that supplies the nation with more than a quarter of its food, including 350 crops ranging from almonds to wheat. There was a time when a farmer could drill a few hundred feet into the ground and have enough water to supply his farm for years, but, in the last decade, Central Valley farmers began to hire rigs that could bore 1,000 feet or more into the earth at a cost of hundreds of thousands of dollars. Extracting water from these ancient groundwater aquifers

dries out and compacts layers of geologic sediments, causing the ground above them to sink. Once compacted, the capacity of the material to store water is permanently reduced. "Everybody is starting to panic," a well driller named Steve Arthur told the *San Jose Mercury News* in March 2014. "Without water, this valley can't survive."

It's the same story being heard all across the western United States. In April 2015 California's Governor Jerry Brown ordered municipalities throughout the state to reduce water use by 25%. The city of Santa Cruz issued a moratorium on outdoor watering. Lake of the Woods, north of Los Angeles, has taken to trucking in their drinking water. Las Vegas, Nevada, is drilling a new tunnel underneath the Hoover Dam because water levels in Lake Mead are dropping below existing outlets.

"Water, water, every where,
Nor any drop to drink."

Samuel Taylor Coleridge, *The Rime of the Ancient Mariner*, 1798

For thousands of years, humans have built structures to divert and store precious **freshwater** for drinking, sanitation, agriculture, and industry. However, there is a growing mismatch between human needs—not to mention human wants—and the existing water supplies around the world. More than 1.1 billion people lack access to clean drinking water and more than 2.6 billion people, nearly 40% of the global population, lack sufficient freshwater for basic sanitation. These shortages promise to worsen since, as we saw in Chapter 5, the global population is projected to increase by another 2 billion by the year 2050.

Water shortages across Earth present one of the most daunting environmental challenges of the 21st century. Damming rivers and diverting water can often solve water problems in the short term, but can lead to larger problems in the future. And any use of water by humans can harm aquatic and terrestrial ecosystems, which provide crucial services such as reducing sediment or purifying runoff before it reaches the ocean.

freshwater Water with a salt content, or salinity, below that of brackish water (i.e., salinity less than 500 mg/l).

Central Question

How can we meet human needs for freshwater, while avoiding or reducing environmental impact?

(NASA)

6.1–6.2 Science

Imagine you are visiting Earth from a distant planet. As you approach, you see a blue world covered mostly by vast oceans and enveloped by clouds of water vapor. The continents are dotted by lakes and ponds and laced with rivers and streams. You also see extensive areas near the poles covered by snow and ice. If you had ground-penetrating sensors, they would also detect vast deposits of underground water. Earth is awash in immense amounts of water, but the problem is we can't always access water when we need it, and much of it is too salty for most uses. What can't be seen from this sky-high view is that water is constantly moving over time, cycling around Earth.

6.1 The hydrologic cycle moves water around Earth

Pick a single drop of water and follow it from a mountainside spring down a freshwater river and into the salty ocean. Warmed by the Sun, this drop of water evaporates and its molecules rise high into the atmosphere, where they can fall to Earth again in the form of rain or snow. The process by which Earth's water moves among the oceans, atmosphere, terrestrial and freshwater environments, and back to the ocean is called the **hydrologic cycle.** This is one of the most critical of Earth's natural processes because it sets limits within which we must work to sustain supplies of freshwater.

The Distribution of Earth's Water

The amount of water cycling through Earth is so vast that we can't talk about it in terms of liters or gallons. The typical unit of volume at this scale is the cubic kilometer (km^3) which is equal to 1 trillion liters (**Figure 6.1**). If it were spread out, 1 km^3 of water would form a band of water 1 meter deep and 25 meters wide that would

hydrologic cycle The movement of Earth's water between the oceans, atmosphere, and terrestrial and freshwater environments.

reservoir A body of water, ranging in size from a pond to an ocean, including below-ground deposits of water; constructed dams retain water in artificial reservoirs, which are commonly used to store and divert water for human use.

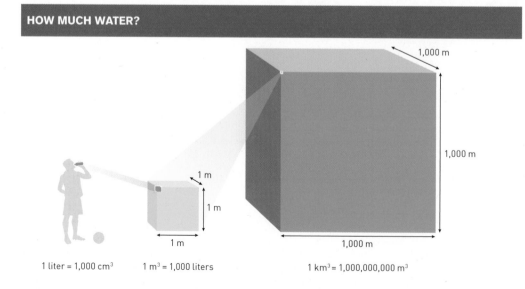

HOW MUCH WATER?

1,000 m
1,000 m
1,000 m

1 m
1 m
1 m

FIGURE 6.1 Thinking quantitatively about any feature of nature—whether distance, mass, or volume—requires an appropriate scale. Familiarity with a range of metric units of volume makes it easier to represent and discuss the hydrologic cycle quantitatively.

1 liter = 1,000 cm³   1 m³ = 1,000 liters   1 km³ = 1,000,000,000 m³

completely encircle Earth at the equator. From a human perspective, this is an immense quantity of water, yet it is tiny in comparison to the water flowing through the hydrologic cycle. The total amount of liquid freshwater on Earth's surface and in the ground amounts to more than 10 million km³. When you add all the water in the atmosphere, oceans, and ice caps, we are talking about a total of 1.4 billion km³.

Earth's water resides in pools called **reservoirs.** The atmosphere, oceans, lakes, rivers, soils, glaciers, snowfields, and groundwater are Earth's main reservoirs of water. The oceans make up the largest of Earth's water reservoirs, containing nearly 97% of all Earth's water. All of the water in the oceanic reservoir is, of course, saltwater. The largest freshwater reservoir, containing over two-thirds of all the freshwater on Earth, is frozen in glaciers and the polar ice caps. Most of the remaining third may be found in **groundwater,** located in the pore spaces in rock and sediment beneath Earth's surface. Groundwater is the source of water into which wells tap and from which spring-fed oases emerge. Lakes, wetlands, and rivers collectively contain only 0.3% of Earth's freshwater. And the atmosphere, one of the smaller of the reservoirs, holds only 0.04%, or about 13,000 km³. The remainder is held in soils, as soil water or ice (**Figure 6.2**).

A geologic formation containing a deposit of groundwater is called an **aquifer.** The amount of water present in an aquifer results from the balance of water flowing in via infiltration and draining out through groundwater flow. The **water table** is the uppermost level of the groundwater, forming the boundary between the groundwater and overlying soil or rock. The layers of rock below the water table, in which the pore spaces in the geologic formation are saturated with groundwater, are called the **saturated zone;** the layers above the water table, which are not saturated with water, are referred to as the **unsaturated zone.** Groundwater gets *recharged* as water from rain or melting snow infiltrates the ground, descends under the force of gravity, and eventually flows to the water table. The rate of groundwater recharge depends on the amount of precipitation, rate of evaporation, and permeability of the soil and rock. Infiltration rates are higher through coarse-grained, sandy soils than through fine-grained, clay soils. Precipitation that does not infiltrate the soil or evaporate into the air moves across the land as **runoff.**

When the soil is topped with leaf litter and other organic matter, as well as broken up by tree roots and the burrows of animals such as earthworms, there will be a significant amount of infiltration. However, in areas with few plants and animals, precipitation rates can easily exceed rates of infiltration, resulting in high amounts of runoff. One example of this is the desert, where flash floods are common during infrequent but intense storms.

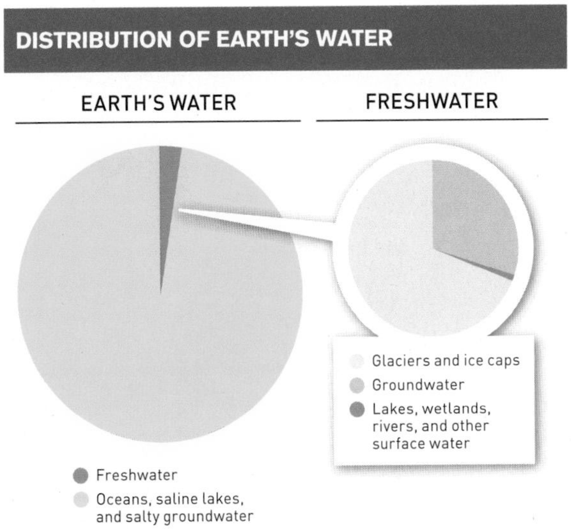

DISTRIBUTION OF EARTH'S WATER

EARTH'S WATER · **FRESHWATER**

- Glaciers and ice caps
- Groundwater
- Lakes, wetlands, rivers, and other surface water

- Freshwater
- Oceans, saline lakes, and salty groundwater

FIGURE 6.2 The oceans contain the bulk of water on Earth. Meanwhile, most freshwater is presently frozen in the planet's glaciers and ice caps or found under Earth's surface as groundwater.

Groundwater flows from areas of recharge to areas of groundwater **discharge,** such as rivers, lakes, and springs, or directly into oceans. Aquifers made up of coarse, sandy materials generally support higher groundwater flow rates. The time required for groundwater to flow from recharge to discharge areas can range from days to thousands of years, depending on flow rates and the distance traveled (**Figure 6.3**).

A **watershed,** or **catchment,** is the area of land collecting water, falling as precipitation, that flows into an aquifer or a river system. For instance, the Mississippi River watershed is the fourth largest in the world, and it includes parts of 31 states and 2 Canadian provinces. The surface water in this watershed eventually passes through the Mississippi Delta and into the Gulf of Mexico.

Water Moves Among Reservoirs

Water moves from one reservoir to another through processes such as precipitation, evaporation, and runoff. This **flux** of water among reservoirs, which is the basis of the hydrologic cycle (**Figure 6.4**), is powered by the Sun, which causes water to evaporate. As water vapor rises into the atmosphere, it cools and condenses, forming clouds. Much of the snow and rain on land originates from the evaporation of moisture from soils and from the loss of water from trees and other plants. More than one-third of this precipitation flows back to the oceans as surface runoff in rivers and streams or as subsurface runoff, as groundwater discharges from land to the sea. This flux represents a very small fraction of the total volume of water circulated annually by the hydrologic cycle (less than 0.0003%), which is replaced each year by precipitation that originates as evaporation from the oceans and falls onto land (see Figure 6.4).

Would the water deposited by precipitation move differently through an urban environment than through a temperate forest environment? How and why?

groundwater Water found in the pore spaces in rock and sediments beneath Earth's surface; feeds wells, springs, and desert oases, and is Earth's second largest reservoir of freshwater.

aquifer A geologic formation containing groundwater; gains water through the process of infiltration and loses water through groundwater flow.

water table The uppermost level of groundwater, which forms the boundary between the saturated and unsaturated zones.

saturated zone The layers of rock below the water table, in which the pore spaces in the geologic formation are saturated with water.

unsaturated zone The layers above the water table, which are not saturated with water.

runoff The amount of water falling as precipitation that flows off the land as surface and subsurface flow.

discharge In an aquifer, the movement of water from the groundwater to a body of surface water (e.g., a river or lake).

watershed (catchment) The area of land that collects water, which falls as precipitation and flows into an aquifer or river system.

flux The rate of flow of materials or energy across a given area (e.g., the flow of water vapor from the ocean's surface to the atmosphere or the flow of radiant energy between an organism and its surroundings).

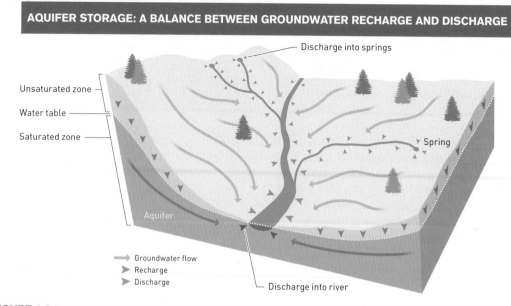

AQUIFER STORAGE: A BALANCE BETWEEN GROUNDWATER RECHARGE AND DISCHARGE

FIGURE 6.3 Recharge adds to groundwater in an aquifer, whereas discharge removes water. Groundwater flows slowly between zones of recharge and zones of discharge.

This tiny fraction is the freshwater primarily available for human use.

⚠ Think About It

1. How does Figure 6.4 explain why the 40,000 km³ of annual runoff is our renewable freshwater supply, which is a far smaller amount than the total volume of freshwater in groundwater, lakes, rivers, and streams?

2. How many liters of water per person does annual runoff equate to, assuming a global population of 7 billion?

3. Why is the number you calculated in response to Question 2 not a practical indicator of the amount of water available to human populations?

4. Can we treat groundwater and surface water as separate resources and manage water sustainably?

?

Is the movement of water through the hydrologic cycle influenced by physical processes only? Explain.

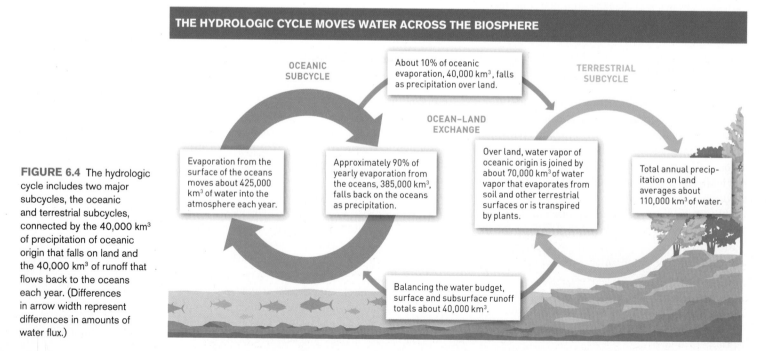

THE HYDROLOGIC CYCLE MOVES WATER ACROSS THE BIOSPHERE

OCEANIC SUBCYCLE

TERRESTRIAL SUBCYCLE

About 10% of oceanic evaporation, 40,000 km³, falls as precipitation over land.

OCEAN–LAND EXCHANGE

Evaporation from the surface of the oceans moves about 425,000 km³ of water into the atmosphere each year.

Approximately 90% of yearly evaporation from the oceans, 385,000 km³, falls back on the oceans as precipitation.

Over land, water vapor of oceanic origin is joined by about 70,000 km³ of water vapor that evaporates from soil and other terrestrial surfaces or is transpired by plants.

Total annual precipitation on land averages about 110,000 km³ of water.

Balancing the water budget, surface and subsurface runoff totals about 40,000 km³.

FIGURE 6.4 The hydrologic cycle includes two major subcycles, the oceanic and terrestrial subcycles, connected by the 40,000 km³ of precipitation of oceanic origin that falls on land and the 40,000 km³ of runoff that flows back to the oceans each year. (Differences in arrow width represent differences in amounts of water flux.)

6.2 The El Niño Southern Oscillation causes periods of dry years and wet years

Every three to five years, on average, the equatorial Pacific Ocean off the coast of South America warms up by a few degrees, leading to significant changes in rainfall around a large part of the globe. This phenomenon, known to the fishermen of Peru for centuries, brought with it torrential rains, flooding on land, and poor fishing at sea. Because the periodic warming generally occurred during Christmastime, the local fishermen named it "El Niño," the Spanish name for Jesus Christ as a child. Other observers noted that droughts also seemed to come and go in cycles.

Today, we recognize the **El Niño Southern Oscillation** as one of the most important influences on atmospheric circulation and surface temperatures across the Pacific Ocean. During typical conditions in the tropical Pacific, the *trade winds,* which blow from the east, push warm surface water westward. As this water moves westward, cold water from the bottom of the ocean on the west coast of South America is drawn up to the surface, in a process called **upwelling.** However, as **El Niño** conditions develop, the trade winds weaken and the warm surface water flows in the opposite direction—eastward. When it reaches South America, this warm water moves north and south along the coast, pushing the normally cool surface water downward.

Such changes in atmospheric circulation and ocean surface temperatures influence the hydrologic cycle globally. The warmer than average sea surface temperatures during an El Niño are accompanied by low atmospheric pressure in the eastern Pacific Ocean. These physical conditions favor the production of storms in the eastern Pacific Ocean, as the water vapor in the warm rising air condenses to form storm clouds. When this happens, conditions are wet and cool across the southern portion of the United States and northern Mexico during December through February. Meanwhile, Southeast Asia and Australia are dry. So, while the southeastern United States faces torrential rains and flooding, Southeast Asia and Australia brace for drought and wildfires. The climatic conditions associated with El Niño generally persist for a year, sometimes two, before the climate swings back toward average conditions or beyond, to the opposite climatic extreme.

The opposite of the El Niño is now known as La Niña, which develops about as frequently as El Niño—approximately every 3 to 5 years. During a **La Niña,** the trade winds are stronger than usual and push warm water even farther west across the tropical Pacific. Meanwhile, thanks to increased upwelling, sea surface temperatures are lower than average in the eastern Pacific Ocean. Combined with cooler sea surface temperatures, La Niña brings higher atmospheric pressures. Under such conditions, cloud formation is inhibited over the eastern Pacific Ocean, whereas on the western Pacific Ocean more storms occur. December through February of a La Niña is characterized by dry and warm conditions across the southern United States and northern Mexico and wet conditions in Southeast Asia and northern Australia (**Figure 6.5**).

It took scientists most of the 20th century to piece together the global impact of El Niño and La Niña. Based on tree ring data, which correlates with precipitation amounts, and other evidence, scientists estimate that El Niño-like conditions have been occurring periodically for tens of thousands of years.

TEMPERATURE AND ATMOSPHERIC PRESSURE DIFFERENCES ACROSS THE PACIFIC OCEAN DURING EL NIÑOS AND LA NIÑAS

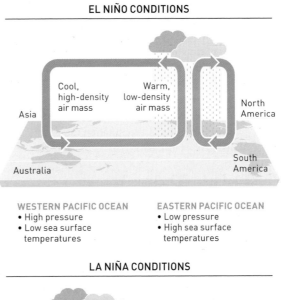

EL NIÑO CONDITIONS

WESTERN PACIFIC OCEAN
- High pressure
- Low sea surface temperatures

EASTERN PACIFIC OCEAN
- Low pressure
- High sea surface temperatures

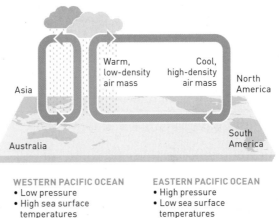

LA NIÑA CONDITIONS

WESTERN PACIFIC OCEAN
- Low pressure
- High sea surface temperatures

EASTERN PACIFIC OCEAN
- High pressure
- Low sea surface temperatures

FIGURE 6.5 The El Niño Southern Oscillation moves atmospheric circulation associated with storm generation to the eastern Pacific Ocean during El Niños and to the western Pacific Ocean during La Niñas.

What are the disadvantages of living in a region strongly connected to the El Niño Southern Oscillation system? Might there be advantages?

El Niño Southern Oscillation An oscillating climatic system involving variation in ocean surface temperatures and barometric pressures across the Pacific Ocean.

upwelling The movement of cold subsurface water to the ocean's surface when warmer surface waters move offshore under the influence of prevailing winds.

El Niño A period of warmer than average sea surface temperatures and lower barometric pressure in the eastern Pacific Ocean, favoring the production of storms in the eastern Pacific Ocean.

La Niña A period of lower than average sea surface temperatures and higher barometric pressures in the eastern Pacific Ocean, resulting in reduced storm activity in the eastern Pacific Ocean.

Today, scientists can predict the onset of an El Niño year about nine months in advance.

There is at present no clear consensus among climate scientists as to how climate warming (see Chapter 14, page 443) will influence the intensity and frequency of El Niño and La Niña. While some climate models predict higher frequencies and intensities of these events, other climate models predict the opposite. Regardless, the long geologic history of this system suggests that the El Niño will continue to affect Earth's weather. From a water supply perspective, those regions under the influence of the El Niño experience much greater fluctuation in precipitation than do regions outside of its influence.

⚠ Think About It

1. The onset of an El Niño or La Niña can now be predicted several months in advance. How might this predictive capability be used to the advantage of water planners?

2. How might water planners and managers adjust to the challenges presented by the climatic variation associated with El Niño and La Niña?

6.1–6.2 Science: Summary

The hydrologic cycle, which is powered by the Sun, moves Earth's water across the biosphere and strongly influences sustainable water supplies. The oceans form the largest reservoir of water on Earth, followed by polar ice caps and glaciers, groundwater and freshwater lakes and rivers, and the atmosphere. Approximately one-tenth of the water evaporated from the oceans annually (40,000 km^3) falls as precipitation on land. Another 70,000 km^3 of precipitation on land originates as water that evaporates from soils and plants. El Niño and its opposite, La Niña, are extremes in an oscillating climatic system involving variation in ocean surface temperatures and atmospheric pressures across the Pacific Ocean. Conditions during an El Niño favor the production of storms in the eastern Pacific Ocean that cause the southern United States and northern Mexico to experience periods of extreme wetness. During a La Niña, conditions favoring the production of storms move away from the eastern Pacific Ocean, drying out the United States and Mexico and delivering moisture to Southeast Asia and Australia.

6.3–6.6 Issues

Approximately 5,000 years ago, two Mesopotamian city-states, Lagash and Umma, went to war over water. They eventually signed the first known peace treaty, but their skirmishes continued for the next 150 years. Conflicts over water in Africa and the Middle East have broken out periodically over the last century and, in 1995, then Vice President Ismail Serageldin of the World Bank declared, "The wars of the next century will be over water." No other resource is more important to human welfare than water. The quality of our lives is directly linked to water and the services it provides, but we are faced with the fundamental problem of water scarcity in the face of a growing human population. Every liter of freshwater used for someone's crops, livestock, showers, and bottles of soda potentially reduces the amount available to neighbors—and to natural ecosystems.

6.3 Access to adequate water supplies as a human right

The environmental scientist Peter H. Gleick pointed out in 1999 that water has not been included as a right in most international conventions on human rights. Although these conventions recognize the human right to life and to adequate food, both of which depend on access to water, they have managed to omit the life-giving liquid itself.

Daily Water Needs

Defining the minimum amount of water a person needs to live is not easy. First off, it depends on the climate. People living in warmer climates will obviously need more water than people living in cooler climates. But it also depends on the individuals, how old they are, how much they weigh, and how active they are. An adult living in a moderate climate and engaging in moderate levels of activity needs to consume about 3 to 5 liters of water per day. Daily water requirements increase with temperature, activity level, and body size. Furthermore, drinking is only one of many ways we use water. We use water to brush our teeth, wash our hands and clothing, and clean our dishes. Allowing for these purposes, Gleick, the founder of the Pacific Institute in Oakland, California, has suggested a volume of about 50 liters (13.2 gallons) per day, budgeted as follows: 20 liters for sanitation services (mainly sewage disposal), 15 liters for bathing, 10 liters for food preparation, and 5 liters for drinking (**Figure 6.6**). The World Health Organization estimates that daily access to between 50 and 100 liters of water is the minimum to ensure adequate sanitation and health.

Relative to the amount of water used in the United States and other developed countries, 50 to 100 liters isn't much (**Figure 6.7**). Even so, according to the United Nations, some 2.6 billion people around the world lack access to sufficient water to meet their basic sanitation needs, and 900 million don't have safe drinking water. Even in places where there is theoretically enough water for residents, they often have no good way to obtain it. The average distance that women and children in Asia walk to obtain water is 6 kilometers (3.7 miles), making less time available for other work and for education.

Water as a Human Right: The Challenges

In 2010 the UN General Assembly at last adopted a resolution recognizing "the human right to water and sanitation." The resolution passed by a vote of 122 nations in favor, none against, with 41 nations

AVERAGE PER CAPITA WATER USE AROUND THE WORLD

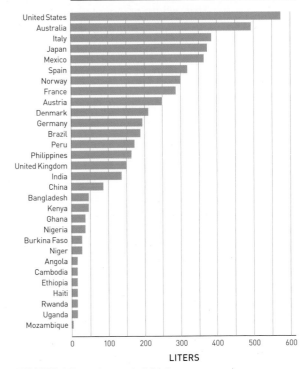

FIGURE 6.7 While people living in some countries use hundreds of liters per day, elsewhere they make do with much less. (Data from Data 360, www.data360.org/dsg.aspx?Data_Set_Group_Id=757)

If you were limited to 50 liters of water per day, how would you cope with everyday tasks?

WHAT IS THE MINIMUM HUMAN NEED FOR WATER?

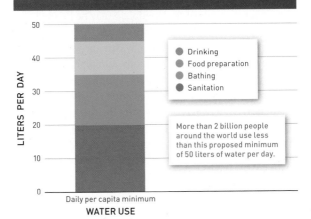

FIGURE 6.6 Peter H. Gleick has suggested 50 liters per day as a minimum water right for all individuals on Earth, based on estimates of the minimum volumes required for drinking, food preparation, bathing, and sanitation. (Data from Gleick, 1999)

abstaining. Although the resolution passed, the debate over recognizing water as a human right continues.

Many of the countries that abstained during the UN vote to recognize water as a human right—including the United States, Canada, and Australia—have an economic stake in water, viewing it as a commodity that should be managed like any other natural resource, such as wood, minerals, and soils. They fear that declaring water as a human right will interfere with the multibillion-dollar water industry and with international water trading, creating thorny legal and political questions.

What if, for example, some nations were required to transfer water across international boundaries. Would Canada then be obligated to transfer water to meet the needs of a growing U.S. population? Another concern is that if access to water is a human right, who bears the responsibility of providing water and paying for the infrastructure to do so? Meanwhile, those in favor of water as a human right counter that the right to water is part of the public trust that belongs to all species. As we move into the future, the proponents of these opposing views will have to engage in an international dialogue, which may lead to a mix of governmental and private business approaches that address the water needs of more people.

⚠ Think About It

1. If water is guaranteed as a human right, how much water should people be guaranteed?

2. Should water-rich regions contribute water to heavily populated arid or semi-arid regions? Explain why or why not.

3. If water can be considered a human right, should this right be addressed at the expense of other species or the natural ecosystems of a region—for example, endangered fish and amphibian species living in arid regions?

Should water be used for growing crops in the desert that might be grown elsewhere?

irrigation A system for artificially delivering water to crops so that they can grow in areas with too little precipitation to support them otherwise.

instream uses Benefits, such as dilution of sewage and recreational fishing and boating, that result from water flowing in river or stream channels.

dam A structure that blocks the flow of a stream or river; may be used to reduce downstream flooding or to store water in a reservoir.

flood A river or stream overflowing its banks and inundating the surrounding landscape known as the floodplain.

floodplain The area of land that stretches from a water channel to the valley walls.

drought An extended period of dry weather during which precipitation is reduced sufficiently to damage crops, impair the functioning of natural ecosystems, or cause water shortages for human populations.

6.4 Humans already use most of the world's accessible freshwater supplies

Even though the total amount of freshwater flowing down mountainsides and into rivers and lakes should be enough to meet our global needs, much of it is not practical to access. Half of all surface runoff each year occurs during heavy rains and flows straight out to the sea. Another fifth of Earth's freshwater runoff occurs in remote, sparsely populated regions. The Amazon River in South America and the Congo River in Africa carry a large fraction of the world's freshwater runoff, but there's no easy way for humans to make use of it. This leaves about one-third of the world's runoff, about 12,500 km³ per year, accessible to human population centers.

Currently, humans use more than half of this accessible runoff. The largest fraction of this, more than 20%, is for crop **irrigation,** a system for artificially delivering water to crops so that they can grow in areas with too little precipitation to support them otherwise. The next largest fraction of accessible runoff is used for diluting sewage and other wastes, maintaining rivers as shipping channels, providing for recreation, and sustaining fish populations. We call these **instream uses,** that is, uses within the river or stream channel. Industries and municipalities combined use about 10% of accessible runoff. Finally, a little more than 2% of accessible runoff is lost through evaporation from flood control and water storage reservoirs (**Figure 6.8**).

Arid Regions Have Maxed Out Their Water Supplies

Arid and semi-arid regions make up approximately one-third of Earth's land surface and support about one-fifth of the global population. Much of the water supply in these areas comes from **dams,** which capture runoff from streams and rivers and store it in reservoirs to ensure a consistent supply of water. This water is diverted to cities to provide water for drinking, sanitation, and industrial

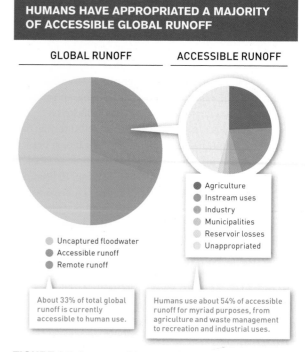

HUMANS HAVE APPROPRIATED A MAJORITY OF ACCESSIBLE GLOBAL RUNOFF

GLOBAL RUNOFF | ACCESSIBLE RUNOFF

- Agriculture
- Instream uses
- Industry
- Municipalities
- Reservoir losses
- Unappropriated

- Uncaptured floodwater
- Accessible runoff
- Remote runoff

About 33% of total global runoff is currently accessible to human use.

Humans use about 54% of accessible runoff for myriad purposes, from agriculture and waste management to recreation and industrial uses.

FIGURE 6.8 As we consider renewable water supplies, we need to factor in the practical consideration of location of runoff relative to human populations. Much of the total runoff from land comes in the form of unpredictable flooding and in flow in remote rivers, such as those flowing into the Arctic Sea. (Data from Jackson et al., 2001)

uses and to farms for irrigation. As the human populations in water-scarce regions continue to grow, these regions grapple with water supply problems (**Figure 6.9**).

Droughts and Floods Make Water Planning Difficult

A **flood** occurs whenever a river or stream overflows its banks and inundates the part of the surrounding landscape called the **floodplain,** the area of land that stretches from the water channel to the valley walls. The higher portions of the floodplain may be flooded once per century or even less frequently. Though they deliver abundant water, floods may damage water supply systems, including dams, diversions, and pipelines, as well as contaminate drinking water supplies. At the other extreme, a **drought** is an extended period of dry weather that damages crops, impairs the functioning of natural ecosystems, and causes water shortages for human populations. When the weather cycles between the extremes of flood and drought, it is often difficult to deliver enough water to cities and farms in different parts of the country (**Figure 6.10**).

Colorado River Basin

The Colorado River Basin encapsulates the challenges created by drought cycles and water shortages in arid

EARTH'S ARID REGIONS

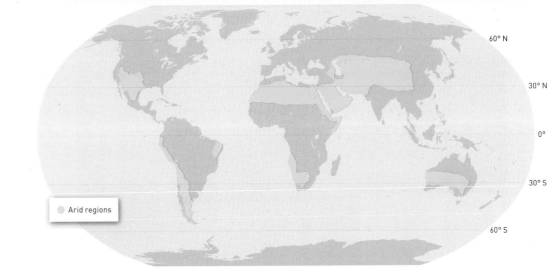

(Reto Stöckli, NASA Earth Observatory)

FIGURE 6.9 Water supplies are severely restricted for the people living in arid regions, which cover a large fraction of Earth's land surface.

regions. At the beginning of the 20th century, the Colorado River, which drains an area of nearly 650,000 km², or about 8% of the 48 contiguous United States, was a wild river, little modified by human activity or structures. Today, 265 dams greater than 5 meters in height interrupt the flow of the Colorado River, making it possible to provide water for millions of people living over a vast area

(**Figure 6.11**). This system supplies water for people living in seven U.S. states and two states in Mexico, including irrigation water for approximately 1.5 million hectares (3.7 million acres) of farmland and water to support the estimated 25 million people living in the region. The Colorado River has become so heavily drawn upon that it now rarely reaches the Gulf of California.

DROUGHT CAN BE HIGHLY REGIONAL

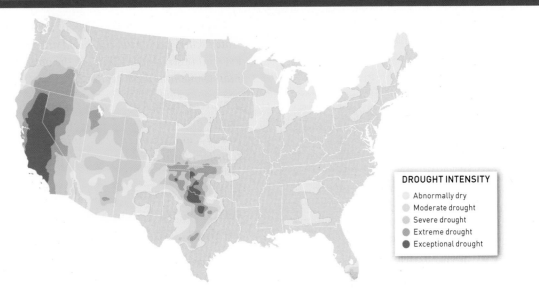

DROUGHT INTENSITY
- Abnormally dry
- Moderate drought
- Severe drought
- Extreme drought
- Exceptional drought

FIGURE 6.10 There was a great amount of variation in drought conditions across the United States during the drought of 2015, which was the third of three consecutive record droughts in California. (Data from United States Drought Monitor http://droughtmonitor.unl.edu/)

FIGURE 6.11 The Colorado River carving the lands of western North America over the course of millions of years has produced some of the most spectacular landscapes on Earth, the best known of which is the Grand Canyon. Spectacular landscapes and a pleasant climate have attracted a large and growing human population, which is on a collision course with water shortages.

THE COLORADO RIVER BASIN: AN ARID REGION WITH A RAPIDLY GROWING HUMAN POPULATION

(NPS Photo)

How would you allocate the water of the overappropriated Colorado River flow?

A complex combination of legal documents, including international treaties, regulates the appropriation of water within the basin among the affected U.S. states,

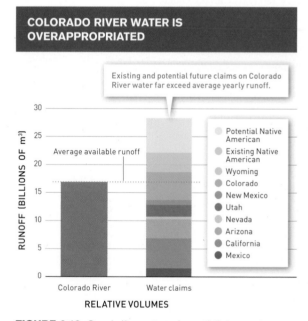

COLORADO RIVER WATER IS OVERAPPROPRIATED

Existing and potential future claims on Colorado River water far exceed average yearly runoff.

FIGURE 6.12 Over half a century of negotiations, court decisions, and international treaties established the present recognized claims to Colorado River water. Potential future claims and changes in annual runoff may require further negotiations of how to allocate Colorado River water, especially as forecasts of future available runoff become more accurate. (Data from Gelt, 1997; Bureau of Land Management, 2008)

Native American tribes, and Mexico. Because early appropriations were based on estimates of average runoff that were made during an unusually wet period in the Colorado River Basin, there is a disparity between supply and allocation (**Figure 6.12**). Conflicts over these limited water supplies are expected to increase with the growing population in the region and continued climate change.

⚠ Think About It

1. It is clear that water supplies in the historically water-scarce regions of the world are becoming even more stretched. What are the options for water management in these regions?

2. Water is removed from the Colorado River to supply the needs of arid areas such as Las Vegas, Nevada, and southern California. Critique this transfer of water, including both the pros and cons.

6.5 Groundwater is being depleted faster than it is replenished

When there is no easily accessible runoff, people drill wells and tap into the groundwater. Approximately one-fourth of Earth's population relies on groundwater, which accounts for approximately 99% of liquid freshwater on Earth (**Figure 6.13**). In parts of North Africa, groundwater is being pumped at twice the rate it is being replenished

IN MANY REGIONS, WE ARE "MINING" GROUNDWATER

(NASA/GSFC/METI/ERSDAC/JAROS & U.S./Japan ASTER Science Team)
(California Department of Water Resources)

Irrigated fields, including crop circles, in Kansas

High-volume groundwater pump

FIGURE 6.13 Agricultural production in many arid and semi-arid regions has been dependent on pumping of groundwater faster than it is recharged. This photo shows the circular irrigated fields in one such region in Kansas.

by recharge, resulting in rapidly falling water tables. As a consequence, groundwater-fed oases in the Sahara Desert that people have relied on for centuries are drying out. In 2000 the depth to groundwater (i.e., the water table) in parts of the North China plain was dropping by as many as 4 meters (13 feet) per year; in southern India, it was dropping by 2 to 3 meters (6.5 to 10 feet) per year. In the central Great Plains of the United States, groundwater levels in some areas have fallen more than 70 meters (230 feet) since the 1940s, an average of about 1 meter per year.

The problem is that while groundwater deposits are vast, approximately three-fourths of present-day groundwater was deposited long ago and is renewed on timescales ranging from hundreds to thousands of years. This "fossil" water, once used, will not be replaced for a very long time (**Figure 6.14**). Pumping and utilization of these nonrenewable supplies of groundwater represent the mining of a limited resource similar to the extraction of fossil fuels such as petroleum and coal.

Ogallala Aquifer

In the United States, some of the most productive farmland in the High Plains depends on a massive groundwater resource, the Ogallala Aquifer. This aquifer underlies approximately 450,000 km^2 (174,000 mi^2) of the High Plains, more than the combined areas of Nebraska and Kansas. Pumping of water from the Ogallala Aquifer began mainly after the 1940s, when farmers began to drill deep wells into the aquifer, with the number of wells increasing rapidly to approximately 170,000 by 1978. From 1949 to 1978, the yearly amount of water pumped from the Ogallala increased from 4.9 km^3 to 28.4 km^3. That's nearly twice the average yearly flow of the Colorado River during the 20th century.

Unfortunately, the rate of withdrawal is approximately 2.5 times greater than the annual recharge rate (see Figure 6.14). As a result, the water table across the Ogallala Aquifer has dropped an average of 4.3 meters

GROUNDWATER RECHARGE AND WITHDRAWAL IN AQUIFERS AROUND THE WORLD

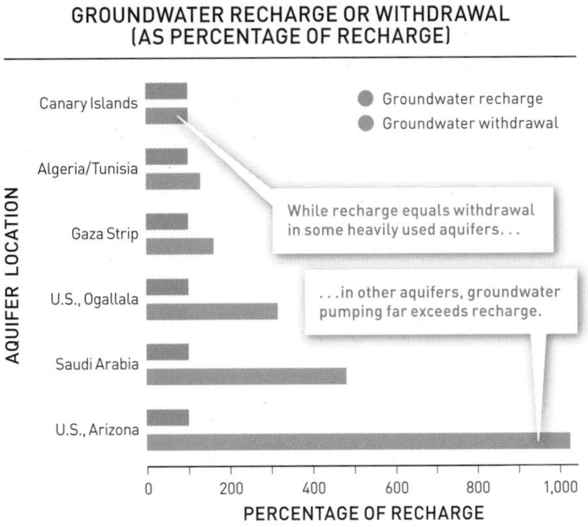
GROUNDWATER RECHARGE OR WITHDRAWAL
(AS PERCENTAGE OF RECHARGE)
Groundwater recharge
Groundwater withdrawal
AQUIFER LOCATION
Canary Islands
Algeria/Tunisia
Gaza Strip
U.S., Ogallala
Saudi Arabia
U.S., Arizona
While recharge equals withdrawal in some heavily used aquifers...
...in other aquifers, groundwater pumping far exceeds recharge.
0
200
400
600
800
1,000
PERCENTAGE OF RECHARGE

FIGURE 6.14 The sustainability of groundwater depends on the relative rates of recharge and pumping. Where pumping rates are greater than recharge, groundwater use is not sustainable. (Data from Gleick, 2000)

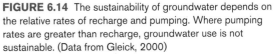

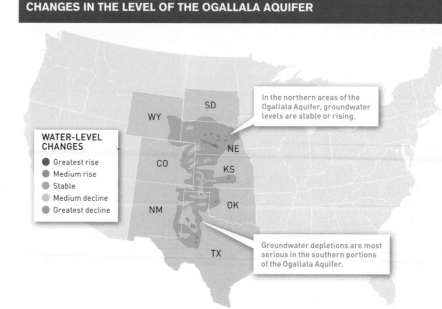

CHANGES IN THE LEVEL OF THE OGALLALA AQUIFER

WATER-LEVEL
CHANGES

● Greatest rise
● Medium rise
● Stable
● Medium decline
● Greatest decline

In the northern areas of the Ogallala Aquifer, groundwater levels are stable or rising.

Groundwater depletions are most serious in the southern portions of the Ogallala Aquifer.

FIGURE 6.15 From predevelopment to 2005, the water table in the Ogallala Aquifer has fallen more than 70 meters (230 feet) in some areas, while it has risen more than 25 meters elsewhere. The greatest water table rise has occurred along the Platte River Valley in Nebraska, which received exceptionally high rainfall from 1980 to 1999 and where the geology is highly permeable to infiltration.

(14 feet). In some areas, such as Texas and Kansas, the water table has dropped more than 30 meters (100 feet). As a result of the imbalance between rates of groundwater pumping and recharge, the Ogallala aquifer may run out of water within a century (**Figure 6.15**). We discuss changes in agricultural practice that could greatly increase the expected life of the Ogallala Aquifer in Chapter 7 (see page 220).

Subsidence and Depletion

One common physical consequence of excessive groundwater withdrawal is **subsidence** of the overlying land surface into the spaces left as water is pumped out. Such subsidence can substantially reduce the storage capacity within the aquifer and damage surface structures in rural and urban areas alike. The U.S. Geological Survey estimates that an area in the United States roughly the size of New Hampshire and Vermont has been impacted by subsidence. Particularly dramatic and sudden subsidence occurs in Florida, with the appearance of house-swallowing sinkholes. In some cases, hundreds of sinkholes have formed following the installation and pumping by a single irrigation well.

The highest rates of groundwater depletion and land subsidence in the United States in recent years have occurred in California's San Joaquin Valley. Faced with an epic drought, farmers in the San Joaquin Valley were pumping massive amounts of groundwater to compensate for the lack of water in storage reservoirs. In response, water tables in some parts of the valley fell 60 meters

(200 feet) in just two years, and land was subsiding at the unprecedented rate of 30.5 centimeters (1 foot) per year. This rate of subsidence was greater than even the historic extent of land sinking in the San Joaquin Valley (**Figure 6.16**). An unsustainable reliance on groundwater is a global-scale problem. A 2010 analysis of the global extent of **groundwater depletion,** the amount of groundwater pumped in excess of recharge, found that yearly depletion of groundwater more than doubled from 126 km³ in 1960 to 283 km³ in 2000.

⚠ Think About It

1. In response to the severe drought of the 1930s, many farms in Nebraska and elsewhere in the Great Plains were abandoned. By contrast, many fewer farms were abandoned during the 1950s drought. Why?

2. Based on Figure 6.15, which regions drawing water from the Ogallala Aquifer appear to be closest to using it sustainably? Which are using it at the least sustainable rate?

6.6 Managing water for human use threatens aquatic biodiversity

Building dams for irrigation, draining wetlands for agriculture, or filling a marsh to build an airport are

subsidence A settling or sudden sinking, in the case of sinkhole formation, of a land surface as a result of processes such as groundwater withdrawal or loss of organic matter in soil.

groundwater depletion The amount of groundwater pumped from an aquifer in excess of recharge. Groundwater depletion can result in land subsidence, which reduces the capacity of an aquifer to store water and can damage buildings and other infrastructure.

GROUNDWATER DEPLETION CAN CHANGE THE FACE OF THE LAND

FIGURE 6.16 Central California's San Joaquin Valley, subject to massive groundwater withdrawals, has set records for land subsidence in the United States.

obvious examples of how humans destroy aquatic ecosystems. Invasive species, such as Asian carp (see Figure 3.22, page 79), can compete with and displace native fish species. Straightening river channels to make them better suited for shipping has also reduced the diversity of aquatic habitats along approximately half a million kilometers of river channels around the world.

Dams and Aquatic Biodiversity

To manage water supplies, we have built dams on rivers the world over. While dams help solve problems of variable water supplies and can protect against flood

damage, they change the environment in many ways and often threaten biodiversity (**Figure 6.17**). During periods of drought, dammed rivers may dry up entirely. Even where water continues to flow below a dam, the river environment is never the same. Because reservoirs trap sediments and nutrients, they reduce the amount available to the river below the reservoir.

Dams also alter the temperature of rivers. When frigid water is released from gates at the bottom of a reservoir during the summer, the river temperature downstream can drop by several degrees. Many species require floods or low flows to complete their life cycles and are thus harmed when dams prevent those conditions. River ecologists estimate that dams and water diversions have altered more than 75% of the 139 largest rivers in the Northern Hemisphere. Within the United States alone, there are more than 75,000 dams.

Dams usually prevent fish from moving up and down the length of the river, which is especially harmful

DAMS SIGNIFICANTLY CHANGE RIVER ENVIRONMENTS

Grand Coulee dam

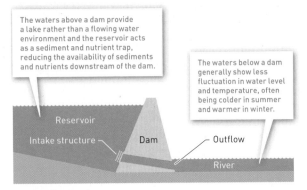

FIGURE 6.17 Dams, such as the Grand Coulee on the Columbia River, interfere with the movements of migratory fish, such as salmon and sturgeon. Dams can also alter river environments in more subtle but ecologically significant ways.

HUMAN IMPACTS ON FRESHWATER ENVIRONMENTS ENDANGER MANY SPECIES

(Ben Kiefer/UDWR)

Endangered Colorado pikeminnow

(USFWS photo by Andy Roberts)

Endangered North American freshwater mussels

(Xu Jian/Nature Picture Library)

Extinct Yangtze River dolphin

FIGURE 6.18 The Colorado pikeminnow, the largest minnow native to North America, has become endangered as a result of dam building, altered temperatures, and the introduction of non-native species to its habitat. In addition, more species of freshwater mussels are native to North America than anywhere else in the world. However, up to 70% of them are extinct or threatened by overharvesting, dam building, sedimentation, and competition with invasive species. Finally, the Yangtze River dolphin, or Baiji, one of very few freshwater dolphin species in the world, was driven to extinction in the late 20th century as a result of unintentional killing by various fishing methods, pollution, and habitat alteration.

oxbow lake A crescent-shaped lake formed on a river's floodplain by rerouting the main river channel, generally during a flood.

channelize To engineer a change to the natural form of a stream or river, including straightening, deepening, or widening the channel.

riparian The transition zone between a river or stream and the terrestrial environment, generally inhabited by a biological community distinctive from adjacent aquatic and upland communities. Riparian zones naturally flood periodically and usually have shallow water tables.

to migratory fish such as salmon (see Chapter 8). Nonmigratory species might also decline as a result of dam building. For instance, many of the native fish populations of the Colorado River, including the Colorado pikeminnow (*Ptychocheilus lucius*), declined as the river was modified for water management. The Colorado pikeminnow was more monster than minnow, weighing up to 80 pounds and reaching lengths of 5 feet. This fish had evolved into the top predator in the Colorado River over a period of millions of years but was brought to the brink of extinction in less than a century by dam building. The dams impact the pikeminnow in several ways: They create the lake-like habitats of reservoirs to which this river fish is not adapted; they lower the water temperature below dams to levels unsuitable for reproduction; and they provide refuge for non-native predatory fish species, such as striped bass, that feed on young pikeminnows.

The pikeminnow is just one example of how humans have harmed aquatic organisms through their activities (**Figure 6.18**). Approximately 20% of the freshwater fishes of the world are threatened with extinction or are already extinct. Within the United States, nearly half of the federally listed endangered vertebrate and invertebrate animals are freshwater species.

Water Management and Wetland Biodiversity

When rivers flood their banks, water spills out onto the floodplain. While floods may harm crops and destroy houses and other structures on a floodplain, they are essential to the health of wetlands. Floods disperse valuable nutrients into the soils of the floodplain and rearrange the landscape; for example, they isolate old river channels, forming riverside habitats called **oxbow lakes.** Likewise, wetland ecosystems act as natural water purifiers capable of removing or reducing the concentrations of various contaminants (**Figure 6.19**).

Building dams and regulating river flow directly harm these wetlands by preventing natural floods. A common manipulation of rivers is to **channelize** them, an engineered change to the natural form of a stream or river, including straightening, deepening, or widening the channel. While a channelized river may make navigation easier, it disrupts the natural connection between a river and its floodplain (**Figure 6.20**). In addition, the sediment-free water released from dams has a higher ability to carve away at the river channel below, causing the water table to fall—sometimes below the rooting zone for riverside trees and other plants, often resulting in the deaths of the trees. Finally, digging drainage canals and pumping out groundwater cause the water table to be lower than the adjacent river or reservoir, further reducing the chance that a river will flood and refresh its floodplain.

Riparian areas, which form the transition zone between a river or stream and the terrestrial environment, have been heavily impacted by dam building and flow regulation. These areas usually have shallow water tables and depend on regular floods in order to support a biological community distinctive from adjacent aquatic

FLOODING HELPS SUSTAIN THE HEALTH OF RIVERS AND ASSOCIATED WETLANDS

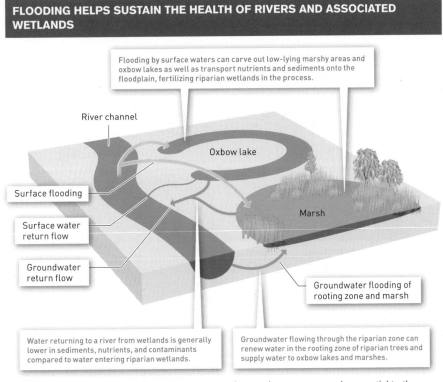

Flooding by surface waters can carve out low-lying marshy areas and oxbow lakes as well as transport nutrients and sediments onto the floodplain, fertilizing riparian wetlands in the process.

River channel

Oxbow lake

Surface flooding

Surface water return flow

Groundwater return flow

Marsh

Groundwater flooding of rooting zone and marsh

Water returning to a river from wetlands is generally lower in sediments, nutrients, and contaminants compared to water entering riparian wetlands.

Groundwater flowing through the riparian zone can renew water in the rooting zone of riparian trees and supply water to oxbow lakes and marshes.

FIGURE 6.19 Flooding, both by surface water and groundwater seepage, is essential to the natural functioning of rivers and associated wetlands.

Many wetlands were drained to reduce the incidence of mosquito-transmitted diseases, such as malaria. In such situations, how would you balance disease control and wetland diversity?

and upland communities. Connections between rivers and wetlands are especially critical in arid climates, where riparian wetlands support unusually high levels of biodiversity, compared with the surrounding landscape. However, dams and channels in those areas have reduced the frequency and intensity of flooding. Where the

demand for water to supply municipalities and agriculture has been especially great, water diversions have commonly dried river channels entirely (**Figure 6.21**). Concerns about such impacts on biodiversity and productivity have stimulated efforts at riparian and wetland restoration (see Solutions, page 181).

HUMANS HAVE GREATLY SIMPLIFIED THE STRUCTURE OF RIVERS

Channelized Rhine River, Dusseldorf, Germany

Unmodified river in Kobuk Valley National Park, Alaska

FIGURE 6.20 As we have straightened river channels to ease navigation and built riverside levees to control flooding, we have reduced the availability of habitats on which many species depend.

RIBBONS OF GREEN THROUGH DESERT LANDSCAPES

San Pedro River, Arizona

(Annie Griffiths Belt/Getty Images)

Salt River, Phoenix, Arizona

(Tim Roberts Photography/Shutterstock)

FIGURE 6.21 Riparian forests and wetlands are generally areas of exceptionally high primary production and species richness in arid and semiarid landscapes, such as here along the San Pedro River in southern Arizona. Periodic flooding is essential for maintaining healthy riparian areas in arid lands, since many key riparian plant species require flooding for reproduction. By contrast, in many urbanized landscapes in arid and semiarid regions, such as the Salt River in Phoenix, riparian wetlands and forests have been nearly eliminated as watercourses have been dried and natural habitat replaced by engineered structures.

⚠ Think About It

1. Should the value of ecosystem services lost as a consequence of water management be considered as part of the price of water delivered to consumers? Explain.

2. Municipalities and landowners generally have a legal right to use a certain amount of water per year. Should there be legislation limiting human water use and requiring minimum river flows to protect endangered species?

3. In the event of dwindling water supplies, what type of use should be given higher priority and which should be given lower priority? Why?

6.3–6.6 Issues: Summary

Approximately 2.6 billion people around the world lack access to enough water to meet their basic needs. Continued population growth, coupled with natural variation in water supplies, will intensify the potential for conflict among water users, especially in arid regions. Only about one-third of freshwater runoff is accessible to human population centers, of which a majority has been appropriated for human uses, including agriculture, waste dilution and disposal, shipping, recreation, and residential and industrial uses. While there are extensive deposits of groundwater, such reservoirs are slowly renewed and are not generally a sustainable supply on timescales that match human needs.

Alteration of freshwater environments around the world threatens many aquatic species, ranging from mollusks to fish. Dams, built to store water and control floods, change river ecosystems in many ways, including blocking the passage of migrating fish, altering historical patterns of flow, changing water temperatures, and reducing the availability of nutrients and sediments in river sections below dams. By reducing flooding, dams also decrease the connection between rivers and riverside wetlands and forests, which make particularly valuable contributions to biodiversity and primary production in arid and semi-arid landscapes.

6.7–6.10 Solutions

How would you like to drink a nice, tall glass of toilet water? In May 2014 the city of Wichita Falls, Texas, began recycling 5 million gallons of sewage, treating it, and sending it right back into the faucets of its residents. The program was a last-ditch effort by the city to cope with a devastating drought, but it's not the first city to do so—that honor goes to Windhoek, Namibia—and it certainly won't be the last. With rivers overallocated, groundwater being depleted, and aquatic ecosystems dying off, finding innovative solutions to our water needs represents a fundamental challenge. The good news is that there are fixes, large and small, that can have a major impact on water consumption. And they don't all involve drinking toilet water. In this section, we focus on a variety of ways to meet domestic water supply needs; in Chapter 7, we address the water efficiency of agriculture, perhaps the biggest challenge of all.

6.7 Water conservation can increase water use efficiency substantially

It pays to fix dripping faucets. Water losses from leaks in water supply systems can range anywhere from 10% to 30% of the total water volume pumped. These losses occur in both modern and old water systems and in cities small and large. The water supply system of Mexico City loses enough water through leaks to supply Rome with all the water it needs!

Water Conservation in a Large City

If all the water supply and wastewater lines in New York City were laid out, they would stretch to California and back, twice (**Figure 6.22**). Keeping them in shape

TRY TO IMAGINE THE NETWORK OF WATER LINES

(Songquan Deng/Shutterstock)

FIGURE 6.22 The size and structural complexity of New York City present the managers of its water, which is delivered to every residence and business on every floor of every building, with a substantial challenge.

is a monumental task. In the 1970s, flows of water and wastewater in the New York City system began to exceed safety limits and push pollution maximums. In response, the city decided to conserve water.

First on their list of priorities was a public education campaign to inform water users of the need for, and means to achieve, conservation. The campaign included over 200,000 door-to-door informational visits, during which residents were offered free low-volume showerheads and a free leak inspection. In addition, the city offered rebates on low-volume flush toilets. Through this effort, the city eventually replaced over 1.3 million high-volume, conventional toilets—which use about 13.2 liters (3.5 gallons) per flush—with low-volume toilets, which use 6 liters (1.6 gallons) per flush. As a follow-up to the education program, the city installed water meters in all unmetered residences to help the city and individual residents track water use. Another major effort focused on detecting leaks in the water distribution system. The key to this effort was installing leak detectors on all the main water lines in the city.

New York City's conservation efforts saved nearly 1.2 billion liters (320 million gallons) of water per day (**Figure 6.23**). Surprisingly, the greatest volume of water saved was associated with installing water meters in unmetered residences. It appears that when given the means to keep track, many residents reduced their water use.

What are some implications of water conservation achieved by simply making water meters accessible to individual water users?

How could you personally contribute to water savings in your community? Prioritize the steps you would take to conserve water.

Gauging Conservation Progress

Other municipal conservation programs around the world have resulted in impressive reductions in water use. From 1988 to 2013, water conservation efforts in Albuquerque, New Mexico, reduced daily per capita

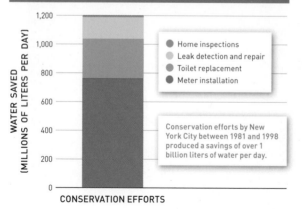

FIGURE 6.23 New York City's water savings were achieved through a combination of installing water meters and low-flow toilets in residences, finding and repairing leaks, and home inspections. (Data from U.S. EPA, 2002)

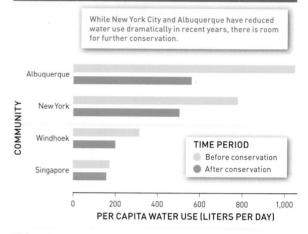

FIGURE 6.24 Four communities (New York City; Albuquerque, New Mexico; Windhoek, Namibia; and Singapore) that have ongoing programs promoting water conservation have achieved significant water savings. (Data from http://www.cabq.gov/albuquerquegreen/green-goals/water, http://www.nyc.gov/html/dep/html/home/home.shtml, http://www.windhoekcc.org.na/index.php, http://www.pub.gov.sg/Pages/default.aspx)

water use from 1,056 liters (279 gallons) to 560 liters (148 gallons). Meanwhile, the residents of Windhoek, the bone-dry capital of Namibia in southwestern Africa, decreased their consumption from 309 liters (82 gallons) to 196 liters (52 gallons) per capita per day (**Figure 6.24**).

While the results of these conservation efforts are impressive, water consumption in these three cities far outpaces that in the Republic of Singapore, a tiny, densely populated island faced with severe water constraints. Its 20 major reservoirs supply only half of its water needs, and it must purchase the rest of its water from nearby Malaysia. Consequently, it has focused its efforts on conservation and reduced per capita water consumption from 172 liters (45.4 gallons) per day in 1995 to 154 liters (40.7 gallons) per day in 2013. The Singapore water authority believes that further progress is possible and aims to reduce consumption to 147 liters per day by 2020 through its water conservation campaign. One suggestion made by the Singaporean Department of Environment and Water is for all residents to save water by reducing their length of showers by 1 minute.

Conservation by Commercial and Institutional Buildings

Commercial and institutional buildings, including office buildings, hotels, commercial laboratories, and university buildings, are major users of water. Conservation efforts have produced large water savings and reduced operating costs in all these types of buildings. Of these, water conservation by hotels may be the best documented and the most familiar. The hotel industry is responsible for 15% of commercial and institutional water consumption in the United States. Three-fourths of hotel water use is for restrooms, laundry, landscaping, and kitchens. Faced with limited water supplies and increasing prices of water, the hotel industry has taken steps to be more efficient in their water use. A first step, as in private residences, has been to install more water-efficient toilets, showerheads, and faucets in guest rooms and to install more efficient laundry and dishwashing equipment. Another essential element in hotel conservation efforts is to encourage guests to conserve water by, for example, taking shorter showers and reusing bath towels. Hotels are also replanting their landscapes with more drought-tolerant plants and installing more efficient watering systems.

These efforts have produced impressive water savings. The Kalaloch Lodge in the Olympic National Park in Washington set a goal of reducing its water use by 40% by the year 2020. In addition to making its water system more efficient, Kalaloch Lodge issued a 5-minute shower challenge to guests and equipped all rooms with a timer set to 5 minutes. Their innovation paid off, producing a 46% reduction in water use between 2011 and 2014. Similarly impressive results have been achieved in large hotels in urban centers. For instance, by simply installing water-efficient fixtures, the 470-room Hilton Palacio del Rio Hotel in San Antonio, Texas, reduced its annual water use by 49% between 2004 and 2011. As a result, the hotel reduced its yearly water use by 26 million gallons and saved about $160,000 annually in water, sewer, and energy costs.

⚠ Think About It

1. How do the levels of water consumption in the four communities shown in Figure 6.24 compare with the proposed minimum consumption proposed as a water right (see page 163)?

2. What are the water use issues in your community and how are they being addressed?

3. How are water conservation programs aimed at private residences and at hotels similar? How might they differ to be most effective?

6.8 Reclamation and recycling are saving water throughout the world

Every liter of water that is reused reduces the amount of water drawn from surface water or groundwater supplies. When communities practice **water recycling,** they substitute engineered treatment, or purification,

of water for the processes that occur naturally during the hydrologic cycle. Treated wastewater can be used for a variety of purposes, including industrial processes, irrigation, recharging groundwater supplies, restoring wetland ecosystems, or even drinking water. In general, the process of treating wastewater to make it safe for reuse is called **water reclamation.** The more rigorous the treatment, the wider range of uses to which recycled water can be put (**Figure 6.25**).

Santa Rosa, California

Santa Rosa, California, is a city of 170,000, about an hour's drive north of San Francisco and 50 kilometers (31 miles) inland from the Pacific Ocean. The heart of the Sonoma Wine Country, the region around Santa Rosa has a Mediterranean climate, with cool rainy winters and warm dry summers. During the dry summers in the area, the flows of rivers and streams around Santa Rosa drop to very low levels. Because environmental regulations do not permit Santa Rosa to discharge its wastewater into rivers at a volume of more than 1% of river flow, the city had to find another way to handle its wastewater in summer: recycling (**Figure 6.26**).

The Santa Rosa water reclamation facility provides **tertiary treatment** of its wastewater, including disinfecting the water by exposing it to ultraviolet light to kill any pathogens that it may contain. This reclaimed water, which is used for a variety of purposes, meets rigorous standards set by the California Department of Health for full contact with the human body, irrigation of food crops and landscaping, and consumption by livestock. Some of the water is used to irrigate 2,590 hectares (6,400 acres) of land, including pastures, vineyards, vegetable fields, and

water recycling Using treated wastewater for beneficial purposes, including industrial processes, irrigation, recharging groundwater supplies, restoring wetlands and aquatic ecosystems, and augmenting drinking-water supplies.

water reclamation Any process of treating wastewater to make it safe for reuse or recycling.

tertiary treatment Advanced treatment of wastewater, which follows primary and secondary treatment, that removes dissolved organic chemicals, nitrogen, phosphorus, several other dissolved salts, and pathogens.

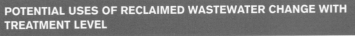

POTENTIAL USES OF RECLAIMED WASTEWATER CHANGE WITH TREATMENT LEVEL

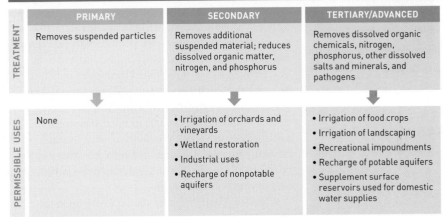

	PRIMARY	SECONDARY	TERTIARY/ADVANCED
TREATMENT	Removes suspended particles	Removes additional suspended material; reduces dissolved organic matter, nitrogen, and phosphorus	Removes dissolved organic chemicals, nitrogen, phosphorus, other dissolved salts and minerals, and pathogens
PERMISSIBLE USES	None	• Irrigation of orchards and vineyards • Wetland restoration • Industrial uses • Recharge of nonpotable aquifers	• Irrigation of food crops • Irrigation of landscaping • Recreational impoundments • Recharge of potable aquifers • Supplement surface reservoirs used for domestic water supplies

FIGURE 6.25 Wastewater receiving secondary treatment is suitable for uses where there is little chance of direct human contact. Tertiary treatment creates the possibility of much wider uses because it is free of dissolved organic chemicals and pathogens.

RECLAMATION IS CENTRAL TO WATER MANAGEMENT IN SANTA ROSA, CALIFORNIA

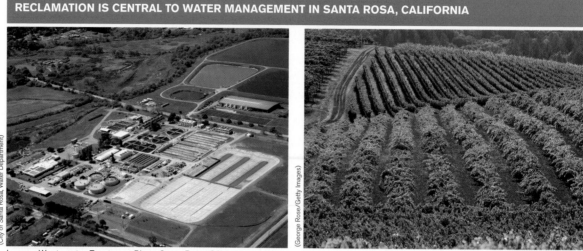

Laguna Wastewater Treatment Plant, Santa Rosa, CA

(City of Santa Rosa, Water Department)

Vineyards near Santa Rosa, CA

(George Rose/Getty Images)

FIGURE 6.26 The Laguna Wastewater Treatment Plant in Santa Rosa, California, provides tertiary treatment to 66 million liters of wastewater per day. Vineyards such as these are some of the crops irrigated by water reclaimed at the Santa Rosa wastewater treatment facilities.

urban landscapes. During the moist winter months when irrigation is unnecessary, Santa Rosa discharges reclaimed water into a tributary to the nearby Russian River and into storage reservoirs.

In 2003 Santa Rosa began pumping water through a pipeline to the Geysers steam field for injection into steam wells to generate electricity from geothermal energy (see Figure 10.17, page 306). In 2008 nearly 50,000 m³ of recycled water was delivered each day to the Geysers steam field, enough to generate sufficient electricity to supply 100,000 households. The water recycling program in Santa Rosa is one of many in California, where the use of recycled water has grown rapidly (**Figure 6.27**).

Windhoek, Namibia

The country of Namibia occupies some of the driest regions of southwest Africa. Most of its territory lies within the Namib Desert, one of the most spectacular deserts on Earth (**Figure 6.28**). Because of its arid climate and perennial water shortages, Namibia has had to be innovative in its use of its limited freshwater supplies. The capital, Windhoek, was founded on a natural spring, but it is a rapidly growing city whose size has increased five-fold from just over 60,000 to 300,000 between 1970 and 2008 (**Figure 6.29**). Windhoek obtains its water primarily from local aquifers and reservoirs that trap the occasional flows of nearby intermittent rivers, which flow only during heavy rainstorms.

When local water supplies started to dry up in the 1960s, Windhoek began to work seriously on plans for the sustainable management of its local water. The nearest abundant source of freshwater is the Okavango

River, 650 kilometers (400 miles) north of the city. As an alternative to expensive, long-distance water diversion, Windhoek turned to water reclamation and recycling to help alleviate its water shortage. The first water reclamation plant began producing clean, recycled water in 1968. The capacity of that first reclamation plant was 1.7 billion liters of water per year. Upgraded in 2002, the plant can now produce 7.7 billion liters of

PAST, PRESENT, AND PROJECTED RECLAIMED WATER USE IN CALIFORNIA

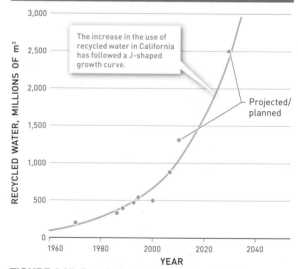

The increase in the use of recycled water in California has followed a J-shaped growth curve.

Projected/planned

FIGURE 6.27 Recycled water is viewed as critical to sustaining California's rapidly growing population and its exceptionally valuable agricultural productivity, which depends heavily on irrigation. (Data from Gleick, 2000; www.owue.water. ca.gov/recycle/)

reclaimed water per year. Like the city of Santa Rosa, Windhoek puts its reclaimed water through a system of tertiary treatment, including disinfection, and tests it to make sure it is free of disease-causing bacteria, viruses, and toxic chemicals. But Windhoek goes beyond Santa Rosa in using reclaimed water as drinking water. During drought periods driven by periodic El Niños, reclaimed water can supply up to 30% of domestic water needs.

Drinking treated wastewater sounds unappealing, but treated wastewater has long been released into rivers by one upstream city only to enter the drinking water supply of the next city downstream. For instance, wastewater from Dallas and Fort Worth, Texas, ends up in the Trinity River, which feeds Houston's water supplies.

Industrial Water Reclamation

Worldwide, industrial water use exceeds that of municipalities (see Figure 6.8, page 164), suggesting significant opportunities for water reclamation and reuse by industries. In fact, many industries are doing just that, including the paper, textile, and food and beverage industries. Their primary goal is to save money, but an important by-product is reduced pressure on water resources. For instance, the Bear Republic Brewing Company, a beer brewery in water-stressed California, recently partnered with Cambrian Innovation, a bioenergy and water provider. Breweries

NAMIBIA IS A COASTAL NATION IN SOUTHWEST AFRICA

(Galyna Andrushko/Shutterstock)

FIGURE 6.28 Most of Namibia is highly arid desert, with few freshwater resources. The massive dunes shown here are in an area of the Namib Desert called the "sand sea."

use large volumes of water to produce beer and generate substantial amounts of solid and liquid waste. Traditionally, breweries separated their solid wastes, consisting mainly of spent grains, from liquid waste. They

NAMIBIA: A PIONEER IN THE USE OF RECYCLED WATER

(© Friedrich Stark/Alamy)

FIGURE 6.29 Rapid population growth in Namibia's capital city, Windhoek, motivated the local government to develop wastewater reclamation and reuse to make more efficient use of its limited water resources.

SCIENCE　　ISSUES　　SOLUTIONS

give the spent grains to farmers to feed their livestock, thus avoiding waste disposal fees.

The liquid waste, however, is less desirable. It contains large concentrations of dissolved organic compounds, which is costly to treat on site or to dispose of through a waste treatment facility. Now, Cambrian Innovations has a method to convert this liquid into two components: methane gas and clean water. The gas can be burned for electricity. The water can be used to clean brewery equipment. The result is a savings of water, energy, and money. Similar opportunities for water reclamation and reuse are being exploited in many other industries.

⚠ **Think About It**

1. How are water-recycling systems and the hydrologic cycle the same? How are they different?

2. Reclaimed and recycled water in most places is mainly put to uses other than drinking, even where recycled water more than meets health standards and is superior in quality to other water sources. The reluctance to drink recycled water has been referred to as the "yuck factor." To what do you attribute this resistance to drinking recycled water?

3. Relative to Question 2, how would you go about overcoming this resistance to drinking recycled water? How would you safeguard public health?

6.9 **Desalination taps Earth's largest reservoir of water**

It's a frustrating fact of nature that humans cannot survive on seawater. Although salt is an essential mineral, the problem is that seawater is approximately 3 times as salty as our blood, and drinking it would quickly overwhelm our kidneys and lead to death. While **seawater** may have 35,000 milligrams per liter of salt, drinking water should contain less than 500 milligrams per liter (**Figure 6.30**). **Desalination,** the process of turning saltwater into freshwater, is one solution to finding drinking water where natural sources are scarce. In 2012 there were more than 16,000 desalination plants in operation, generating over 28 km³ of freshwater per year. That's only 1% of freshwater used by humans from all sources, but the small flow of freshwater that it does produce is critical for sustaining some human communities.

The earliest approach to desalination involved **distillation,** a process that uses heat to evaporate water from seawater or **brackish water** (natural waters with a salt content intermediate between freshwater and seawater) and then condenses (and captures) the

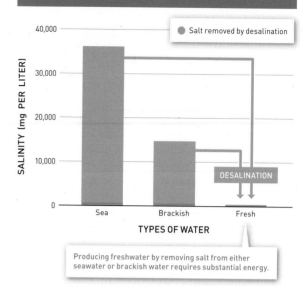

FIGURE 6.30 The red area of each bar shows the amount of salt that must be removed to convert (average) seawater or (median) brackish water to freshwater with salinity less than 500 milligrams per liter. (Data from Art, 1993)

resulting salt-free water vapor. The outputs from the distillation process are freshwater and concentrated salt brine (**Figure 6.31**). In 2012 distillation accounted for approximately 15% of water produced by desalination worldwide. Because it requires a lot of energy, this approach to desalination is generally expensive. However, various engineering techniques have improved the efficiency of the process a great deal.

For example, it is possible to use the heat produced as a by-product of electrical generation as a source of heat energy for desalination. This approach is called **cogeneration,** a term generally referring to the use of a single source of energy for multiple purposes. Cogeneration can be combined with other techniques, such as reducing the atmospheric pressure in the distillery. Because water boils at a lower temperature at lower atmospheric pressure, less energy is needed to evaporate it. Still, because distillation processes are so energy-expensive, they are used principally in energy-rich, water-scarce countries of the Middle East and North Africa. These countries are responsible for more than half the annual production of desalinated water.

Reverse osmosis is a more energy-efficient desalination process that uses semipermeable membranes to separate salts and water. The membranes are permeable to water, allowing water molecules to pass across, but they are not permeable to salts. As in distillation, reverse osmosis creates two streams of water: (1) freshwater, which can

---

seawater   Ocean water and the water of seas, such as the Caribbean and Mediterranean seas. The salinity of seawater averages about 34,000 mg/l (34 g/l), but ranges from 30,000 to about 40,000 mg/l (30 to 40 g/l).

desalination   The process of removing salts from seawater or brackish water to form freshwater.

distillation   A desalination process that uses heat to evaporate water from seawater or brackish water and condenses the resulting salt-free water vapor to produce freshwater.

brackish water   Natural waters with a salt content intermediate between freshwater and seawater, commonly occurring near the mouths of rivers where freshwater and seawater mix.

cogeneration   Generally refers to the use of a single source of energy for multiple purposes.

reverse osmosis   A desalination process that uses selectively permeable membranes and pressure to separate salts and water.

DESALINATION BY DISTILLATION: EVAPORATION AND CONDENSATION OF WATER

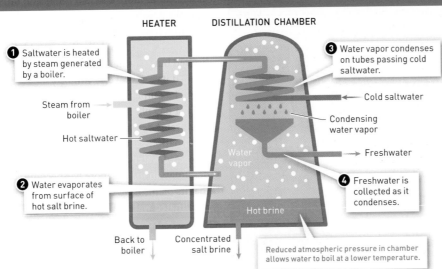

FIGURE 6.31 Distillation is the oldest method of desalination. The basic technology is well developed and dependable. Its major shortcoming is that the process requires a lot of energy. Lowering the atmospheric pressure in distillation chambers reduces energy costs for distillation. To increase water yield, several distillation chambers may be connected in series, each with a lower atmospheric pressure.

be up to 99.7% pure water that is piped into the water distribution system, and (2) higher-salinity water that is discharged back into the environment. The energy needed to drive the process is used mainly for pressurizing the

water fed into the system (**Figure 6.32**). As filters have improved, the pressure needed to run these systems has declined substantially, thereby reducing energy costs. However, there are also environmental impacts from the disposal of the salt brine, which has about twice the salinity of seawater, and the potential to harm marine life that gets sucked into the intakes.

The Tampa Bay Desalination Plant

Tampa, Florida, has turned to desalination to supplement its traditional water supplies, opening the largest reverse osmosis plant in the United States in 2007 (**Figure 6.33**). Now fully operational, the plant has the capacity to produce 95 million liters (25 million gallons) of freshwater per day, approximately 10% of freshwater consumption for the Tampa Bay region.

Some unique local and regional factors reduce the financial costs of the Tampa Bay desalination plant. First, because the salt content of the water in Tampa Bay averages approximately 25% less than full-strength seawater, the energy cost is reduced. Second, the plant is located next to an electrical generation plant that supplies low-cost power. The power plant also shares a variety of infrastructure and services with the desalination facility. For instance, the salt brine produced by the plant is diluted in the cooling water released by the power plant at a 70:1 ratio. As a result, no increases in salinity have been observed in Tampa Bay, even when the desalination plant is operating at full capacity.

OSMOSIS MOVES WATER ACROSS A SELECTIVELY PERMEABLE MEMBRANE

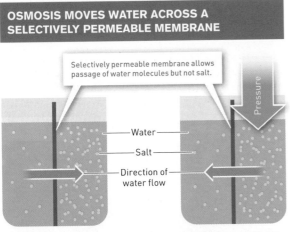

FIGURE 6.32 During the natural process of osmosis, water moves from the side of a selectively permeable membrane on which the concentration of salts is lower to the side on which the concentration of salts is higher. Reverse osmosis changes ("reverses") this normal direction of flow by applying physical pressure to the side of the membrane with higher salt content, forcing water to move toward the side of the membrane with the lower salt content.

LAKE POWELL STORES SEVERAL YEARS OF COLORADO RIVER FLOW

(Airphoto–Jim Wark)

FIGURE 6.36 The Colorado River backing up behind Glen Canyon Dam formed Lake Powell, which largely eliminated spring floods and traps the sediment they once carried through the Grand Canyon.

of 1996, with others following in 2004, 2008, and 2012. During controlled flooding, river managers leave the floodgates open for about 3 to 5 days and time the floods to follow inputs of sand from tributaries of the Colorado River. For example, the Paria River, which flows into the Colorado River downstream from Glen Canyon Dam, washed more than 1 million tons of sediment into the river a few months before the 2004 controlled flood. Controlled flooding has been successful at restoring sandbars to the Colorado River shoreline within the Grand Canyon (Figure 6.37).

Restoring River and Wetland Structure

The Kissimmee River in central Florida was altered dramatically in the 1940s during efforts to control flooding on private property. Flood control efforts converted the meandering channel to a straight, 90-kilometer-long canal, which was 9 meters (30 feet)

deep by 100 meters (328 feet) wide (Figure 6.38a). Because the canal was completely cut off from the floodplain, the exchange of nutrients and organic matter between the river and floodplain was eliminated. The impact on biodiversity was dramatic. For instance, the number of waterfowl declined by 90% and the number of nesting bald eagles fell by 75%. In addition, the canal became a sink for organic matter and dissolved oxygen levels (the amount of oxygen molecules in the water) fell, leading to the decline of a once thriving sport fishery for largemouth bass, which require higher oxygen levels.

The Kissimmee restoration project, which began in 1992, is the largest ecosystem restoration attempted to date. When completed in 2015, the project will have restored 100 km² (40 mi²) of river–floodplain ecosystem, containing over 8,000 hectares (20,000 acres) of productive wetlands. Restoration goals also include reestablishing 61 kilometers (40 miles) of meandering river channel (Figure 6.38b).

The response of the biological community to restoration has been rapid and impressive. Numbers of wading birds in the restored river and floodplain areas have increased five-fold. Numerous duck and shorebird species that had been absent from the landscape for decades have returned. Organic deposits in the riverbed have been reduced by over 70% and dissolved oxygen levels have increased. Those increases in dissolved

CONTROLLED FLOODS IN THE GRAND CANYON ACHIEVED PREDICTED RESULTS

(USGS/Matt Kaplinski/Northern Arizona University)

FIGURE 6.37 One of the major goals of controlled flooding below the Glen Canyon Dam was to sustain sandbars in the Grand Canyon, where they provide nursery habitat for native fish and camping sites for rafters and backpackers. The sandbar shown in this photo was formed during the controlled flood in the spring of 2004.

KISSIMMEE RIVER STRUCTURE: LOST AND RESTORED

a. Channelized Kissimmee River

b. Restored reach of Kissimmee River

FIGURE 6.38 Flood control on the Kissimmee River involved replacing the complex channel structure of the river with a straight canal, which severed the former connection between the river and its floodplain. Restoration of the Kissimmee River included filling sections of the channelized river and rerouting river flow through the original channels, which once again overflow onto the surrounding floodplain and wetlands during heavy rains.

oxygen and habitat complexity have been accompanied by a resurgence of populations of largemouth bass and other sunfish.

⚠ Think About It

1. The cost of restoring the Kissimmee River has been projected at about $500 million. What are the potential benefits?

2. Why do water managers need to coordinate controlled flooding with sediment inputs from Colorado River tributaries below the Glen Canyon Dam?

3. How might dam building and prevention of the historical floods on the Colorado River give an advantage to invasive non-native fish species in competition with native fish species?

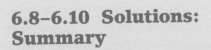

6.8–6.10 Solutions: Summary

Water conservation can be very effective at reducing water use—at both the individual and city scale. A water conservation campaign in New York City has saved over 1 billion liters per day. Water recycling is also saving water around the world. Wastewater can be made safe for recycling through a variety of treatments collectively called water reclamation. Depending on the level of treatment, recycled water is being put to a wide range of uses around the world, including industrial processes, irrigation, recharging groundwater supplies, restoring or augmenting wetland or aquatic ecosystems, or even drinking water. Desalination, which removes salt from water, can be used to convert seawater or brackish water into freshwater. Although desalination is energy-expensive, developments in cogeneration and reverse osmosis are rapidly increasing efficiency and decreasing costs.

Restoring aquatic ecosystems involves establishing minimum stream flows and reintroducing periodic flooding. For example, controlled flooding is being used on the Colorado River to sustain critical sandbar habitat within the Grand Canyon. The Kissimmee River in central Florida was "channelized" to prevent flooding, which had disastrous consequences for the functioning of the river–floodplain ecosystem and associated biodiversity. Restoration has produced rapid recovery of the Kissimmee River system.

Answer the following questions for each chapter section and then answer the Central Question.

Central Question: How can we meet human needs for freshwater, while avoiding or reducing environmental impact?

6.1–6.2 Science

- How does the hydrologic cycle move water around Earth?

- What effect does the El Niño Southern Oscillation have on the hydrologic cycle?

6.3–6.6 Issues

- How much water do humans require to live and should that be considered a basic human right?

- How do humans use water and what are factors that affect water's availability?

- What is the current state of groundwater and what affects its availability?

- How does managing water affect aquatic biodiversity?

Water Resources and You

Water is essential to our survival and well-being—it literally flows through our individual lives on a daily basis. As a result, we all have abundant opportunities to contribute to solving the many water-related issues we face.

☐ *Learn about water supply, consumption, and issues in your community.*

Learn about your community water supply in detail. What is the source of water in your community? Is it mainly groundwater, surface water, or a fairly even mix of the two? How does water consumption in your community compare with other communities in your region and elsewhere across the country and world? What are the major water supply issues in your community? What are the major water supply issues in your region? If possible, tour the water treatment facilities in your community to learn about where the water you use goes and how it is treated.

☐ *Actively support water conservation programs in your community.*

As you inform yourself regarding water supply and infrastructure, you will build a foundation for becoming involved in local and regional water issues. As opportunities arise, you may lend your voice to initiatives designed to save water, such as water recycling, upgrading and maintenance of the community water distribution system, or planting of drought-resistant landscaping plants in public spaces. If no such

water-saving initiatives now exist and you see a need for them, you might consider suggesting one or more through local media or by writing to the mayor or city council in your community.

☐ *Commit yourself to water conservation in your daily life.*

Consider how you might conserve water in your everyday life. As a student, you may be able to find opportunities to encourage the installation of low-flow toilets or showerheads on campus, if your campus has not already done so. Wherever you live (e.g., in a dormitory, apartment, or house), you can report or fix leaky faucets or bathroom fixtures. The average time spent in the shower in the United States is 8 minutes, which expends 68 liters (18 gallons) of water when using a standard 8.3 liters (2.2 gallons) per minute showerhead. Use a timer on your smartphone or shower timer to limit the time you shower to 5 minutes—this can save over 25 liters (6.6 gallons) of water every time you shower!

☐ *Become involved in river or wetland restoration.*

Explore potential restoration opportunities with federal, state, or local agencies or with nongovernmental conservation organizations in your area. Your active participation in restoration as a volunteer will help reduce impacts on aquatic biodiversity as we supply water for human populations. It will also be an opportunity to explore key aspects of aquatic biodiversity firsthand.

6.7–6.10 Solutions

- **What are some water conservation approaches and how do they affect water availability?**

- **Where have water reclamation projects been implemented and what effect are they having?**

- **How does desalination work and what are some prominent desalination projects?**

- **What are some of the key ways in which water conservation can protect aquatic ecosystems?**

Answer the Central Question:

_____

_____

_____

_____

_____

_____

Chapter 6

Review Questions

1. How many liters are contained in 1 km³ of water?
a. 100,000 liters
b. 1,000,000 liters
c. 1,000,000,000 liters
d. 1,000,000,000,000 liters

2. How do El Niño and La Niña affect Australia?
a. El Niño brings wet conditions; La Niña brings dry conditions.
b. El Niño brings dry conditions; La Niña brings wet conditions.
c. El Niño brings wet conditions, while La Niña has no effect.
d. El Niño has no effect, while La Niña brings dry conditions.

3. How many countries included in Figure 6.7 have per capita water consumption rates less than Gleick's proposed 50-liter-per-day human right?
a. One country   c. Nine countries
b. Seven countries   d. Eleven countries

4. How much of global runoff is now appropriated for human use?
a. About 17%   c. A bit over 50%
b. About 31%   d. Nearly 100%

5. The arid regions of the world support what percentage of the global population?
a. Less than 1%
b. Approximately 10%
c. Approximately 20%
d. Over 50%

6. Which of the following are problems with groundwater pumping around the world?
a. Falling water tables
b. Land subsidence
c. Low recharge rates
d. All of the above

7. What fraction of large rivers in the Northern Hemisphere has been altered by dams and water diversions?
a. About one-tenth
b. About one-third
c. Approximately half
d. Approximately three-fourths

8. What aspect of New York City's water conservation program resulted in the greatest water savings?
a. Meter installation
b. Installing low-flow toilets
c. Detecting and repairing leaks
d. Home inspections

9. What factor most limits the use of desalination as a means of supplementing water supplies?
a. A sufficient source of saline water
b. The lack of effective technology
c. The cost of energy to run the process
d. A general lack of interest in the process, even in water-scarce regions

10. Although there are remarkably successful examples of river and wetland restoration, what factors make such restoration very difficult in many situations?
a. Severe pollution
b. Intensive urban development
c. Local or regional groundwater depletion
d. All of the above

Critical Analysis

1. Use the information in this chapter and other resources, such as Singapore's national water agency, the Public Utilities Board (PUB, www.pub. gov.sg/Pages/default.aspx) to discuss the ways in which Singapore's water supply issues and solutions are similar to or very different from those of Earth as a whole.

2. Use the information in the text to trace the potential paths a water molecule travels from the oceanic subcycle to the terrestrial subcycle, back to the oceanic subcycle.

3. Apply the general principles of the hydrological cycle and the regional patterns of water table fall or rise (see Figure 6.15) to develop a long-term conceptual plan for sustainable use of the Ogallala Aquifer, which occurs mainly under the temperate grassland biome.

4. Using a variety of sources, design a sustainable water-management plan for a region, such as the American Southwest, Northwest, or Southeast, in which precipitation and temperatures are strongly influenced by El Niño and La Niña.

5. Compare river or wetland restoration projects that have been successful with those that have failed. Propose the best predictors of success or failure. An Internet search will yield abundant examples.

Find additional resources and links online at www. macmillanhighered.com/launchpad/molles1e.

Central Question: How can we produce food and forest products while minimizing environmental impact?

Describe how the physical environment and biodiversity influence the availability of terrestrial resources.

SCIENCE

Sustaining Terrestrial Resources

Analyze the environmental impacts of harvesting terrestrial resources.

Discuss tactics for minimizing the impact of farming, ranching, and forestry.

ISSUES

SOLUTIONS

(Yuriy Chertok/Shutterstock)

High rates of erosion, which began around 7,000 BCE, have removed the soil cover of many landscapes in Greece.

Terrestrial Resources Depend on Fertile Soils

Over 2,000 years ago, people understood that tending to the health of soils is essential to sustaining the welfare of human communities.

Sometime around 2,400 years ago, the Greek philosopher Plato gazed up at the dry, barren hills surrounding his native Athens and came to a startling conclusion about the impact of human agriculture over the preceding centuries. A once-extensive forest had been cleared, and the landscape had subsequently grown parched. "What now remains compared to what then existed," he wrote, "is like a skeleton of a sick man, all the fat and the soft soil having wasted away, and only the bare framework of the land being left."

Like a detective, Plato gathered evidence to test his theory by scouting the countryside. He noticed that forests growing some distance from Athens included enormous, old trees and deep, fertile soils. These forests, he reasoned, are what Athens must have been like during some past era. In fact, Plato noticed that in the center of the city where large trees no longer grew, there were buildings with wide, wooden roof beams extracted from these long-lost forests. In addition, he observed that many religious shrines in the area, which ancient Greeks typically sited near perennial springs and streams, had no flowing water nearby. In the past, he reasoned, the combination of forested hills and deep soils would have captured precipitation and slowed its passage through the landscape. Deforestation had increased rates of soil loss, which meant that water rapidly ran off the land during rainstorms.

> "A good part of agriculture is to learn how to adapt one's work to nature, to fit the crop-scheme to the climate and to the soil. . . . To live in right relation to the natural conditions is one of the first lessons that a wise farmer or any other wise man learns."
>
> Liberty Hyde Bailey, *The Holy Earth*, 1915

Athens was hardly unique. The changes Plato noticed in his backyard are just one part of the global story of how humans have altered the world we live in. In the early days of our species, we lived by gathering wild plant materials for food and shelter and hunting wild animals that gave us furs and meat. In other words, we lived entirely off the production of natural ecosystems. Then about 10,000 years ago, human communities around Earth began to domesticate plants and animals, manipulating natural ecosystems to serve our needs.

Managing terrestrial resources through ranching, forestry, and farming has increased the amount of food and forest products (e.g., wood) available to human populations, as well as reduced variation in supplies of these resources. A more reliable food supply increased the rate of human population growth (see Chapter 5, page 130), which led to the development of cities. Forestry cleared land for crops and created pastures for ranching while also supplying building material and fuel. The benefits of agriculture and forestry, however, were accompanied by unsustainable environmental costs as humans irrigated lands, replaced natural vegetation with crops, and cleared forests for land to support human populations. The need for careful management of Earth's terrestrial resources brings us to the central question of this chapter.

Central Question

How can we produce food and forest products while minimizing environmental impact?

7.1–7.3 Science

Using photosynthesis, plants turn energy from the Sun, carbon dioxide from the atmosphere, and moisture and nutrients from the soil into leafy, green biomass. The amount of biomass produced—called *primary production* (see Chapter 2)—varies widely across natural terrestrial ecosystems. Through agriculture and forestry, humans use primary production in various ways, including for direct human consumption, fodder for livestock, and forest products for building materials. In this chapter, we explore how we can harvest these necessities from terrestrial ecosystems sustainably. However, this extractive view of these ecosystems should not distract us from the huge value of the other services they provide, which are discussed in Chapter 4 (see page 112). Three of the most significant influences on terrestrial primary production, and therefore on the availability of terrestrial resources, are climate, nutrients, and biodiversity.

7.1 Climate, biodiversity, and nutrients influence terrestrial primary production

The Amazonian rain forest has a very different climate than arctic tundra. The rain forest is moist and warm, with lush growth of countless varieties of plants; the tundra is dry and cold, with sparse growth of far fewer plant species. Clearly, the climate of an area, especially its prevailing temperature and precipitation, affects the biomass and the type of vegetation that grows there. Climate is also one of the main factors influencing variation in primary production (**Figure 7.1**). As any gardener knows, most plants grow best when they have plenty of water and sunlight, so long as temperatures are not so hot that the plants wilt or so frigid that they freeze.

Species-Richness Effects

Natural ecosystems add layers of complexity atop these environmental variables. More specifically, they contain a variety of plant species that interact with one another, and scientists have long suspected that a link exists between the number of species and the productivity of an ecosystem—the idea being that each plant species has slightly different growth requirements and strategies. Together, they can take advantage of every beam of sunlight, drop of water, and soil nutrient.

To test this hypothesis, in the early 1990s, an ecologist named David Tilman prepared 147 plots—10 feet by 10 feet—in the Minnesota prairie. He and his colleagues seeded these plots with anywhere from 1 to 24 species of native grasses. As predicted, they found that plots with more species had higher primary production levels. In fact, a long-term study by Tilman's research group found that primary production in the most diverse study plots was over 340% higher than in plots with a single species.

Plant growth patterns and physiology suggest causal mechanisms for these findings. Some root systems plunge deep into the soil, whereas others crawl along just under the surface, which means that each plant takes up nutrients and moisture from different parts of the soil column. Also, some plant species make the environment more favorable for other species by, for example, adding nitrogen to the soil or providing shade for species that prefer the community understory. These positive effects

THE GLOBAL DISTRIBUTION OF NET PRIMARY PRODUCTION CLOSELY MATCHES THE DISTRIBUTION OF EARTH'S CLIMATIC ZONES AND BIOMES

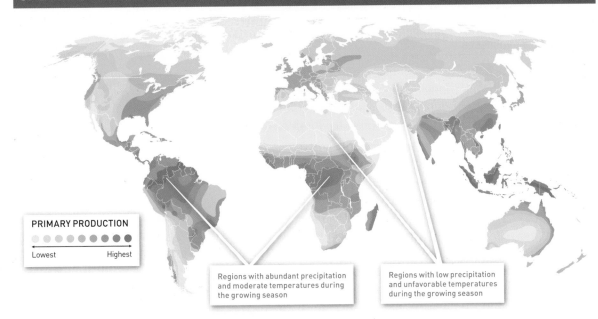

PRIMARY PRODUCTION

Lowest   Highest

Regions with abundant precipitation and moderate temperatures during the growing season

Regions with low precipitation and unfavorable temperatures during the growing season

FIGURE 7.1 Primary production is highest in equatorial regions where the natural vegetation is tropical forest. Temperate forest regions support the next highest levels of primary production. Meanwhile, the lowest rates of primary production occur in the cold dry tundra and in deserts, which may be cold or hot but are always dry. Between these end members are boreal forests, temperate grasslands, and savannas, which support medium levels of primary production. (After Del Grosso et al., 2008)

can facilitate greater productivity where multiple species grow together.

Soil Nutrients

Although climate and plant diversity have substantial effects on levels of terrestrial primary production, gardeners also know that the amount of fruits and vegetables they harvest is affected by soil fertility. Soil fertility is especially connected to levels of certain key elements, such as the availability of nitrogen, the soil nutrient that most commonly limits terrestrial primary production. Consequently, retaining nutrients in soil is critical for sustaining primary production. During one of the field experiments conducted by Tilman and his

colleagues, they learned that study plots with more plant species were better able to take up and retain nitrate, a chemical form of nitrogen useful to plants but subject to leaching through the soil by water. More plant species on a single plot retained more nitrogen, and therefore overall primary production was higher (**Figure 7.2**).

The Nitrogen Cycle

Nitrogen, like other essential nutrients on Earth, cycles from soil to water to air. The chief reservoir of nitrogen, an essential component of protein and nucleic acids and thus critical for life, is the atmosphere, where elemental nitrogen, N_2, makes up 78% of atmospheric gases. However, most organisms cannot use elemental

FIGURE 7.2 Seeding with greater numbers of grassland species in experimental field plots resulted in lower concentrations of nitrate, an easily leached form of soil nitrogen, below the reach of roots in grassland plots. (Data from Tilman et al., 1996) These results indicate lower nutrient loss by leaching from the topsoil of higher-diversity experimental grassland ecosystems. Tilman's study included 147 experimental plots covering an area of 9 square meters (100 square feet), seeded with a range of plant species native to North American prairies: 1, 2, 4, 6, or 8 species (20 plots each), 12 species (23 plots), or 24 species (24 plots).

BIODIVERSITY APPEARS TO INFLUENCE NUTRIENT RETENTION BY ECOSYSTEMS

(David Tilman, UMN)

David Tilman's field study site at Cedar Creek, MN

Study plots seeded with a greater diversity of plants had lower nitrate concentrations below the plant rooting zone.

NITRATE (mg PER kg SOIL)

NUMBER SPECIES SEEDED

FIGURE 7.3 Nitrogen makes up 78% of dry air. However, only a few microorganisms are capable of nitrogen fixation, a process that requires breaking the strong bonds that join the two atoms in N_2 molecules. Once incorporated into the molecules, such as amino acids and nucleic acids, of which these nitrogen fixers are made, nitrogen can be cycled through an ecosystem.

nitrogen cycle The process whereby nitrogen passes through and between ecosystems, involving several key actions by microorganisms, including nitrogen fixation, decomposition, ammonification, nitrification, and denitrification.

nitrogen fixation Incorporation of atmospheric nitrogen, N_2, into nitrogen-containing compounds by bacteria, living in association with plants or free living.

ammonification The process by which decomposers break down proteins and amino acids, releasing nitrogen in the form of ammonia (NH_3) or ammonium ion (NH_4^+).

nitrification The conversion of ammonia or ammonium to nitrites (NO_3^-) by nitrifying bacteria.

nitrogen assimilation The incorporation by plants of nitrate and ammonium into essential nitrogen-containing organic compounds.

denitrification The process by which specialized bacteria in soil and water convert nitrate ions back into nitrogen gas (N_2), which returns to the atmosphere.

weathering The fragmentation and decomposition of mineral materials as a result of chemical, biological, and mechanical processes, resulting in the release of nitrogen, phosphorus, and other elements.

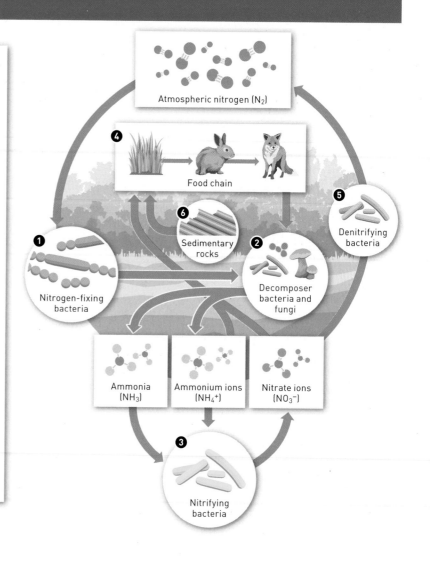

THE NITROGEN CYCLE

KEY PROCESSES

❶ NITROGEN FIXATION
Specialized bacteria in soil, root nodules, and water convert atmospheric nitrogen (N_2) into ammonia. A small amount of nitrogen is fixed by lightning.

❷ DECOMPOSITION and AMMONIFICATION
Decomposer bacteria and fungi consume animal, plant, and microbial waste products and remains, thereby releasing ammonia and ammonium ions.

❸ NITRIFICATION
Nitrifying bacteria convert ammonia and ammonium ions into nitrate ions.

❹ NITROGEN ASSIMILATION
Plants absorb ammonium ions and nitrate ions and incorporate them into essential molecules such as amino acids, proteins, and nucleic acids.

❺ DENITRIFICATION
In the absence of oxygen, specialized bacteria in soil and water convert nitrate ions back into nitrogen gas (N_2), which returns to the atmosphere.

❻ WEATHERING
In ecosystems on nitrogen-rich sedimentary rocks, weathering can be a significant source of nitrogen for plants.

Atmospheric nitrogen (N_2)

Food chain

Denitrifying bacteria

Sedimentary rocks

Nitrogen-fixing bacteria

Decomposer bacteria and fungi

Ammonia (NH_3)

Ammonium ions (NH_4^+)

Nitrate ions (NO_3^-)

Nitrifying bacteria

---

nitrogen to make essential nitrogen-containing compounds. As a consequence, they are dependent on nitrogen-containing compounds in soils and water that are released during the course of the **nitrogen cycle.**

Six major processes drive the nitrogen cycle (**Figure 7.3**). Elemental nitrogen from the atmosphere enters the cycle through the process of **nitrogen fixation,** during which specialized nitrogen-fixing bacteria convert N_2 to ammonia, NH_3, which is incorporated into organic molecules such as amino acids, the building blocks of proteins. A small amount of nitrogen is also fixed by lightning. Nitrogen fixation is a critical link between the atmospheric pool of nitrogen, which cannot be used by most organisms, and soil and aquatic pools of nitrogen.

Ammonia and ammonium ions are released from decomposing plant, animal, and microbial biomass in a process called (2) **ammonification.** Some ammonium and ammonia are converted to nitrates through (3) **nitrification,** a two-step process in which bacteria first produce nitrites (NO_2^-) and then nitrates. Primary

producers can take up the nitrates and ammonium in terrestrial or aquatic ecosystems.

Once within the plant or alga, nitrate and ammonium enter a process called (4) **nitrogen assimilation,** in which they are incorporated into essential nitrogen-containing organic compounds, such as DNA and amino acids. When plants, or the consumers that eat them, decompose, the nitrogen in their bodies returns to soil or water, completing the nitrogen cycle.

Nitrogen can also be lost from ecosystems through the process of (5) **denitrification.** Denitrification takes place in poorly drained, poorly aerated soils, or in low- or no-oxygen environments in lakes and marshes, where denitrifying bacteria convert nitrates into elemental nitrogen gas. If nitrogen is not replenished by nitrogen fixation, denitrification may deplete the available nitrogen within an ecosystem. In ecosystems developing on nitrogen-rich sedimentary rocks, (6) **weathering** can make significant contributions to the amount of cycled nitrogen.

⚠️ Think About It

1. If you were to travel northward from the deserts of the American Southwest, North Africa, or Central Asia, primary production would first increase and then decline as you continued north. Why?

2. Can you envision a scenario where two species growing together might decrease primary production? How does that fit into what we know from David Tilman's experiments?

3. How might different leaf shapes among plant species cause diverse ecosystems to maximize primary production?

4. How would life on Earth change if all nitrogen-fixing organisms suddenly became extinct?

7.2 Agriculture, forestry, and grazing systems are built on the natural biomes

Humans establish their **terrestrial harvest systems,** such as forestry, ranching, and farming, within the boundaries of natural biomes (**Figure 7.4**). That means the same factors that influence primary production in nature—namely, climate, nutrients, and biodiversity—will also play a role in these systems. With careful stewardship, all the biomes can sustain production of resources useful to humans.

Farming

The temperate forest biomes, with their moderate climates and fertile soils, can generally support intensive agriculture. This is particularly true for temperate deciduous forests, which have been cleared of oaks,

terrestrial harvest systems
Ways of extracting production from ecosystems, ranging from hunting and gathering in unmanaged natural ecosystems to nomadic herding and small-scale subsistence farming to industrialized agriculture.

SYSTEMS THAT HUMANS HAVE DESIGNED FOR HARVESTING LAND-BASED RESOURCES ARE BUILT ON EARTH'S TERRESTRIAL BIOMES

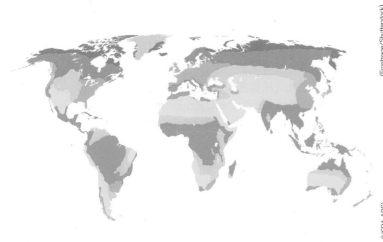

TROPICAL FOREST Timber harvesting, clearing for livestock production and grain and soybean farming

TEMPERATE FOREST Timber harvesting, mixed crop and livestock farming

TEMPERATE GRASSLAND Large-scale grain and soybean farming, grazing in drier regions

DESERT Irrigated agriculture in river valleys, livestock grazing on uplands

TROPICAL SAVANNA Livestock grazing, limited crop production

MEDITERRANEAN SCRUB Orchard and vineyard crops, mixed livestock farming

TAIGA Timber harvesting, small-scale farming

TUNDRA Nomadic herding and hunting

FIGURE 7.4 People have found ways to acquire the resources needed to maintain themselves in all of Earth's terrestrial biomes, each presenting a unique set of opportunities and challenges.

maples, and other trees to make way for farms. For example, the highly productive rice farming of China and Japan is concentrated in the temperate forest biome. Some of the most fertile soils occur in temperate grasslands, which is why the prairie regions of North America and Eurasia have been centers of wheat and maize farming. Faming in prairie and Mediterranean regions often requires irrigation, which creates a challenge to sustainability (see Figure 6.13, page 167).

Ranching

Why have the indigenous herding cultures of both tundra and deserts been largely nomadic?

Large grazing and browsing animals, such as elk, kangaroo, and musk ox, naturally exist on every continent, suggesting that **ranching,** the practice of raising domesticated livestock for meat, leather, wool, and other products is possible almost everywhere. Historically, migratory herding people inhabited the tundra, desert, and semi-desert biomes because those biomes have the lowest levels of primary production and are unsuitable for farming. The practice of irrigation has greatly increased the productivity potential of arid lands, allowing ranchers to settle in one place; however, it is not without environmental cost and controversy. Temperate grasslands, especially those receiving greater amounts of precipitation, are more fertile than deserts and, in their native state, may provide some of the most productive grazing habitats for livestock.

Forestry

ranching The practice of raising domesticated livestock for meat, leather, wool, and other products.

forestry The management of forests and woodlands for the harvest of timber or fuelwood.

Tropical biomes generally sustain high levels of primary production (see Figure 7.1) and are home to valuable hardwoods such as mahogany and rosewood. However, because many tropical forests grow on old, highly oxidized soils from which abundant rains have washed most plant nutrients, reforestation can be a frustratingly slow process that makes sustainable

harvesting more difficult (see Figure 7.6). In temperate coniferous forests, soils are generally less fertile than soils in deciduous forests because conifer needles are acidic. Taiga also has acidic soils and short growing seasons. Thus, these biomes are better suited to supporting **forestry,** the management of forests and woodlands for the harvest of timber or fuelwood, than agriculture.

⚠ Think About It

1. The United States is large and encompasses several biomes with different potentials for agriculture and forestry. What problems might smaller countries face in meeting their food and timber needs?

2. Are the biomes with the highest level of primary production always the best places for agriculture? Why or why not?

3. When humans replace native vegetation with agricultural crops or a second growth forest, how might primary production differ from the original plant community?

7.3 Soil structure and fertility result from dynamic processes

Soils are more than just dirt. Plants depend on them for physical support, water, and nutrients. They are also home to diverse underground organisms. Creatures living in soils range from microscopic bacteria to the largest organism known—a honey fungus, *Armillaria ostoyae,* in eastern Oregon, which is 3 times the size of New York's Central Park (**Figure 7.5**). Soil development

---

TEEMING WITH LIFE, SOILS ARE HOME TO A VAST ARRAY OF BIODIVERSITY

Clover root nodules housing nitrogen-fixing bacteria

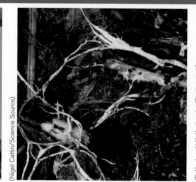

Soil fungi associated with plant roots

Earthworm

Pocket gopher

FIGURE 7.5 Many thousands of bacterial species can be identified from a single gram of soil. Shown here are just a few of the vast number of life forms that inhabit soils.

COMBINED EFFECTS OF CLIMATE AND ORGANISMS PRODUCE MAJOR DIFFERENCES AMONG THE TYPICAL SOILS OF EARTH'S BIOMES

TROPICAL FOREST
Soils of tropical rain forests generally infertile due to heavy rainfall and high decomposition rates; low organic matter content, which reduces nutrient holding capacity, reflected in light soil color; soils of tropical seasonal forests usually more fertile.

TEMPERATE FOREST
Usually fertile and neutral to slightly acidic; medium rates of decomposition foster accumulation of nutrients and organic matter in soil.

TEMPERATE GRASSLAND
Neutral to slightly basic, high to moderate fertility; can be deep and high in organic matter; color ranges from brown to black.

DESERT
Low fertility and little organic matter; often high in salt content; may have a layer of stony material that impedes water infiltration.

TROPICAL SAVANNA
Generally low fertility, often with impermeable layer that retains water near the soil surface.

MEDITERRANEAN SCRUB
Generally low to moderate fertility, organic matter content moderate to low; fragile and subject to erosion.

TAIGA
Shallow and highly acid; low decomposition rates slow soil building and tie up nutrients in plant litter.

TUNDRA
Rich in peat and humus and usually with a layer of permafrost; freezing and thawing create netlike surface features.

FIGURE 7.6 Differences in climate and dominant organisms in terrestrial biomes are reflected in variations in amounts of organic matter, depth, color, and fertility of their soils.

and soil structure vary around the world largely because of differences in climate and the types and abundance of organisms present (**Figure 7.6**).

Soil Structure

If you were to dig into a mature soil, you would find a series of layers, called *soil horizons*. In a temperate deciduous forest, these are known as the O, A, E, B, C, and R horizons (**Figure 7.7**).

The surface layer, or the **O horizon,** is a site of active decomposition of organic matter, such as leaves, twigs, and bark. Burrowing organisms and physical processes, such as freezing and thawing, mix this decomposing organic matter with the inorganic clays and sands found in the lower soil horizons, producing the crumblike structure characteristic of many fertile soils.

The **A horizon,** which sits beneath the the O horizon, is typically referred to as the **topsoil.** It is rich in essential plant nutrients, such as nitrogen, phosphorus, and potassium, on which much plant production is dependent. Although it contains significant amounts of organic matter, which is generally dark in color, the A horizon is predominantly an inorganic layer, consisting of a mixture of various proportions of sand, silt, and clay.

FIGURE 7.7 Mature soils under temperate deciduous forests commonly have six horizons: the O, A, E, B, C, and R horizons.

BASIC SOIL STRUCTURE CONSISTS OF A VERTICAL SEQUENCE OF SOIL HORIZONS

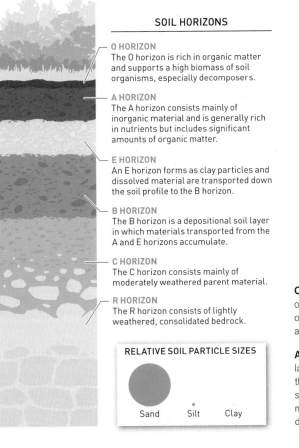
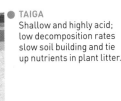
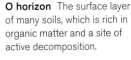

SOIL HORIZONS

O HORIZON
The O horizon is rich in organic matter and supports a high biomass of soil organisms, especially decomposers.

A HORIZON
The A horizon consists mainly of inorganic material and is generally rich in nutrients but includes significant amounts of organic matter.

E HORIZON
An E horizon forms as clay particles and dissolved material are transported down the soil profile to the B horizon.

B HORIZON
The B horizon is a depositional soil layer in which materials transported from the A and E horizons accumulate.

C HORIZON
The C horizon consists mainly of moderately weathered parent material.

R HORIZON
The R horizon consists of lightly weathered, consolidated bedrock.

RELATIVE SOIL PARTICLE SIZES
Sand   Silt   Clay

O horizon The surface layer of many soils, which is rich in organic matter and a site of active decomposition.

A horizon (topsoil) Soil layer immediately below the O horizon that includes significant amounts of organic matter, generally expressed by dark color.

Why are soils with a deep A horizon generally considered good for farming?

soil texture The relative fineness or coarseness of a soil, which is determined by its proportions of sand, silt, and clay.

loam A soil consisting of approximately equal proportions of sand, silt, and clay.

E horizon Soil layer between the A and B horizons, from which clays and dissolved materials are transported down the soil profile to the underlying B horizon.

B horizon A depositional soil layer in which materials transported from the A and E horizons accumulate.

C horizon The deepest soil layer, consisting mainly of lightly weathered parent material.

parent material The bedrock or unconsolidated deposits, such as windblown sand or silt, from which soil develops.

R horizon The base of a soil profile composed of consolidated bedrock, immediately below the C horizon.

erosion A process that removes geologic materials, ranging from clay-sized particles to boulders, from one part of a landscape to be deposited elsewhere; increased rates of soil erosion due to human activity can reduce soil fertility.

FIGURE 7.8 Soils form as a consequence of climate and organisms acting on parent material over long periods of time. The sequence here sketches soil development in a climate that would support temperate deciduous forest at maturity.

The proportions of sand (coarse and gritty), silt (intermediate grain size), and clay (small grain size) in the topsoil determine **soil texture.** Soils dominated by one of the three types of mineral soil particles are called sandy, silty, or clay soils. Meanwhile, a soil consisting of approximately equal proportions of sand, silt, and clay is a **loam.** A loam soil, which has properties intermediate between sandy and clay soils, is considered one of the most desirable for agriculture.

In well-developed soils, there can be an **E horizon** below the A horizon. The light-colored E horizon results from clays and dissolved organic matter flowing downward to deeper soil layers, leaving pale-colored sand and silt. The subsoil, or **B horizon,** is a depositional layer, rich in materials that leached out of the E horizon. The **C horizon,** the deepest soil layer, typically contains unconsolidated weathered rocks from the **parent material** from which soil develops. Parent material can be made up of rock, windblown or water-transported sand, or organic matter, such as peat. At the base of a soil developed on rock is the **R horizon,** which is partially weathered bedrock.

Soil Development

Soil forms as the environment interacts with parent material. The factors important to soil formation include climate, organisms, the nature of the parent material, the topography (or form) of the land surface, and time (**Figure 7.8**).

Climate directly influences soil development through temperature and precipitation. Temperature fluctuations, freezing and thawing in cold climates, and heating and cooling in hot climates, promote weathering of rocks. Weathering begins with fracturing and fragmentation of large rock, which eventually reduces even great rock formations to small soil particles. The rate of soil development reaches maximum levels in the warm, humid tropics. Climate also indirectly affects soil development through its influences on the activity of soil organisms and plant roots.

Wind and rain add nutrients as they deposit dust on a landscape (**Figure 7.9**). Nitrogen-fixing bacteria and plants, such as legumes, produce most of the biologically available nitrogen in soils. Runoff from rainfall can cause **erosion** of soils from the landscape, whereas rainfall that percolates into soils can remove nutrients and carry them down the soil profile and into groundwater.

Though all soils are subject to erosion by wind or water, those on steep slopes are more vulnerable. As soils erode from higher ground, the soil remaining on the slopes is thin and prone to drying. Soils washed into valleys and other low points in a landscape, such as swales, cause a thickening and moistening in these depositional areas.

⚠ Think About It

1. Why might some plants grow well on soils in the early stages of development but not in later stages (see Figure 7.8)?

2. What properties might lead farmers to consider loam to be an ideal soil?

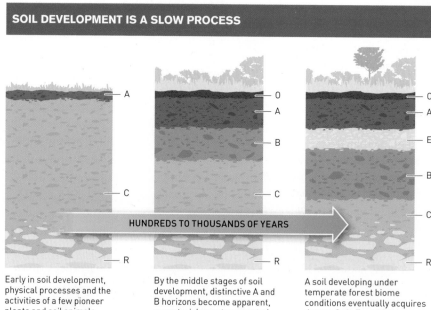

SOIL DEVELOPMENT IS A SLOW PROCESS

HUNDREDS TO THOUSANDS OF YEARS

Early in soil development, physical processes and the activities of a few pioneer plants and soil animals begin to break down parent material and add organic matter to the surface layer.

By the middle stages of soil development, distinctive A and B horizons become apparent, as materials are transported from the A to B horizons; a thin surface organic layer gradually builds up.

A soil developing under temperate forest biome conditions eventually acquires deeper O, A, E, and B horizons as a consequence of continued transport of materials down the soil profile to the B horizon.

SOIL STRUCTURE AND FERTILITY RESULT FROM A DYNAMIC INTERPLAY OF PROCESSES

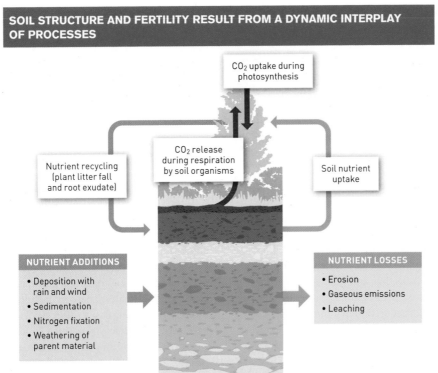

FIGURE 7.9 Avenues of soil nutrient loss oppose several sources of nutrient addition. Soil nutrients taken up by plant roots and carbon dioxide from the atmosphere are incorporated into plant tissues during photosynthesis. Then as plant litter and other organic matter decompose (e.g., fallen leaves, dead roots, or shed bark), carbon dioxide is released to the atmosphere and soil nutrients are returned to the soil.

3. How would differences in the main factors influencing soil development (climate, organisms, parent material, topography, and time) influence the store of nutrients and organic matter in soils as depicted in Figure 7.9?

4. If global climate changes such that the optimal zone for grain production moves far to the north, how might soils limit agriculture in this new "climatically optimal" zone?

7.1–7.3 Science: Summary

Climate, which includes temperature and precipitation, is one of the most important factors influencing the amount of biomass an ecosystem produces. Careful experimental research has revealed that biodiversity also has a significant positive influence on productivity,

independent of climate. These documented relationships may help in the design of sustainable agricultural systems.

Different climatic conditions support a wide range of biomass production and different soil types. This variation in climate and soils is linked to Earth's terrestrial biomes, within which human populations have developed systems of farming, ranching, and forestry to harvest primary production for human use.

In a mature temperate deciduous forest, soils have a distinct sequence of layers called the O, A, E, B, C, and R horizons. The main factors important to soil formation include climate, organisms, the nature of the parent material, the topography (or form) of the land surface, and time. The supply of essential plant nutrients (e.g., nitrogen and phosphorus) and organic matter in soils is not static, but rather the result of a dynamic interaction between several processes, such as erosion, deposition, and decomposition.

7.4–7.9 Issues

In the early days of our species, humans lived off the products of natural ecosystems, foraging for nuts, berries, and other wild plant materials, and hunting wild animals, such as mammoths and musk ox. At the end of the last ice age, some 10,000 years ago, a radical shift occurred as human communities began to domesticate food plants: corn and squash in the Americas, wheat and barley in the Middle East, and rice in China. We also began breeding livestock for a variety of purposes, including meat, milk, leather, and wool. As beneficial as such technologies have been to human societies, they have inevitably had an impact on the environment, particularly with the rise of industrial agriculture developed to meet our rapidly growing urban populations.

7.4 Industrial agriculture, which increases production, comes with environmental impacts

At first, humans performed the work of cultivating and harvesting crops with the help of simple tools and animal power. Haphazard gardens near settlements might have included a **polyculture** of multiple domesticated crops intermixed with useful wild species. With little or no knowledge of soil chemistry, early farmers learned to maintain soil fertility by rotating crops on two-, three-, or four-year cycles; they alternated between grains, which deplete nitrogen, and plants such as legumes, which enrich it. Formal experiments with natural fertilizers (e.g., bat guano and bone meal), which began in the 18th century, soon reduced the need for **crop rotation.** By the 1920s, most farmers in the United States and Europe were cultivating **monocultures,** vast fields of single crops, which could be maintained year after year with the application of natural and synthetic fertilizers, along with pesticides. These monocultures also had the benefit of being easily tilled, seeded, fertilized, and harvested using tractors powered by fossil fuels, which vastly decrease the amount of labor required on the farm (**Figure 7.10**).

The gains in production made with industrial agriculture spread to the developing world with the Green Revolution, which was spearheaded by a dedicated plant breeder named Norman Borlaug (**Figure 7.11**). Working with wheat in Mexico, Borlaug made several thousand crosses between genetic varieties, producing high-yielding strains that could be grown in a wide range of ecological conditions and that were resistant to many diseases that infect wheat. The results were dramatic. In just 25 years, the national average wheat yield in Mexico increased four-fold, from 750 kilograms (1,650 pounds) per hectare to 3,000 kilograms (6,600 pounds) per hectare.

polyculture The growing of multiple domesticated crops that may be intermixed with useful wild species.

crop rotation A method farmers use to maintain soil fertility and reduce the buildup of pests by rotating crops on two-, three-, or four-year cycles.

monoculture A planting of a single variety of crop, generally over a large area, that creates an attractive target for pests and pathogens of the crop.

A QUEST FOR HIGHER PRODUCTION AND EFFICIENCY REDUCED BIODIVERSITY IN MODERN AGRICULTURE

Polyculture: wheat and barley production under date palms

Monoculture of wheat

FIGURE 7.10 The apparent biodiversity in traditional polycultures contrasts sharply with the large tracts of land planted to monocultures in contemporary industrial agriculture.

NORMAN BORLAUG SHOWING THE RESULTS OF WHEAT BREEDING EXPERIMENTS, WHICH WERE KEY TO THE GREEN REVOLUTION

FIGURE 7.11 Borlaug's work at hunger relief drew on early life experiences. His childhood was spent working on his family's farm in Iowa, where he developed a basic understanding of farming methods and a feel for the problems faced by farmers. His work during the Great Depression brought him in contact with people suffering the ravages of hunger, an experience that motivated him to live a life dedicated to reducing hunger around the world.

The revolution that Borlaug began in Mexico spread, first to other Latin American countries and, by the 1960s, to India and Pakistan, which were facing famine. In 1970 Borlaug received the Nobel Peace Prize in recognition of his life-saving work there. But because the Green Revolution involved intensive agriculture, with its range of attendant problems (see Figure 7.24, page 207), it has come under criticism. Even Borlaug recognized that one of the ways to continue to increase production, with reduced environmental impact, may be to return, at least partly, to the past, by incorporating crop rotation and a greater diversity of crops into agricultural systems.

⚠ **Think About It**

1. In the mid-1960s, some scientists were predicting widespread deaths due to famine by the 1970s. Luckily, this did not occur. How did the Green Revolution alter the predicted course of history?

2. Because predictions of famine made half a century ago did not materialize, is there no cause for concern about future famines?

7.5 **Common farming, grazing, and forestry practices deplete soils**

In his 2005 best-selling book *Collapse: How Societies Choose to Fail or Succeed,* Jared Diamond describes how soil depletion has contributed to the fall of ancient civilizations, such as the Maya in Central America. Today, damage to soils, either through depletion of nutrients or physical loss of soil, continues to be a major global environmental concern (**Figure 7.12**).

Although the Green Revolution emphasized the planting of single varieties of wheat and other grains, why did it depend ultimately on biodiversity to make its gains in production possible?

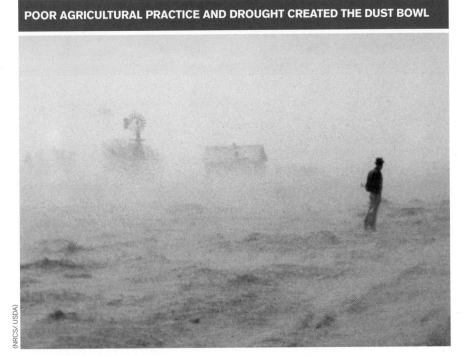

POOR AGRICULTURAL PRACTICE AND DROUGHT CREATED THE DUST BOWL

FIGURE 7.12 Dust (essentially blowing topsoil) from this 1930s environmental disaster filled the air in communities as far away from the Great Plains as New York City and set millions of refugees in motion seeking a better life across the continent.

How might studies of undisturbed natural ecosystems help design sustainable agricultural systems?

Soil Loss by Erosion

Erosion is a natural land-forming process that moves soil from one place to another. However, human activities can accelerate it—to devastating effect. In undisturbed temperate forest, soils are formed at a rate of about 1 ton per hectare per year, which is 100 to 1,000 times greater than the natural erosion rate. However, when people cut forests or churn up soils during agriculture or road-building, rates of soil loss skyrocket. Over the past 150 years, erosion has removed more than 50% of the topsoil from the prairies of Iowa, which include some of the most productive farmland in North America. Overall, soil losses in North America and Europe average approximately 17 metric tons per hectare (ha) per year. Soil losses in the croplands of South America, Asia, and Africa are even greater, around 30 to 40 tons per hectare annually.

Losses of topsoil to erosion can be broken down into losses of mineral particles, organic matter, or nutrients. In one landmark 1992 study, Cornell University researchers reported that the 17 metric tons of topsoil lost per hectare in the United States contain an average of approximately 14.5 metric tons of mineral soil, 2 metric tons of organic matter, and nearly 0.5 metric ton of inorganic nutrients (**Figure 7.13**).

Those nutrients include potassium, phosphorus, and nitrogen, which are critical to agriculture and to maintaining soil fertility. Most replacement potassium and phosphorus comes from mining activities, which have a significant impact on the environment and require substantial amounts of fossil fuels. In contrast, replacement nitrogen is derived from the atmosphere in a process that also requires substantial use of fossil

fuels (through industrial synthesis). The losses of soil organic matter can also reduce farm productivity because the crumblike structure of soil organic matter is key to promoting the infiltration of water, aeration, retention of nutrients, and resistance to erosion.

Conventional-Tillage Agriculture

Soil erosion is a potential problem associated with farming and ranching, and each has unique challenges. **Conventional-tillage agriculture** involves tilling a field to break up soil clumps and smooth the soil surface before planting, as well as weeding using specialized machinery. Although conventional tillage can be very effective for seeding crops and controlling weeds, it exposes large areas of bare soil to wind and water, often resulting in erosion of topsoil (**Figure 7.14**).

Overgrazing of Rangelands and Desertification

Grazing by livestock causes erosion in the same way as conventional tillage, by reducing plant cover and disturbing topsoil. Livestock are generally heavy animals that consume a lot of plant material and compress soils with their hooves. Overgrazed rangelands may lose up to 100 tons of soil per hectare per year. Rangelands in hot, dry regions are particularly sensitive to erosion because they naturally support sparse plant cover, which can be easily overgrazed. Severe erosion can occur during torrential rains and flash floods (**Figure 7.15**).

The impacts of overgrazing in arid and semiarid rangelands commonly lead to **desertification,** a process of degradation of once fertile lands to a desertlike condition of reduced plant cover and primary production. Desertification is a major problem in central

conventional-tillage agriculture Tilling a field to break up soil clumps and smooth the soil surface before planting, as well as weeding using specialized machinery.

desertification A process of degradation of once fertile lands to a desertlike condition of reduced plant cover and primary production.

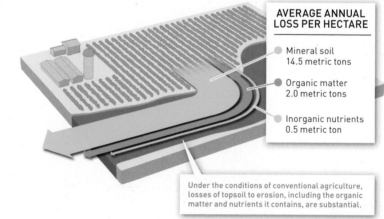

SOIL EROSION REMOVES MINERAL SOIL, ORGANIC MATTER, AND INORGANIC NUTRIENTS

AVERAGE ANNUAL LOSS PER HECTARE

- Mineral soil 14.5 metric tons
- Organic matter 2.0 metric tons
- Inorganic nutrients 0.5 metric ton

Under the conditions of conventional agriculture, losses of topsoil to erosion, including the organic matter and nutrients it contains, are substantial.

FIGURE 7.13 The average loss of soil on each hectare of agricultural land in the United States as a result of erosion amounts to 17 metric tons annually. (Data from Pimentel et al., 1992)

CONVENTIONAL-TILLAGE AGRICULTURE EXPOSES BARE SOIL TO EROSIVE FORCES

(USDA Media by Lance Cheung)

FIGURE 7.14 Intensive annual cultivation, which is central to conventional-tillage agriculture, can result in massive soil losses through erosion by both wind and water.

Asia, much of China, and of northern Africa, particularly in the Sahel region (**Figure 7.16**).

⚠ Think About It

1. How might a soil profile on a farm in a region that would naturally support either temperate

grassland (or temperate forest) show evidence of soil erosion? (See Figure 7.6.)

2. Why are the soils of mountain landscapes more subject to erosion? (Hint: What force besides water and wind power is especially influential in mountain landscapes?)

OVERGRAZING HAS RESULTED IN HIGH RATES OF EROSION ON MANY RANGELANDS

(Lynn Betts/USDA)

FIGURE 7.15 The formation of gullies, such as the one shown here on an overgrazed pasture in southern Iowa, is one of the most severe forms of erosion.

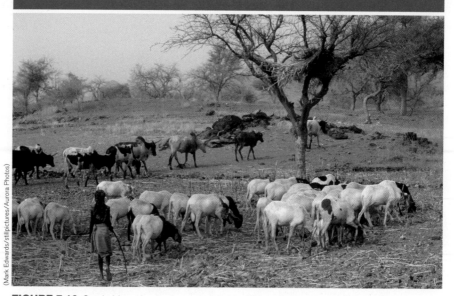

OVERGRAZING HAS CONTRIBUTED TO DESERTIFICATION

(Mark Edwards/stillpictures/Aurora Photos)

FIGURE 7.16 Semiarid grazing lands around the world have been converted to desertlike ecosystems through desertification, a process that is increasing the extent of low-production wastelands. The contribution of livestock to desertification of the Sahel region of northern Africa, pictured here, is well studied.

3. Why haven't the soils of the tundra and taiga biomes been subject to higher levels of erosion, compared with the soils of other biomes, at least to this point in their history?

7.6 Deforestation and some forestry management practices deplete soils and increase flooding danger

The Mekong River originates on the Tibetan plateau and flows nearly 3,000 miles through the spectacular tropical forests of southeast Asia—a biodiverse region that supports the livelihoods of some 70 million people. Over the last 40 years, however, these forests have been shrinking dramatically: Cambodia has lost one-fifth of its forests, Laos and Myanmar have lost one-quarter, and Thailand and Vietnam have lost just under half. The people of Southeast Asia will not only be coping with a scarcity of timber as a building material and cooking fuel, but they will also see changes in the functioning of their ecosystems. In a region known for heavy monsoon rains, the loss of forest cover will stimulate increased erosion, landslides, and degradation of soils. Rainwater normally collected and purified in forested basins will be cloudy with sediment, harming the freshwater fish that feed so much of the population.

The situation in the Greater Mekong Basin is part of a larger story of the loss of forests around the world for timber and agriculture. According to the United

Nations Food and Agriculture Organization's Global Forest Resources Assessment, updated in 2015, forests cover approximately 4 billion hectares, or 31% of the world's land area, and are being lost at a rate of 13 million hectares per year. This rate of deforestation is slightly lower than it was in the 1990s, but forests are still vanishing at an alarming rate. The highest rates of deforestation are occurring in places with the highest levels of biodiversity, including Southeast Asia, South America, and Africa. However, deforestation is also occurring at very high rates in temperate and boreal forests. In Russia, over 400,000 km² of boreal forest was cleared from the late 1990s to the early 2000s—that's slightly more than the entire state of Montana.

Forest Harvest and Clearing

The practice of forestry involves the cutting and removal of trees and other plant material for timber, fuelwood, and paper pulp. In temperate regions, foresters typically employ **clear-cutting,** an economically efficient technique whereby an entire area is cleared of its trees. Because clear-cutting is often done with heavy equipment in steep terrain, it produces high levels of soil disturbance, which leaves the bare mountainside exposed to erosion (**Figure 7.17**). As erosion removes topsoil from these forested landscapes, their potential to produce future crops of timber is reduced.

Many studies have been done to quantify the effects of logging on soil loss, which is often measured as the amount of soil transported by streams flowing out of forested stream basins, generally standardized as "sediment

clear-cutting An economically efficient technique whereby an entire area is cleared of its trees.

LOGGING ACTIVITIES COMMONLY PRODUCE SIGNIFICANT SOIL DISTURBANCE

(© Robert McGouey/Landscape/Alamy)

FIGURE 7.17 Exposure of bare soil and disturbance of topsoil during logging, particularly due to road building and log skidding, create conditions conducive to increased soil erosion.

yield" per square kilometer. One such study in the mountains of southern British Columbia, Canada, compared sediment yields in side-by-side stream basins. One of the two basins, Redfish, had the trees removed from approximately 10% of its area, whereas the other, Laird, was unlogged. In addition, 19 kilometers of roads had been built in the Redfish Basin during logging operations, whereas no roads had been built in the Laird Basin. Despite the small area logged, logging in the Redfish Basin was associated with a 50% increase in soil losses from erosion (**Figure 7.18**).

Finally, **slash-and-burn** is a common technique used in tropical countries to rapidly convert forestlands into temporary farms. Rather than harvesting the timber, trees are burned, thereby releasing some of their nutrients into the poor tropical soils, which remain fertile for several years. Slash-and-burn has been an effective form of agriculture for regions with low population densities. However, it is not an effective or sustainable practice for large populations at an industrial scale because widespread slash-and-burn practice would result in massive erosion and loss of soil fertility.

Fire Suppression

Forest ecosystems do not have to be logged to be harmed. In the western United States, pine stands once featured an open understory of widely spaced trees (**Figure 7.19a**), which was a consequence of frequent, low-intensity fires, caused by lightning or Native Americans, who traditionally cleared vegetation to

improve habitat for game animals. To protect valuable timber resources, the U.S. Forest Service began a policy of fire suppression in 1910, and tree densities have since increased to hundreds or thousands per hectare (**Figure 7.19b**).

When fires occur in these forests, which are now much denser with woody growth, they generally burn larger areas

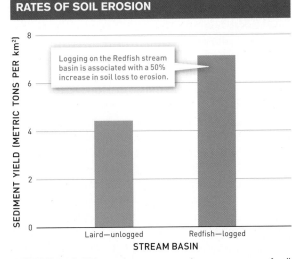

STUDIES SHOW THAT LOGGING INCREASES RATES OF SOIL EROSION

Logging on the Redfish stream basin is associated with a 50% increase in soil loss to erosion.

FIGURE 7.18 This graph compares a nine-year average of soil losses, as measured by sediments transported in the associated streams, in two stream basins. One of the basins, Redfish, had been logged over 10% of its area and included 19 kilometers of logging roads, whereas the other basin, Laird, was unlogged and had no roads. (Data from Jordan, 2006)

slash-and-burn A common technique used in tropical countries to rapidly convert forestlands into temporary farms.

FIRE PROTECTION IN FORESTS OF THE WESTERN UNITED STATES HAS INCREASED THE DENSITY OF TREE POPULATIONS

a. Bitterroot National Forest, Montana, 1909

b. Bitterroot National Forest, Montana, 1997

FIGURE 7.19 Historically, ponderosa pines in Arizona and elsewhere in the American West grew at low densities, producing a parklike landscape. After decades of fire protection, however, the densities of ponderosa pines in these landscapes increased to unsustainable levels.

Where livestock have been introduced and fires suppressed, North American grasslands have gradually changed to shrublands or woodlands. Why?

with greater intensity. The catastrophic 2011 Wallow Fire in eastern Arizona that burned nearly 2,100 km² (817 mi²) of forest, the most extensive in the history of the state, is an example of such a fire (**Figure 7.20a**). These high-intensity fires result in massive losses of soil carbon and nutrients from the fire itself as well as from subsequent erosion (**Figure 7.20b**). Ironically, fire suppression has made the forests of western North America more vulnerable to fires and more susceptible to soil and nutrient loss.

⚠ **Think About It**

1. What are the consequences of the loss of forest in the Mekong River Basin?

2. What are some of the environmental damages resulting from clear-cutting of forests?

3. How have fire suppression efforts altered forests in the western United States?

OFTEN THE LONG-TERM CONSEQUENCE OF FIRE PROTECTION HAS BEEN, IRONICALLY, CATASTROPHIC FIRE

a. Wallow Fire, Arizona, 2011

b. Soil erosion following forest fire

FIGURE 7.20 The Wallow Fire of 2011 burned with such intensity that it consumed more than 200,000 hectares (500,000 acres) of forest in Arizona before firefighters could contain it. One consequence of intense forest fires, such as the Wallow Fire, is massive soil loss to erosion.

THREE MAIN FORMS OF CROP IRRIGATION IN USE TODAY

a. Flood irrigation

b. Central-pivot sprinkler irrigation and crop circles

c. Drip irrigation

FIGURE 7.21 Flood irrigation depends on proper leveling of agricultural fields to move water at just the right speed to irrigate the rooting zone of crop plants without wasting too much water. Central-pivot sprinkler irrigation, an automated system of spray irrigation, is widely used in western North America, where it produces the circular agricultural plots visible to air travelers. Drip irrigation, which delivers water to the rooting zone of plants, is another automated irrigation system that finds its widest use in the production of high-value crops in regions where water is especially scarce.

7.7 Irrigation can damage soils

One of the common ways to increase the agricultural productivity of farmlands in arid areas is through irrigation, which is a system for artificially delivering water to crops. Digging ditches and other simple forms of irrigation date back thousands of years. Irrigation can be sustainable, but it also has the potential to waste freshwater resources and damage soils.

Irrigation Systems

As we saw in Chapter 6, irrigation places one of the largest demands on Earth's supplies of freshwater. By current estimates, irrigation accounts for nearly 70% of water withdrawals from surface and groundwater around the world and threatens biodiversity by reducing the availability of water in freshwater ecosystems. This is a particularly serious concern in arid regions, where freshwater ecosystems are most vulnerable to competing water demands such as municipal and industrial use. In addition, although irrigation can increase crop production, it can also damage soils in ways that reduce crop production.

There are many ways of applying irrigation water to fields, but today the three main techniques are flood, sprinkler or spray, and drip irrigation. Flood irrigation moves water across the surface of a field guided by gravity and berms (raised barriers) or by a series of furrows (small channels in the ground) (**Figure 7.21a**).

The advantage of flood irrigation is that it does not require expensive equipment. It can, however, waste a lot of water.

Sprinkler, or spray, irrigation applies water to agricultural fields by pumping pressurized water through sprinkler or spray nozzles. The most common type of sprinkler irrigation, the center-pivot system, in which a pipe bearing the sprinklers pivots around a central point, provides fairly even irrigation coverage. However, the equipment is costly and water losses through evaporation can be high (**Figure 7.21b**).

Drip irrigation, which delivers water and often nutrients directly to the rooting zone of crop plants, is perhaps the most precise and efficient irrigation system in common use (**Figure 7.21c**). But because this system is expensive, farmers use drip irrigation only on high-value crops, such as strawberries and tomatoes, in water-starved regions.

Irrigation and Waterlogged and Saline Soils

Applying water to an agricultural field faster than it drains, as can happen with flood irrigation, may produce **waterlogged soil,** where the water table is at or near the soil surface (**Figure 7.22**). Terrestrial plants require sufficient soil moisture, but because waterlogged soil has its pore spaces filled with water instead of air, it therefore deprives plant roots of the oxygen they need.

Irrigation combined with inadequate soil drainage can also result in soil **salinization,** a process of salt buildup in a soil. If irrigation raises the water table, salts are not

What criteria would you use for dividing up water supplies during wet periods and during droughts?

waterlogged soil A condition in which the water table is at or near the soil surface.

salinization The process of salt buildup in a soil.

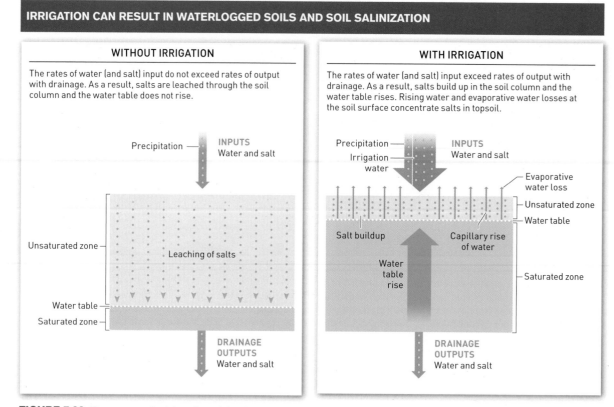

FIGURE 7.22 The amount of salt in soils and the depth to the water table are the result of a dynamic balance between water and salt inputs with precipitation and, in the case of irrigated soils, irrigation and outputs of water and salts with groundwater drainage.

Why is soil salinization generally more of a problem in hot, dry climates?

pesticide Generally a chemical substance used to kill destructive organisms, including insects (insecticide), fungi (fungicide), weeds (herbicide), and rodents (rodenticide).

flushed away but rather deposited in topsoil as water evaporates from the soil surface.

Waterlogged and saline soils are a global agricultural problem. Many millions of hectares of irrigated agricultural land, approximately one-third of the world's total, have been damaged by improperly managed irrigation, and the world loses an area larger than Ireland to salinization every decade (**Figure 7.23**).

⚠ Think About It

1. Avoiding soil salinization in irrigated fields depends on maintaining adequate soil drainage so that salts are leached down the soil profile and drained away with groundwater. How might this practice affect the environment?

2. Salinization of soils has taken place in many landscapes that have never been irrigated. How can salinization in such places be explained?

3. How are waterlogging and salinization of soils, in fact, different aspects of the same general problem?

7.8 Intensive agriculture can cause pollution and promote pesticide resistance

Many approaches to improving agricultural production, such as the addition of fertilizers and **pesticides,** can cause pollution (**Figure 7.24**). In Chapter 13, we examine the impacts of agricultural pollution, particularly pollution resulting from concentrated animal feeding operations and intensive agriculture, on aquatic ecosystems and populations; in Chapter 11, we examine them from environmental risk and human health perspectives. Here, we focus our attention on pollution by chemical pesticides and the evolution of pesticide-resistant organisms.

Pest Control and Pollution

The leaf beetle *Leptinotarsa decemlineata* is about the size of a pencil eraser and has a bright orange head with ten brown and yellow stripes along its back. It was largely unknown until 1859, when an outbreak occurred on potato fields near Omaha, Nebraska. As the leaf beetle population expanded across the United States

WATERLOGGED AND SALINE SOILS SUPPORT REDUCED CROP PRODUCTION

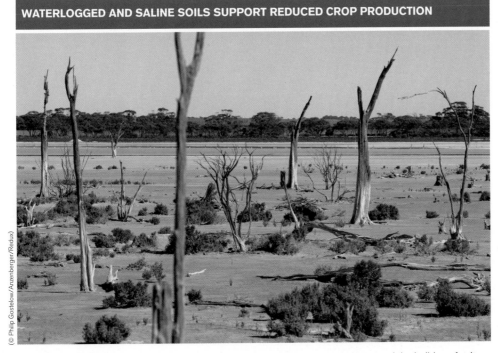

FIGURE 7.23 The filling of soil pores with water reduces root access to oxygen, and the buildup of salts in soils puts most crop plants under physiological stress. Both of these impacts are reflected in the dead trees and scrub vegetation in this once productive agricultural land in Australia.

and Canada to Europe and Asia, decimating crops in its wake, it earned its common name, the Colorado potato beetle. Since then, hundreds of chemicals have been tested and developed to fight this beetle, making it a key focus during the development of the modern pesticide industry.

Today, U.S. farmers apply 500 million kilograms (1.1 billion pounds) of pesticides annually to protect crops valued at about $40 billion from pests, pathogens, and plant competitors. In spite of the application of massive quantities of pesticides, insects, pathogens, and weeds still reduce potential annual crop production in the United States by nearly 40% (**Figure 7.25**).

The benefits of pesticides in crop protection are accompanied by various costs, however. The annual price tag for pesticides in the early 2000s was

INTENSIVE AGRICULTURE IS A SOURCE OF SEVERAL MAJOR FORMS OF WATER AND AIR POLLUTION

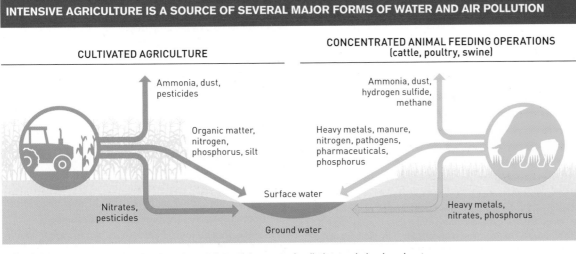

FIGURE 7.24 Intensive agriculture can be a substantial source of pollution to air, land, and water.

MANY ORGANISMS REDUCE HUMAN FOOD SUPPLIES

Corn borer consuming corn kernels   Root-knot nematode attacking tomato root   Weedy grass stunting soybean growth   Brown rats infesting stored wheat

FIGURE 7.25 Herbivorous insects, such as this corn borer, attack the above-ground parts of crop plants, while soil nematodes attack the roots of crop plants. Weeds can reduce crop production by competing with crop plants for nutrients, water, and light, while rats and many insects consume stored grains.

approximately $10 billion. Additional environmental and social costs resulting from pesticide use totaled $10 billion, including the costs associated with the poisonings of domestic animals, wild birds, fish, and honeybees and other pollinators (**Figure 7.26**). Around the world, pesticide poisoning results in the hospitalization of approximately 3 million people and more than 200,000 deaths each year, the majority of which occur in developing countries. The estimated annual cost of human pesticide poisoning in the United States alone is over $1 billion.

PESTICIDE COSTS TAKE A BITE OUT OF PROFITS

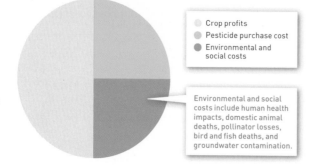

- Crop profits
- Pesticide purchase cost
- Environmental and social costs

Environmental and social costs include human health impacts, domestic animal deaths, pollinator losses, bird and fish deaths, and groundwater contamination.

pesticide resistance
An evolved tolerance to a pesticide by a pest population as a result of repeated exposure to a pesticide, ultimately rendering the chemical ineffective.

natural enemies Predators and pathogens that attack herbivorous insects and other pest organisms.

FIGURE 7.26 The value of crops protected by pesticides in the United States at the beginning of the 21st century totaled $40 billion. Half of this value was consumed by the cost of the pesticides applied to protect them. (Data from Pimentel, 2005)

Pesticide Resistance and Loss of Insect Predators

Pesticides have not proved to be a silver bullet in saving crops from insect damage. In 1952 farmers noticed that the widely used pesticide dichloro-diphenyl-trichloroethane (DDT) was no longer effective at killing the Colorado potato beetle. In other words, the beetle had evolved **pesticide resistance.** Soon farmers reported resistance to other chemicals, and today the beetle has evolved resistance to over 50 different pesticides. The lesson from the potato beetle has been that intensive use of chemicals on crops often creates environmental conditions that favor the evolution of pesticide resistance and outbreaks of crop pests (**Figure 7.27**).

For instance, growing potatoes or other crops in extensive monocultures creates an attractive target for pests and pathogens. A large crop field is an easy target for insect pests to find, colonize, and multiply their population. Also, the physical homogeneity of monocultures reduces the amount of habitat available to support the **natural enemies,** the predators and pathogens that attack herbivorous insects and other pest organisms. Then, as we apply pesticides, we kill not only pests, but also the spiders and predaceous insects that help control them.

In a final irony, applying chemical pesticides exerts strong natural selection for the evolution of resistance to those same pesticides—the individuals that are resistant to the pesticide survive and populate the next generation of insects, increasing the prevalence of pesticide resistance. In a biological arms race, farmers are forced to apply more or different pesticides, increasing their costs

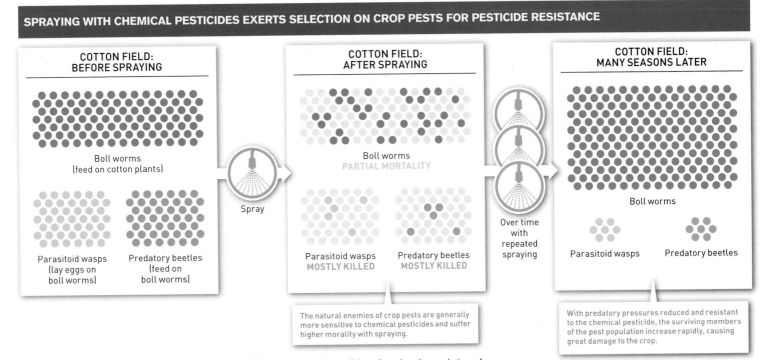

SPRAYING WITH CHEMICAL PESTICIDES EXERTS SELECTION ON CROP PESTS FOR PESTICIDE RESISTANCE

COTTON FIELD:
BEFORE SPRAYING

Boll worms
(feed on cotton plants)

Parasitoid wasps
(lay eggs on
boll worms)

Predatory beetles
(feed on
boll worms)

Spray

COTTON FIELD:
AFTER SPRAYING

Boll worms
PARTIAL MORTALITY

Parasitoid wasps
MOSTLY KILLED

Predatory beetles
MOSTLY KILLED

Over time
with
repeated
spraying

COTTON FIELD:
MANY SEASONS LATER

Boll worms

Parasitoid wasps

Predatory beetles

The natural enemies of crop pests are generally more sensitive to chemical pesticides and suffer higher morality with spraying.

With predatory pressures reduced and resistant to the chemical pesticide, the surviving members of the pest population increase rapidly, causing great damage to the crop.

FIGURE 7.27 Intensive agriculture commonly creates environmental conditions favoring the evolution of pesticide resistance and outbreaks of crop pests.

of production, creating additional environmental and social costs, and exerting further selection for pesticide resistance in the pest population.

⚠ Think About It

1. According to Figure 7.26, how much of the value of crops protected by pesticides is profit, once the costs of pesticides have been paid?

2. How might research on alternatives to chemical pesticides be affected if farmers and the chemical industry paid all costs, including environmental and social costs, of pesticide use?

3. What role does the intensity of pesticide application likely play in the evolution of pesticide resistance among agricultural pest populations?

7.9 Genetically modified crops are sources of controversy and agricultural potential

The pace of developing new varieties of plants and animals has quickened with the arrival of **biotechnology.** Biotechnology uses engineering techniques to modify organisms genetically for a particular purpose, ranging

from producing varieties that provide a more dependable source of medicines to devising crops that are more nutritious. A **genetically modified (GM) organism,** or **GMO,** has one or more new genes introduced into its genetic makeup by biotechnology, using an array of engineering methods.

When the new genes come from other species, as is generally the case, GMOs are called **transgenic organisms.** For example, genes from the bacterium *Bacillus thuringiensis* have been inserted into the DNA of several crops, including corn, to increase the corn plant's resistance to insects who chew on the plants.

In this process, a gene from *B. thuringiensis,* which codes for the insect-killing crystalline substance **Bt,** is inserted into the corn's DNA. Because the transgenic, or GM, corn now carries this bacterial gene, it incorporates Bt crystals in its tissues as it grows. As a result, insects are poisoned when they feed on one of these GM corn varieties (**Figure 7.28**). Meanwhile, natural enemies of these pest insects, along with humans and other vertebrates, are seemingly unharmed by eating Bt-containing tissue, which is only toxic at the high pH levels that are found in the pest's gut.

Status of GM Crops

Most commercially grown GM crops have been engineered with three traits in mind: (1) the capacity to produce insect-killing chemicals (e.g., Bt), thereby increasing crop plants' resistance to pests; (2) resistance

biotechnology The application of engineering techniques to modify organisms genetically for a particular purpose.

genetically modified (GM) organism (GMO) An organism into which one or more genes have been incorporated using the techniques of biotechnology.

transgenic organism A GM organism that contains genes from another species.

Bt Insect-killing crystalline substance produced by the bacteria species *Bacillus thuringiensis.*

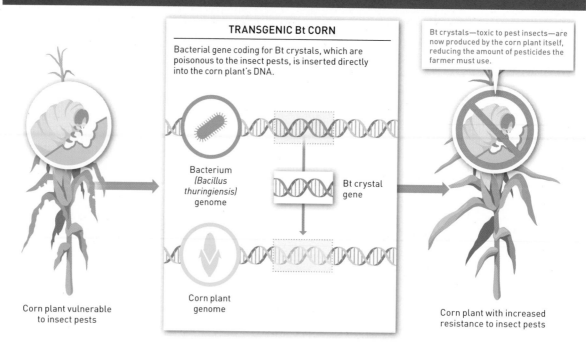

BACTERIAL GENES HAVE BEEN IMPLANTED IN SEVERAL CROP PLANTS TO MAKE THEM MORE RESISTANT TO INSECT ATTACKS

TRANSGENIC Bt CORN

Bacterial gene coding for Bt crystals, which are poisonous to the insect pests, is inserted directly into the corn plant's DNA.

Bacterium (Bacillus thuringiensis) genome

Bt crystal gene

Corn plant genome

Corn plant vulnerable to insect pests

Bt crystals—toxic to pest insects—are now produced by the corn plant itself, reducing the amount of pesticides the farmer must use.

Corn plant with increased resistance to insect pests

FIGURE 7.28 Corn engineered to contain spores of the bacterium *Bacillus thuringiensis* (Bt) kills herbivorous insects feeding on its tissues, but it is not known to harm humans.

to chemical herbicides, allowing farmers to control weeds with herbicides without harming crops; or (3) resistance to plant viruses, reducing losses of crop plants to these pathogens (**Table 7.1**).

The United States leads the world in the adoption of GM crops. For example, we grow more GM crops than in any other nation (see Table 7.1). Also, in 2010, farmers in the United States planted nearly 67 million hectares (165 million acres) in GM crops, dwarfing the plantings in other nations (**Figure 7.29**). In addition, U.S. farmers show a high level of acceptance of GM crop varieties. According to the U.S. Department of Agriculture, by 2010, most of the corn (86%), cotton (93%), and soybeans (93%) grown in the United States were GM varieties. However, more and more farmers outside the United States are also adopting these GM varieties.

GM Crops: The Potential and the Controversy

Scientists are now developing many other GM crops, with traits including greater drought tolerance, improved storage potential, and better nutritional quality. One of the best-known examples of a GM crop with improved nutritional quality is "golden rice," which produces β-carotene, a precursor to vitamin A. Since the first golden rice variety was developed, scientists have increased the β-carotene content of golden rice 23-fold. This concentration can make a difference in the lives of millions in developing countries who suffer from vitamin A deficiency, which can cause premature blindness

(**Figure 7.30**). GM crops may have the potential to improve human health and provide food for a growing human population, but they have also engendered a great deal of controversy.

In addition to improved nutrition, proponents of GM crops indicate several ways in which they benefit the environment. A 2010 U.S. National Academy of Sciences study reports that insecticide use has decreased with increased planting of Bt corn and cotton. Herbicide-resistant crops can improve soil health, particularly where reduced tilling of fields lowers soil compaction

Why do you think the United States grows and consumes so many more GM crops than the rest of the world?

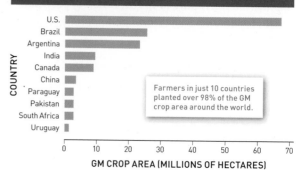

GROWING OF GM CROPS IS CONCENTRATED IN 10 COUNTRIES

Farmers in just 10 countries planted over 98% of the GM crop area around the world.

GM CROP AREA (MILLIONS OF HECTARES)

FIGURE 7.29 Among these countries, the greatest allocation of land to GM crops is found in just three: the United States, Brazil, and Argentina. (Data from ISAAA, 2011)

TABLE 7.1

MAJOR GENETICALLY MODIFIED (GM) CROPS PLANTED AROUND THE WORLD IN 2010

GM Crop	Bioengineered Traits	Primary Uses of Crop	Where GM Varieties Are Grown
Alfalfa, *Medicago sativa*	Herbicide tolerance, insect resistance	Forage for livestock: hay, silage, grazed, or fed as chopped greens	U.S.
Canola, *Brassica napus*	Herbicide tolerance	Cooking oil, salad oil, industrial lubricants	Australia, Canada, Chile, U.S.
Corn (maize), *Zea mays*	Herbicide tolerance, insect resistance	Livestock feed, human food (e.g., cereal, meal, oil)	Argentina, Brazil, Canada, Chile, Czech Republic, Egypt, Honduras, Philippines, Poland, Portugal, Romania, Slovakia, South Africa, Spain, Uruguay, U.S.
Cotton, *Gossypium hirsutuma*	Herbicide tolerance, insect resistance	Textiles, paper, oil, livestock feed	Argentina, Australia, Brazil, Burkina Faso, China, Colombia, Costa Rica, India, Mexico, Myanmar, Pakistan, South Africa, U.S.
Papaya, *Carica papaya*	Viral resistance	Fresh fruit	China, U.S.
Poplar, *Populus nigra*	Insect resistance	Paper, lumber	China
Potato, *Solanum tuberosum*	Insect resistance	Fresh consumption, snack foods	Czech Republic, Germany, Sweden
Soybeans, *Glycine max*	Herbicide tolerance, insect resistance, increased mono-unsaturated fatty acid	Oil, textured vegetable protein, tofu, livestock feed	Argentina, Bolivia, Brazil, Canada, Chile, Costa Rica, Mexico, Paraguay, South Africa, Uruguay, U.S.
Squash, *Cucurbita pepo*	Viral resistance	Fresh consumption	U.S.
Sugar beet, *Beta vulgaris*	Herbicide tolerance	Sugar	Canada, U.S.
Sweet pepper, *Capsicum annuum*	Viral resistance	Fresh consumption, flavoring, coloring	China
Tomato, *Lycopersicon esculentum*	Viral resistance	Fresh consumption, or preserved by canning or drying	China

Data from ISAAA, 2011.

GOLDEN RICE: BIOTECHNOLOGY THAT TARGETS A GLOBAL SOURCE OF MALNUTRITION

CREATING TRANSGENIC GOLDEN RICE

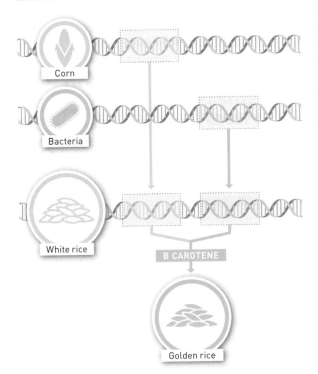

and erosion (see page 215). These benefits, coupled with improved yields in the face of plant diseases and droughts, have the potential to improve food security.

On the other hand, skeptics of GM crops have identified several areas of serious concern. Pest and weed populations have developed resistance to Bt toxins and herbicides through natural selection. If the genes that confer herbicide resistance are transferred to weeds, they may become more difficult to control or may spread to non-GM crops and contaminate them genetically. Genes from GM corn have, in fact, been documented in the native varieties of corn in Mexico. Selection for herbicide tolerance has also been documented in "superweeds" that have become too costly to control. In addition, hundreds of insect populations have evolved resistance to Bt tolerance, and when pests targeted by Bt varieties decline, secondary pests that thrive in the absence of Bt-sensitive pests can increase.

Use of herbicide-resistant GM crops also means that farmers can use high levels of herbicides to control weeds without harming crops, but this can lead to unintended consequences. For example, widespread herbicide application has decreased milkweed abundance

FIGURE 7.30 Recent varieties of golden rice contain high levels of β-carotene, which the body converts to vitamin A. The developers of golden rice propose that it can help alleviate vitamin A deficiency, which causes an estimated 250,000 cases of blindness among children each year.

an extra ton of corn is only one-sixth the cost of shipping from the United States. Sanchez points out that investing in local farmers to help improve their farm production may be the surest and least expensive way to reduce the level of hunger in the world. He notes, encouragingly, that such assistance in some of the regions of Africa most afflicted by hunger has more than doubled corn yields on small farms.

Crop Diversity for Productivity and Sustainability

Intensive agriculture, which was at the center of the Green Revolution, focused on producing food through large-scale monocultures of low genetic diversity, with high production maintained by high inputs of fertilizers, pesticides, and energy. In contrast, many of today's agricultural scientists are turning back to traditional practices to develop sustainable, high-production systems that require lower inputs of chemicals and energy.

Genetic diversity, one of the basic components of biodiversity (see Chapter 3, page 62), has proved a powerful tool for increasing rice yields in China, echoing the results of Tilman's study, covered earlier in the chapter. Youyong Zhu of Kunming University in southwestern China led a team that investigated the potential for reducing the loss of rice production—caused by a serious fungal disease called rice blast, *Magnaporthe grisea*—by increasing the genetic diversity of rice plantings. Sticky or glutinous rice varieties are sought after for making specialty dishes and therefore draw a higher selling price. These sticky varieties seem especially vulnerable to infection by rice blast. To combat rice blast, the researchers copied a traditional planting scheme used by local farmers. They increased the genetic diversity of rice in fields by planting susceptible, tall, sticky rice and resistant, short hybrids in rows in a 1 : 4 ratio: Two rows of short rice on the left, one row of tall rice in the middle, two rows of short rice on the right.

The reduction in rice blast infection in these mixed fields was dramatic. The rate of infection in monocultures of the sticky rice was 20% compared with 1% in the mixed plantings (**Figure 7.33**). The increased distance between susceptible plants and modification of temperatures, humidity, and light were likely responsible for creating conditions less favorable for the pathogen than those in monocultures of sticky rice. Furthermore, it was not necessary to spray fungicidal chemicals to control rice blast in the mixed field, thus saving money. Reducing rice blast infections resulted in an 89% increase in yield of the valuable sticky rice varieties and a 40% higher gross economic return per hectare.

Rice farmers across China took notice. In Yunnan and 10 other Chinese provinces, the area in mixed plantings increased to 1.57 million hectares between 2000 and 2004. Across this vast area, rice yields increased by an average of 675 kilograms (1,488 pounds) per hectare,

What are the potential costs of planting two or more varieties of a crop in close proximity?

intercropping Growing two or more crops in the same field.

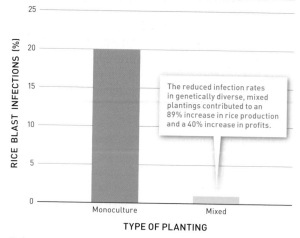

POTENTIAL POSITIVE EFFECTS OF BIODIVERSITY ON CROP YIELDS BEGINS WITH GENETIC DIVERSITY

The reduced infection rates in genetically diverse, mixed plantings contributed to an 89% increase in rice production and a 40% increase in profits.

FIGURE 7.33 Increasing the genetic diversity of rice plantings in China has reduced losses to pathogens and has increased yields and profits. (Data from Zhu et al., 2000)

and $259 million were gained either through increased income or reduced costs. Agricultural scientists have demonstrated similar boosts in yield resulting from higher genetic diversity in other crops.

Some of the benefits of biodiversity to crop production can be realized by growing different crops sequentially on the same field generally over a period of three to five years, a practice called *crop rotation*. For example, a farmer might plant corn in one field, and then plant a different crop in that same field the next year, and the year after. The benefits of crop rotation include increased yields, and lower insect and disease infestations. Crop rotation can also improve soil aeration when deep-rooted crops are part of the rotation, as well as improve soil fertility, as when a nitrogen-fixing legume, such as alfalfa or soybeans, is grown.

Biodiversity can be incorporated into agricultural systems by growing two or more crops in the same field, a technique called **intercropping.** Intercropping has been practiced for centuries. For example, Native Americans developed a traditional intercropping system in which they grew corn, beans, and squash together, a combination of crops commonly referred to as "the three sisters" (**Figure 7.34**).

In this system, beans enrich the soil by fixing nitrogen, the corn provides a surface for attachment by the climbing bean plants, and the squash shades the soil, thereby reducing temperature fluctuations and water loss. The three crops also provide complementary sources of nutrition: Combining the amino acids of beans, corn delivers all the essential amino acids required by humans, while squash provides a rich source of vitamin A.

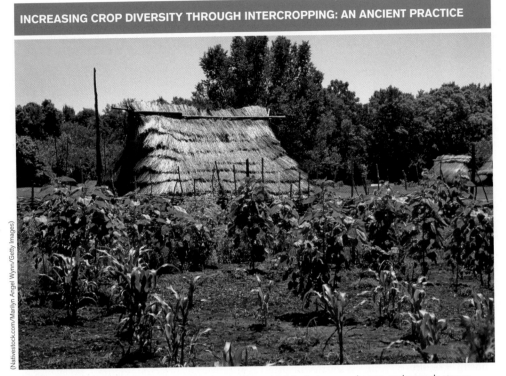

INCREASING CROP DIVERSITY THROUGH INTERCROPPING: AN ANCIENT PRACTICE

(Nativestock.com/Marilyn Angel Wynn/Getty Images)

FIGURE 7.34 The three sisters intercropping system, which includes corn, beans, and squash, was a traditional planting system used by Native Americans.

Modern researchers and farmers are rediscovering the benefits of intercropping. Faced with the daunting task of feeding its 1.3 billion people on limited agricultural lands, China has been investing heavily in agricultural research. The ultimate prize in such research would be the development of low-input (low-cost), high-yield (high-profit) agricultural systems. Examples of Chinese intercropping systems include growing sugarcane and potatoes or wheat and broad beans together on the same field. Intercropping has increased yields from 33% to 84%, reduced the incidence of crop diseases, and increased the income of farmers.

⚠ **Think About It**

1. Aside from monetary costs, what is the difference in relieving hunger by buying and shipping food from developed countries versus helping local farmers improve their farm production?

2. What are the benefits of crop genetic diversity? Besides rice, do you know of any other crops threatened by a lack of diversity? Cite examples.

3. Why does intercropping or crop rotation generally give a bigger boost to crop production when one of the crops involved is a legume, such as soybeans or alfalfa?

7.11 Sustainable farming, forestry, and ranching practices can reduce soil losses and improve soil fertility

Ultimately, the sustainable production of terrestrial resources depends critically on a healthy environment, especially healthy soils. Farming, ranching, and forestry aimed at sustaining production also address many significant environmental issues.

Managing Tillage in Agriculture

Cultivating soil parallel to the slope of the land (straight up and down the slope) leads to rapid soil erosion. Simply by cultivating on the contours (across the slope) instead, farmers can greatly reduce soil erosion in hilly terrain and create interesting landscapes in the process (**Figure 7.35a**). In very steep terrain, terraces can retain sufficient soil to grow valuable crops. In some places, properly maintained terraces have sustained agricultural productivity on very steep slopes for many centuries (**Figure 7.35b**). Where wind erosion is a potential problem, farmers have planted windbreaks, which reduce the force of the wind (**Figure 7.35c**).

Recent developments in agriculture are greatly reducing the intensity of cultivation and, as a consequence, the rate

What characteristics does intercropping share with natural ecosystems?

SEVERAL FARMING PRACTICES CAN REDUCE SOIL EROSION ON AGRICULTURAL LANDS

a. Contour plowing

b. Terracing

c. Tree windbreaks

FIGURE 7.35 Cultivating and planting across the slope in hilly farmland, such as this farm in Pennsylvania, helps reduce soil erosion. Terracing has prevented soil loss on very steep terrain where rice has been farmed for centuries in Southeast Asia. Windbreaks of trees are used extensively to reduce soil loss due to wind erosion.

of erosion. These techniques, called **low-till** (*till* is short for "tillage") or **no-till agriculture,** create less soil disturbance and soil compaction and leave crop residues on the field. Crop residues are parts of the crop that are not harvested for use, such as corn or wheat stalks; their roots provide structure to hold together the soil while leaf parts left on the soil protect it from erosion in the off-season.

These methods also allow the soil to maintain more moisture. Low- and no-till agriculture techniques require specialized seeding equipment, such as seed drills, to effectively plant a crop (**Figure 7.36**) by placing seeds at the proper depth in the soil and the correct distance from neighboring seeds. In the absence of conventional tillage, farmers control weeds with herbicides or by planting cover crops of plants called *green manure.* While growing, green manure suppresses weed growth, prevents soil erosion, and where nitrogen-fixing plants are used as a cover crop, they add nitrogen to soils. When a farmer is ready to plant, the

NO-TILL AGRICULTURE REQUIRES SOME SPECIALIZED EQUIPMENT

no-till (low-till) agriculture
An approach to growing crops involving reduced or no cultivation; creates less soil disturbance and leaves crop residues on the field.

FIGURE 7.36 Agricultural engineers have invented equipment designed to address the special challenges presented by no-till agriculture; for example, the seed drill shown here plants a crop through a dense cover of crop residue and live noncrop plants.

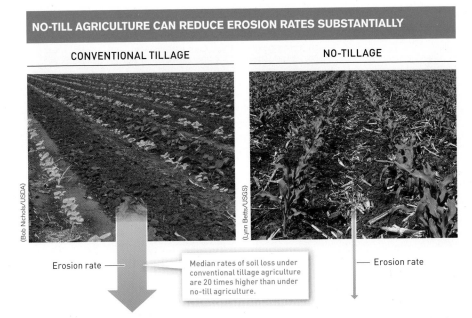

NO-TILL AGRICULTURE CAN REDUCE EROSION RATES SUBSTANTIALLY

CONVENTIONAL TILLAGE

NO-TILLAGE

(Bob Nichols/USDA)

(Lynn Betts/USGS)

Erosion rate

Erosion rate

Median rates of soil loss under conventional tillage agriculture are 20 times higher than under no-till agriculture.

FIGURE 7.37 A comparison of median rates of soil loss by erosion from fields under conventional tillage shows that no-till agriculture lowers erosion rates dramatically. (Data from Montgomery, 2007)

green manure is mown and left to mulch the soil, which adds organic matter and improves soil fertility.

Reduced tillage produces several improvements over conventional tillage, including lower rates of erosion, because crop residues physically shelter the soil surface from wind and rain. No-till agriculture has an average erosion rate one-twentieth of that associated with conventional tillage (**Figure 7.37**).

Organic Farming

Organic techniques can be used to produce food without expensive and unsustainable inputs of pesticides and herbicides. In the United States, organic food can be certified when the producer does not use synthetic fertilizer, pesticides, or GMOs. Critics of organic foods claim that organic crops are lower yielding because of higher loss to pests or decreased productivity from not using pesticides or synthetic fertilizers. However, a global study showed that organic methods can produce an almost equal amount of food as non-organic methods, indicating that organic agriculture can be a viable part of a sustainable agricultural solution.

Another trend related to organic farming is emphasizing local food production such as farmer's markets and locally sourced products. These markets provide locally grown produce—that is usually also organic—directly to consumers. Advantages include reduced transportation and storage cost, since the food travels fewer miles, and eating foods just when they are locally in season means that those foods do not need to be grown in a greenhouse or imported when the product is out of season.

Livestock Stocking Rates and Resting the Land

Raising livestock on rangelands can lead to desertification and soil loss. Sustainable ranching depends on putting the right number of livestock on a plot of land, based on the climate, soil type, and plant growth (**Figure 7.38**). Carefully managing stocking rates

INTENSITY AND TIMING OF GRAZING ARE KEY TO SUSTAINING RANGELAND PRODUCTIVITY

(USDA photo by Jack Dykinga)

FIGURE 7.38 As shown on this healthy semiarid rangeland, carefully managing stocking rates can maintain plant cover, which greatly reduces rates of soil erosion.

FIGURE 7.39 Closing roads and reestablishing vegetation on the natural slope of the land, as shown here, greatly reduce soil loss on logged forests.

ROADS IN MANAGED FORESTS ARE A MAJOR SOURCE OF SOIL EROSION

(U.S. Forest Service)

Initial stage of forest road restoration

(U.S. Forest Service)

Land contour restored; seed, fertilizer, and mulch applied

How might no-till farming affect the biodiversity of soil organisms?

shelterwood harvesting
Removes the tallest trees in a series of partial cuts, leaving behind enough of a forest canopy to provide shelter for speedy regrowth of shade-tolerant trees (e.g., red oak, American beech).

selective logging The clearing of land for lumber that focuses on the most mature, high-value trees, leaving the forest ecosystem largely intact.

also reduces soil damage by compaction, which crushes soil pore spaces critical for plant growth. Studies have shown that "resting the land," which involves removing livestock for some period, can reverse soil compaction. For example, two years of rest following 11 years of continuous grazing by sheep on pastures in Oregon was sufficient to reverse the effects of soil compaction, including the restoration of soil pore spaces, which are critical for soil aeration.

Forest Harvest, Landscape Restoration, and Fire Management

Forestry inevitably increases soil erosion. However, a number of practices can reduce soil loss. First, foresters should minimize the disruption of the land surface from heavy equipment and road building during logging. Timber harvests can be conducted in ways that reduce soil losses. More environmentally sensitive variations on clear-cutting involve leaving a few standing dead trees,

for wildlife habitat or cutting only strips out of forests to reduce high winds and erosion. **Shelterwood harvesting** removes the tallest trees in a series of partial cuts, leaving behind enough of a forest canopy to provide shelter for the speedy regrowth of shade-tolerant trees, such as red oak or American beech.

Another alternative to clear-cutting for timber is **selective logging,** which focuses on the most mature, high-value trees, leaving the forest ecosystem largely intact. Once an area is logged, restoring vegetative cover can greatly reduce soil losses. However, the key to reducing erosion on logged landscapes is to remove logging roads and restore the natural contours of the landscape, particularly at stream crossings (**Figure 7.39**).

High-intensity forest fires often lead to catastrophic erosion (see Figure 7.20b, page 204), whereas low-intensity fires generally do not. Consequently, thinning the understory can reduce the intensity of forest fires, helping sustain forest soils and the forest itself. The 2007

REDUCING TREE DENSITY AND CONTROLLED BURNING OF DEAD WOOD CAN LOWER FOREST FIRE INTENSITY

FIGURE 7.40 The first photo shows an area burned during the 2007 Angora Fire that had not been treated by thinning and fuel reduction and where no trees survived the fire. Compare this to the second photo of a treated area burned in the fire, where 90% of the trees survived and where the fire burned mainly along the ground and not in the canopy. (Data from Safford et al., 2009)

(H. D. Safford/U.S. Forest Service)

Trees not thinned or fuel reduced prior to Angora fire

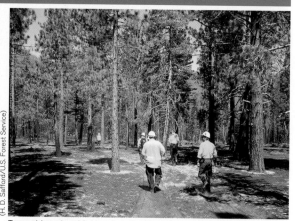

(H. D. Safford/U.S. Forest Service)

Trees thinned and fuel reduced prior to Angora fire

AN ANCIENT PRACTICE SUSTAINED FERTILITY OF RAIN FOREST SOILS

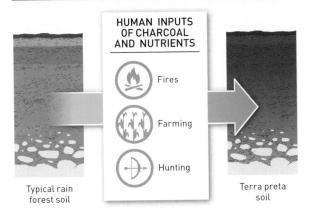

FIGURE 7.41 Fertile terra preta soils have developed in the Amazon Basin as a consequence of human activity associated with small urban centers dating as far back as 7,000 years before the present.

Angora Fire, which burned 1,243 hectares (3,072 acres) of forest in the Lake Tahoe Basin on the border between California and Nevada, provided a dramatic demonstration of how vegetation management can reduce forest fire intensity. By coincidence, the Angora Fire burned 194 hectares (479 acres) of forest that had been thinned to reduce the danger of intense fire, while the remainder of the fire swept through forest that had not been thinned. The difference in fire intensity and tree mortality was stark (**Figure 7.40**).

In areas where the forest had been thinned and the density of fuel reduced, tree survival was over 85%, compared with only 22% survival in areas that had not been thinned. The lower-intensity fire that resulted from

forest management ultimately left more living trees, which reduced the potential for soil losses from erosion. Periodic controlled burning of forests can reduce buildup of fuels capable of creating severe fire conditions and can thus reduce the chance of massive soil loss following an intense wildfire.

An Ancient Practice That Sustains Soil Fertility

Though tropical forests in the Amazon River Basin are lush, their soils tend to be infertile, making them largely unsuitable for agriculture (see Figure 7.6, page 195). However, soil scientists and archaeologists have noticed that the Amazon Basin is dotted with large patches of dark, fertile soils called **terra preta,** which literally means "dark soil" in Portuguese. Many areas of terra preta were in continuous cultivation for centuries, and although abandoned long ago, they are still capable of producing twice the crop yield per hectare compared with nearby, unaltered rain forest soils.

How were these islands of fertility created and by whom? It turns out that native populations of the Amazon Basin added organic matter and nutrients in the form of wastes from fish and game harvested from the surrounding landscape, along with human wastes and an abundance of charcoal (**Figure 7.41**). Rather than slash-and-burn agriculture, which releases a significant amount of carbon dioxide into the atmosphere, these people conducted slash-and-char agriculture, using low-heat fires to produce charcoal structures, which are ideal for holding nutrients within soils so that they are not washed out by heavy tropical rains.

In fact, these charcoal deposits in terra preta soils have inspired scientists around the world to begin experimenting with the use of biomass to produce charcoal, so-called biochar, as a soil additive (**Figure 7.42**).

AN OLD TECHNIQUE IS BEING USED TO SUSTAIN SOIL FERTILITY

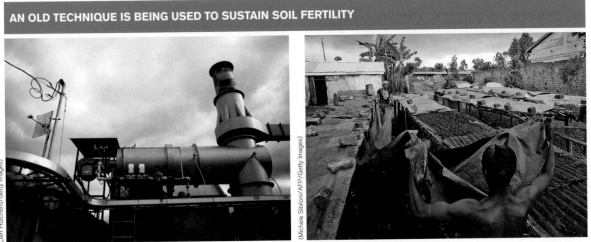

Industrial-scale biochar production system

Simple biochar production system

terra preta Dark, fertile soils high in charcoal and nutrient content, created by native populations in the Amazon River Basin before the arrival of Europeans.

FIGURE 7.42 Modern technologies for producing biochar are developing rapidly. However, simple small-scale systems are also capable of producing this effective soil additive.

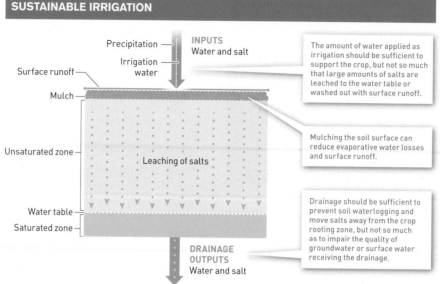

CAREFUL WATER AND SALT MANAGEMENT ARE ESSENTIAL FOR SUSTAINABLE IRRIGATION

FIGURE 7.43 Preventing waterlogging and salinization of irrigated soils depends on applying water at rates that do not exceed rates of drainage.

⚠ Think About It

1. Name ways we can practice agriculture and still sustain our soils.

2. How would you balance the benefit of reduced soil erosion under no-till agriculture against the potential environmental costs associated with applying chemical herbicides?

3. While wildfires are generally considered a destructive force, explain how excluding fire from some ecosystems can also be considered a disturbance and a destructive policy.

4. What other soils might benefit from the ancient practice of incorporating charcoal similar to the terra preta soils of the Amazon Basin? (Hint: Consider Figure 7.6, page 195.)

7.12 Sustainable irrigation requires careful management of water and salts

Our food security depends on freshwater stored in manmade reservoirs, natural underground aquifers, and water transported across vast distances in aqueducts or other waterways. In 2010 approximately 43% of global cereal production (mainly wheat, rice, and maize) occurred on irrigated land. Without irrigation,

mulch A natural or synthetic covering to the soil surface that conserves moisture, reduces soil temperature variation, and decreases growth of weeds.

production on these lands would decrease by nearly half, reducing global production of cereals by 20%. However, agricultural production is much more dependent on irrigation in some regions than in others. Southern Asia, for example, would lose an estimated 45% of cereal production, and Northern Africa, 66%. Europe, by contrast, receives enough regular rainfall that its cereal production would be unaffected. This means that food security over the long term will be achieved only if irrigation is sustainable, which requires careful management of water and salts.

Water Management

Efficient use of water is one of the most important elements of sustainable irrigation. Techniques to improve water use efficiency contribute to sustainable irrigation directly by reducing withdrawals from groundwater or surface water. For example, careful leveling and sloping of fields watered by flood irrigation (see Figure 7.21a, page 205) increase the uniformity of irrigation, thereby decreasing the amount of water needed to achieve high levels of crop production. Delivering just the right amount of water to the rooting zone of growing crop plants with a precision technique such as drip irrigation (see Figure 7.21c, page 205) can save even more water.

More efficient water use can also be achieved by reducing water losses from crops and the soil surface. One way to decrease evaporative water losses is to cover the soil surface with **mulch,** a natural or synthetic covering to the soil surface that conserves moisture, reduces soil temperature variation, and decreases the growth of weeds.

Salt Management

Salts are added to soil with every drop of irrigation water. This means farmers must ensure that soils are well drained in order to avoid raising the water table, which causes waterlogging and salinization (see Figure 7.22, page 206). Using highly efficient irrigation systems to deliver just the amount of water needed to sustain agricultural production, while not producing excessive saline drainage water, can reduce the salts exported to groundwater and surface water (**Figure 7.43**).

⚠ Think About It

1. What practices can reduce the incidence of waterlogged soils in irrigated landscapes?

2. How could irrigating only when a crop needs water reduce the amount of drainage water that needs to be managed?

3. Development of the Colorado River Basin (see Figure 6.11, page 166) included converting approximately 1 million hectares (2.5 million acres) of land in the basin to irrigated agriculture. A significant rise in the salinity of Colorado River water, particularly downriver near Mexico, has been linked to this development. Explain the link between these events.

7.13 Integrated approaches to pest control can reduce pesticide pollution and evolution of pesticide resistance

Plants in natural ecosystems are generally protected by a combination of their own chemical and physical defenses, as well as by predators and pathogens of their insect enemies. As farmers cope with failing pesticides and devastating insect outbreaks, they turn to natural ecosystems for inspiration in guiding pest control.

Integrated Pest Management

Pesticide resistance, as we saw in the case of the Colorado potato beetle, has been one consequence of intensive agriculture, which has been sustained by the application of larger quantities and varieties of pesticides. To avoid getting caught in an endless cycle, agriculturalists developed **Integrated Pest Management,** or **IPM,** which incorporates multiple sources of information to contain pest damage within acceptable limits, while trying to minimize harm to people, property, and the environment. Practitioners begin by investing in preventative measures, such as planting pest-resistant crop varieties, rotating crops to prevent the buildup of pests, and removing diseased or infested plants. During the growing season, they monitor for early signs of both pest populations and natural enemies of the pests, which could keep those pests in check (**Figure 7.44**). Finally, they begin killing pest insects using highly targeted techniques, such as trapping

Integrated Pest Management (IPM)
An approach to managing pests (e.g., insects, pathogens, weeds) that incorporates multiple sources of information to contain pest damage within acceptable limits while trying to minimize harm to people, property, and the environment.

INFORMATION IS ESSENTIAL TO SUCCESSFUL INTEGRATED PEST MANAGEMENT

(USDA photo by Jack Dykinga)

FIGURE 7.44 Monitoring pest populations to determine whether control measures are warranted is a key element of IPM.

INTEGRATED PEST MANAGEMENT RECOMMENDS A GRADED RESPONSE TO PEST CONTROL

(© Alistair Scott/Alamy)

FIGURE 7.45 As pest populations begin to build, IPM begins with control methods that have the lowest environmental impact. For example, traps such as this one containing chemicals that attract specific pest species, can be effective at controlling some problem species and have little environmental impact.

and mechanical controls, before resorting to broad-spectrum pesticides (**Figure 7.45**).

GM crops such as Bt corn have not ended the need for IPM because pests can still evolve resistance to these insect-killing crops. The key to controlling Bt-resistant pests is to make sure that each generation of the crop pest includes susceptible individuals. To do so, farmers plant a certain proportion of their fields in non-Bt varieties of their crop. Almost all the insect pests living on these non-Bt crop refuges will be susceptible to Bt. The mates for the small number of resistant pests growing to maturity on the Bt variety of the crop will be almost entirely Bt-susceptible individuals (**Figure 7.46**). Consequently, the offspring of the resistant individuals mating with nonresistant individuals will be killed when they feed on Bt crop tissue, and the resistance traits will not increase in frequency in the pest population.

Herbicide-resistant GM crops have been widely adopted by farmers, making up, for example, 91% of the soybeans grown in the United States. However, with the widespread use of herbicides, farmers are observing the evolution of herbicide-resistant "super weeds." The key to avoiding herbicide resistance is for farmers to vary the form and strength of evolutionary selection on the weed population. Applying the same herbicide repeatedly to a weed population will exert strong directional selection on the population for resistance to that herbicide (see Figure 3.6, page 65). In such circumstances, if resistance traits exist in the weed population, they will likely increase in frequency.

Rather than applying the same herbicide repeatedly, farmers can combine herbicide control with old-fashioned mechanical weeding. Alternatively, a farmer can switch between two types of herbicides with different modes of action. To be successful, sustainable agriculture requires paying close attention to the crop environment and having a sophisticated understanding of multiple natural processes, including evolution by natural selection.

THE REFUGE TECHNIQUE

By flooding the environment with susceptible individuals, the refuge technique greatly reduces the probability that two resistant pests will mate with each other, but will instead mate with susceptible individuals emerging from the non-Bt refuge.

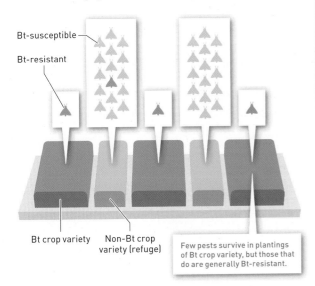

Bt-susceptible

Bt-resistant

Bt crop variety

Non-Bt crop variety (refuge)

Few pests survive in plantings of Bt crop variety, but those that do are generally Bt-resistant.

FIGURE 7.46 The refuge technique for preventing evolution of Bt resistance depends on producing abundant Bt-susceptible mates for any resistant variants of the pest that emerge in each generation. The success of this approach depends on the offspring of matings between susceptible and resistant pests being poisoned by the Bt toxin.

⚠ Think About It

1. How does Integrated Pest Management (IPM) address some of the most severe environmental impacts associated with pesticide use?

2. How do you weigh the concerns over the planting of GM crops against their benefits when used in a program of IPM?

3. Would the refuge method for preventing the evolution of Bt resistance in crop pests (see Figure 7.46) work if Bt-resistant pests preferred to mate with other Bt-resistant individuals?

7.10–7.13 Solutions: Summary

The most sustainable approach to combating malnutrition and undernourishment around the world may be to help local farmers produce adequate quantities of nutritious food. Biodiversity in agriculture increases agricultural production and reduces chemical and energy inputs.

To minimize loss of healthy, fertile soils, cultivating with the contours in sloping lands or constructing terraces can reduce soil erosion. Growing crops with little tillage (low-till) or without tillage (no-till) can further reduce soil losses by decreasing soil disturbance and leaving crop residues on the field. Selective logging, restoring vegetative cover, and removing logging roads after logging help reduce soil losses from managed forests. Managing fuel and fire can reduce the chance of intense, catastrophic forest fires that lead to rapid soil erosion. On rangelands, proper stocking rates and periodic rest of the land can prevent desertification. Native populations of the Amazon Basin sustained agricultural production in highly weathered tropical forest soils by adding nutrients and charcoal rather than practicing nomadic, slash-and-burn agriculture.

Sustainable irrigation requires careful management of water and salts. More precise irrigation is an effective way to conserve water and control salts.

The key elements of Integrated Pest Management (IPM) are information about the state of pest populations and populations of their natural enemies, preventative practices to discourage pest populations, and staged control measures, beginning with the least risky. IPM increasingly involves GM crops, especially those that have been engineered for herbicide and insect resistance.

Answer the following questions for each chapter section and then answer the Central Question.

Central Question: How can we produce food and forest products while minimizing environmental impact?

7.1–7.4 Science

- How do climate and biodiversity influence terrestrial primary production?
- How is soil formed?
- What types of agricultural systems are built on natural biomes?

7.5–7.9 Issues

- How do farming, grazing, and forestry practices affect soil?
- In what ways does irrigation alter the quality of soil?
- What are the problematic consequences of intensive agriculture?
- In what ways are genetically modified (GM) crops a source of controversy and potential?

Terrestrial Resources and You

Terrestrial resources provide us with food, fuel, and building material. Because these resources are so central to our lives, there are many opportunities to use them more sustainably in everyday life.

☐ *Keep informed.*

Learn about the sources of food in your community. In addition to large grocery stores, are there farmer's markets where you can buy directly from local growers? (Check out localharvest.org to find markets near you.) What water resources are available for agriculture in your area? If there are forestlands in your area, learn how they are managed.

☐ *Eat lower on the food chain.*

In Chapter 6, we considered ways to reduce the amount of water we use in our daily lives for sanitation and other uses. However, as we have seen, the largest user of water around the world is agriculture. Despite that, not all agricultural products require the same amount of water for their production. Among the largest users of water is the production of animal products, especially beef. Recent estimates indicate that shifting to a vegetarian diet would reduce water consumption related to food production by 36%. Even if you are not a vegetarian, one of the most effective ways to reduce the amount of water used to produce the food you eat is to eat fewer animal products. Even one meat-free meal a week for a family of four can save over 100 pounds of meat over the course of a year.

☐ *Grow some of your own food.*

Consider planting a garden, if you have sufficient space, or growing herbs and vegetables in a small raised bed planter or in pots. Many communities offer the opportunity to garden through community gardens. Growing even a small amount of your own food connects you to the process of food production and helps you understand the problems faced by farmers.

☐ *Buy and consume sustainably produced terrestrial resources.*

As a consumer, you have the power to influence the way the products you purchase are produced. In respect to terrestrial resources, you can choose to purchase products that are certified to have been produced using sustainable agricultural and forestry practices. For example, the Rainforest Alliance certifies consumer goods ranging from agricultural and forestry products to home and office supplies. You can find information on where to purchase sustainably produced foods from a variety of sources, such as the online Eat Well Guide. When purchasing animal products, such as meat, eggs, or milk, consider products endorsed by an organization such as Animal Welfare Approved.

- What are the benefits of increasing genetic and crop diversity?

- How can sustainable farming, forestry, and ranching practices reduce soil loses and improve soil fertility?

- What factors must be managed to irrigate sustainably?

- What is Integrated Pest Management and what are its benefits?

Answer the Central Question:

Chapter 7

Review Questions

1. Which of the following ecosystems is most likely to support the highest level of primary production?
a. Tropical forest
b. Temperate forest
c. Hot desert
d. Temperate grassland

2. Which of the following soil horizons generally contains the lowest amount of organic matter?
a. O horizon
b. A horizon
c. B horizon
d. C horizon

3. Why do tropical forest soils generally contain low amounts of organic matter?
a. Low levels of primary production by tropical forests
b. Lack of decomposition of organic matter in tropical forests
c. High rates of decomposition of soil organic matter
d. An absence of soil animals, such as ants, in tropical forests

4. Which of the following is a major source of erosion in forests?
a. Road building
b. Wildfires
c. Soil disturbance by heavy equipment
d. All of the above

5. In which of the following situations would irrigation most likely produce soil salinization?
a. In soils with high rates of drainage and low levels of irrigation water inputs
b. In soils with low rates of drainage and high levels of irrigation water inputs
c. In soils with a low (deep) water table
d. In sandy soils with high rates of leaching

6. The individuals carrying genes for pesticide resistance are generally rare in a pest population prior to exposure to a pesticide. What does this suggest about the pesticide-resistant individuals in the population?
a. Pesticide-resistant individuals are at a competitive disadvantage relative to other individuals in the population in the absence of the pesticide.
b. Pesticide-resistant individuals are at a competitive advantage relative to other individuals in the population in the presence and absence of the pesticide.
c. Pesticide-resistant individuals are at a competitive disadvantage relative to other individuals in the population in all environments.
d. There are no differences in competitive ability between pesticide-resistant individuals and pesticide-sensitive individuals.

7. What determines whether a genetically modified organism (GMO) is transgenic?
a. It carries genes different from wild ancestors.
b. It is more resistant to attacks by insects.
c. It grows at faster rates than other varieties.
d. It carries a gene from another species.

8. Which of the following was not part of the Green Revolution?
a. Use of pesticides
b. Conventional tillage
c. Genetically modified crops
d. Intensive plant breeding programs

9. Which of the following approaches to farming will generally have the lowest fuel costs?
a. Conventional tillage farming
b. Low-till farming
c. No-till farming
d. All approaches to farming have similar fuel costs.

10. Which of the following is not part of IPM?
a. Closely monitoring the size of pest populations
b. A graded response to increasing pest populations, beginning with trapping
c. Preventative measures such as planting pest-resistant crop varieties
d. Eventually eradicating pest populations with chemical sprays

Critical Analysis

1. Why do some environmental observers say that the impact of intensive agriculture is really a population problem rather than a problem with agriculture?

2. At the height of the Dust Bowl (see Figure 7.12, page 199), U.S. President Franklin D. Roosevelt said, "The nation that destroys its soils destroys itself." Explain his rationale.

3. How are the concepts of evolutionary biology, community ecology, and ecosystem ecology increasingly informing today's agriculture?

4. Refer to Figure 4.3 (page 99), Figure 7.4 (page 193), and Figure 7.6 (page 195), summarizing the global distribution of biomes, climates, and soils. Discuss the influence that these factors may have had on the development of agriculture.

5. Develop a detailed conceptual design for farming sustainably on land requiring irrigation for crop production.

Find additional resources and links online at www.macmillanhighered.com/launchpad/molles1e.

Central Question: Can we sustainably manage fisheries and aquaculture?

Describe the ways in which fish and shellfish are harvested and how the physical environment influences the availability of aquatic resources.

SCIENCE

CHAPTER 8

Sustaining Aquatic Resources

Analyze the impacts of humans
on aquatic resources.

Discuss approaches for developing more
sustainable aquaculture and fisheries.

ISSUES

SOLUTIONS

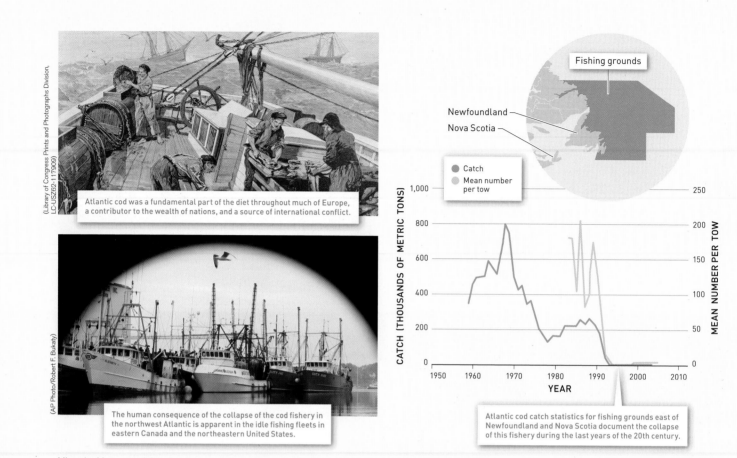

Atlantic cod was a fundamental part of the diet throughout much of Europe, a contributor to the wealth of nations, and a source of international conflict.

(Library of Congress Prints and Photographs Division, LC-USZ62-117909)

The human consequence of the collapse of the cod fishery in the northwest Atlantic is apparent in the idle fishing fleets in eastern Canada and the northeastern United States.

(AP Photo/Robert F. Bukaty)

Fishing grounds

Newfoundland
Nova Scotia

Catch
Mean number per tow

Atlantic cod catch statistics for fishing grounds east of Newfoundland and Nova Scotia document the collapse of this fishery during the last years of the 20th century.

Historical images of the cod fishery of the northwest Atlantic, a fishery that produced abundant harvests for five centuries and then collapsed. (Data from Lilly et al., 2006; cited in Stares et al., 2007)

A Tale of Overharvesting and Fishery Collapse

When Europeans first began fishing for Atlantic cod off North America, they encountered an astonishing abundance of fish. Now, five centuries later, that natural wealth is nearly gone.

The Atlantic cod (*Gadus morhua*) is a bottom-dwelling, predatory fish that can reach 200 pounds. It is a top, or apex, predator that prowls the ocean floor, hunting for smaller fish and crustaceans. Beginning in the 15th century, people began to prize it for its flaky, white flesh and for the fact that it could be preserved for long periods of time by drying and salting. It didn't take long before cod became the focus of one of largest and most valuable

fisheries on the planet. The worldwide cod harvest peaked in the 1960s, when massive bottom trawlers—nets dragged along the sea floor—scooped up 3.9 million tons of codfish in a single year. The ocean's bounty seemed endless.

But in the spring of 1992, fishing boats returned to ports from New Bedford, Massachusetts, to Halifax, Nova Scotia, with their hulls nearly empty. Cod stocks dropped to less than 1% of historic levels, and regulators shut down the prime fishing grounds on Georges Bank, an elevated plateau off the Atlantic coast of the United States and Canada. Retailers everywhere switched white fish recipes from cod to Alaskan pollock and New Zealand hoki. The cod **fishery** had collapsed.

"If the oceans don't make it, neither will we."

Jackson Browne, "If I Could Be Anywhere," 2010

Fisheries collapse occurs when annual catches of the species in question decline below 10% of the historic catch. Today, following the collapse of the cod fishery in the northwest Atlantic and the subsequent closing of the fishing grounds, the fishing fleets that harvested hundreds of thousands of tons of cod each year now sit idle. What factors led to the collapse of this, and other, fisheries in the United States? What, if anything, can be done to restore valuable fisheries here and around the world? In this chapter, we will explore the science behind fisheries and what steps we need to take to restore them to their former health.

The collapse of the rich cod fishery off New England and Canada is not unique. Similar crashes have occurred in other fisheries around the world due to the rise of industrial fishing in the 20th century. In the United States, the National Oceanic and Atmospheric Administration (NOAA) considers 40 fish stocks to be overfished and 28 subject to continued overfishing. Once a major fish exporter, the United States now imports the majority of its seafood. These fishery declines threaten marine species with extinction, as well as the food webs and productivity of marine ecosystems, which in turn affects the human communities that depend on them for ecosystem services and economic livelihood.

fishery A population of fish or shellfish, and the economic system involved in harvesting the population, often identified by the geographic area where the fish or shellfish are harvested.

fisheries collapse The decline in a certain species' annual catch below 10% of its historic catch.

Central Question

Can we sustainably manage fisheries and aquaculture?

(Bill Dewey, Taylor Shellfish Farms)

8.1–8.3 Science

We often think of early humans as hunters of large land mammals, but fishing has long provided subsistence to populations that lived near rivers, lakes, and especially the ocean. For thousands of years, humans have harvested an abundance of freshwater fish and seafood, leaving leftovers from these ancient seafood meals in great heaps called *middens,* which have been identified by archaeologists along coastlines around the world. The future productivity of the world's fisheries depends not only on how we harvest them, but also on the health of the earth's ecosystems.

8.1 Commercial fish populations are heavily harvested and actively managed

Around the world, fisheries employ approximately 40 million people who regularly harvest some 1,500 species. The total value of fisheries and aquaculture around the world is estimated at $217.5 billion. Target species include molluscs, such as scallops and clams; crustaceans, such as lobster, crab, and shrimp; freshwater fish, such as catfish and trout; and marine fish, such as tuna and anchovy. Out of the 90 million metric tons (99 million tons; 1 metric ton equals 1,000 kilograms, or 2,204 pounds) of fish captured around the world in 2011, about 90% comes from marine fisheries; the remaining 10% comes from inland rivers and lakes, according to the United Nations Food and Agriculture Organization (FAO).

An additional 60 million metric tons come from **aquaculture,** the controlled growing of aquatic

organisms, including fish, shellfish, algae, or plants, as a crop, mainly for food. From 1950 to 2011, global fish consumption increased from less than 20 million metric tons per year to about 150 million metric tons. In addition to the fish specifically targeted by fishers, other marine species have also been impacted through **bycatch,** the killing and discarding of noncommercial fish, birds, dolphins, sea turtles, and other wildlife that occurs as a result of contact with fishing gear. Because bycatch can affect so many nontarget species, it can have widespread negative effects on the marine food web.

Types of Fishing

The earliest fishers practiced **subsistence fishing,** catching enough for themselves and their families, including a small amount to be bartered or sold. Some of the simplest techniques involve catching fish by hand in shallow waters, spearing or trapping them, or simply scooping them out of the water with a net or other vessel. Today, subsistence, or noncommercial fishing, is still practiced in many indigenous and rural communities around the world. For instance, the state of Alaska allows residents to catch a certain number of salmon each year using a handheld dipnet or a snag hook—a hook that allows fishers to yank fish out of the water without the use of bait. Alaskan residents with the proper permits are allowed to subsist on a wide range of other species, including halibut, crab, and clam. In general, subsistence fisheries tend to be small and have a limited impact on fish populations.

In contrast to subsistence fishing, **commercial fishing** involves catching fish for profit, and it represents the vast

aquaculture The controlled growing of aquatic organisms (e.g., fish, shellfish, algae, or plants) as a crop, mainly for food; carried out in marine, brackish water, or freshwater environments.

bycatch Discarded catch and mortality of any organism (e.g., fish, invertebrate animals, birds, dolphins, sea turtles) as a result of contact with fishing gear.

subsistence fishing The practice of catching enough fish for one's family plus a bit more for bartering or selling.

commercial fishing Catching fish for profit; represents the vast majority of the fish captured around the world.

majority of the fish captured around the world. More than 90% of commercial fishing around the world is done by **small-scale fishers,** who use minimal gear, such as handlines or hand nets, and may fish from small boats such as motorized skiffs or nonmotorized canoes. They typically stay close to the coastline and fish for only a few hours or days at a time.

Industrial fishermen use more expensive and technologically advanced gear to catch fish and may often travel for weeks at a time, processing and refrigerating or freezing their catch on board. **Bottom trawlers** drag weighted nets along the ocean floor in order to catch groundfish, including cod and flounder, along with scallops, shrimp, and crab. **Longline fishing** involves laying out a very long line with hundreds or thousands of baited hooks, used to catch tuna (near the surface) or groundfish (e.g., halibut, cod). **Gillnetting** involves placing a net with large mesh in the water column to selectively catch fish, including salmon, in the Pacific Northwest. The size of fish caught depends on the net mesh size. When a fish cannot pass all the way through the gillnet, it is ensnared by its gill covers when trying to retreat. Other methods, such as baited **pot-traps,** are used to catch lobster and crab.

Finally, **sport** or **recreational fishing** may include fly-fishing on a stream in Montana or hiring a tourist charter boat to catch trophy-sized fish, including sharks, swordfish, and tuna. Some sport-fishers eat their catch, whereas others engage in **catch-and-release fishing,** in which they release the fish back into the water where they caught it. However, care must be taken to ensure that the captured fish set free will survive the encounter.

Fisheries Management

The major question for fisheries regulators, such as Alaska Fish & Game or the National Marine Fisheries Service in the United States, is determining the level of sustainable harvest of a particular stock. A **stock** is loosely defined as a discrete subpopulation of a species, which is reproductively isolated from other stocks. Some wide-ranging species like southern bluefin tuna consist of a single stock, whereas dozens of salmon stocks are known, depending on the specific river to which they return for breeding. Fisheries managers perform a **stock assessment** to estimate the size of the fish stock, the rate at which it is growing, and the rate of sustainable harvest. A key piece of data for making stock assessments is the **catch-per-unit effort,** which is a measure of how many fish can be caught using a specific piece of gear—a net or a line—for a certain period of time. In "mark and recapture" studies, fisheries biologists may also release tagged fish and then try to catch them again to get a better idea of the size of a population or the boundaries of a particular stock.

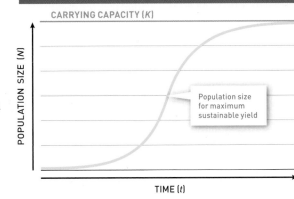

MAXIMUM SUSTAINABLE YIELD

CARRYING CAPACITY (K)

POPULATION SIZE (N)

Population size for maximum sustainable yield

TIME (t)

FIGURE 8.1 Theoretically, the maximum sustainable yield of a population is achieved when a population is growing at a maximum rate. In a population growing logistically, maximum growth rate occurs when the population size is one-half of the carrying capacity. Consequently, fisheries managers generally attempt to maintain fish populations near this size but not significantly lower or higher.

Although fisheries statistics can be complex, the basic principles are simple. When fishing pressures are high, fish populations decline. When fishing pressures are reduced, fish populations grow. However, scientists have noted that when fish exist at high densities, their level of reproductive success is lower. Consequently, harvesting fish to prevent them from reaching high densities can make the fishery more productive. However, if fish populations are reduced to extremely low levels, the fish have trouble finding suitable mates, and fishery productivity may thus decrease.

The goal of fisheries managers has long been to manage fish populations so that harvest rates are at or near a theoretical level called **maximum sustainable yield (MSY),** which is the maximum harvest of a renewable natural resource that does not reduce future yields. If we assume S-shaped, or logistic, population growth (see Figure 3.11, page 69), the maximum sustainable yield is expected when the population size is approximately one-half of the carrying capacity (**Figure 8.1**). At this size, the rate of population growth is highest, and recovery from harvest should be most rapid. In well-managed fisheries, ongoing scientific study provides information to help managers and fishers avoid depleting stocks below these levels by reducing or sometimes closing a fishery to continued harvest. When a fish stock drops below the population size that provides the MSY, it is considered overfished.

⚠ Think About It

1. Why is it important to determine the MSY for each stock?

small-scale fishers
Commercial fishers who use minimal gear and fish from small boats or nonmotorized canoes.

industrial fishermen
Commercial fishers who may travel for weeks at a time and use expensive, technologically advanced gear to process and refrigerate or freeze their catch on board.

bottom trawlers
Weighted nets dragged along the ocean to catch groundfish (e.g., cod, flounder, scallops, shrimp, crab).

longline fishing
The practice of laying out a very long line with hundreds or thousands of baited hooks; used to catch tuna (near the surface) or groundfish (e.g., halibut, cod).

gillnetting
The practice of placing panels of large mesh net in the water column to catch fish, the size of which depend on the mesh size; fish that cannot pass all the way through the gillnet are ensnared by their gill covers when trying to retreat.

pot-traps
Baited traps used to catch lobster or crab.

sport (recreational) fishing
The practice of fishing for pleasure (e.g., fly-fishing, hiring a tourist charter boat to catch trophy-sized fish).

catch-and-release fishing
The practice of releasing fish back into the water after catching them.

stock
A discrete subpopulation of a species, which is reproductively isolated from other stocks.

stock assessment
Estimated size of a fish stock, the rate at which the population is growing, and the rate of harvest.

catch-per-unit effort
A measure of how many fish are caught using a specific piece of gear—a net or a line—for a certain period of time.

maximum sustainable yield (MSY)
The maximum harvest of a renewable natural resource that does not reduce future yields (e.g., the sustainable annual catch from a fish population).

2. What are some potential reasons for the extremely rapid cod fishery collapse after being fished for hundreds of years?

8.2 Nutrient availability influences primary production in marine environments

The large fish we prize in the developed world, such as cod, salmon, and tuna, are usually apex predators that feed at or near the top of the food web. These fish and all other consumers in the food web would not exist without organisms at the bottom of the food web that convert solar energy into the chemical energy of sugars. The photosynthetic organisms of the oceans—including seaweeds, reef-building corals, and phytoplankton, the microscopic algae that drift with the ocean currents—account for roughly half of global primary production. Aquatic primary production varies widely across natural aquatic ecosystems, and it depends on climate and other forces that affect the global distribution of nutrients.

As wind blows across the surface of the ocean, it pushes and pulls water, creating ocean currents and influencing the nutrients available to fish stocks. On Earth, we have **prevailing winds** that blow

consistently from one direction, but they do not end up moving directly north or south. Rather, the rotation of Earth creates a deflection in the winds called the **Coriolis effect,** which deflects winds to the right in the Northern Hemisphere and to the left in the Southern Hemisphere. The result is the global pattern of prevailing winds: northeast trade winds, westerlies, and polar easterlies in the Northern Hemisphere; southeast trade winds, westerlies, and polar easterlies in the Southern Hemisphere (**Figure 8.2**). As the prevailing winds blow across the oceans, they set in motion oceanic currents. The Coriolis effect acts on these currents to create large-scale patterns of oceanic circulation that move to the right in the Northern Hemisphere and to the left in the Southern Hemisphere (see Figure 8.2). As a result, each hemisphere of each ocean basin has a large circular current called a *gyre,* which is centered under subtropical, high-pressure regions.

Oceanic currents exert major influences on regional climates by transporting heat or, in some situations, cooling waters from one region to another. The Gulf Stream in the Atlantic Ocean, for instance, transports heat from the tropics to higher latitudes, extending temperate climates much farther north in northwest Europe than would be the case otherwise (**Figure 8.3**). Meanwhile, the Labrador Current in the western Atlantic Ocean cools northeastern North America. The currents also modify the distribution of marine environments (**Figure 8.4**). For example, the currents extend cool

prevailing winds Winds that blow consistently from one direction (e.g., the northeast trade winds blow from the northeast).

Coriolis effect A deflection in the winds from a straight north–south path as a consequence of Earth's rotation on its axis from west to east; deflects winds to the right of their direction of travel in the Northern Hemisphere and to the left in the Southern Hemisphere.

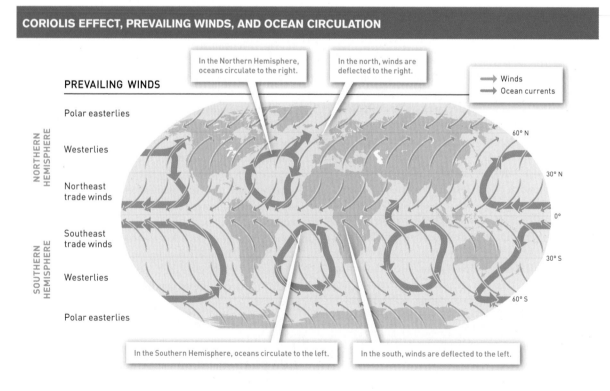

CORIOLIS EFFECT, PREVAILING WINDS, AND OCEAN CIRCULATION

FIGURE 8.2 The Coriolis effect deflects prevailing winds and ocean currents to the right of their direction of travel in the Northern Hemisphere and to the left in the Southern Hemisphere.

THE GULF STREAM OFF THE EAST COAST OF NORTH AMERICA

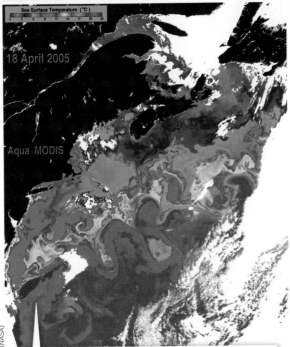

Sea Surface Temperature (°C)

18 April 2005

Aqua MODIS

(NASA)

The Gulf Stream, which transports warm tropical waters to northern latitudes, remains a visually distinctive water mass for thousands of kilometers and has been called a "river in the ocean."

FIGURE 8.3 Captured here in a satellite photo, the Gulf Stream is one of the best known of the major ocean currents. It was first mapped in 1770 by Benjamin Franklin and his cousin Timothy Folger, a whaling captain, using whalers' observations of water temperatures, color, and ocean life as they pursued their prey.

marine waters northward along the southwest coast of Africa and southward along Africa's northwest coast. This transport of cool surface waters significantly narrows the band of warm, tropical marine waters in the eastern Atlantic Ocean, compared with the western Atlantic.

Light, Nutrients, and Primary Production

Although light penetrates water, it grows weaker with depth, extending to a maximum depth of 200 meters and restricting photosynthesis to a surface layer of the oceans and lakes known as the **euphotic zone** (Figure 8.5). As organic matter produced in the euphotic zone sinks through the water column, it carries with it various elements essential for photosynthesis, such as nitrogen, phosphorus, and iron. Warmer surface layers are less dense than deeper

In March 2011 a tsunami washed massive amounts of debris, including entire houses, into the Pacific Ocean off Japan's east coast. Explain why several months later that debris began washing up on the western shores of North America.

euphotic zone A surface layer of the oceans and deep lakes where there is sufficient light to support photosynthetic aquatic organisms.

MAJOR MARINE ENVIRONMENTS

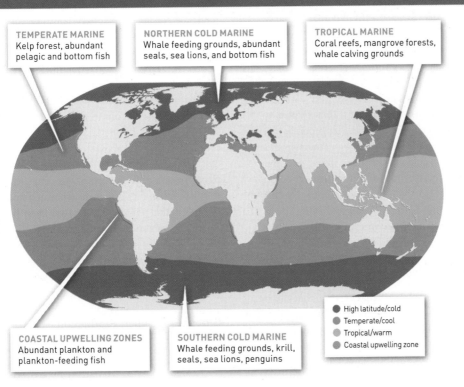

TEMPERATE MARINE
Kelp forest, abundant pelagic and bottom fish

NORTHERN COLD MARINE
Whale feeding grounds, abundant seals, sea lions, and bottom fish

TROPICAL MARINE
Coral reefs, mangrove forests, whale calving grounds

COASTAL UPWELLING ZONES
Abundant plankton and plankton-feeding fish

SOUTHERN COLD MARINE
Whale feeding grounds, krill, seals, sea lions, penguins

● High latitude/cold
● Temperate/cool
● Tropical/warm
● Coastal upwelling zone

FIGURE 8.4 Average ocean temperature defines the major marine environments. Boundaries between marine environments were mapped in the Northern Hemisphere using a February thermal image and in the Southern Hemisphere using an August thermal image. (National Virtual Oceanographic Data System [NVODS], http://ferret.pmel.noaa.gov/NVODS/)

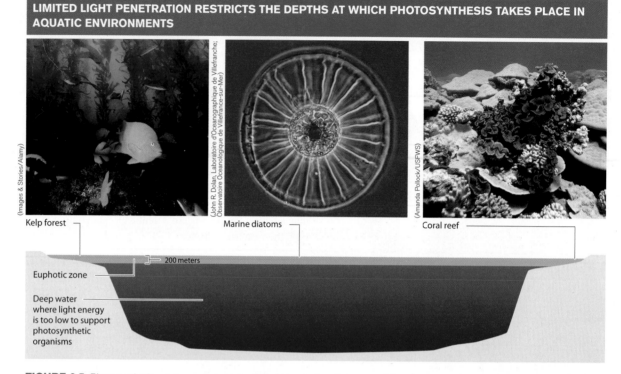

FIGURE 8.5 Photosynthetic marine organisms, such as kelp, marine diatoms, and the algae associated with reef-building corals, are limited to the euphotic zone, the surface layer of the world's oceans.

layers, which means there is little vertical mixing with deeper cool water. As a consequence, essential chemical nutrients depleted from warm surface waters build up in deeper cool-water layers as sinking organic matter decomposes, and primary production in the euphotic zone gradually declines. This means that any mechanism that promotes renewal of nutrients in surface waters, such as the vigorous mixing of deep and surface waters, will increase rates of primary production.

The process of **upwelling** does just that. Driven by prevailing or seasonal winds, upwelling generally occurs where winds blow warmer surface water away from shore and replace it with colder, subsurface water. As shown in Figure 8.2, extensive areas of upwelling occur along the west coasts of North and South America, North and South Africa, southwestern Europe, and along shores of the northwest Indian Ocean, where upwelling is driven by seasonal monsoon winds.

The Phosphorus Cycle

Phosphorus is one of the critical elements that can be brought to the surface via upwelling. While the nitrogen and carbon cycles include a major atmospheric reservoir, another critical biogeochemical cycle—the phosphorus cycle—does not. As **Figure 8.6** shows, phosphorus enters the cycle through the weathering of rock, so it begins its journey on land before becoming important

in the ocean. Phosphorus released by weathering is taken up from the soil by plants and incorporated into plant tissues, where it is used to form cell membranes, nucleic acids, and the energy-bearing molecule **ATP** (**adenosine triphosphate**). Herbivores feeding on plant tissues ingest phosphorus, as do carnivores feeding on herbivores. Phosphorus is then released by these animals when feces or urine is returned to the soil; phosphorus is also released from dead and decaying organic matter by detritivores and decomposers. Once released into soils, this phosphorus can be taken up again by plants and recycled within the ecosystem or exported to the ocean by rivers or wind, where it participates in similar cycling patterns with algae, zooplankton, and fish.

When a fish dies and is incorporated into marine sediment, that sediment often ultimately becomes a rock that integrates phosphorus into a mineral, closing the cycle. Other minerals in the ecosystem that, like phosphorus, do not have gaseous forms either (e.g., iron, potassium) undergo similar cycles, with only minor variations in some of the details.

Global Patterns in Production

The highest levels of marine primary production lie along the margins of the continents, especially where upwelling brings nitrogen and phosphorus-rich deep waters to the surface euphotic zone (**Figure 8.7**). However, there are

upwelling The movement of cold subsurface water to the ocean's surface when warmer surface waters move offshore under the influence of prevailing or seasonal winds.

ATP (adenosine triphosphate) An energy-bearing molecule containing phosphorus used to transport energy within cells.

THE PHOSPHORUS CYCLE

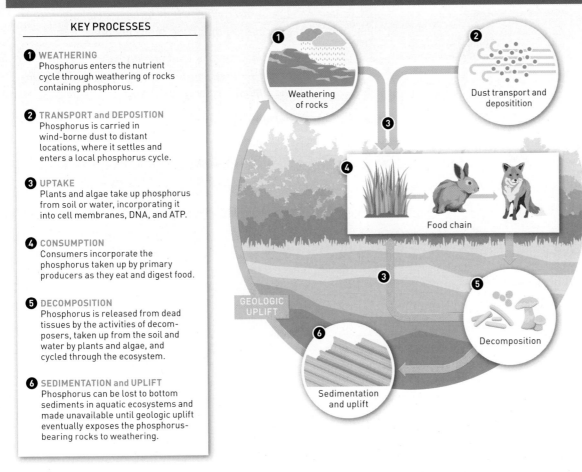

KEY PROCESSES

❶ WEATHERING
Phosphorus enters the nutrient cycle through weathering of rocks containing phosphorus.

❷ TRANSPORT and DEPOSITION
Phosphorus is carried in wind-borne dust to distant locations, where it settles and enters a local phosphorus cycle.

❸ UPTAKE
Plants and algae take up phosphorus from soil or water, incorporating it into cell membranes, DNA, and ATP.

❹ CONSUMPTION
Consumers incorporate the phosphorus taken up by primary producers as they eat and digest food.

❺ DECOMPOSITION
Phosphorus is released from dead tissues by the activities of decomposers, taken up from the soil and water by plants and algae, and cycled through the ecosystem.

❻ SEDIMENTATION and UPLIFT
Phosphorus can be lost to bottom sediments in aquatic ecosystems and made unavailable until geologic uplift eventually exposes the phosphorus-bearing rocks to weathering.

❶ Weathering of rocks

❷ Dust transport and deposition

❸

❹ Food chain

❸

❺ Decomposition

GEOLOGIC UPLIFT

❻ Sedimentation and uplift

FIGURE 8.6 Phosphorus is essential to all living systems as a component of energy-carrier molecules, such as ATP, and also of DNA. Unlike the carbon and nitrogen cycles, the phosphorus cycle does not include a gaseous form occupying a major atmospheric pool.

other coastal areas of high primary production, such as those along the eastern coasts of North America, South America, Africa, and Asia. Here, waters are shallow enough to renew nutrient levels in surface waters during periods of intense mixing by winds or storms. In these waters, nutrients are also elevated by runoff from land.

VARIATION IN PRIMARY PRODUCTION ACROSS THE WORLD'S OCEANS

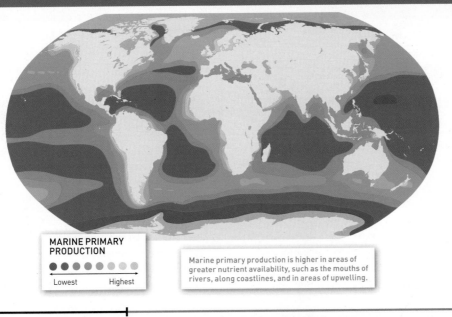

MARINE PRIMARY PRODUCTION

Lowest ———— Highest

Marine primary production is higher in areas of greater nutrient availability, such as the mouths of rivers, along coastlines, and in areas of upwelling.

FIGURE 8.7 High levels of primary production are limited to approximately 10% of the world's oceans. (Data from Ryther, 1969; Field et al., 1998)

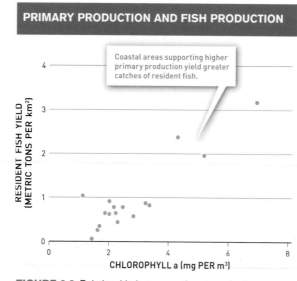

FIGURE 8.8 Relationship between primary production, as measured by chlorophyll a concentration in surface waters, and fish caught along the Pacific Coast of North America, from southern California to Alaska. (Data from Ware and Thomson, 2005)

Why are so many of the world's very productive fishing grounds found in cool upwelling waters?

Marine areas of highest primary production, such as the west coast of North America, support the highest biomass of fish and shellfish (**Figure 8.8**). Similarly, higher levels of phytoplankton production off the Faroe Islands in the northeast Atlantic are what sustained the historic production of Atlantic cod.

The lowest levels of primary production tend to occur in deep, mid-ocean environments, particularly in tropical oceans where warm surface water rarely mixes with nutrient-rich deeper layers of cool water. Here, nutrients that could stimulate primary production are trapped below the euphotic zone.

⚠Think About It

1. How do surface winds influence global patterns of primary production in the ocean?

2. What processes tend to reduce the concentrations of nutrients essential for marine production from the euphotic zone of the oceans?

3. How does upwelling increase the nutrient content of surface waters?

4. How would you expect marine primary production to respond if changing environmental conditions caused widespread extinctions of marine phytoplankton species?

8.3 El Niño and other large-scale climatic systems affect fisheries

During the 20th century, scientists uncovered the influence of the El Niño Southern Oscillation (ENSO) (see Chapter 6) on fisheries. During an El Niño, the waters off the west coast of South America are warmed; during a La Niña, they are cooled (see Figure 6.5, page 161). The warm waters brought by El Niño to the coast of South America have been long associated with crashes in commercially important fish populations, such as anchovies, that also trigger widespread mortality among fish-eating seabirds and sea mammals. Why should the raising of sea surface temperatures by only a few degrees decimate these marine populations? A critical clue is provided by the most common cause of death: starvation.

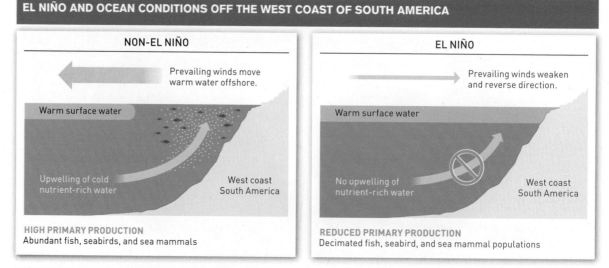

FIGURE 8.9 The effect of El Niño on the coastal ecosystems of western South America results from its creating a barrier to upwelling, which delivers nutrients to the euphotic zone.

In the absence of El Niño, there is strong upwelling along the coast of western South America (see Figure 8.4, page 233), which supports high levels of primary production (see Figure 8.7, page 235) and one of the world's most productive fisheries. However, the coming of El Niño warms the waters off western South America, and essentially shuts off the upwelling of nutrient-rich waters to the euphotic zone (Figure 8.9). As a result, primary production declines drastically, with consequences throughout the entire food web. For example, there is little plankton to feed the small fish, which are essential as food for larger fish, seabirds, and sea mammals. As a result, these consumers, living higher in the marine food web, suffer massive starvation and reproductive failure, and the fishing economy of the region also suffers.

⚠ Think About It

1. How does El Niño suppress marine primary production along the west coast of South America?

2. Do the El Niño/La Niña phenomena exert density-dependent or density-independent controls on sea mammal populations along the west coast of South America? Explain (see Chapter 3, page 71).

8.1–8.3 Science: Summary

The goal of fisheries managers is to estimate how much we can harvest from a fish stock while still keeping it productive.

Prevailing winds blow across the oceans, driving oceanic currents, which influence patterns of primary production and fish stock productivity by transporting warm or cooling waters from one region to another. Upwelling further modifies marine environments by bringing nutrient-rich cool water to the surface. Limited penetration of water by light restricts photosynthesis to the euphotic zone.

Many commercially significant fish stocks vary as a consequence of oscillations in large-scale climate systems. For example, the El Niño Southern Oscillation influences fisheries productivity by altering physical and chemical conditions that can influence fish populations directly or indirectly through its effects on rates of primary production.

Many of the feeds used in aquaculture incorporate fish meal made from forage fish caught along the west coast of South America. How should the pricing of these feeds vary with the El Niño/La Niña cycle?

8.4–8.6 Issues

Peruvian anchovy populations crashed in the 1970s. Blue walleye went extinct in the Great Lakes in the 1980s. Atlantic cod collapsed in the 1990s. The dire state of all these commercial stocks demonstrated that we have the technical capacity to deplete what was once thought to be inexhaustible. Fish stocks around the world have been harmed not only by overharvesting, but by pollution, dams, and the changing climate.

8.4 Tragedy of the Commons: Intensive harvesting has resulted in overexploitation of many commercially important marine populations

Once humans were able to navigate the open waters of the entire planet and developed techniques for catching and processing massive harvests at sea, they soon had the means to decimate entire populations of marine organisms.

Depletion of Whale Populations

Humans have been hunting whales for more than 3,000 years—a fact we know from scenes carved on whalebones. Early whale hunters likely had little impact on whale numbers, since they worked to supply food to relatively small local populations. But that changed with the appearance of commercial whaling when the demand for whale products, particularly oil, soared in the 19th century. The main targets of whalers in the North Atlantic Ocean, from the 16th through 19th centuries, were North Atlantic right whales, *Eubalaena glacialis,* and bowhead whales, *Balaena mysticetus* (Figure 8.10), which are slow swimmers that float when killed. Using open rowing skiffs and hand-thrown harpoons, whale hunters slaughtered an estimated 120,000 right whales and bowhead whales, jeopardizing their very existence.

For many years, the blue whale, *Balaenoptera musculus,* and the fin whale, *B. physalus,* remained beyond the reach of early whaling technology because they were too fast, too strong, and sank when killed. This changed with the invention of the harpoon gun, explosive harpoons,

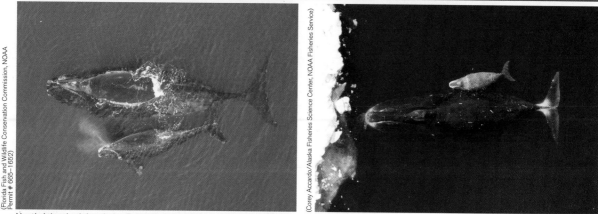

TWO SPECIES OF WHALES HEAVILY EXPLOITED BY EARLY WHALING IN THE NORTH ATLANTIC

(Florida Fish and Wildlife Conservation Commission, NOAA Permit # 665–1652)

(Corey Accardo/Alaska Fisheries Science Center, NOAA Fisheries Service)

North Atlantic right whale, *Eubalaena glacialis*

Bowhead whale, *Balaena mysticetus*

FIGURE 8.10 The depletion of populations of North Atlantic right whales and bowhead whales by early whalers showed for the first time the capacity of humans to deplete marine resources once thought to be inexhaustible.

What in the life histories (see Chapter 3, page 71) of large marine species would make them more vulnerable to overharvest than species of smaller body size?

and steam-powered winches and catcher boats, which introduced the age of modern whaling. The populations of blue and fin whales in the Southern Hemisphere declined from about 400,000 in 1920 to a few thousand in 1960.

The history of commercial whaling is a good example of how the Tragedy of the Commons leads to overexploitation of resources (see Chapter 2, page 49). During the time of peak commercial whaling, no international agreements or regulations limited harvest. Because whales mostly live in areas away from international borders, whalers were free to harvest as many whales as they could sell to meet

commercial demand. Ultimately, the collapse of many whale populations and rising popular awareness of the overharvest of whales sparked international agreements to ban whaling in 1982. Those bans are still in place today, with a few notable and controversial exceptions.

An Ecosystem Upturned: Atlantic Cod

As the cod fishery in the northwest Atlantic collapsed (**Figure 8.11**), marine scientists discovered that other species, such as hake, haddock, and pollock, were also in trouble. All these fish were historically dominant predators, and their absence due to overharvest

THE NORTHWEST ATLANTIC COD FISHERY HARVEST AND COLLAPSE

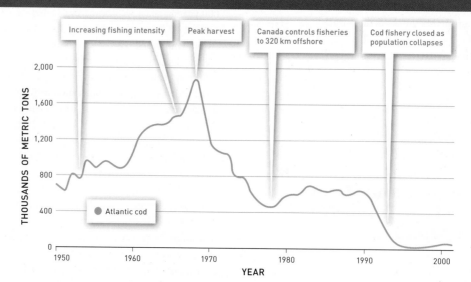

Increasing fishing intensity

Peak harvest

Canada controls fisheries to 320 km offshore

Cod fishery closed as population collapses

THOUSANDS OF METRIC TONS

Atlantic cod

YEAR

FIGURE 8.11 While commercial harvest of the cod populations off Canada spanned over 500 years, intensive modern harvesting increased after 1950, with the population collapsing four decades later. (Data from FAO, 2005)

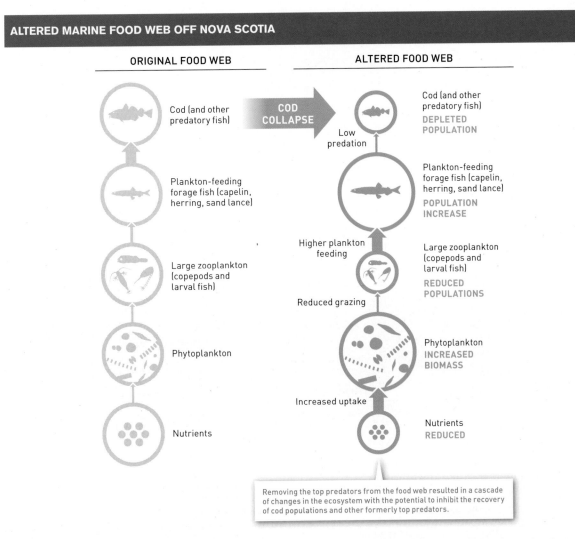

ALTERED MARINE FOOD WEB OFF NOVA SCOTIA

ORIGINAL FOOD WEB

Cod (and other predatory fish)

Plankton-feeding forage fish (capelin, herring, sand lance)

Large zooplankton (copepods and larval fish)

Phytoplankton

Nutrients

ALTERED FOOD WEB

COD COLLAPSE

Low predation

Cod (and other predatory fish)
DEPLETED POPULATION

Plankton-feeding forage fish (capelin, herring, sand lance)
POPULATION INCREASE

Higher plankton feeding

Large zooplankton (copepods and larval fish)
REDUCED POPULATIONS

Reduced grazing

Phytoplankton
INCREASED BIOMASS

Increased uptake

Nutrients
REDUCED

Removing the top predators from the food web resulted in a cascade of changes in the ecosystem with the potential to inhibit the recovery of cod populations and other formerly top predators.

FIGURE 8.12 The collapse of the cod population off Nova Scotia resulted in a radical change in the structure of the marine food web. (Information from Frank et al., 2011)

transformed the ocean ecosystem in many ways (**Figure 8.12**). Bottom-dwelling invertebrates, such as snow crab, that were common prey for cod doubled in number. Similarly, small forage fish, such as herring and capelin, no longer faced heavy predation pressures by cod and other predators, so their population numbers increased by a factor of 9. All these small fish preyed heavily on zooplankton, causing that population to decline. Because zooplankton feed on phytoplankton, fewer zooplankton meant that phytoplankton populations increased, and their greater abundance ultimately reduced the nutrient content of surface waters. The effect of the collapsed cod fishery rippled throughout the food web, much like the extensive effects of wolves on the Yellowstone food web (see Chapter 4, page 103). These dramatic changes led to speculation that recovery of the apex predators like cod might not be possible for a very long time.

Fisheries Collapse: A Global Problem

Similar to the history of whaling, the collapse of the cod fishery off Canada and New England is a sobering example

of a Tragedy of the Commons (see page 49). But many other exploited fish populations have also collapsed, including the sardine fishery off California and the blue tuna fishery in the Atlantic. Recent estimates indicate that more than 25% of commercially important fish stocks have suffered declines in numbers sufficient to be classified as a "collapse" in the fishery. **Figure 8.13** shows the patterns of population decline under exploitation for two of these fish stocks: the South Atlantic snowy grouper and the South Atlantic black sea bass.

One factor in these declines has been the over-expansion of fishing fleets and a system that creates a competitive, "race-to-fish" approach to harvest. Rather than regulating the amount of fish being caught, regulators in the United States traditionally limit the type of gear that can be used and the number of days a fishing vessel can spend at sea. Regulators also monitor the number of fish being caught and shut down fishing for the season when too many fish are caught. Naturally, fishers scramble to harvest as many fish as possible before regulators cut off fishing, and this system unintentionally incentivizes illegal harvest (i.e., poaching) by unregistered

COLLAPSE OF TWO FISHERIES

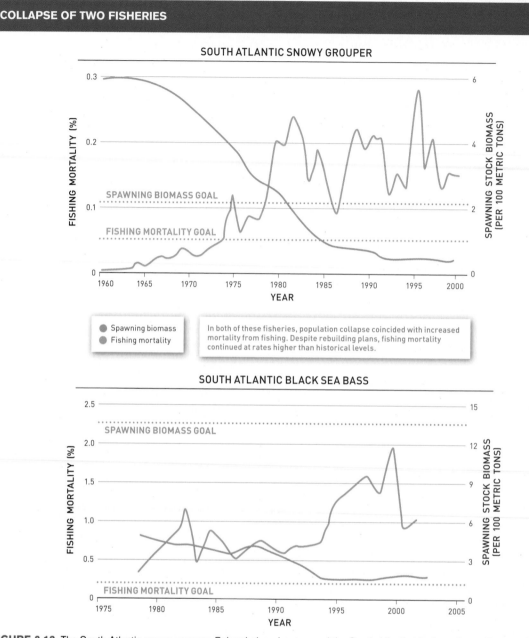

FIGURE 8.13 The South Atlantic snowy grouper, *Epinephelus niveatus,* and the South Atlantic black sea bass, *Centropristis striata.* (Data from Rosenberg, Swasey, and Bowman, 2006)

vessels. Under such conditions, it is difficult to avoid overfishing.

Remaining Uncertainty

While our understanding of the status and biology of commercially important fish populations grows rapidly, significant gaps in our understanding remain. In 2013, of the 230 stocks of commercially important marine fish under U.S. jurisdiction, the status of 23% was uncertain or undetermined. Meanwhile, significant percentages of commercially important fish stocks off northwest Europe and New Zealand also had uncertain or undetermined

status. Clearly, increased information on the status of these stocks would help with their management, especially in regard to regulating fishing pressure.

⚠ Think About It

1. How did technological development influence the overharvesting of whale and fish populations?

2. In what ways does the collapse of the commercial whaling industry or the northwest Atlantic cod fishery represent a Tragedy of the Commons?

8.5 Dams and river regulation have decimated migratory fish populations

Building dams on rivers can stabilize water supply and protect people and infrastructure from flooding. However, dam construction can also displace human populations dependent on rich floodplain resources, forcing them to make a living in less productive environments. Dam construction has many benefits and costs, but here we focus on how river modification by dams threatens populations of commercially important migratory fish, especially salmon.

The Columbia River

In the early 19th century, some 8 to 10 million adult salmon would swim up the Columbia River every year to spawn in the river and its tributaries. But the construction of over 100 large hydroelectric dams converted the once large, free-flowing river to a series of long reservoirs (**Figure 8.14**). Salmon no longer have access to an estimated 45% of their historic spawning areas. Even where salmon can get around dams through fish ladders—constructed stair-steps of water

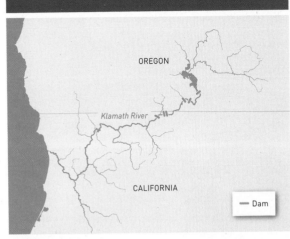

KLAMATH RIVER BASIN

FIGURE 8.15 Dams on the Klamath River prevent the passage of migratory salmon and other migratory fish species to the upper river basin, greatly reducing the area available for spawning.

that allow a fish to swim over a dam—they face severe environmental challenges, including migrating through lakes rather than along a river and returning to degraded habitat. Because of population declines following damming, the U.S. Fish and Wildlife Service has designated most of the salmon populations in the Columbia River system as threatened or endangered.

The Klamath River

Another important salmon river, the Klamath, flows through northwest California and southeast Oregon (**Figure 8.15**). It was once the third most productive salmon river of the U.S. West Coast, with half a million fish returning to spawn each year. However, Copco 1, a hydroelectric dam built in 1918, made most of the upper Klamath River basin inaccessible to salmon and other migratory fish. Three additional hydroelectric dams, built on the Klamath from 1925 to 1962, prevented migratory salmon from reaching approximately 970 kilometers (600 miles) of spawning streams in the upper Klamath River system, reducing the potential of the river system to produce salmon.

These Klamath River dams have also had indirect impacts on salmon populations. Water diversions for agriculture have reduced flows in the river, and drainage from irrigated agricultural fields has introduced excess nutrients and organic contaminants such as pesticides into the river. Nutrients coming from upstream agricultural areas foster blooms of algae in the reservoirs and below them. Decomposing algae trigger oxygen depletion, which stresses salmon physiologically and leads to the spread of salmon diseases. In 2002, for instance, pathogens killed at least 33,000 adult salmon

How do you think we should weigh the relative environmental costs against the economic benefits of dams?

A FREE-FLOWING RIVER NO MORE

Dams and reservoirs developed for hydroelectric power and water storage on the Columbia River system have disrupted salmon migration and have severely reduced the productivity of a major salmon fishery.

FIGURE 8.14 This map shows the locations of major dams on the lower Columbia River system, including one of the Columbia River's main tributaries, the Snake River. Hydroelectric development has transformed most of the entire length of the Columbia and Snake rivers from a flowing river to a series of reservoir pools.

FIGURE 8.16 In 2002 degraded water quality made salmon migrating up the Klamath River more vulnerable to attack by pathogenic organisms, and tens of thousands of them died as a consequence.

SALMON KILL ON THE KLAMATH RIVER

(AP Photo/The Herald and News, Ron Winn)

How are the collapse of the Klamath River salmon and the collapse of the cod fishery similar? How are they different?

in the Klamath (**Figure 8.16**). In addition, water in the reservoirs behind the dams warms to temperatures unsuitable for salmon, which are a cold-water adapted fish that will die when temperatures exceed 24°C (75°F).

As a result of the combined effects of reduced spawning area, low water quality, and the ravages of pathogens, salmon and steelhead runs in the Klamath River have been reduced by approximately 90%. The decline in these stocks led to the closure of 1,000 kilometers (700 miles) of the West Coast of the United States to commercial salmon fishing from 2008 to 2011.

Mekong River Dam

The Mekong River is one of the world's longest rivers, flowing 2,700 miles from the Tibetan plateau through Southeast Asia and into the South China Sea. Fish from these waters provide much-needed protein for more than 40 million people who live along it, making it one of the most important inland fisheries in the world. However, China has built seven dams along the upper stretches of the Mekong and is now financing several massive hydroelectric dams in Laos and Cambodia,

which would alter the river's flow and threaten its 850 fish species, including many migratory ones. For instance, the 107-foot-tall Xayaburi Dam in northern Laos may drive the Mekong giant catfish—an economically important species and one of the world's largest freshwater fish— to extinction. Many other small-scale and subsistence fishers will also have their livelihoods threatened, and locals are fighting this and other dams on the Mekong through protests and lawsuits.

⚠ Think About It

1. Should China consider how dam construction in that country will affect river productivity in other downstream countries? Explain why or why not.

2. Although dams harm migratory fish populations, some nonmigratory species of lake fish benefit from dams. Is this a balanced tradeoff?

8.6 Aquaculture can pollute aquatic environments and threaten wild fish populations

Wild fish stocks have declined and the commercial harvest of fish from oceans has reached a plateau, so the seafood industry has turned to aquaculture (see page 230) to meet increasing demand for seafood (**Figure 8.17**). Over the past three decades, aquaculture production has increased by about 9% annually, a growth rate exceeding that of all other forms of food production. By 2011 this resulted in more than 40% of total fisheries production (**Figure 8.18**).

Aquaculture occurs in marine, brackish, or freshwater environments all over the world. Farmers raise fish in ponds or in mesh cages suspended in water, while shrimp and crabs are generally grown in ponds or tanks. In intensive aquaculture, farmers feed fish and crustaceans specially formulated diets for optimum growth. By contrast, filter-feeding shellfish such as oysters and mussels can be suspended in the water column on racks or attached to lines, where they feed by filtering plankton and other organic matter from the surrounding water. Properly managed aquaculture systems can provide an environmentally friendly source of dietary protein. However, just like agriculture, aquaculture has the potential to threaten the biodiversity and health of ecosystems in several different ways.

Problems with Cultured Fish and Shellfish

Critics of aquaculture point out that in all but the most secure systems, some of the organisms being cultured will escape into the wild. For example, cage-

AQUACULTURE YIELDS A DIVERSITY OF VALUABLE CROPS

(Brian O'Hanlon/Open Blue)
Cobia growing in open-water cages

(Hoang Dinh Nam/AFP/Getty Images)
Farm-grown shrimp in Viet Nam

Shellfish farming in Tomales Bay, California
(NOAA Photo/Office of Aquaculture, National Marine Fisheries/Diane Windham)

FIGURE 8.17 Aquaculture produces a wide variety of aquatic resources, including fish, shrimp, and oysters.

grown fish can escape if the netting confining the fish is torn. Pond-raised fish or crustaceans may jump or crawl into nearby waterways. Escapees may become invasive species (see Chapter 3), such as the Asian carp, which escaped from aquaculture ponds to colonize the Mississippi River system and which now threaten the Great Lakes. Even native species pose threats: When domesticated individuals mate with wild, locally adapted individuals, it generally reduces the fitness of wild, locally adapted populations. Domesticated fish have genetic characteristics that have been selected for confined rearing, whereas wild fish have genetic characteristics optimized for their native environments. Therefore, when aquaculture fish breed with wild fish, they introduce genes that decrease adaptation of wild populations to their environment.

Currently, there are no genetically modified varieties of fish used in aquaculture. One company, AquaBounty Technologies based in Massachusetts, is seeking approval from the U.S. Food and Drug Administration (FDA) for its faster-growing, genetically modified (GM) Atlantic salmon. Despite protective measures the company has taken, some scientists and conservationists believe that these fish will threaten wild fish populations if they escape.

Water Pollution

Aquaculture can also be a significant source of water pollution. As in agriculture on land, runoff from fish and shellfish feeding operations contains nutrients, especially nitrogen and phosphorus, which can produce noxious algal blooms that impair water quality and can kill wild fish. Aquaculture waste may also contain enough organic matter to deplete oxygen supplies in waters receiving the waste, again potentially resulting in fish kills.

Clearing Mangrove Forests

Shrimp farms are increasing pressure on mangrove forests, one of the most valuable and endangered tropical ecosystems on Earth. Mangrove forests grow in coastal waters, where they protect coastal areas from storm surge and may protect against tsunami damage. The roots of mangroves also provide protection from predation for young fish, so they additionally act as nurseries that enhance the productivity of coastal tropical fisheries. Mangroves also happen to be desirable locations for shrimp

If GM Atlantic salmon are approved for human consumption, what safeguards should be used to prevent their escape and thus potential mating with wild salmon?

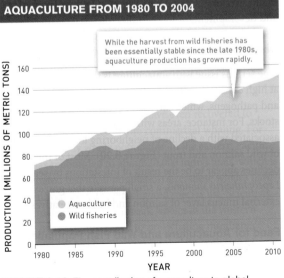

PRODUCTION OF WILD FISHERIES AND AQUACULTURE FROM 1980 TO 2004

While the harvest from wild fisheries has been essentially stable since the late 1980s, aquaculture production has grown rapidly.

PRODUCTION (MILLIONS OF METRIC TONS)

- Aquaculture
- Wild fisheries

YEAR

FIGURE 8.18 The contribution of aquaculture to global production of aquatic resources tripled in 25 years. (Data from FAO, 2005, 2013)

INFLUENCE OF INDIVIDUAL TRANSFERABLE QUOTAS (ITQS) ON FISHERIES SUSTAINABILITY

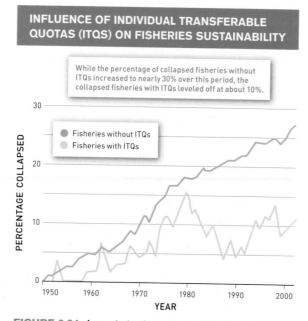

While the percentage of collapsed fisheries without ITQs increased to nearly 30% over this period, the collapsed fisheries with ITQs leveled off at about 10%.

- ● Fisheries without ITQs
- ● Fisheries with ITQs

PERCENTAGE COLLAPSED

YEAR

FIGURE 8.24 An analysis of more than 11,000 fisheries shows that individual transferable quotas greatly reduce the probability of fisheries collapse. (Data from Costello et al., 2008)

FIGURE 8.25 Biodiversity helps sustain economically valuable services provided by aquatic ecosystems, including salt marshes, kelp forests, mangrove forests, and riparian wetlands.

eventually attain its former abundance. Still, these results suggest that the radical change in ecosystem structure following the collapse of the Atlantic cod populations of the northwest Atlantic may be reversing itself. These results create a sense of cautious hope for the eventual recovery of other collapsed fish populations.

⚠ **Think About It**

1. What are the implications of the observation that collapsed fish or whale populations have commonly recovered after harvest has been reduced or eliminated, whereas others have failed to recover?

2. Establishing individual transferable quotas (ITQs) has seemed to put many fisheries on a path to sustainability. Why?

8.8 Biodiversity contributes to the productivity and stability of fisheries

Many vital services by marine and freshwater ecosystems, such as higher production and stability of aquatic resources, are sustained by biodiversity (**Figure 8.25**). One explanation for higher production is that diverse ecosystems make more efficient use of nutrients and light (**Figure 8.26**). In addition, higher genetic and species diversity leads to higher stability in the face of disturbance and faster recovery following disturbance. For instance, genetic diversity in populations of seagrass, *Zostera marina*, is associated with greater resistance to disturbance by grazers and faster recovery following heat-induced mortality (**Figure 8.27**). When the seagrass is healthy, so, too, are the fisheries that depend on the seagrass. The fact that diverse ecosystems have high

AQUATIC ECOSYSTEM SERVICES

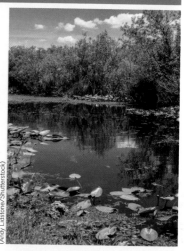

SALT MARSHES
Salt marshes act as barriers to waves and storm surges; filter pollutants, including sediments, nutrients, and pesticides; provide nesting, nursery, and feeding grounds for a diversity of wildlife.

(Randolph Femmer/USGS)

KELP FORESTS
Kelp forest net primary production is among the highest of any ecosystems on Earth, providing food and shelter for an abundance of fish, as well as shore protection.

(Claire Fackler, NOAA National Marine Sanctuaries)

MANGROVE FORESTS
Mangrove forests protect coastlines from storm waves and tsunami, while providing nursery grounds for juvenile fish, harvestable fish and shellfish, and sources of wood products.

(David Burdick/NOAA)

RIPARIAN WETLANDS
Riparian wetlands are hotspots of biodiversity and productivity; during flooding, particularly along large rivers, many species of fish move into riparian wetlands for feeding and spawning.

(Andy Lidstone/Shutterstock)

INFLUENCE OF MARINE PRIMARY PRODUCER DIVERSITY ON ECOSYSTEM FUNCTION

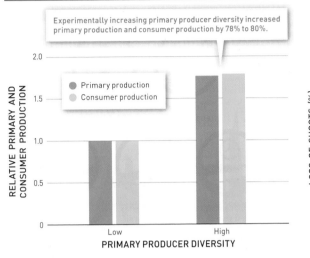

> Experimentally increasing primary producer diversity increased primary production and consumer production by 78% to 80%.

FIGURE 8.26 A survey of several experimental studies, which manipulated primary producer diversity, demonstrated the strong positive effect of primary producer diversity on primary and consumer production. (Data from Worm et al., 2006)

GENETIC DIVERSITY AND ECOSYSTEM STABILITY

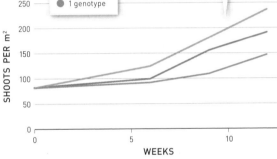

> Eel grass plots with greater genetic diversity were more resistant to grazing by geese.

> Recovery following dieback during a heat wave was faster in eel grass plots containing higher genetic diversity.

FIGURE 8.27 Plots of eel grass, *Zostera marina,* containing more genotypes lost fewer shoots to grazing geese, indicating higher resistance to this form of disturbance. (Data from Hughes and Stachowicz, 2004) Eel grass plots containing more genotypes also recovered more rapidly following heat-related mortality, indicating higher resilience following disturbance. (Data from Reusch et al., 2005)

productivity and buffer communities from disturbance is a key reason they provide important ecosystem services to humans. Biodiverse and stable ecosystems with complex physical structure sustain higher species and population numbers, which humans can harvest indefinitely with sustainable fisheries management.

Biodiversity and Bristol Bay Sockeye Salmon

The fishery for sockeye salmon (*Oncorhynchus nerka*) around Bristol Bay, Alaska, has produced high-quality dietary protein and has brought a stable income to the communities around Bristol Bay for more than a century (**Figure 8.28**). How has this fishery been sustained, while

What mechanisms may connect diversity among aquatic primary producers to higher diversity among aquatic consumers?

SOCKEYE SALMON, *ONCORHYNCHUS NERKA,* AN IMPORTANT AQUATIC RESOURCE

Sockeye salmon swimming up a river to spawn

The Bristol Bay Alaska salmon fishing fleet catching salmon before they enter their spawning rivers

FIGURE 8.28 Sockeye salmon, *Oncorhynchus nerka,* are native to the North Pacific region, where they spawn in river and lake ecosystems from the Klamath River of California and Oregon to Alaska, Siberia, and the island of Hokkaido, Japan. The species is an important source of income and nutrition to local communities throughout its range. For example, in recent years the sockeye salmon fishery of Bristol Bay, Alaska (shown in action here), alone has been valued at over $100 million annually.

SCIENCE   ISSUES   SOLUTIONS

SUCCESSFUL RIVER RESTORATION

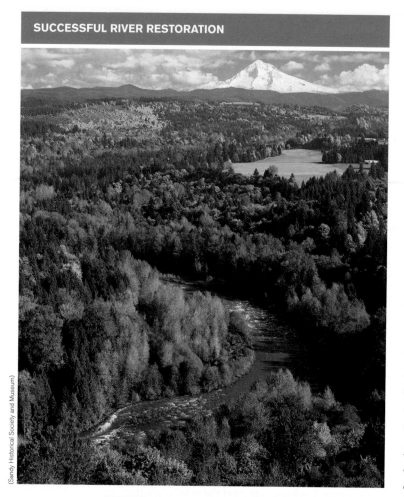

(Sandy Historical Society and Museum)

FIGURE 8.35 Flowing from Mount Hood to the Columbia River, the Sandy River, Oregon, is a wild river once more.

that removing the dams will revitalize the regional salmon fishery, which could generate tens of millions of dollars annually for nearby fishing communities.

Based on consideration of these issues and extensive scientific and engineering studies, the diverse interest groups involved converged on an agreement in early 2010. The agreement included a plan for sharing water in ways that would both restore salmon runs to the upper Klamath River system and protect the interests of farmers. Three years later, then U.S. Secretary of the Interior Ken Salazar recommended removal of the four major dams on the Klamath, an action requiring authorization by the U.S. Congress. Although removal of the four mainstream hydroelectric facilities would eliminate 169 gigawatts (GW) of hydroelectric power, renewable power sources such as wind or solar could replace it while fisheries populations recover, leading to a win–win scenario for long-term sustainability. If the plan is approved, it will be the largest dam removal project ever attempted.

⚠ Think About It

1. What do the flushing of sediments and spawning success by salmon following dam removal on the Sandy River suggest about the resilience of rivers and salmon populations?

2. What criteria should be used when deciding a course of action in a very complex economic and cultural situation such as that on the Klamath River?

?

What factors likely contributed to reaching an agreement involving so many potentially conflicting interests in the Klamath River system?

integrated multi-trophic aquaculture (IMTA) An approach to aquaculture that involves raising several species of aquatic organisms with complementary feeding habits in close proximity.

PGE followed up in 2008 by removing Little Sandy Dam, which opened up the entire Sandy River system to migrating salmon. For the first time in nearly 100 years, the Sandy River was flowing freely from its headwaters on Mount Hood to the Columbia River (**Figure 8.35**). Scientists are carefully following developments in this very large-scale experiment in returning an entire river system to a free-flowing state.

Klamath River Dam Removal

Restoration of a highly modified river system such as the Klamath requires that we consider factors far beyond the impact of dams on salmon populations. For example, removing the dams will eliminate the 50 jobs associated with operating the hydroelectric system on the Klamath River. At the same time, dam removal would support an estimated additional 450 jobs across the region over the following 50 years. The existing reservoirs behind the dams support recreational fishing, which generates income to the local community. However, fisheries scientists project

8.10 Aquaculture can provide high-quality protein with low environmental impact

As the world recognizes the limits of the oceans, many hope that aquaculture can take some pressure off wild populations. However, as aquaculture continues to expand, one of the challenges is making it more efficient and environmentally friendly.

Reducing Pollution by Aquaculture

One of the most effective ways to reduce pollution from aquaculture is through **integrated multi-trophic aquaculture (IMTA),** which, in the marine environment, is sometimes called *integrated mariculture.* In this process, several species of aquatic organisms with complementary feeding habits are raised in close proximity. The waste product of one species is food for the others, thereby reducing impacts on the environment. For example, in Sungo Bay, China, aquaculturists grow a combination of fish, abalone, seaweeds, and scallops in a

CHINESE INTEGRATED AQUACULTURE SYSTEM

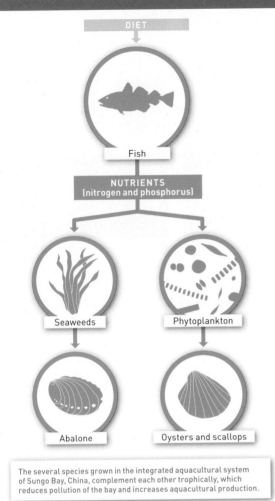

The several species grown in the integrated aquacultural system of Sungo Bay, China, complement each other trophically, which reduces pollution of the bay and increases aquacultural production.

FIGURE 8.36 Nutrients released from food, fed to fish cultured in cages, is absorbed by either phytoplankton, which is food for oysters and scallops, or by seaweeds, which are eaten by abalone.

sequence of cultures extending for 8 kilometers offshore. In this system, waste from the cage-cultured fish, both dissolved nutrients and particulate matter, is consumed by the other members of this integrated aquaculture system. Filter feeding species can directly consume wastes from the fish feeding operation, while primary producers benefit from dissolved nutrients (mainly nitrogen and phosphorus). The result is a substantial reduction in pollution and greater harvests of seafood (**Figure 8.36**).

Integrated multi-trophic aquaculture is being tested around the world. A system of integrated aquaculture involving Atlantic salmon, mussels, and kelp has also been developed for the Bay of Fundy, Canada. Again, the integrated system reduces pollution from the caged salmon and increases profits. Other systems involving multiple species are in place in Africa, South America, Australia, and Europe.

Aquaculture Wastes and Constructed Wetlands

Biological approaches are also being used to make land-based aquaculture systems more sustainable. All intensive aquaculture systems have the potential to generate pollution. While IMTA can be used to reduce nutrient pollution in marine systems, other techniques are required for freshwater aquaculture in the terrestrial landscape. There are many possible engineering solutions to the problem of aquaculture wastewater treatment, but many are expensive both in terms of energy and money.

Increasingly, **constructed wetlands,** which are artificial wetland ecosystems constructed in areas where wetlands do not occur naturally, can be used to treat wastewater from freshwater aquaculture (**Figure 8.37**). Constructed wetlands foster the complex biological functions of nutrient removal by plants and microbes, and they can also provide habitat for wildlife. While they can be effective and self-sustaining if designed properly, they require long-term monitoring to ensure that they continue to function as designed. Because the problem of treating wastewater from aquaculture is slightly different from treating other forms of wastewater, we will defer a detailed discussion of constructed wetlands until Chapter 13.

Shrimp Farming and Mangrove Conservation

There are several ways to reduce the impact of shrimp farming on mangrove forests. One problem is containment of the nutrient-rich water to minimize coastal pollution. Experts in the design of shrimp ponds point out that the soils where mangroves grow are generally too permeable to contain water. In those cases, the ponds may need to be lined with plastic or an impermeable clay so that waste water can be properly managed. Another problem associated with shrimp farming is that removing mangroves to build shrimp ponds increases the risk that the ponds, which are very expensive to build, will be damaged by storms.

Recognizing this danger, many shrimp ponds are now sited behind protective mangrove forests, and the necessary ocean water is delivered to the inland ponds through a canal or pipeline. Finally, because ponds dedicated to intensive shrimp culture need to be completely drained periodically, they are increasingly built above the high-tide level and therefore inland from mangrove forests. In these cases, reducing direct impact on mangrove forests allows them to continue to provide ecosystem services of fish nursery habitat and flood protection.

Reducing the Impact of Sea Lice

One of the most direct approaches to reducing the impact of sea lice on wild salmon is to monitor and treat lice infestations on captive salmon using pesticides. However,

How might IMTA increase profits and reduce costs in an aquaculture system?

constructed wetlands
Artificial wetland ecosystems, used in the treatment of wastewater, that are constructed in areas where wetlands may not occur naturally.

A CONSTRUCTED WETLAND RECLAIMS A HISTORICAL TREASURE IN HANGZHOU, CHINA

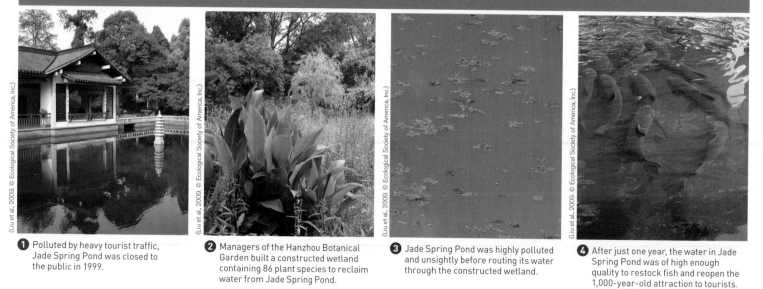

1 Polluted by heavy tourist traffic, Jade Spring Pond was closed to the public in 1999.

2 Managers of the Hanzhou Botanical Garden built a constructed wetland containing 86 plant species to reclaim water from Jade Spring Pond.

3 Jade Spring Pond was highly polluted and unsightly before routing its water through the constructed wetland.

4 After just one year, the water in Jade Spring Pond was of high enough quality to restock fish and reopen the 1,000-year-old attraction to tourists.

FIGURE 8.37 A similar approach, using constructed wetlands, is being used as a low-cost way to treat wastes from aquaculture.

there are concerns that sea lice will evolve resistance to such treatments and that nontarget organisms will be killed—as has occurred with chemical control of agricultural pests (see Chapter 7). Another potential control measure includes restricting salmon farms to areas away from the migration paths of wild fish to minimize their contact. Canada and parts of Europe have already taken the precaution of excluding salmon aquaculture from some areas supporting valuable wild salmon stocks. Growing salmon in land-based tanks rather than sea pens and treating the effluent is another way to reduce infection of wild salmon.

Decreasing the Use of Fish Meal

Reducing the impact of aquaculture on wild forage fish populations hinges on finding substitutes for fish meal and fish oil. Much progress has been made in replacing fish meal with plant protein (e.g., soy). Fish oils have also been partially replaced by plant oils, including canola, soy, sunflower, or olive oil. Fish nutrition experts predict that three-fourths of fish oils currently used in aquaculture feeds could be replaced by plant oils with no loss in growth performance by fish or shellfish. Others have suggested that shifting aquaculture from large, carnivorous species, which require high-protein diets, down the food chain to omnivores or herbivores will also reduce the amount of fish meal used in aquaculture. For example, tilapia and catfish grow well on predominantly plant-based diets. Plant-based feeds are also growing in popularity among shrimp farmers as they have increasingly shifted from growing carnivorous tiger shrimp to growing omnivorous western white shrimp, *Litopenaeus vannamei.*

⚠ Think About It

1. How are integrated aquaculture systems converging on the structure of natural food webs (see Chapter 2, page 40)?

2. How can knowledge of the influence of biodiversity on ecosystem processes help reduce the environmental impact of aquaculture?

3. What are the potential contributions of using plant-based feeds to aquaculture sustainability?

8.7–8.10 Solutions: Summary

One key to properly managing fisheries is a management plan built on sound science. In the case of whales, an international ban helped species recover. Fisheries are generally regulated through gear restrictions and limits to the amount of time and the period when a fishing vessel can be at sea. Competitive approaches to fishing can be avoided by granting fishing quotas to individuals, cooperatives, or communities. Fishing quotas also appear to encourage more cooperation between fishers and fisheries scientists and managers, and marine protected areas improve fisheries while simultaneously increasing marine biodiversity.

Thousands of projects around the world, large and small, are aimed at restoring rivers and wetlands, and this increasingly involves dam removal. Aquaculture now accounts for nearly 40% of total fisheries production. With the massive increase in aquacultural production comes the potential for massive environmental damage. One of the most effective ways to reduce pollution from aquaculture is through integrated multi-trophic aquaculture (IMTA). Many shrimp farms are now being sited and managed in ways that protect nearby mangrove forests. More and more, wastewater from land-based aquaculture is being treated effectively and economically using constructed wetlands. In addition, fish farmers are switching to plant-based diets and raising fish lower on the food chain.

Answer the following questions for each chapter section and then answer the Central Question.

Central Question: Can we sustainably manage fisheries and aquaculture?

8.1–8.3 Science

- **How are commercial fish populations harvested and managed?**

- **What is the relationship between nutrient availability and primary production in the marine environment?**

- **How does El Niño affect fisheries along the west coast of South America?**

8.4–8.6 Issues

- **How is overexploitation of whale and marine fish stocks related to the Tragedy of the Commons?**

- **What impact do large dams have on migratory fish populations?**

- **In what ways does aquaculture affect the aquatic environment and wild fish populations?**

Aquatic Resources and You

Fishery and aquaculture products are a major part of the diets of people around the world. Demand for them continues to increase, creating even more pressures on marine and freshwater environments already threatened by overexploitation. One way that each of us can help foster more sustainable use of these resources is by making informed choices.

☐ *Keep informed.*

Keep up to date on developments in sustainable fishing and aquaculture both in your region, if you live in an area with commercial fishing or aquaculture, and globally. Explore issues, such as the status of GM salmon and what is being done to prevent aquaculture from threatening wild fish stocks. Learn about the sources of fish and shellfish in your area. If you live in a fish-producing region, gather information about the importance of these industries to the regional economy and about the markets for the products from your region.

☐ *Buy and consume sustainably managed aquatic resources.*

Several organizations rate seafood on its level of sustainability. The Monterey Bay Aquarium provides online evaluations of the most sustainable sources of fish and shellfish for individual consumers and businesses. Its website (www.seafoodwatch.org) rates all types of seafood and categorizes them into "green," "yellow," and "red" seafood categories (there's also an app!). When shopping for seafood at grocery stores and markets, look for the blue check mark labeling "Certified Sustainable Seafood MSC" used by the Marine Stewardship Program, which reviews fisheries and certifies those that are being managed sustainably.

☐ *Speak up!*

Talk to restaurant owners and supermarket managers and ask them to serve and supply seafood that is sustainably harvested. You can also write to local and federal policy makers and encourage them to establish laws and initiatives that protect marine areas and aquatic health.

8.7–8.10 Solutions

- **What management strategies are being used to make harvests of aquatic resources more sustainable?**
- **How does biodiversity contribute to the productivity and stability of aquatic resources?**
- **How are migratory salmon responding to removing dams from spawning streams?**
- **What tactics are being used to reduce the environmental impacts of aquaculture?**

Answer the Central Question:

_____
_____
_____
_____
_____
_____
_____

Chapter 8
Review Questions

1. Where in the world's oceans would you likely encounter the lowest level of primary production by phytoplankton?
a. Along the west coast of a continent
b. In the central tropical Pacific Ocean
c. In an area of active upwelling
d. Offshore from a major river

2. How does El Niño impact commercial fish production along the west coast of South America?
a. The warm water brought by El Niño speeds up fish production.
b. The increased nutrients in surface waters associated with El Niño are toxic to phytoplankton.
c. The warm surface water of an El Niño directly kills large numbers of fish.
d. The warm surface water of an El Niño stops upwelling, which reduces nutrient renewal in the euphotic zone.

3. How would total salmon catch in the Bristol Bay fishery be affected if pollution from mining, for example, eliminated a salmon population in a major river that drains into Bristol Bay (see Figures 8.29 and 8.30)?
a. Catches would likely not change.
b. Catches would likely be more stable over time due to less competition among salmon.
c. Catches would likely be more variable over time due to lower salmon population diversity.
d. Catches would likely increase.

4. Approximately what percentage of the world's fish populations has been depleted to the point where those numbers are now considered "collapsed"?
a. 90%   c. 50%
b. 75%   d. 25%

5. Following dam building, the number of salmon returning to spawn in the Klamath River averaged about 35,000, approximately 10% of the historical average. How many salmon historically returned to the Klamath River to spawn?
a. 90,000   c. 3,500,000
b. 350,000   d. 35,000,000

6. Which of the following is a way in which intensive aquaculture affects aquatic ecosystems?
a. Intensive aquaculture can pollute aquatic ecosystems.
b. Intensive aquaculture can enhance the genetic diversity of wild populations.
c. Intensive aquaculture increases the intensity of fishing for wild fish.
d. Intensive aquaculture decreases the disease and parasite load on wild fish populations.

7. Which of the following was an indicator of recovery of the Atlantic cod food web off Nova Scotia?
a. A decline in abundance of zooplankton
b. An increase in abundance of phytoplankton
c. An increase in the abundance of forage fish
d. A decrease in the abundance of forage fish

8. River restoration can involve which of the following?
a. Planting of native riparian vegetation
b. Restoring the natural form of river channels
c. Removing dams
d. All of the above

9. What percentage of the 150 million metric tons of aquatic organisms harvested in 2011 came from aquaculture?
a. 25%   c. 40%
b. 60%   d. 80%

10. How do marine protected areas contribute to sustaining fishery production?

a. They do not contribute to sustaining fishery production but instead preserve marine ecosystems for tourism.
b. Marine protected areas reduce the number of large predaceous fish, so prey fish populations grow.
c. By reducing the abundance and diversity of marine invertebrates in an area, marine protected areas allow marine fish to flourish.
d. Marine protected areas support higher biomass and numbers of fish, providing an abundance of juvenile and adult fish that move into areas that can be fished.

Critical Analysis

1. According to the Food and Agriculture Organization of the UN, the people in the nations on the west coast of Africa consumed much more fish per capita than did the people in the nations along the east coast of Africa. Can you explain this difference in terms of availability?

2. Using Internet resources, explore and summarize the role played by international treaties in the management of Atlantic cod and Pacific halibut. Compare the complexity and success of these treaties at sustaining fish harvests.

3. Explore the information and perspectives on the issue of genetically modified (GM) Atlantic salmon and formulate a cogent argument _for_ or _against_ approval of GM salmon for human consumption.

4. Use Figure 8.12 as a basis for reviewing and explaining the changes in the marine food web off Nova Scotia caused by the depletion of Atlantic cod, the top carnivore in that ecosystem. If the cod population does recover, how will the ecosystem change as it returns to something resembling its former state?

5. Discuss the potential cultural changes associated with a shift from "race to fish" to individual transferable quotas (ITQs) as a basis for regulating commercial fishing. How does this regulatory change potentially affect the ways in which fishers view and harvest fish and shellfish populations?

Find additional resources and links online at www.macmillanhighered.com/launchpad/molles1e.

Central Question: How can we manage nonrenewable energy resources in a way that reduces environmental harm?

Describe the main fossil fuels utilized by modern society.

SCIENCE

CHAPTER 9

Fossil Fuels
and Nuclear Energy

Explain how fossil fuel extraction
and nuclear power use can
damage the environment.

Analyze the tactics for mitigating
the environmental impacts of consuming
fossil fuels and using nuclear power.

ISSUES

SOLUTIONS

(Courtesy U.S. Coast Guard)

(Saul Loeb/AFP/Getty Images)

The Deepwater Horizon drilling rig being consumed by flames (left) and pelicans coated with crude oil from the resulting oil spill (right).

Deepwater Horizon Up in Smoke

A devastating oil spill highlights our dangerous dependence on fossil fuels.

On the night of April 20, 2010, a slurry of methane gas, mud, and seawater shot up the drilling apparatus of an oil rig in the Gulf of Mexico and geysered into the air like a shaken bottle of champagne. With a resounding bang, the gas erupted in flames and the rig quaked back and forth violently. The commotion woke 23-year-old Christopher Choy—one of 126 workers onboard—who climbed out of his bed as fire alarms blared in the hallways. When he got outside to the inferno, he watched men leap 50 feet off the deck into the dark, roiling ocean below. "I'm fixing to die. This is it," he thought, as he later told a news reporter. "We're not gonna get off of here."

In fact, Choy was one of the lucky ones who escaped to safety on a life boat—but that night 11 of his coworkers perished. They died in what would become the largest accidental oil spill in history. The oil rig was called the Deepwater Horizon because it had previously pushed the limits of human engineering by penetrating more than 10,700 meters (35,000 feet) into the ocean floor in search of the precious "black gold" that fuels our energy-hungry economy. On the night of the disaster it was drilling at a more modest 4,000 meters (13,000 feet) deep. Even after the flames were extinguished 36 hours later, thick, black crude gushed from the wellhead on the seafloor for the next three months. All told, approximately 780 million liters (206 million gallons) of oil were released into the Gulf of Mexico.

Following the spill, dozens of diseased dolphins washed up on Gulf coast beaches. Seabirds were coated in thick, black tar. And Louisiana's alligator-filled wetlands smelled like a corner gas station. The fishing and tourism industries along the U.S. Gulf Coast from Louisiana to Florida were devastated. In July 2015, British Petroleum (BP), which was responsible for the well, agreed to pay out $18.7 million in fines and compensation—the largest environmental settlement in U.S. history—and estimates it has incurred more than $40 billion in spill-related costs overall.

> "It is evident that the fortunes of the world's human population, for better or for worse, are inextricably interrelated with the use that is made of energy resources."
>
> M. King Hubbert, 1956

It's been 150 years since the drilling of the world's first commercial oil well—the Drake Well in northeastern Pennsylvania—and we're more dependent than ever on this fossil energy source. Fossil fuels and other **nonrenewable energy** sources, including uranium for nuclear power, supply 87% of the planet's power needs. Such nonrenewable energy sources require millions of years to form through biological, geological, and chemical processes and will eventually be depleted. In the next chapter, we examine the quest to replace them with **renewable energy** sources, such as solar power and biofuels, which can be replenished in a relatively short period of time and are not exhausted by use. But for now we focus on the nature of nonrenewable energy sources and why they continue to underlie the global economy. We discuss some solutions to the environmental challenges they create in the concluding section of the chapter.

The U.S. Energy Information Agency predicts that energy demand will increase approximately 50% over the next 30 years. This rising energy demand, coupled with the rapid pace at which we are depleting nonrenewable energy sources and the extent to which their use impacts the environment, brings us to the central question of this chapter.

nonrenewable energy Sources of energy, including coal, petroleum, natural gas, and nuclear fuels, that are not renewable on timescales meaningful to human lifetimes and that can be depleted with continued use.

renewable energy Sources of energy, including solar, wind, hydrologic, geothermal, and biomass, that can be replenished in a relatively short period of time. Use does not deplete renewable energy sources.

Central Question

How can we manage nonrenewable energy resources in a way that reduces environmental harm?

(Dado Galdieri/Bloomberg via Getty Images)

9.1–9.3 Science

Think about the last time there was a power outage in your city. If it lasted just a few hours, it was likely a diversion as you sat around and shared stories with your family in the dark as the ice cream in your freezer melted. Now imagine if the whole world went dark and the global supply of fuel was exhausted. Airports would shut down. Factories would close their doors. The food supply would diminish. Your only source of news would be the neighbors. Life as we know it would come to a halt. It's not such a far-fetched nightmare. The fact is our modern lives are almost entirely dependent on nonrenewable energy resources, and one day, inevitably, they are going to run out.

fossil fuels Fossilized organic material, mainly the remains of ancient photosynthetic organisms that converted the Sun's radiant energy into chemical energy (e.g., coal, oil, natural gas).

9.1 Fossil fuels provide energy in chemical form

Coal, oil, and natural gas are called **fossil fuels** because they are the fossilized remains of ancient photosynthetic organisms that converted the radiant energy in sunlight to chemical energy. That energy is stored in the form of molecular chains made up of hydrogen and carbon. Burning these fuels breaks up those molecular chains and liberates smaller molecules, including water and carbon dioxide, along with pollutants and ash particles known as black carbon. By burning them, the stored energy is released, primarily in the form of heat.

COAL FORMATION

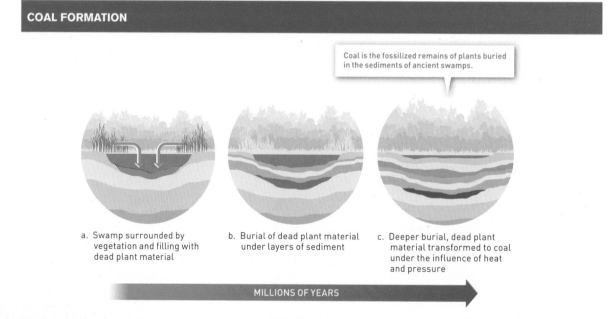

Coal is the fossilized remains of plants buried in the sediments of ancient swamps.

a. Swamp surrounded by vegetation and filling with dead plant material

b. Burial of dead plant material under layers of sediment

c. Deeper burial, dead plant material transformed to coal under the influence of heat and pressure

MILLIONS OF YEARS

FIGURE 9.1 Coal, the fossilized remains of plants buried in the sediments of ancient swamps, was formed by processes taking place over millions of years.

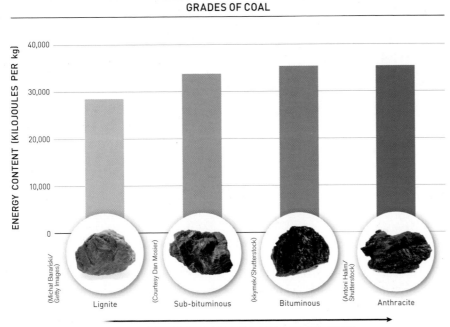

MAJOR TYPES OF COAL VARY IN THEIR CARBON AND ENERGY CONTENT

GRADES OF COAL

ENERGY CONTENT (KILOJOULES PER kg)

40,000

30,000

20,000

10,000

0

(Michal Barariski/ Getty Images)

Lignite

(Courtesy Dan Mosier)

Sub-bituminous

(kkymek/Shutterstock)

Bituminous

(Antoni Halim/ Shutterstock)

Anthracite

INCREASING CARBON CONTENT AND HARDNESS

FIGURE 9.2 The energy content of coal increases from lignite to bituminous coal; anthracite has slightly lower energy content than bituminous coal.

Coal

As the leaves, branches, and trunks of ancient plants accumulated in freshwater swamps some 100 to 300 million years ago, they formed organic-rich bottom sediments (**Figure 9.1a**). Inorganic material, including clays and sand, were then deposited in these ancient swamps, first mixing with and then burying the organic-rich layer. Over the course of millions of years, the rock and soil overlying the swamp sediments increased to great depths (**Figure 9.1b**). With the increasing pressure and heat that resulted from the thick accumulation of rock and soil, the organic-rich material that began as buried swamp sediment was gradually converted into the sedimentary rock that we call **coal** (**Figure 9.1c**).

Geologists classify coal into four major grades, which differ in carbon and energy content (**Figure 9.2**). The differences among coal grades are mainly the result of variations in their age and the amount of heat and pressure to which they were exposed during development. Lignite, the youngest of the coal grades with the lowest carbon and energy content, was subjected to less heat and pressure during development than were the other grades. The other coal grades, listed in order of increasing carbon content, are sub-bituminous coal, bituminous coal, and anthracite. Energy content increases from lignite to bituminous coal, then the increase in energy drops slightly from bituminous coal to anthracite.

Humans have used coal as a source of heating since prehistoric times. Native Americans once burned it in their pottery kilns, and the steamships and railroads of the Industrial Revolution were powered by coal boilers. Today, coal is most commonly used to generate electricity. For example, more than 90% of coal mined in the United States is used in coal-fired electrical power stations. Coal also runs various industrial processes, such as the manufacture of paper or cement. Coal serves as a raw material to manufacture plastics, tar, fertilizer, and medicines. When baked in a hot furnace, coal is transformed to coke, a key material used in the manufacture of steel.

In 2008 energy experts estimated that about 93% of the world's known coal reserves occur in three regions of the world: Northern Eurasia, Asia Pacific, and North America. Meanwhile, Africa, South and Central America, and the Middle East collectively are estimated to hold just 7% of world reserves. Closer analysis shows that proven coal reserves are even more restricted geographically. Just nine nations harbor more than 90% of the world's coal reserves (**Figure 9.3**). At 28% of the world total, the coal reserves in the United States exceed all others.

Petroleum

Petroleum, or **crude oil,** formed in the oceans from the accumulated remains of algae and zooplankton deposited on the sea floor over millions of years (**Figure 9.4**). These deposits mixed with sands and silt, and eventually this organic-rich sediment was buried under deep layers of sedimentary rock. Trapped under a cap of heavy rock,

coal Sedimentary or metamorphic rock high in carbon and energy content formed over millions of years under conditions of high pressure and temperature (lignite, sub-bituminous coal, bituminous coal, anthracite).

petroleum (crude oil) A mixture of hydrocarbons contained in sedimentary rocks of marine origin; developed from the accumulated remains of algae and zooplankton deposited on the sea floor over millions of years.

FIGURE 9.3 Regional distribution of known coal reserves: Three geographic regions possess 95% of the world's coal reserves. Countries with the largest known coal reserves: Just nine countries contain more than 90% of global coal reserves. (2008 Data from U.S. Energy Information Agency, www.eia.gov/cfapps/ipdbproject/IEDIndex3.cfm?tid=1&pid=7&aid=6)

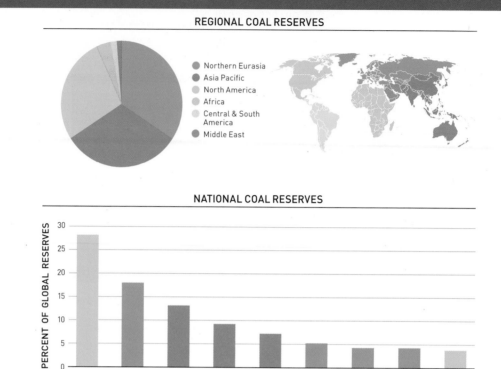

REGIONAL AND NATIONAL COAL RESERVES

REGIONAL COAL RESERVES

- Northern Eurasia
- Asia Pacific
- North America
- Africa
- Central & South America
- Middle East

NATIONAL COAL RESERVES

What do the rich oil fields in places like Texas and North Dakota suggest about the geologic history of these regions?

the developing petroleum was subjected to increasing pressures and heat, which gradually transformed the organic materials into a waxy substance called **kerogen.** With increasing heat and temperature applied over the course of millions of years, kerogen was converted to crude oil.

Chemically speaking, crude oil is a mixture of **hydrocarbons,** molecular chains consisting of only carbon and hydrogen. The smallest hydrocarbon, made up of four hydrogen atoms bonded to a single carbon atom, is methane (CH_4). To be useful and safe, crude oil must be refined, a process that involves separating the

kerogen A waxy substance found in shale and other sedimentary rocks that yields oil when heated; occurs during an intermediate stage of petroleum formation.

hydrocarbon An organic molecule made up of carbon and hydrogen only; the simplest hydrocarbon is methane (CH_4), the main component of natural gas.

FORMATION OF PETROLEUM

Petroleum is derived from the fossilized remains of marine algae and animals.

Marine organisms, like algae and animals, sink to the ocean's bottom after death

Burial of dead marine organisms by sediments

Under deep layers of rock, oil formed through the application of heat and pressure

MILLIONS OF YEARS

FIGURE 9.4 Petroleum, or crude oil, is derived from the fossilized remains of marine organisms over the course of millions of years.

REFINING CRUDE OIL AND ITS MAJOR PRODUCTS

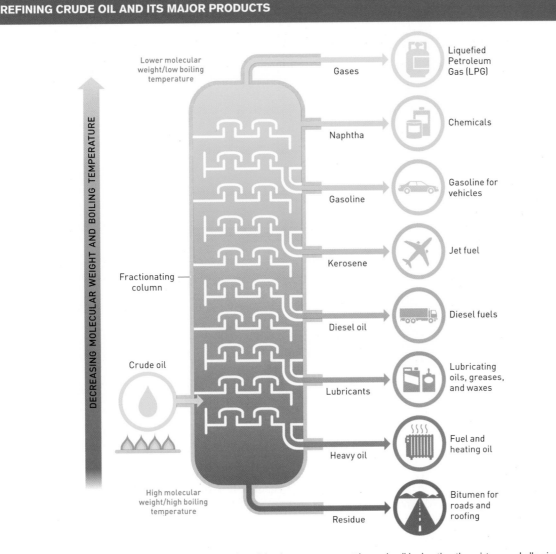

FIGURE 9.5 Oil refineries separate the wide range of useful substances present in crude oil by heating the mixture and allowing the substances to separate themselves by molecular weight, with the lightest compounds rising to the top of the fractionating column.

hydrocarbons into fractions containing hydrocarbons with the same number of carbon atoms.

The main principle behind the refining of oil is that different sizes of hydrocarbons have different molecular weights and will boil (or condense) at different temperatures. To separate these hydrocarbons, heated crude is pumped into the refinery column, as shown in **Figure 9.5**. Because the temperature within the column decreases from bottom to top, the heaviest hydrocarbons (e.g., heating oil) condense and flow out in the lower layers. Meanwhile, the lightest hydrocarbons (e.g., methane) rise to the very top of the column, where they are collected. These gases are typically pressurized until they turn to liquid, resulting in liquefied petroleum gas, or LPG.

Petroleum ends up in our cars and trucks in its most familiar forms as gasoline (petrol) and diesel fuel.

However, various oil products are in high demand to lubricate engines, tar our roads, or heat our homes (Figure 9.5). For many of these applications, alternatives do not work as well or are currently more costly.

The Middle East has oil reserves that account for over half of the world total and dwarf those of all other regions. After the Middle East, South and Central America, North America, Africa, and Northern Eurasia, each control from approximately 7% to 16% of global reserves. Within the world's regions, just 15 countries control more than 90% of the world's known oil reserves (**Figure 9.6**).

Natural Gas

Natural gas is a mixture of gaseous hydrocarbons—primarily methane, but also ethane, propane, and butane

FIGURE 9.6 The oil-rich Middle East Region possesses just over half of proven reserves. Just 15 countries hold more than 90% of the world's known oil reserves. (2011 Data from U.S. Energy Information Agency, www.eia.gov/countries/index.cfm?view=reserves)

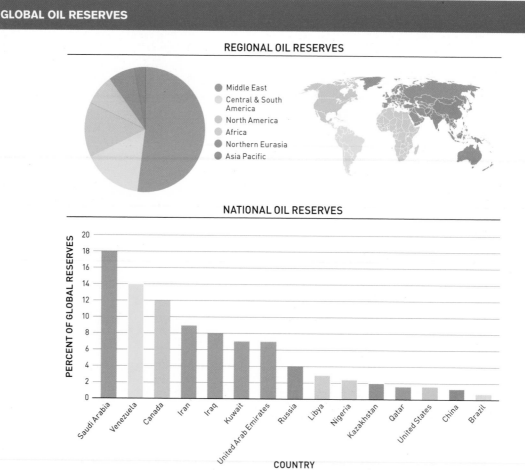

GLOBAL OIL RESERVES

REGIONAL OIL RESERVES

- Middle East
- Central & South America
- North America
- Africa
- Northern Eurasia
- Asia Pacific

NATIONAL OIL RESERVES

PERCENT OF GLOBAL RESERVES

COUNTRY: Saudi Arabia, Venezuela, Canada, Iran, Iraq, Kuwait, United Arab Emirates, Russia, Libya, Nigeria, Kazakhstan, Qatar, United States, China, Brazil

What does the shift in the view of natural gas from "waste product" to valuable resource suggest about changes in energy supply and demand over time?

(Figure 9.7). Natural gas forms in both petroleum deposits and coal beds. The long hydrocarbons in petroleum break down into shorter hydrocarbons of natural gas at temperatures above 100°C. These volatile gases then migrate up through porous rocks until they reach a nonporous impermeable rock, called a *cap rock* (Figure 9.8). Significant deposits of natural gases occur, especially where dome-shaped cap rocks trap a reservoir of gas overlying crude oil deposits. If drilling

ruptures such cap rock, the natural gas under it flows to the surface under pressure. In coal beds, natural gas, mainly methane, is formed from coal either under heat and pressure (as in crude oil) or by methane-producing bacteria operating at lower temperatures.

Natural gas associated with oil deposits and coal beds was once considered a hazard, because it could explode during drilling, and a waste product, because it was difficult to store and transport. Now, however, natural gas is used

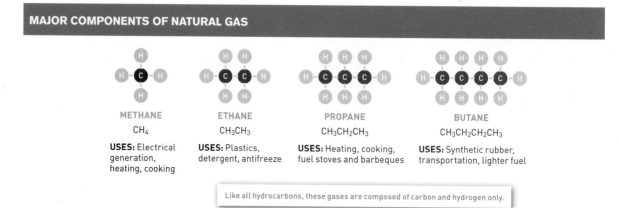

MAJOR COMPONENTS OF NATURAL GAS

METHANE
CH_4
USES: Electrical generation, heating, cooking

ETHANE
CH_3CH_3
USES: Plastics, detergent, antifreeze

PROPANE
$CH_3CH_2CH_3$
USES: Heating, cooking, fuel stoves and barbeques

BUTANE
$CH_3CH_2CH_2CH_3$
USES: Synthetic rubber, transportation, lighter fuel

Like all hydrocarbons, these gases are composed of carbon and hydrogen only.

FIGURE 9.7 Natural gas is a mixture of several kinds of hydrocarbons, but the major gas in the mixture is generally methane. The hydrocarbons in natural gas are all gases at room temperature.

ASSOCIATION OF NATURAL GAS DEPOSITS WITH OIL AND COAL BEDS

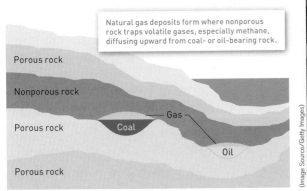

Natural gas deposits form where nonporous rock traps volatile gases, especially methane, diffusing upward from coal- or oil-bearing rock.

Porous rock

Nonporous rock

Gas

Porous rock

Coal

Oil

Porous rock

FIGURE 9.8 Natural gas deposits form where nonporous rock traps volatile gases, especially methane, diffusing upward from coal or oil-bearing rock.

NATURAL GAS, A HANDY AND EFFECTIVE ENERGY SOURCE

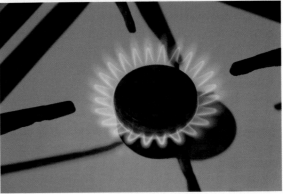

(Image Source/Getty Images)

FIGURE 9.9 Natural gas is widely used in homes for cooking and heating. It is also used for many industrial processes, including increasingly for generating electricity.

widely in residential, commercial, and industrial settings (**Figure 9.9**), where it accounts for about one-fourth of the energy used in the United States. As was the case with oil and coal, natural gas is unevenly distributed around the world. Just three countries, Russia, Iran, and Qatar, control over 50% of the known reserves of natural gas (**Figure 9.10**).

GLOBAL RESERVES OF NATURAL GAS

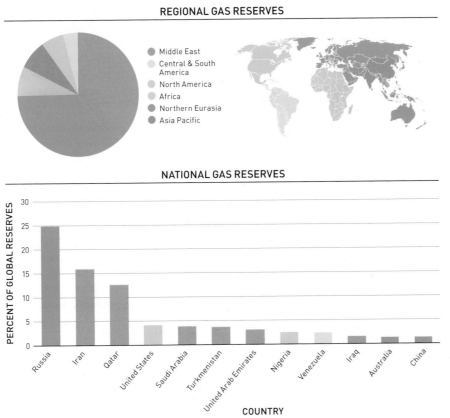

REGIONAL GAS RESERVES

- Middle East
- Central & South America
- North America
- Africa
- Northern Eurasia
- Asia Pacific

NATIONAL GAS RESERVES

FIGURE 9.10 Regional distribution of known natural gas reserves: The Middle East and Northern Eurasia have 75% of the known reserves. Countries with the largest known natural gas reserves: 75% of known reserves occur in 12 countries; just three countries own over half of the world's known gas reserves. (2011 Data from U.S. Energy Information Agency, www.eia.gov/cfapps/ipdbproject/IEDIndex3.cfm?tid=1&pid=7&aid=6)

SCIENCE   ISSUES   SOLUTIONS

⚠ Think About It

1. When converting some source of energy (e.g., natural gas) into electricity, why is the amount of electrical energy always less than the amount of energy used in its generation? (Hint: Consider the second law of thermodynamics discussed in Chapter 2, page 38.)

2. The same processes that produced fossil fuels, such as coal and petroleum, are still occurring on Earth today. Why, then, are fossil fuels considered "nonrenewable"?

3. Explain why the author Thom Hartmann was correct when he referred to fossil fuels as "ancient sunlight."

9.2 Power plants and vehicles burn fossil fuels to generate electricity and movement

During the early days of the Industrial Revolution, most industries powered their grain mills and other mechanical contraptions directly, using wind mills and water wheels. With the invention of the coal-powered steam engine in the late 1700s, factories could be located anywhere fuel was available, whether or not a wind or water power source was nearby. By the end of the 1800s,

BASIC ELEMENTS AND OPERATION OF AN ELECTRICAL GENERATOR

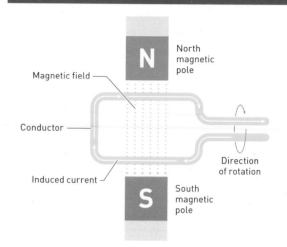

FIGURE 9.12 Passing a conductor through a magnetic field induces an electrical current in the conductor, commonly called *electricity* or *electrical energy.*

scientists had learned how to transform the mechanical energy of these steam engines into electricity.

Electricity, which is the flow of electrons through materials known as conductors, is one of the most useful forms of energy, in part because it can be transmitted over long distances from a source of production to users (**Figure 9.11**). Today, when you cook a meal in your microwave or charge your laptop, it's easy to forget the long path that the electricity may have taken to reach you.

Generating Electrical Power

A device capable of producing electricity, commonly called a *generator,* includes two basic components: a magnet and a conductor (**Figure 9.12**). The magnetic component of a generator creates a magnetic field. The movement of a conductor, generally rotating copper wire, through the generator's magnetic field induces a flow of electrons in the conductor. That flow of electrons can be harnessed to do work ranging from lighting a room to running an electric car. However, we need to remember that it takes energy to move the conductor and generate a flow of electrons. Globally, fossil fuels are the source of approximately two-thirds of electrical generation (**Figure 9.13**).

Power Plants

Generating electricity is as easy as boiling a pot of water. OK, it's not *that* easy—but any fuel source that can heat water enough to turn it into steam may be used to generate electricity. Currently, the most common source of this heat is the burning of fuels, especially coal and natural gas.

Let's take a look at the workings of a typical coal-fired power plant (**Figure 9.14**). Coal is first pulverized into

ELECTRICAL TRANSMISSION LINES

(Vlad Turchenko/Shutterstock)

FIGURE 9.11 One of the convenient aspects of electrical power is that it can be transmitted long distances with relatively small energy losses.

ENERGY SOURCES FOR PRODUCTION OF ELECTRICAL ENERGY

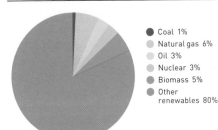

GLOBAL ELECTRICAL GENERATION

- Coal 41%
- Natural gas 21%
- Oil 5%
- Nuclear 13%
- Biomass 2%
- Other renewables 18%

U.S. ELECTRICAL GENERATION, 2013

- Coal 39%
- Natural gas 27%
- Oil 1%
- Nuclear 19%
- Biomass 1%
- Other renewables 12%

BRAZIL ELECTRICAL GENERATION

- Coal 1%
- Natural gas 6%
- Oil 3%
- Nuclear 3%
- Biomass 5%
- Other renewables 80%

FIGURE 9.13 Recently, the energy sources for electrical generation across the globe and in the United States were dominated by nonrenewable energy sources. Meanwhile, the energy for electrical generation in Brazil came mainly from renewable resources. (2011 Data from International Energy Agency, www.iea.org; 2013 Data from U.S. Energy Information Agency, www.eia.gov/electricity/; and Brazilian Ministry of Mines and Energy, www.mme.gov.br)

tiny pieces so it burns more completely. The heat of combustion is used to boil water, which produces steam. The steam creates pressure that turns a turbine attached to an electrical generator. After passing through the turbine, the steam is cooled, condenses back to liquid water, and returns to the boiler, where it is heated to form steam once again. The electricity produced can then flow through the power grid and into homes and businesses.

Each step in the process results in a loss of usable energy. On average, about 35% of the energy content of coal or other combustible fuel is converted into electrical energy. The remainder is lost to the environment as heat (see Figure 2.17, page 47).

Electrical energy can be generated using other combustible fuels, including natural gas, oil, and biomass,

in the form of wood or agricultural wastes. Although the physical nature of these materials requires different handling, the principle of their use is the same as in coal-fired power plants: The chemical energy released during burning is used to produce steam that turns a turbine.

Trains, Planes, Automobiles, and More

With the increased availability of petroleum in the late 1800s, it became possible to develop a more compact engine that didn't require a separate steam chamber and furnace. In an **internal combustion engine,** combustion directly drives a set of pistons or turbines hooked up to a crank arm. Most internal combustion engines are used in vehicles, such as automobiles, diesel-powered trains, jet airplanes and boats. They are also used in power tools

What costs, beyond the price of fuel, should be taken into account when calculating the cost of delivering electrical power to a home or business?

internal combustion engine Engine (most commonly used in cars, boats, and jet airplanes) in which combustion directly drives a set of pistons or turbines hooked up to a crank arm.

ELECTRICAL GENERATION USING COMBUSTIBLE MATERIALS

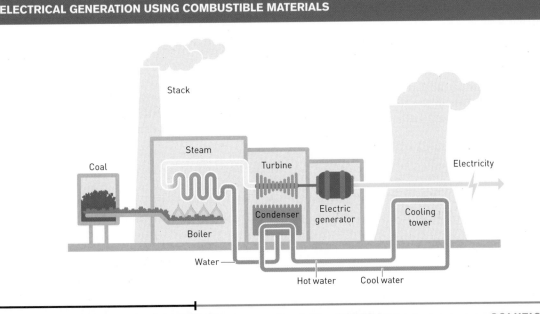

FIGURE 9.14 Heat from the burning of coal, natural gas, oil, or biomass can be used to generate the steam used to spin the turbines of an electric power plant. The power plant featured here uses coal as its source of energy.

such as lawnmowers, leaf blowers, and chainsaws, as well as for portable diesel and gasoline power generators.

Combined cycle power plants combine a type of internal combustion engine known as a **gas turbine engine** with a steam power plant. When operated on its own, a gas turbine engine burns natural gas, sending a hot, high-pressure stream of gas through a turbine connected to an electrical generator. A combined cycle power plant takes advantage of the potentially wasted heat of combustion to boil water and spin an independent steam turbine. Combined cycle power plants can increase the efficiency of power generation from 35% to 60%.

⚠ Think About It

1. Why is electricity often referred to as a "secondary" source of energy?

2. What economic and environmental benefits potentially result from increasing the efficiency of electrical power plants?

3. Is it possible to create a perfectly efficient power plant? Explain.

9.3 Nuclear energy is released by atomic fission and fusion

Soon after the development of nuclear weapons in the middle of the 20th century, we found ways to transform this new, potent source of energy into electricity. **Nuclear energy** is released when the bonds holding the protons and neutrons making up the nucleus of an atom break, a process called **nuclear fission** (**Figure 9.15**). Nuclear energy is also released when the nuclei of two atoms fuse together under high temperature and pressure, a process called **nuclear fusion.** Nuclear fusion provides the fuel for stars, including our own Sun, and is the basis for one type of atomic weapon.

Nuclear bonds hold much more energy than chemical bonds. Therefore, the amount of energy released during nuclear fission is much greater than that released during the combustion of fossil fuels. For example, the energy content of a gram of uranium-235, the most common nuclear fuel, is approximately 3 million times the energy content of a gram of coal.

Uranium, the fuel used in today's nuclear power plants, is a nonrenewable resource. Uranium ore is mined from the ground and processed to produce a powdery, concentrated form known as yellowcake. The isotope used for fission, uranium-235 (U-235), makes up only 0.71% of the uranium in nature. Through a process known as **enrichment,** uranium-235 is separated from less valuable uranium-238. The highly radioactive

combined cycle power plants Power plants that combine a gas turbine engine with a steam power plant.

gas turbine engine Engine that burns natural gas, sending a hot, high-pressure stream of gas through a turbine connected to an electrical generator.

nuclear energy A form of energy released when the nucleus of an atom breaks apart (nuclear fission), or when the nuclei of two atoms fuse (nuclear fusion).

nuclear fission A process in which the bonds holding the protons and neutrons that make up the nucleus of an atom are broken, resulting in the release of a large quantity of energy.

nuclear fusion A process in which the nuclei of two atoms fuse to form a new type of atom, releasing large amounts of energy.

enrichment A nuclear process in which uranium-235 is separated from less valuable uranium-238.

NUCLEAR REACTIONS

NUCLEAR FISSION

A neutron colliding with the nucleus of uranium-235 splits the nucleus, forming atoms of barium-139 and krypton-94 and releasing three neutrons and a large amount of energy.

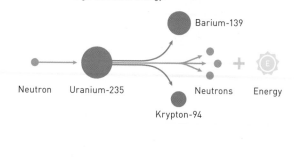

NUCLEAR FUSION

At very high temperatures, two hydrogen nuclei fuse, forming an atom of helium and giving off energy.

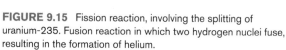

FIGURE 9.15 Fission reaction, involving the splitting of uranium-235. Fusion reaction in which two hydrogen nuclei fuse, resulting in the formation of helium.

enriched uranium can be used as nuclear fuel, whereas the depleted uranium is typically placed in storage.

Approximately 97% of the known reserves of uranium have been found in just 15 countries (**Figure 9.16**). The largest known reserves, nearly one-third of the world total, occur in Australia. The United States, with 100 commercial reactors, is the largest producer of nuclear power in the world, and nuclear energy accounts for 19% of the electricity produced in the country.

In contrast to nuclear fission, fusion doesn't require uranium or other rare, radioactive materials, but rather a seemingly limitless fuel: hydrogen, which can be extracted from water (Figure 9.15). Presently, there are no commercial fusion reactors for power generation, but we will discuss its potential in the Solutions section of this chapter.

Nuclear Power Generation

In nuclear power plants, the heat energy released during the fission of uranium can be harnessed to generate electricity. This heat produces steam that drives a turbine connected to an electrical generator (**Figure 9.17**), similar to what we saw in the coal-fired power plant. However, using nuclear energy safely requires a number of special precautions.

Secure operation of a nuclear power station depends on controlling the speed of neutrons released during

GLOBAL URANIUM RESERVES

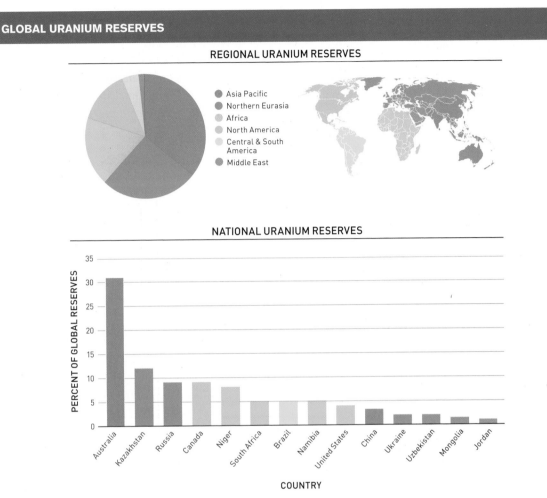

REGIONAL URANIUM RESERVES

- Asia Pacific
- Northern Eurasia
- Africa
- North America
- Central & South America
- Middle East

NATIONAL URANIUM RESERVES

PERCENT OF GLOBAL RESERVES

Australia, Kazakhstan, Russia, Canada, Niger, South Africa, Brazil, Namibia, United States, China, Ukraine, Uzbekistan, Mongolia, Jordan

COUNTRY

FIGURE 9.16 Regional distribution of recoverable uranium: Four regions possess over 90% of known reserves of recoverable uranium. Countries with the largest known reserves of recoverable uranium: More than 97% of known reserves of recoverable uranium occur in 14 countries. (2012 Data from World Nuclear Association, www.world-nuclear.org/info/Nuclear-Fuel-Cycle/Uranium-Resources/Supply-of-Uranium/#.UYrrT4KHamE)

nuclear fission. Uranium-235 is formed into small pellets that are contained in tubes called **fuel rods.** In the most commonly used nuclear reactors, the fuel rods are submerged in pressurized water, which acts as a **moderator** by reducing the speed at which neutrons travel (see Figure 9.17). By slowing down the neutrons

released during fission, the moderator increases their chance of triggering fission of other uranium-235 atoms, which releases more energy and yet more neutrons in a chain reaction.

The increasing number of fission reactions is an example of positive feedback (see Chapter 2, page 48). If

fuel rods Tubes containing small pellets of uranium-235 used as an energy source in nuclear reactors.

moderator A substance (most commonly pressurized water) used in a nuclear reactor to reduce the speed at which neutrons travel.

DIAGRAM OF A NUCLEAR POWER PLANT

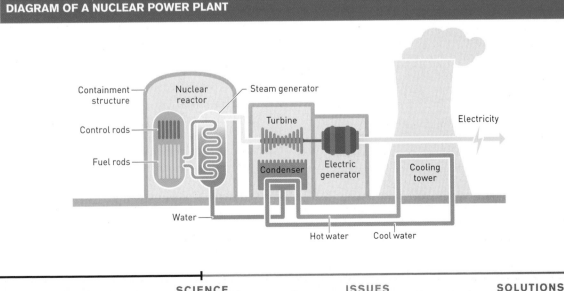

Containment structure · Nuclear reactor · Steam generator · Turbine · Electricity · Control rods · Fuel rods · Condenser · Electric generator · Cooling tower · Water · Hot water · Cool water

FIGURE 9.17 The design of a nuclear power plant is aimed at controlling the rate of nuclear fission and heat production to produce a steady source of steam for driving the electrical generator, while avoiding a reactor meltdown.

Nuclear power is more expensive than power generated using coal or natural gas as energy sources. Why, then, have so many countries around the world invested heavily in the development of nuclear power plants?

control rods Long rods made of neutron-absorbing substances, used to control the rate of fission in a nuclear reactor.

containment structure A steel and concrete enclosure designed to prevent the release of radioactive material in the case of a serious nuclear reactor accident.

uncontrolled, a chain reaction could generate so much heat that it would lead to a meltdown of a nuclear reactor. The chain reaction in nuclear reactors, and thus the amount of heat generated, must therefore be regulated using **control rods,** which are made of substances that absorb neutrons. Because they absorb neutrons, control rods can be inserted into a reactor core to slow the rate of nuclear fission or to shut down a reactor completely, either in an emergency or for repairs.

The pressurized water circulating through the reactor is heated by the energy released during fission reactions. This heat produces the steam, which spins the turbine attached to a generator. Cooling water circulated through the turbine chamber cools the steam, condensing it back to liquid water. This water is returned to the steam generator, where it is again converted to steam. Notice that the three streams of water in the nuclear power plant—pressurized water in the reactor, water in the steam chamber, and cooling water in the condenser—do not mix. For safety, the nuclear reactor and steam generator are enclosed in a **containment structure** with steel and concrete walls generally 1 to 2 meters (3 to 6 feet) thick.

⚠ **Think About It**

1. How are the processes that yield energy during chemical reaction different from the processes that yield nuclear energy?

2. What are some implications of uranium-235 having approximately 3 million times the energy content compared to coal?

3. How does increasing or decreasing the number of control rods inserted in a nuclear reactor influence the amount of heat generated in a nuclear reactor?

9.1–9.3 Science: Summary

The most commonly used sources of nonrenewable energy are fossil fuels in the form of coal, oil, and natural gas, which were formed over millions of years. Any source of heat sufficient to convert liquid water to steam can be used to generate electricity. Currently, the most common source of this heat is the combustion of fossil fuels, especially coal and natural gas. Oil can also be used to power internal combustion engines, which are found in vehicles and portable power generators. In the mid-20th century, we learned how to harness nuclear energy through nuclear fission. Nuclear fission is a very powerful source of heat, which can be used to generate electricity, but nuclear energy requires special precautions for safe use.

9.4–9.6 Issues

In the aftermath of the Deepwater Horizon oil spill, the United States has not shied away from the environmental risks from oil drilling. It has redoubled its efforts. In the last three years, oil production has increased by over 30%, and, in 2014, President Barack Obama signed a bill to reopen the Eastern Seaboard to offshore oil exploration. There's even talk of lifting the nation's four-decade-old ban on exporting crude oil. Yet the central challenge of nonrenewable resources such as fossil fuels is that their supply is limited. In our continuing quest for energy, supplies of which are dwindling, we have the potential to disrupt the environment at scales small and large.

9.4 Global energy use grows as energy shortages loom

The global economy is powered by the extraction of raw natural resources and a steady supply of cheap energy. But what if the flow of energy begins to slow and the prices rise? Because we obtain most of our electricity from nonrenewable resources, it's crucial to examine how much longer those resources will last.

Global Energy Consumption

Cell phones, laptop computers, and other electronic devices were formerly reserved for the wealthiest

WORLD CONSUMPTION OF ENERGY RESOURCES

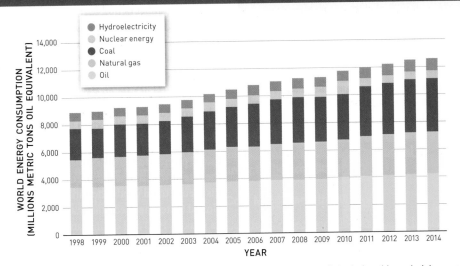

FIGURE 9.18 While oil, nuclear energy, and hydroelectricity consumption have grown little during this period, increases in natural gas and coal consumption have been substantial. (Data from BP 2009, 2011, 2013, 2015)

members of society, but today we see cell phones in some of the most remote corners of the globe. This spread of technology—and the energy to power it—goes beyond personal electronics. Villages where the only form of transportation had been an oxcart are now buzzing with motorbikes, while open-air restaurants are now sealed off and cooled with humming air-conditioning systems.

So it shouldn't be surprising that, with increasing development around the world, global consumption of **primary energy**—mainly, coal, oil, natural gas, nuclear energy, and hydroelectricity—grew by 44% between 1998 and 2014 (**Figure 9.18**). Primary energy is a form of energy that requires only extraction or capture for use;

examples are coal, crude oil, and wind. Although oil is still the single largest primary energy source consumed, in recent years its share declined from 39% in 1998 to 33% of total global energy consumption. Meanwhile, natural gas and coal use have increased.

Today, China, the United States, and the European Union are the three largest consumers of energy on Earth. In 2011 their combined primary energy consumption made up over 53% of the world total. However, their trajectories in energy use over the period of 1998 to 2014 differ dramatically (**Figure 9.19**). While energy consumption in the United States increased by only 3% and in the European Union it declined by 6% over this

What factors likely contribute to oil's falling share of total energy consumption?

primary energy A form of energy that requires only extraction or capture for use (e.g., coal, crude oil, wind).

ANNUAL ENERGY USE BY THE UNITED STATES, EUROPEAN UNION, AND CHINA FROM 1998 TO 2014

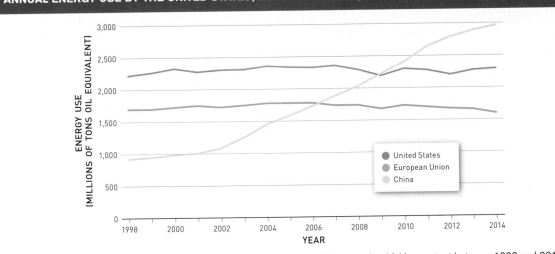

FIGURE 9.19 While energy use in the United States and the European Union remained fairly constant between 1998 and 2014, China's energy use grew by 217%. (Data from BP 2009, 2011, 2012, 2013, 2015)

SCIENCE ISSUES SOLUTIONS

FIGURE 9.23 This approach to mining coal, which removes the tops of some of the oldest mountains on Earth along with their cloak of highly diverse temperate forest ecosystems, is one of the most destructive forms of mining in use today.

MOUNTAINTOP REMOVAL MINING IN WEST VIRGINIA

(© Jim West/Alamy)

How should we weigh the jobs provided by mountaintop removal mining against the massive environmental damage caused by this approach to coal mining?

acid mine drainage A problematic result of strip mining, in which surface flow of groundwater turns acidic as it percolates through mine wastes (tailings).

mountaintop removal mining An extremely destructive coal mining practice that involves clear-cutting of the forests on a mountain and adjacent stream valleys; miners then use explosives to break up the rock overlying the coal deposit, depositing it in the adjacent valleys, which are buried as the coal is exposed.

fly ash By-product of coal burning, stored in open ponds and landfills.

acid mine drainage. Where it occurs, acidification can seriously disrupt aquatic ecosystems (see Chapter 13, page 403).

In steep terrain, such as in the Appalachian Mountains, one of the most destructive coal mining practices ever devised is being used: **mountaintop removal mining.** Large deposits of high-quality coal underlie the Appalachian Mountains. Historically, these deposits were mined by tunneling through overlying geologic strata and digging out the coal seams. Subsurface mining continues in much of the region. However, in some areas, mountaintop removal allows access to these deposits, requiring the removal of up to 100 meters (330 feet) of overburden (**Figure 9.23**). The first step in mountaintop removal mining is to clear-cut the forests on a mountain and adjacent stream valleys. Miners then use large explosive charges to break up the rock overlying the coal deposit. Giant earth-moving equipment called *draglines* deposit this overburden in nearby stream valleys, which are gradually buried as the coal is exposed.

The Appalachian Mountains, which run north–south along the eastern United States, are among the oldest mountains on Earth and home to some of the highest levels of temperate biodiversity in the world. The temperate forests that cloak the flanks of these mountains support an exceptionally high number of tree species and are home to large numbers of endemic amphibian species. The streams that drain the forests are also exceptionally diverse, supporting some of the most species-rich invertebrate communities in the world. During mountaintop removal mining, this wildlife

habitat is destroyed, surface streams are lost, and water and air are polluted.

As of 2010, about 410,000 hectares (1.2 million acres) on 500 mountains in Appalachia had undergone mountaintop removal mining. In the process, hundreds of kilometers of streams have been lost and a globally unique and diverse mountain landscape irreversibly altered.

Coal Sludge and Fly Ash Spills

Extracting, processing, and burning coal produce a tremendous amount of toxic waste. For instance, crushing coal during processing results in coal sludge, a mixture of mineral particles, coal dust, and water. Companies usually store this mixture at the coal-processing site in sludge ponds created behind an earthen dam. A second by-product of coal burning is called **fly ash.** Approximately 140 million tons of fly ash and other waste are produced from coal burning in the United States every year. Like coal sludge, fly ash is stored in open ponds and landfills.

Breaches and leaks from these ponds have led to numerous environmental calamities. For example, on December 22, 2008, a breach of a storage pond at the Kingston, Tennessee, coal-fired power plant (**Figure 9.24**) released 1.1 billion gallons of fly ash slurry. It covered approximately 120 hectares (300 acres) of land, inundated 12 homes, blocked a railroad line, and knocked down electrical power lines. Fortunately, no one was injured. However, it killed a significant number of

(David Luttrell/Tennessee Valley Authority)

FIGURE 9.24 The fly ash slurry spill at the Kingston, Tennessee, coal-fired power plant, shown here, took place on December 22, 2008. The failure of the retaining structure on the fly ash slurry pond released 4 million cubic meters of fly ash slurry, some of which can be seen here.

fish in the nearby Clinch and Emory Rivers and elevated concentrations of several heavy metals, including arsenic and mercury. The U.S. Environmental Protection Agency (EPA) and the Tennessee Valley Authority (TVA) estimated that cleanup of this one spill would require three years at a cost of hundreds of millions of dollars. Another major spill took place in February 2014, when a ruptured pipe at a Duke Energy plant led to the dumping of 82,000 tons of coal ash into a North Carolina river.

However, even without these spills, coal sludge and fly ash can produce serious environmental impacts. These waste products contain large concentrations of heavy metals that can leach into groundwater, contaminating drinking water supplies and poisoning aquatic organisms. Currently, the EPA lists over 70 coal-ash pollution sites around the country. In the aftermath of the Kingston spill, the EPA began drafting regulations to reign in coal power plant pollution. In December 2014, the EPA ruled that no new coal sites could be located in wetlands or in earthquake-prone areas and that coal ash ponds had to be lined to prevent groundwater contamination. The ruling also specified that any active coal ash disposal sites that are causing pollution must be cleaned up.

Oil Spills

Oil spills such as the Deepwater Horizon provide some of the most familiar examples of how petroleum extraction threatens the environment. Although such catastrophic oil spills make the headlines, they only represent a small fraction of the 29 million gallons of oil that leak into marine environments in North America. A 2003 report from the National Academy of Sciences estimated that only 8% of that total comes from tanker and pipeline spills, whereas oil exploration and extraction account for just 3%. The majority of the oil spilled—about 85%—comes from chronic runoff from the parking lots in coastal areas, improper disposal of waste oil, and old two-stroke engines on recreational boats that regularly discharge gasoline.

Mining Oil Sands

Some of the dirtiest oil on Earth lies deep within the boreal forests of Alberta, Canada. These deposits are known as Athabasca oil sands or tar sands, due to the tarlike consistency of the oils they contain, commonly called **bitumen.** Bitumen is a flammable, highly viscous or semisolid mixture of hydrocarbons. When these tar sands occur near the land surface, they are mined by removing overburden and transported offsite, where the heavy oils are separated from the sands and clays with which they are mixed. When the oil sands are found farther from the surface, they are heated in place to allow the oil to flow and are then pumped to the surface. In either case, the tar sands oil is more expensive to extract than so-called conventional oil that occurs naturally in a liquid state. However, as the price of conventional oil has risen, the tar sands have become economical to mine. And as the mining proceeds, the landscape is left in a devastated condition (**Figure 9.25** on the next page).

Various environmentalists, farmers, and landowners in the United States have tried to halt or slow the extraction of tar sands by blocking the approval of the Keystone XL Pipeline, which will transport more than half a million barrels of oil per day from Alberta to the Gulf Coast, where it can be refined and exported. The pipeline will cross the Nebraska Sandhills and the Ogallala Aquifer, which supplies groundwater for drinking and irrigation in eight states from South Dakota to Texas, and many critics fear that an oil spill could be devastating. The TransCanada Corporation, which seeks to build the pipeline, has argued that Keystone will be "the safest pipeline ever built."

The Fracking Boom

After years of dependence on foreign oil, the United States is experiencing a resurgence of cheap energy through a boom in *fracking,* which poses new threats to the environment. **Fracking (hydraulic fracturing)** is a

bitumen A flammable, highly viscous or semisolid mixture of hydrocarbons.

fracking (hydraulic fracturing) An extraction technique that involves drilling horizontally into a rock formation and pumping in a mixture of fluids and sands to fracture it, thus creating a path through which natural gas or oil can flow out.

FIGURE 9.25 Before mining, the landscape overlying the oil sands is made up of a mix of boreal forest, wetlands, and ponds. Mining leaves no vestige of the rich ecosystem that is removed to gain access to the buried fossil fuels.

MINING THE ATHABASCA OIL SANDS

Before mining

After mining

controversial extraction technique used when oil or natural gas is tightly bound in geologic formations, such as the Bakken Shale of North Dakota or the Marcellus Shale of Pennsylvania, and cannot be pumped from the formation using conventional technology. Fracking involves drilling horizontally into a rock formation and pumping in a mixture of fluids and sands to fracture it. The fractures, which are held open by the sand in the fracking fluid, create a path through which natural gas or oil can flow out of the formation. While hydraulic fracturing has been used for more than 50 years, new technology, including horizontal drilling and new chemical mixtures, provide access to previously unavailable natural gas deposits.

The EPA has identified numerous ways that fracking can impact surface or groundwater sources of drinking water, as has occurred in Pennsylvania and West Virginia (EPA, 2015). Fracking operations generally involve pumping millions of gallons of fluid in each well. Thus, the two main concerns are that fracking potentially reduces drinking water supplies, as water is pumped from an aquifer or from a lake or river, and that the chemicals in the fracking fluids could contaminate drinking water (**Figure 9.26**).

The fracking fluids contain 98% water and sand and 2% chemical additives. What are these additives? Companies have used more than 1,000 different chemical compounds in various fracking operations, the most common ingredients being hydrochloric acid, methanol, and petroleum derivatives. But not all the ingredients are known: Companies have refused to disclose the identity of 10% of them for "proprietary business reasons." Regardless, the EPA did gather enough information to evaluate human health effects of 73 of the "mystery ingredients" and found that they included carcinogens and other toxic substances affecting heart, liver, kidney, and reproductive systems (see Chapter 11, page 332).

Furthermore, the EPA identified a number of instances in which contamination of drinking water due to fracking has been verified. But that number was relatively small compared with the number of fracking wells that have been drilled in the country. As a result, the study concluded that fracking has not produced "widespread, systemic impacts on drinking water resources in the United States." In other words, contamination of drinking water by fracking has been a local or regional problem. While the petroleum industry lauded the report as demonstrating that fracking is therefore indeed safe, environmental organizations emphasized that the study confirmed fracking still has the *potential* to contaminate drinking water.

Another area of concern is an uptick in the number of earthquakes in areas where fracking activity has increased. According to the United States Geological Survey (USGS), the frequency of noticeable earthquakes (magnitude 3 or higher) in the central and eastern United States began increasing sharply around 2009. Oklahoma has been particularly hard hit. One earthquake in November 2011 struck near the town of Prague and caused more than $10 million worth of damage. And during 2014 Oklahoma recorded 15 magnitude-4 earthquakes—more than the state had seen in an entire century.

A recent USGS report, however, concluded that fracking was *not* responsible for most of these earthquakes. Rather, it laid the blame on another controversial practice of the energy industry: injecting wastewater into deep wells (see Chapter 12, page 381). Drillers in this region have been pumping up a briny mixture containing oil and gas—the remnant of an ancient sea. After separating the fuel from the saltwater, they inject waste back into deep wells, which can fracture rock formations near major faults. Oklahoma has more than 4,600 of these disposal wells, and lawmakers,

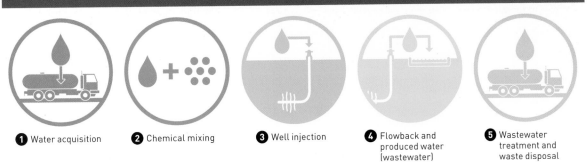

THE FRACKING WATER CYCLE

1 Water acquisition **2** Chemical mixing **3** Well injection **4** Flowback and produced water (wastewater) **5** Wastewater treatment and waste disposal

FIGURE 9.26 The extraction, use, treatment, and disposal of water for fracking have the potential to impact drinking water supplies at each point in the process.

companies, and residents are currently debating whether the practice needs to be more closely regulated.

⚠ Think About It

1. To what extent should the environmental damage caused by fossil fuel extraction (see the discussion of externalities in Chapter 2, page 49) be included in the actual price of an energy source?

2. A company causing environmental damage during the course of fossil fuel extraction or transport is generally required to pay for the damages caused, as well as the cleanup. However, are there ways that the market might positively reward those companies that extract and transport such resources without environmental damage?

3. While the United States fined BP for the Deepwater Horizon oil spill and the company is paying for economic damages, Brazil filed criminal charges that could result in long prison terms for energy-company personnel involved in oil spills off its coast in 2011. How do you think environmental impacts should be penalized? Include an analysis of the costs and benefits of potential penalties.

9.6 Nuclear power development comes with environmental costs

During the early years of the atomic age, nuclear power was championed as a nearly infinite source of cheap and clean energy. "Our children will enjoy in their homes electrical energy too cheap to meter," the head of the U.S. Atomic Energy Commission confidently declared in 1954. But in the 1960s and 1970s, a coalition of environmentalists rallied against the use of nuclear power

because of issues related to uranium mining, radioactive waste disposal, and the risk of a nuclear accident. Even as some pundits today continue to predict a "nuclear renaissance" in response to rising petroleum prices, we have come no closer to solving these environmental issues.

Uranium Mining

Grand Canyon National Park sits right in the middle of the country's richest deposits of uranium ore. Beginning in 1944, over 3.9 million tons of uranium were extracted from over 1,000 mines in the region, and the low-level radioactive waste was left behind in piles scattered across federal, state, and Native American lands, potentially posing threats to human health and the environment. Uranium mining was banned from the region in 1986, but a company called Energy Fuels Resources has been fighting to reopen one particular mine—the Canyon Mine—that lies 6 miles from the park entrance.

Uranium mining, which can be conducted underground or on the surface, poses many of the same environmental threats as coal mining, but with the addition of radioactivity, which can cause cancer and other ailments. Although mine waste typically has a very low level of radioactivity, there are over 15,000 mine sites in the western United States and they are not regularly monitored. Members of the Navajo Nation, where much of this mining has occurred, argue that long-term environmental exposure has led to an increase in cancer in the region. In a study that included Navajo uranium miners from the 1950s, the U.S. National Institute for Occupational Safety and Health found 3 to 6 times the number of lung cancer deaths than expected, along with a significant increase in deaths due to tuberculosis and other lung diseases. The study did not measure the environmental exposures of residents who were not working in the mines.

Uranium mining accidents also have the potential to seriously damage the environment. In July 1979 a waste

THREE MILE ISLAND NUCLEAR POWER PLANT

(Jamous Patrick/Paris Match via Getty Images)

FIGURE 9.27 In 1979 the operators and safety systems at the Three Mile Island nuclear power plant prevented a partial meltdown from developing into a serious nuclear accident at the plant.

pond at a uranium mill in Church Rock, New Mexico, breached its dam. Over 1,000 tons of acidic, radioactive mining waste flowed into the Puerco River, contaminating water supplies used for watering livestock. Church Rock has been called the biggest radioactive accident in U.S. history.

Radioactive Waste Disposal

Each step in the process of turning uranium ore into enriched fuel rods generates a significant amount of waste. However, most of the front-end waste generated during the production of fuel rods has only a low level of radioactivity and poses little risk if disposed of properly. The spent fuel rods, however, are considered high-level waste (HLW) and present a more serious challenge. Whereas fuel rods can no longer be used to generate electricity in a reactor, they remain highly radioactive and must be isolated from the environment.

Even though the United States has been generating nuclear power for decades, we still have no safe system for the long-term disposal of spent fuel rods. After cooling off in water pools for up to 10 years, these rods are currently placed in steel cylinders, encased in concrete, and stored at the nuclear plant. Ideally, these casks will be moved to a safe repository deep underground, where they can be protected from natural disasters or a terrorist attack.

CHERNOBYL, UKRAINE, SITE OF THE WORST NUCLEAR ACCIDENT OF THE 20TH CENTURY

RADIOACTIVE FALLOUT FROM CESIUM-137 AFTER CHERNOBYL [kBq/m^2]

- \> 1,480
- 185–1,480
- 40–185
- 10–40
- 2–10
- < 2

Chernobyl

(Shone/Gamma-Rapho via Getty Images)

FIGURE 9.28 An explosion at the Chernobyl nuclear power plant on April 26, 1986, and a subsequent 10-day fire sent a plume of radioactive materials across Europe.

Unfortunately, no country currently has a viable system in place for long-term disposal. In 1978 the Department of Energy first identified Yucca Mountain in Nevada as a potential nuclear disposal site and experimental tunnels were bored into the rock. However, questions concerning the geologic stability of the site and political opposition to the facility, led by long-term Nevada Senator Harry Reid, have prevented it from opening. In May 2009 the Secretary of the Department of Energy said, "Yucca Mountain as a repository is off the table" (see Chapter 12 for more on nuclear waste disposal).

Nuclear Meltdowns

The worst type of accident that can happen at a nuclear power plant is called a *meltdown*. A meltdown occurs when the core of a nuclear reactor grows too hot and begins to melt, a situation that could lead to the release of radiation into the environment.

This nightmare scenario happened in March 1979 at Pennsylvania's Three Mile Island nuclear power plant. A valve got stuck inside, which led to the loss of coolant from the nuclear reactor (**Figure 9.27**). Fortunately, the reactor was contained in a strong confinement structure and plant operators acted quickly to minimize the release of radioactive material into the environment. This event, which was considered a partial meltdown, stoked public fears about the threats of nuclear technology. The reactor was too badly damaged to resume operations, and cleanup from the accident took 14 years and cost $1 billion.

A second nuclear accident further heightened doubts about nuclear power safety. In the early morning hours of April 26, 1986, a test of emergency systems at the Chernobyl power plant in Ukraine resulted in a steam explosion that tore apart one of its reactors (**Figure 9.28**). The resulting fire burned for 10 days. Because the Chernobyl reactor had no confinement structure, the raging fires spewed radioactive material into the atmosphere, with winds transporting them to the northwest. Significant levels of contamination from the Chernobyl fires extended over 2,000 kilometers, impacting crops and dairy products in countries as far away as Italy, Ireland, and Norway. However, the most contaminated areas were in Ukraine and nearby Belarus. In the aftermath of the accident, there have been approximately 5,000 cases of nonfatal thyroid cancer, and the U.S. National Academy of Sciences has estimated that the accident will cause more than 4,000 extra cancer deaths in the region over the next 50 years.

Nuclear power experts point out that the disaster at the Chernobyl power plant resulted from a flawed reactor design, mistakes made by inadequately trained operators, and insufficient attention to safety. Simply having a structure to contain a reactor explosion, as do all post-Soviet reactors, would have greatly reduced the chance of environmental contamination. Since the disaster, all power plants with Chernobyl's design have been modified

FUKUSHIMA NUCLEAR POWER PLANT DISASTER

FIGURE 9.29 The record magnitude 9.0 earthquake on March 11, 2011, produced a tsunami, measured at 14 meters in height when it struck the Fukushima nuclear power plant, flooding and disabling the plant's backup diesel-powered electrical generators.

to make them safer, and no major nuclear accidents have occurred at these plants since the accident at Chernobyl.

The Fukushima Nuclear Disaster

On March 11, 2011, the strongest earthquake ever recorded off Japan produced a tsunami that inundated a nuclear power plant in Fukushima, Japan, and led to a meltdown (**Figure 9.29**). The nuclear power plant survived the initial earthquake, but the external electrical supply to the power plant was knocked out. Diesel generators in the basement roared into action to provide backup electrical power to run the power plant's cooling system. However, the tsunami that arrived 45 minutes later flooded the generators, disabling them. The plant switched to backup battery power, but it was exhausted after eight hours.

Without electrical power to run the plant's controls and monitor its environment, the situation developed into a serious crisis. With control rods in place, fission within the reactor was reduced greatly but not eliminated entirely. The reactors began to overheat and pressure began building up in them. With no electrical power, power plant operators were forced to vent the reactors manually to relieve the building pressures. However, the venting also released highly unstable hydrogen gas, which reacted explosively with oxygen, blowing the roofs off of two buildings that housed reactors and where spent fuel rods were housed in pools of water to keep them cool.

Because the spent fuel pools were exposed to the atmosphere, operators worried that the water in them

How might the use of nuclear energy as a weapon influence perspectives on its development for electrical generation?

would evaporate. If that had happened, the spent fuel would have eventually heated up sufficiently to burn, spewing radioactive cesium-137 into the atmosphere and contaminating the environment, as occurred around Chernobyl. Fortunately, no spent fuel rods were exposed to the atmosphere, but the worry continued for some time during the crisis. Still, the explosions and fires at the Fukushima plant discharged enough radioactive material into the surroundings to force the evacuation of 80,000 people from a 20-kilometer (12-mile) radius around the power plant.

Eventually, pipes were laid to deliver a constant stream of water to the Fukushima reactors, gradually cooling them down. Finally, on December 16, 2011, nine months after the earthquake, the prime minister of Japan reported that all three damaged reactors at the Fukushima power plant were stable and in a state of cold shutdown. However, it may be decades before anyone can return safely to the evacuation zone.

⚠ Think About It

1. What do the accidents at Three Mile Island, Chernobyl, and Fukushima have in common?

2. What could be done to improve the safety of the power plants at Chernobyl and Fukushima?

3. What would motivate a nation, such as Germany, to phase out nuclear power, while others, such as the United States, continue to use nuclear power as a significant part of their energy mix?

9.4–9.6 Issues: Summary

In the next 20 years, fossil fuel production will likely peak and then begin to decline until we have exhausted Earth's nonrenewable supplies of energy. More intense fossil fuel extraction in response to higher prices and greater demand will continue to be a source of significant environmental damage. Strip mining of coal and oil sands and mountaintop removal mining have already devastated many landscapes. Coal sludge and fly ash ponds have contaminated water supplies with heavy metals. Oil spills have disrupted ecosystems around the world.

Nuclear power is neither as safe nor as cheap as once promised. Three nuclear accidents have changed public perceptions of nuclear power and altered the course of its development. An accident in 1979 at Pennsylvania's Three Mile Island nuclear power plant led to a partial meltdown of one of the plant's reactors. Another accident in 1986 at the Chernobyl power plant in Ukraine spread radioactive material over thousands of square kilometers across Europe. Explosions and fires at the nuclear power plant in Fukushima, Japan, deposited enough radioactive material into the surroundings to force the evacuation of tens of thousands of residents.

9.7–9.9 Solutions

Take a ride on an airplane over the plains of Montana, and you'll see a landscape scarred by thousands of miles of dirt roads, barren oil well pads, and long-abandoned uranium mines. This is what 100 years of energy development and resource extraction can do to Earth, but it doesn't have to be this way. Although the ultimate solution to reducing the impact of nonrenewable energy is to switch to renewable energy, as we shall see in the next chapter, we'll still be using fossil fuels and nuclear power for many more decades to come. In this chapter, we discuss smarter development practices that can minimize the footprint of these industries, reduce the possibility of

catastrophes, and restore the environment to its natural state when these industries depart.

9.7 New laws and technology are cleaning up the oil industry

In the aftermath of the Deepwater Horizon spill, BP spent $14 billion on cleanup, over a four-year period, sifting beach sand for oil and working to restore over 22,500 kilometers (14,000 miles) of shoreline. Overseen by the

U.S. National Oceanic and Atmospheric Administration (NOAA), the cleanup entered the third phase with efforts to restore barrier islands, marshes, and oyster beds. However, in the deep-sea habitat near the well site, there is little anyone can do but wait, and scientists say that it will be decades before the habitat is fully recolonized by marine invertebrates.

Yet it would be unfair to blame BP entirely for the oil spill. As consumers of products ranging from plastics and gasoline to natural gas, we are all, in some ways, complicit in the spill. Ultimately, reducing the impact of the oil industry is going to depend on reducing our own consumption and demanding that companies and politicians do a better job of protecting and restoring ecosystems affected by the oil industry.

In the worst-case scenario, it pays to have a recovery plan. On March 24, 1989, the Exxon Valdez oil tanker ran aground in Prince William Sound, Alaska, releasing 40 million liters (11 million gallons) of petroleum into the marine environment (**Figure 9.30**). The immediate impact of the spill included the deaths of approximately 250,000 sea birds, 2,800 sea otters, 300 harbor seals, and 22 killer whales. It is unknown how many marine invertebrates and fish also died.

Exxon was heavily criticized for its limited response to the spill at the time, but the chemical and mechanical cleanup techniques it employed remain the standard in the industry. The most critical step in the cleanup process is to lay out floating **booms** and other barriers to contain oil slicks and prevent them from entering sensitive coastal areas. Next, a **skimmer** is used to collect the oil from the water's surface. Passive skimmers are either enclosures that skim off only the top layer of oil or are made of materials that mop up oil. Suction skimmers, by contrast, are like vacuum cleaners that suck up the oil and pump it into a storage tank.

A second key element of the cleanup process, which was used widely during the Deepwater Horizon spill, involved using a **dispersant,** a chemical that thins and dissolves the thick crude. Unfortunately, Corexit, one of the most commonly used dispersants, has been implicated as a potential cause of health problems among workers, including liver, kidney, lung, nervous system, and blood disorders.

After the immediate cleanup concluded, Exxon agreed to pay the U.S. and Alaskan governments $900 million for environmental restoration and research. Much of the marine community has recovered, including bald eagles,

AERIAL PHOTO OF THE MARCH 1989 EXXON VALDEZ SPILL IN PRINCE WILLIAM SOUND, ALASKA

(Natalie Forbes/National Geographic/Getty Images)

FIGURE 9.30 The release of approximately 40 million liters of crude oil into this ecosystem continues to impact marine life more than 25 years after the disaster.

boom A barrier used to contain oil slicks and prevent them from entering sensitive coastal areas.

skimmer A device used to collect spilled oil from the water's surface.

dispersant A chemical used in oil spill cleanup that thins and dissolves the thick crude.

salmon, river otters, and sea otters. However, tens of thousands of liters of oil remain buried in the beaches of Prince William Sound and will not be removed by natural processes for a considerable period of time, perhaps a century.

The most important outcome of the Valdez Spill was the passage of the Oil Pollution Act of 1990, which required companies to have a program in place to prevent oil spills and a detailed cleanup plan in the event of a significant accident. The Oil Pollution Act also led to the phase-out of large single-hull tankers by banning them from U.S. waters after 2010. Double-hull tankers have a second, inner hull and are 5 times less likely to spill oil. In the event of a hull breach, double-hulls will spill far less oil.

Regulations for Safer Fracking

In response to growing concerns that fracking could pose a threat to drinking water, draft legislation proposed in 2012 required that companies disclose all chemicals they use for hydraulic fracturing on federal lands in the United States. Three years later, in March 2015, the Bureau of Land Management of the Department of Interior released a set of new rules governing fracking on federal public and Native American tribal lands. A key requirement, which the gas and oil industry had long resisted, was that drilling companies must indeed disclose the chemical composition of any hydraulic fracturing fluids used.

In addition, drillers are required to protect any aquifers penetrated during drilling, using strong concrete barriers to prevent contamination by fracking fluids. There are also requirements to provide more secure storage of waste fracking fluids recovered from wells to reduce the exposure of air, water, and wildlife to the fluids. With more than 100,000 gas and oil wells on federal lands, which encompass over 3 million square kilometers (1.2 million square miles), these new rules apply broadly to gas and oil drilling on federal lands because 90% of the new wells being drilled on those lands employ fracking.

Oil-Free Zones

The only way to guarantee that a spill is not going to occur in a particular region is to ban oil exploration altogether. Following the 1969 oil spill in Santa Barbara, California, that state halted the leasing of offshore tracts within 3 miles of its coastline. Other sensitive areas have also restricted oil exploration and drilling. In 2003 Norway declared the Lofoten Islands oil-free, because of their rich cod and herring fisheries along with their abundant wildlife. One of the most contentious oil-free zones is the Arctic National Wildlife Refuge in Alaska. The refuge is the largest protected wilderness in the United States, but the U.S. Geological Survey estimates that it contains 10.3 billion barrels (433 billion gallons) of oil.

What is implied by the phrase "or to an economically usable state" in the mining law requiring reclamation of mined lands?

reclamation   A process that restores an ecosystem to its natural structure and functioning prior to mining or to an economically usable state.

The status of drilling on the refuge has repeatedly been a hot-button issue during election years. Proponents have claimed that drilling would have no impact on local wildlife and would allow the United States to reduce its dependency on foreign oil. Opponents argue that oil development is particularly harmful in the high Arctic region because the extreme cold and short summer growing season mean that the land would take centuries to recover in the event of a spill.

⚠ Think About It

1. Do you think Exxon and BP were penalized sufficiently, considering the damage that was (and continues to be) caused?

2. What are the pros and cons for maintaining the Arctic National Wildlife Refuge as an oil free zone?

3. Do you think the new governing rules for fracking on federal public lands in the United States should be extended to apply to private lands as well? Explain.

9.8 Ecosystem restoration can mitigate the environmental impacts of fossil fuel extraction

Strip mining for coal, oil sands mining, and mountain top removal mining of coal all have the potential to destroy vast tracts of natural ecosystems. To reduce these environmental impacts, laws in the United States and other countries require miners of these resources and elsewhere to repair damages done to mined lands. In the United States, the federal law mandating restoration is the Surface Mining Control and Reclamation Act (SMCRA), which came into force in 1977. The laws of individual state and local authorities also generally require mine **reclamation,** that is, restoring an ecosystem to its natural structure and functioning prior to mining or to an economically usable state. According to the National Mining Association, more than 900,000 hectares (2.2 million acres) of mined lands have been restored in the United States alone. In many cases, ecosystem restoration has been remarkably successful.

Restoration of Strip-Mined Prairie

The state of Wyoming includes some of the most valuable low-sulfur coal deposits in North America, making them the focus of intensive strip mining. Strip mining of coal produces massive landscape disturbance (see Figure 9.22, page 277) and the Wyoming prairies are no exception.

RESTORATION FOLLOWING STRIP MINING OF WYOMING PRAIRIE

(Bruce Gordon/EcoFlight)
Powder River Basin mine prior to restoration

(Lee Buchsbaum)
Restored grassland in Powder River Basin

FIGURE 9.31 Prior to mining, the undisturbed landscape in the Powder River Basin consisted of rolling hills covered mainly with sage brush and grasses. Strip mining removed all the topsoil and vegetation, but restoration has been remarkably successful in creating productive wildlife grazing areas like this tract of restored grassland.

However, the Jacobs Ranch Mine in the Powder River Basin has become an award-winning case study in successful ecological restoration in the region.

The mine, which is owned and operated by Rio Tinto Energy America, naturally supports a semi-arid plant community dominated by several species of native grasses and shrubs. This ecosystem, which is capable of supporting a rich community of native herbivores or domestic cattle, sets the benchmark for ecological restoration because SMCRA mandates that mined lands be restored to original or better condition. As the Jacobs Ranch Mine is worked, the landscape is stripped bare (**Figure 9.31**). However, as mining proceeds, regulations require that the topsoil fraction of overburden be stockpiled separately from the deeper mineral layers for later use during restoration of the mined site. Following extraction of the coal, the lower layers of overburden are spread across the mined area and worked with heavy machinery to restore the natural contours. Then stockpiled topsoil is spread across the site, followed by reseeding and replanting of native vegetation. Restoration of these lands also includes establishing small reservoirs as water sources for wildlife.

One restored area at the Jacobs Ranch Mine, identified as critical winter habitat for the local elk herd by the Wyoming Department of Game and Fish, has become a showcase for ecological restoration and cooperative work between industry and conservation organizations. In 2004 Rio Tinto Energy America began working with the Rocky Mountain Elk Foundation to create a 405-hectare (1,000-acre) conservation easement at Jacobs Ranch Mine to permanently protect critical winter habitat for elk at the site. These negotiations ended with Rio Tinto Energy

America donating the land to the Rocky Mountain Elk Foundation. The core of this conservation easement, which ensures that the land will be sustained as wildlife use indefinitely, is a 296-hectare (730-acre) parcel of mined land restored to full productivity (see Figure 9.31). The hope is that similar successes can be made as we try to restore the many millions of hectares of land and water impacted by surface mining.

Restoration of Boreal Forest Oil Sands Mining

Sometimes it's not possible to restore a landscape to its natural state. The boreal forest where oil sands extraction takes place is a patchwork of forest, lakes, and several types of wetlands. This patchwork is going to look very different in the future (**Figure 9.32**). The area of forest will increase by 40% and the area of lakes by 177%, while the total area of wetlands will decrease by 36%. Most significant, the area of peat wetlands will decrease by 67%.

Peat wetlands are the result of centuries of development, and they cannot be restored. The loss of these peat wetlands is environmentally significant because they are a major repository of climate-warming gases in the landscape. Disturbing them would release carbon dioxide and methane into the atmosphere and diminish their ability to store carbon in the future (see Chapter 14). Whether or not we should be extracting these resources and irretrievably destroying this landscape is not a question that science can answer—it's one for society to answer. This is one of many reasons why the debate over the Keystone Pipeline has been so heated.

Is it possible to "improve" a landscape over its condition prior to disturbance by mining? If so, what criteria would you use?

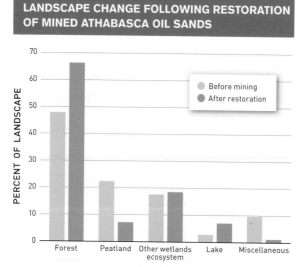

FIGURE 9.32 Restoration of landscapes following mining of Athabasca oil sands results in substantial changes in land coverage. Coverage by forests and lakes increases significantly, while peatland (bogs and fens) cover decreases. (Data from Rooney et al., 2012)

Restoration Following Mountaintop Removal Mining

Once you've blown off the top of a mountain with explosives and dumped the material into adjacent valleys, how do you replace it? You can't.

However, it turns out there's a loophole in the Surface Mining Control and Reclamation Act of 1977. The law requires that land be restored to a natural condition similar to that prior to mining *or to an economically useful condition.* Companies involved in mountaintop removal mining have taken this second course and have not attempted to restore the natural mountain contours, which, if not impossible, would not be economically feasible. Consequently, areas that were once forested peaks are now large, flat or gently sloping patches in an otherwise mountainous landscape (**Figure 9.33**). They have become grazing lands, forestry plantations, or wildlife areas.

The state of Kentucky, for instance, working with several mining companies and the Rocky Mountain Elk Federation, has reintroduced elk, which were extirpated from the region long ago. Kentucky's reintroduced elk mainly use grazing lands on restored mountaintop mining areas, where the population has grown rapidly to over 10,000 individuals and is now the focus of hunting in the region. Other economic developments on reclaimed mountaintop mine lands include golf courses, regional airports, correctional facilities, and industrial parks. Mining companies cite these economic developments as positive benefits to the largely impoverished region.

Critics of mountaintop removal mining assert that there has been too little economic development of the reclaimed mining areas and that what has occurred has been at the expense of one of North America's biodiversity hotspots. Indeed, the soils, one of the foundations of terrestrial ecosystem productivity and health (see Chapter 7, page 194), in areas reclaimed after mountaintop removal mining are deficient in several ways. They are denser and much lower in organic matter content, which reduces water infiltration and favors surface runoff. Many areas remain barren after 15 years or more. In addition, critics argue that grazing lands inhabited by elk cannot compensate for the valleys and headwater streams, home to exceptionally species-rich communities, that have been destroyed or for the pollution that finds its way downstream from these filled valleys.

⚠ Think About It

1. What are the unique challenges and opportunities of restoring Wyoming's surface-mined prairies, the boreal forests landscape overlying the Athabasca oil sands, and the Appalachian mountaintops that have been removed?

2. What should be the goals of ecosystem restoration? Restoring functional properties such as natural levels of primary production (see Chapter 7, page 190) or restoring natural levels of species richness and native species composition (see Chapter 4, page 96), or both?

3. How would you go about evaluating the relative merits of restoration of mined lands to a state of economic usefulness versus their original condition?

9.9 Advances in nuclear power plant operation and design are aimed at improving safety

In the wake of the Fukushima accident, Japan and Germany, the third and fourth largest economies in the world, reexamined the place of nuclear power in their energy mix. Within a year, Japan had shut down all but two of its 50 main nuclear reactors for testing and evaluation of safety systems, with the remaining two scheduled for shutdown and evaluation soon thereafter. Two years after the tsunami shut down the Fukushima nuclear power

What are the relative challenges involved in restoring a prairie grassland versus an old growth forest disturbed by fossil fuel extraction?

RESULTS OF RESTORATION FOLLOWING MOUNTAINTOP REMOVAL MINING

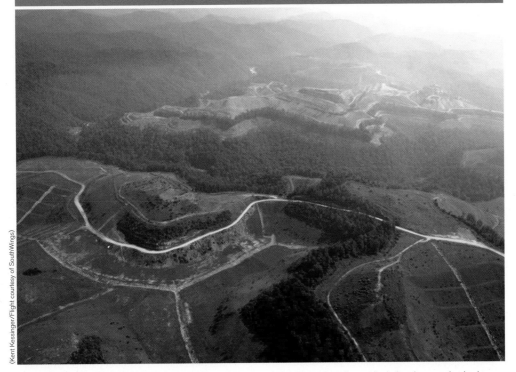

(Kent Kessinger/Flight courtesy of SouthWings)

FIGURE 9.33 Restoration following mountaintop removal mining generally results in level or gently sloping terrain, most commonly vegetated with grasses and other herbaceous plants.

plant, no nuclear reactors in Japan were operating; the first nuclear plant was restarted in August 2015. In 2011 Germany shut down 8 of its 17 reactors and planned to shut down the remainder by 2022. Even France, which generates more than 75% of its electricity using nuclear power, spent the first half of 2013 studying whether to reduce its reliance on this energy source.

Nuclear power has not proven to be as cheap or as clean as some pundits have argued. Nevertheless, nuclear power will continue as a significant source of energy over the next 50 years, and engineers are trying to improve safety at existing nuclear power plants, as well as attempting to make the nuclear power plants of the future safer still.

Improvements to Existing Nuclear Power Plants

Regulatory agencies around the world reassessed the design and operation of nuclear power plants following the accident at Fukushima. In the United States, the Nuclear Regulatory Commission released a list of recommendations for improving nuclear power plant safety. The list included three main measures: (1) Planning for multiple natural disasters or other threats and building the capacity to support the safety functions of all reactors

at a site simultaneously; (2) improved monitoring of pools containing spent fuel rods and robust backup systems for cooling spent fuel rod pools; (3) better systems for venting reactors of the type at Fukushima, in case of power loss.

Another general recommendation was for each nuclear power plant to reanalyze the earthquake and flood risks and to assess how communication and safety equipment would perform with extended power loss. The experiences at Three Mile Island, Chernobyl, and Fukushima are also influencing the design of new power plants, which include safety features that were not part of earlier nuclear power plants.

Making Future Nuclear Power Plants Safer

The latest nuclear power plant designs are simpler and rely more heavily on passive safety systems instead of active ones. The new designs rely more on natural forces, such as gravity and heat convection, and less on mechanical systems, such as valves and pumps. The actions of human plant operators are also less critical to ensuring safety in these newer reactor designs.

At the end of 2011, the U.S. Nuclear Regulatory Commission approved two new power plants that incorporate such principles. One of them, the

What are some reasons related to environmental protection and human health that would encourage countries like France and the United States to not replace their nuclear power plants with fossil fuel–burning power plants?

Is it possible to anticipate and prevent all possible threats to a nuclear power plant? Explain.

THE WESTINGHOUSE AP1000 NUCLEAR POWER PLANT DESIGN

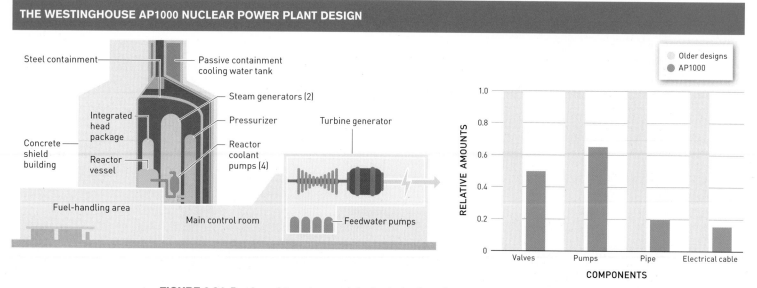

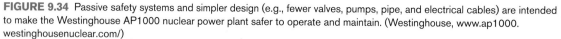

FIGURE 9.34 Passive safety systems and simpler design (e.g., fewer valves, pumps, pipe, and electrical cables) are intended to make the Westinghouse AP1000 nuclear power plant safer to operate and maintain. (Westinghouse, www.ap1000.westinghousenuclear.com/)

Westinghouse AP1000, will be the first new nuclear power plant built in the United States since 1996. One of the advantages of this power plant is that it is much simpler structurally than were earlier plants. For example, the AP1000 has substantially fewer pumps, less control cable, less piping related to safety, and fewer safety valves. The volume of the earthquake-resistant building is also smaller (**Figure 9.34**). The simpler design of the AP1000 is intended to reduce construction and maintenance costs, decrease construction time, and improve safety.

The AP1000 is also designed to shut down safely without the need for any action by human operators and without electrical power or pumps in the event of a design failure, such as a break in a coolant pipe. Because the AP1000 uses gravity, passive circulation, and compressed gases to cool the reactor core and containment structure, there is no potential danger from failures of backup generators and pumps. There are also many active components designed into the AP1000 power plant, but they are not part of critical safety functions or are redundant to the passive safety systems.

Recycling Fuel Rods

In Europe and Japan, spent uranium fuel is reprocessed and reused, which increases efficiency and reduces

nuclear waste. To recycle the fuel, depleted uranium and plutonium must be separated from certain waste materials produced during fission. The uranium and plutonium are then combined to create a fuel known as MOX (Mixed Oxide Fuel), which can be used in nuclear reactors.

MOX accounts for just 2% of the nuclear fuel used today. But if all the spent nuclear fuel in present use were recycled into MOX, it could replace three years' worth of uranium extracted from mines around the world. One factor preventing wider adoption of MOX fuels is that plutonium can be used in nuclear weapons. Some fear that the wider commercial use of plutonium could increase nuclear proliferation. In the United States, no recycling takes place currently.

Nuclear Fusion

Some scientists believe that they can "bottle up the Sun" and develop a controlled nuclear fusion reaction here on Earth. Although they have succeeded in producing small experimental fusion reactors, known as *tokamaks*, the amount of energy required to run the reactors is far greater than the amount of energy they release. The International Thermonuclear Experimental Reactor currently under construction in France could represent the first step toward a working fusion-powered generator. If successful, it will produce 500 megawatts of power. Initial experiments are scheduled to begin in 2020.

⚠ Think About It

1. What are the advantages of designing passive emergency protection systems that are not dependent on an intact electrical system or even a backup electrical supply?

2. How would making nuclear power plants simpler in design contribute to greater reliability and safety?

3. Are the risks associated with nuclear power so great that we should stop trying to make nuclear power plants safer and just abandon the technology entirely? Explain.

9.7–9.9 Solutions: Summary

The impacts of strip mining for coal, oil sands mining, and mountain top removal mining can be reduced by restoring the land to a natural or economically useful condition. The impacts of mountaintop removal mining are not reversible, so mining companies have opted for producing flat areas, suitable for some forms of economic activity, in an otherwise mountainous landscape.

Because it appears that nuclear power will continue as a significant source of energy, engineers have suggested improvements to existing nuclear power plants, including more robust electrical backup in the event of natural disasters, better monitoring of spent fuel rod tanks, and more reliable venting systems for Fukushima-type reactors. The latest designs for new nuclear power plants are simpler than those for older power plants, and they rely more heavily on passive safety systems instead of active ones. The new design features are intended to reduce construction and maintenance costs, decrease construction time, and make the newer nuclear power plants safer to operate.

Answer the following questions for each chapter section and then answer the Central Question.

Central Question: How can we manage nonrenewable energy resources in a way that reduces environmental harm?

9.1–9.3 Science

- What are the different types of fossil fuels?
- How do power plants and vehicles utilize fossil fuels to generate electricity and movement?
- How is nuclear energy released and harnessed?

9.4–9.6 Issues

- What is the nature of global energy consumption and have we reached peak oil?
- How does extracting fossil fuels harm the environment?
- What are the risks of nuclear energy?

Fossil Fuels, Nuclear Energy, and You

Ecosystems, our economic systems, and our very bodies run on energy of one form or another. We are now at a point in history when inexpensive, readily available fossil fuels are approaching peak production. Meanwhile, demand for energy is on the rise. The future of humanity and the biosphere literally depends on how we manage the remaining fossil fuels and transition to sustainable energy sources.

☐ *Stay informed.*

One way that each of us can help foster more sustainable use of these resources is by keeping informed and using whatever influences each of us has to address these challenges. Stay informed regarding rapid developments in the fossil fuel industry, especially in regard to estimates of the sizes of energy reserves and rates of consumption of nonrenewable fossil fuel resources around the world. Pay close attention to issues associated with the development of unconventional sources of oil such as oil sands and tight oil. Stay abreast of developments in renewable biofuels, which will be critical substitutes for crude oil.

☐ *Get involved.*

Explore local, regional, and national energy developments. Try to constructively influence the direction of those developments, either individually or collectively, through organizations that represent your perspective, whatever it may be. Use your knowledge to help others who may be interested to understand developments and issues regarding energy technology and use. As an environmental science student, you will have developed a basis for navigating these complex subjects.

☐ *Consume energy wisely.*

We are all, by necessity, consumers of energy. As such, we can affect the consumption of energy in many ways, from the kinds of energy-consuming devices we purchase to how we control the temperature in our homes and workplace. Residential and commercial buildings are collectively the largest consumers of energy in the United States. For lighting, choose bulbs that use less energy, such as compact fluorescent lights. If you have access and your health permits, set the thermostats in the place where you live or work to conserve energy, meaning a few degrees lower in winter and higher in summer.

☐ *Make energy-wise transportation choices.*

Our transportation choices can also make a significant difference. If you can do so securely, walk or bicycle to school or work. If you can't, take a bus or train. If you must drive, try to carpool. All these alternatives can save money as well as energy. If you purchase a motor vehicle, consider models that are both safe and fuel-efficient. While governmental mandates affect the fuel efficiency of vehicles produced by manufacturers, our choices as consumers can also exert significant influence on manufacturing.

- How can we improve the efficiency and reduce the impact of fossil fuel use?

- What is the process of environmental restoration after fossil fuel extraction?

- What improvements have been made in nuclear power plants?

Answer the Central Question:

_____

_____

_____

_____

_____

Chapter 9
Review Questions

1. What is the origin of the chemical energy of coal and crude oil?
a. The heat energy of molten volcanic lava transferred to sedimentary rocks
b. Photosynthesis by primary producers that lived millions of years ago
c. Sedimentary rocks exposed to high heat and pressure
d. Sedimentary rocks modified by exposure to solar energy

2. Which of the following types of coal has the highest energy content?
a. Lignite
b. Sub-bituminous coal
c. Bituminous coal
d. Anthracite

3. What is the approximate energy efficiency of a coal-fired electrical power plant?
a. 10%
b. 25%
c. 35%
d. 70%

4. What is the basic difference between a coal-fired power plant and today's nuclear power plants?
a. There is no difference; both use chemical energy as a source of heat to produce steam.
b. Coal-fired power plants use chemical energy, whereas nuclear power plants use nuclear fusion as a source of heat to produce steam.
c. Coal-fired power plants rely on nuclear energy, whereas nuclear power plants use chemical energy as a source of heat to produce steam.
d. Coal-fired power plants use chemical energy, whereas nuclear power plants use nuclear fission as a source of heat to produce steam.

5. Which country now consumes the most energy per capita?
a. The United States
b. China
c. The European Union
d. Russia

6. When do experts now predict we will reach peak oil production?
a. In 5 to 10 years
b. In 10 to 20 years
c. In 50 to 60 years
d. Peak oil already occurred in 2000.

7. EPA regulations now require which of the following for coal ash ponds?
a. Coal ash ponds cannot be established on wetlands.
b. New coal ash ponds must be lined to prevent contamination of groundwater.
c. Coal ash ponds cannot be built in earthquake-prone areas.
d. All of the above

8. What is the major target of hydraulic fracturing, or fracking?
a. Extraction of coal
b. Extraction of natural gas
c. Extraction of water in arid regions
d. Extraction of carbon dioxide

9. Which of the following is _not_ currently included in restoration requirements following mountaintop removal mining?
a. Restoring the native plant and animal diversity
b. Restoring top soils to original structure and fertility
c. Restoring original drainage patterns and water quality
d. None of the above

10. Which of the following is the major feature of improved safety for the latest nuclear power plants?

a. Larger numbers of professional operators to ensure safety procedures are followed
b. Larger numbers of pumps and safety switches for plenty of backups
c. Passive safety systems that work without human intervention or an outside power source
d. Larger, more robust housing for the reactors

Critical Analysis

1. What are the connections between growth of the global population and global economy and the increasing demand for energy? Use the laws of thermodynamics in your discussion (Chapter 2, page 37).

2. Some local U.S. communities have passed ordinances prohibiting fracking within their jurisdictions. Meanwhile, some state governments have tried to pass laws to limit or deny such prohibitions by local communities. Discuss the pros and cons of local versus centralized control of fossil energy exploration and extraction.

3. What do the nuclear accidents at Three Mile Island, Chernobyl, and Fukushima have in common? How are they different? What lessons can be learned from these accidents?

4. Should the costs of the ecosystem services, such as air and water purification, impaired by strip mining and mountaintop mining be included in the cost of coal?

5. How would a serious application of the precautionary principle (see Chapter 1, page 12) influence the development of nuclear power.

Find additional resources and links online at www. macmillanhighered.com/launchpad/molles1e.

Central Question: Can we develop renewable energy resources to help sustain a thriving economy without adversely affecting the environment?

Describe the renewable sources and technologies of solar, wind, hydroelectric, hydrokinetic, and geothermal energy.

SCIENCE

Renewable Energy

Identify the environmental and human impacts of renewable energy.

Investigate potential strategies to maximize the sustainability of renewable energy.

ISSUES

SOLUTIONS

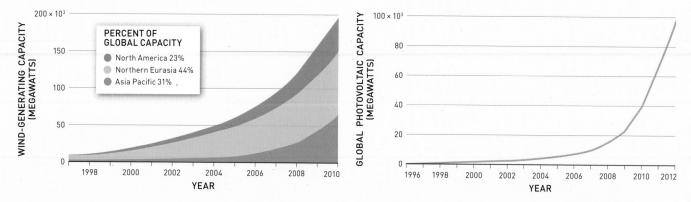

GROWTH IN WIND-GENERATING CAPACITY IN NORTH AMERICA, NORTHERN EURASIA, AND ASIA PACIFIC

PERCENT OF GLOBAL CAPACITY
- North America 23%
- Northern Eurasia 44%
- Asia Pacific 31%

Wind-generation capacity in these regions grew exponentially during the first decade of the 21st century. (Data from BP, 2011)

GLOBAL PHOTOVOLTAIC GENERATION CAPACITY

Photovoltaic generation capacity grew rapidly from 1996 to 2012. (Data from BP, 2013)

Energy Independence

A small island in Denmark is leading the way with renewable energy.

The island of Samsø is a two-hour ferry ride from the eastern coast of Denmark. A small, windswept island in the North Sea with 4,000 residents, it would be unremarkable except for one simple fact: The electrical grid is 100% powered by renewable energy. Twenty-one massive wind turbines rise 150 feet from the pastoral landscape and from the choppy waters of the Kattegat Strait, generating 34 megawatts (MW) of power. On windy days, they produce so much energy that the island sells it to mainland Denmark. In addition to wind power, Samsø has four plants that supply 70% of the island's heat needs using straw waste from barley production and sustainably harvested wood chips from local forests, along with solar energy. The local golf course boasts a solar-powered lawn mower, and one enterprising farmer converts his canola crop into fuel to run his tractors. The island plans to be entirely fossil fuel–free by 2030.

> "I'd put my money on the sun and solar energy. What a source of power! I hope we don't have to wait until oil and coal run out before we tackle that."
>
> Thomas A. Edison, 1931, in *Uncommon Friends* (1987)

By charting their own course and developing a diverse array of renewable energy sources, Samsingers—as the people of Samsø are known—are avoiding many of the problems we have explored that relate to fossil fuels and nuclear power. They demonstrate that it is possible to maintain a thriving economy and high standard of living while reducing their environmental impact. Many other communities around the world are making smaller, but nonetheless significant steps in that direction. According to the Renewable Energy Policy Network (REN21), renewable energy resources powered half of the world's newly installed electrical generating capacity in 2010. Most of these gains are coming from massive hydropower dams in countries such as Brazil and China, but we're also seeing dramatic growth in other renewable energy sectors. For example, biofuels (e.g., algae) production increased by 14% in 2010 and wind power increased 24%, reaching nearly 200 gigawatts (200,000 MW; left graph). The production of electricity by photovoltaic systems has grown at an explosive pace, reaching nearly 100 gigawatts (100,000 MW) in 2012 (right graph) and growing to 177 gigawatts (177,000 MW) in 2014.

Even though the adoption of renewable energy supplies represents a step in the right direction, it comes with its own issues. Solar panels require the extraction of mineral resources and can take up a large footprint on land. Many consider wind turbines an eyesore that creates noise pollution; wind turbines also pose a threat to migrating birds and flying bats. And dams require the flooding of river valleys. The challenge of ensuring that the transition from nonrenewable energy to renewable energy is done sustainably lies at the heart of the Central Question of this chapter.

Because of the size of the human population and the great amounts of energy needed to sustain our economies, we have to consider that the large-scale development of renewable energy resources may threaten the environment in numerous ways.

Central Question

Can we develop renewable energy resources to help sustain a thriving economy without adversely affecting the environment?

FIGURE 10.16 Wave energy is highest along coastlines in far northern and southern latitudes. (Based on research from Barstow et al., 1998)

AVERAGE AVAILABLE WAVE ENERGY ALONG COASTAL AREAS

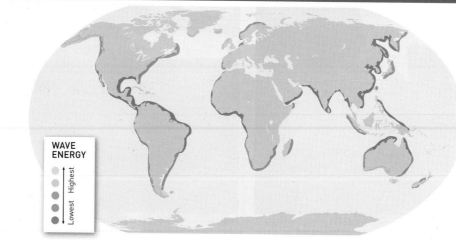

WAVE ENERGY

Lowest — Highest

Could the geothermal resources of Iceland be used to grow tropical plants year round? How?

How is renewable energy availability related to the distribution of natural physical hazards to people and property?

Geothermal Energy

When a volcano erupts, it gives us a peek at the awesome power contained beneath the surface of Earth. The heat energy produced in Earth's core by the radioactive decay of chemical elements, such as uranium, is known as *geothermal energy*. Where accessible near Earth's surface, it becomes a valuable source of renewable energy.

Geothermal energy is used in two principal ways. Where it heats water that flows to Earth's surface as hot springs, geothermal energy can heat buildings. For example, all the buildings in Reykjavik, the capital of Iceland, are heated by water from geothermal springs. The second way geothermal energy can be used is to drive electric generators (**Figure 10.17**).

The regions most suitable for developing geothermal energy are "hot spots" that occur at plate boundaries or at places where the crust is thin enough to let the heat through. These are areas of significant volcanic activity, such as the "ring of fire" that encircles the Pacific Ocean (**Figure 10.18**). Geothermal activity is also intense in areas of sea floor spreading. The mid-Atlantic ridge is a place where two tectonic plates are diverging, producing much volcanic activity—and Iceland has capitalized on it, generating 66% of its energy from geothermal energy. The Great Rift Valley of East Africa, with its great rift lakes and towering volcanic peaks, is another region of substantial geothermal activity.

There are three types of geothermal power plants (**Figure 10.19**). Dry steam plants use the steam from geothermal wells directly to turn generator turbines. The Geysers in northern California, the largest geothermal generating system in the world, is a dry steam plant. Flash steam plants, the most common type in use today, pull deep, high-pressure hot water into lower-pressure tanks. As this hot water enters a lower pressure chamber, it "flashes" into steam, which is used to drive a generator turbine. The third type, the binary-cycle geothermal power plant, can be run with geothermal waters of moderate temperatures, well below the boiling point of water. In a binary-cycle plant, geothermal water is passed through a heat exchanger, where the heat it contains flows to a second fluid with a much lower boiling point than water. This transfer of heat converts the second fluid to vapor, which drives the associated turbine.

⚠ Think About It

1. Using Figures 10.4 to 10.18, determine which regions have the most, and least, diversity of renewable energy resources.

ENERGY FROM EARTH'S CORE

(Eco Images/Getty Images)

FIGURE 10.17 Where hot springs, waters that are heated by geothermal energy, occur near Earth's surface, they can be used to heat buildings or to generate electricity, as at this geothermal power plant in Iceland.

and drainage patterns, and road construction. These threats not only impact wildlife populations but also add to the costs of constructing and operating solar power plants. For example, Bright Source Energy, a company developing solar energy in the Ivanpah Valley in California, spent more than $56 million protecting and relocating desert tortoises. Mitigation efforts included building an 80-kilometer-long (50-mile-long) tortoise-proof fence to prevent the relocated animals from returning.

In addition to damaging habitat and adding to mortality in wildlife populations, energy development will bring new service roads to the deserts of the southwestern United States, fragmenting habitats and impeding the movements of threatened animals. In the Southwest, desert tortoises, desert bighorn sheep, and desert mule deer could be seriously affected (**Figure 10.25**).

Water Consumption by Concentrating Solar Power

Although concentrating solar power stations can reduce our reliance on nonrenewable resources, they can potentially consume a large amount of water and generally operate in arid regions where water is scarce. The most energetically efficient concentrating solar power plants draw on turbine designs developed in coal- and gas-fired power plants and require cooling for condensing steam back to liquid water.

As shown in Figure 10.4 (see page 299), the areas receiving the highest amounts of solar energy are concentrated in arid and semi-arid regions. Water loss is the result of evaporation in the cooling towers where water is cooled before returning to the steam condensers. And, of course, all types of power-generating stations use more water than photovoltaic or wind-generating stations, which do not require cooling (**Figure 10.26**). Consequently, concentrating solar power development faces some fundamental questions. Where would the water for cooling come from and at what cost to other competing uses? Can alternative technologies reduce water consumption by concentrating solar power stations?

Competitive Energy Returns on Investment

A critical test of the sustainability of any energy source is how much energy must be expended to obtain it. For example, to prepare coal to generate electricity, it must be mined, processed, and transported to a power station, all of which require energy. During the production of photovoltaic cells, energy is expended in producing the semiconductor crystals, fabricating the solar cells, and building the inverter, which converts direct current (dc) to the alternating current (ac) used in the electrical grid.

The ratio of energy spent to energy obtained is called the **energy return on energy investment (EROEI)**. For example, an EROEI of 1 means you are just breaking even,

ROADS AND WILDLIFE HABITAT FRAGMENTATION

(NPS Photo by Andrew Cattoir)

FIGURE 10.25 Roads can reduce free movement of many species of wildlife, such as this desert bighorn sheep, across the landscape, subdividing populations into smaller isolated units, each more prone to local extinction.

energy-wise: You are expending the same amount of energy that will ultimately be extracted or produced. The higher the EROEI, the better the energy source. Recent analyses indicate that hydroelectric power and wind energy have a greater EROEI than coal or natural gas (**Figure 10.27**). However, solar energy remains relatively more costly in terms of energy input—on par with nuclear power.

WATER USE BY DIFFERENT GENERATION TECHNOLOGIES

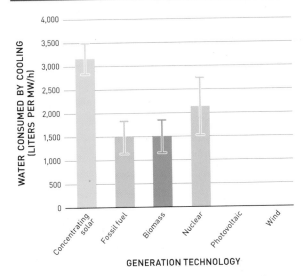

FIGURE 10.26 A comparison of medians and ranges of water consumption by recirculating water cooling systems associated with different generation technologies shows that concentrating solar power generation consumes more water than other technologies. (Data from Carter and Campbell, 2009)

How might efforts to protect desert tortoise habitat benefit other species in the ecosystem?

What factors may contribute to the high rates of water use by concentrating solar power plants operating in deserts?

energy return on energy investment (EROEI) The ratio of the energy content of an energy source (e.g., gasoline) to the amount of energy that must be used in, for example, drilling, transporting, and refining to produce the energy source.

FIGURE 10.27 The EROEI for electrical generation using renewable hydroelectric and wind energy sources is competitive with coal, whereas the EROEI of generating electricity with photovoltaic cells is comparable to generation using natural gas or nuclear energy. (Data from Inman, 2013)

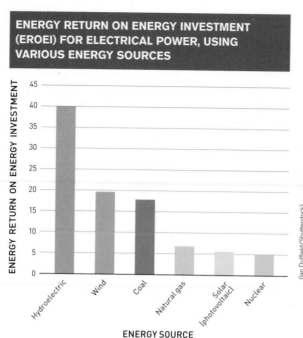

ENERGY RETURN ON ENERGY INVESTMENT (EROEI) FOR ELECTRICAL POWER, USING VARIOUS ENERGY SOURCES

WIND TURBINES AND BIRDS

(Ian Duffield/Shutterstock)

FIGURE 10.28 Most bird mortality associated with wind development results from collisions with turbines and associated structures. Many kinds of birds, but particularly birds of prey—such as this golden eagle, falcons, hawks, and owls—appear to be particularly subject to collisions with earlier-generation wind turbines.

⚠ Think About It

1. How do the wildlife conflicts associated with solar energy development in arid regions challenge the commonly held perspective of these biomes as "empty places"?

2. The largest use of water in arid and semi-arid regions is for irrigation. How do we weigh electrical production against food production in such regions?

10.5 Wind turbines and transmission lines kill birds and bats

Wind turbines work best when mounted on tall towers where winds are the strongest. The electricity generated, in turn, requires high-voltage lines to transmit the generated power to the population centers. All these structures can maim or kill birds and bats that collide with them.

Wind Turbines

Between 2005 and 2007, the 5,000 wind turbines at the Altamont Pass Wind Resource Area (APWRA) in northern California killed approximately 9,000 birds each year. Eagles, hawks, falcons, and other birds of prey proved to be particularly vulnerable, accounting for about one-fourth of this mortality (**Figure 10.28**).

But to put this problem in perspective, the estimated number of birds killed by wind turbines around the country is just 0.04% of the total annual collision deaths.

Somewhere between 300 million and 1 billion birds die each year from colliding with buildings (**Figure 10.29**). Collisions with transmission lines (10 million to 154 million) and communication towers (4 million to 50 million) also kill vast numbers of birds. In addition,

ESTIMATED ANNUAL BIRD DEATHS (MEDIANS AND RANGES) IN THE UNITED STATES, RESULTING FROM COLLISIONS WITH VARIOUS STRUCTURES

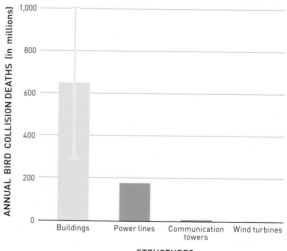

FIGURE 10.29 Estimates of annual bird deaths resulting from collisions with wind turbines were estimated at 573,000, compared with an estimated 300 million to 1 billion deaths resulting from collisions with buildings. (Data from American Bird Conservancy, 2015)

TWO BAT SPECIES VULNERABLE TO WIND ENERGY DEVELOPMENT

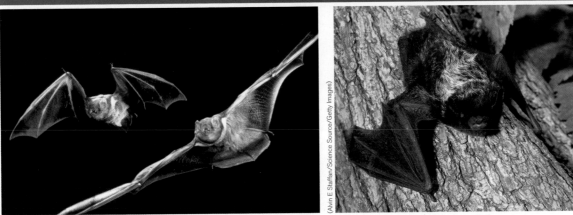

Eastern red bats, *Lasiurus borealis*

Silver-haired bat, *Lasionycteris noctivagans*

FIGURE 10.30 Bat mortality at wind turbines appears to be the result of both collisions and barotrauma, that is, lung damage resulting from bats entering low-pressure micro-environments created by the moving turbine blades. Two particularly vulnerable species are eastern red bats, *Lasiurus borealis,* and silver-haired bats, *Lasionycteris noctivagans.*

each year in the United States, cars, pesticides, and cats kill an estimated 60 million, 70 million, and 100 million birds, respectively. But wind power represents a growing threat: By 2030, when the number of wind turbines is projected to reach 100,000, the number of birds killed by wind turbines in the United States may reach 1 million annually.

While early environmental concern related to wind energy development was focused on birds, bats are also vulnerable. Surprisingly, most bat deaths at wind farms occur at low wind velocities of less than 22 kilometers per hour (13 miles per hour). Ongoing research indicates that at least some of bat mortality is the result of barotrauma, resulting from rapid expansion of their lungs as they enter areas of low air pressure near moving turbine blades.

The most vulnerable bats are tree-roosting migratory species such as eastern red bats, *Lasiurus borealis,* and silver-haired bats, *Lasionycteris noctivagans* (**Figure 10.30**). By 2020 more than 100,000 bats could be killed annually in the Mid-Atlantic Highlands region of the United States alone. This level of mortality could also affect a significant ecological service bats regularly provide: eating insects. A 2011 study (Boyles et al., 2011) estimated the value of bats' contribution to controlling insect pests in the United States alone at nearly $23 billion annually. Reduced bat populations could translate into more insects attacking crops and farmers using more pesticides to counter them.

Transmission Lines

The high-voltage transmission lines leading into and out of a wind farm, or any other type of power plant, are a significant source of bird mortality. There are approximately 800,000 kilometers (500,000 miles) of high-voltage transmission lines in the United States

today (**Figure 10.31**). The additional transmission lines required by utility-scale wind and solar power plants will add significantly to bird mortality.

Some bird species are particularly subject to collisions with transmission lines. Unfortunately, planning for the routing of transmission lines has sometimes failed to consider such vulnerabilities. For instance, a proposed switching facility in eastern New Mexico would place electrical transmission lines across the Middle Rio Grande of central New Mexico, obstructing the daily flight path of wintering greater

Should there be more efforts to reduce fatal bird collisions with buildings?

U.S. TRANSMISSION LINE GRID

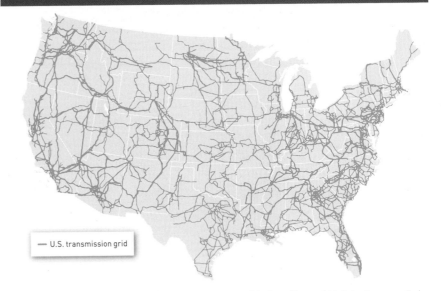

— U.S. transmission grid

FIGURE 10.31 Transmission line networks are responsible for millions of bird deaths annually in the United States alone.

VULNERABILITY TO ELECTRICAL TRANSMISSION LINES

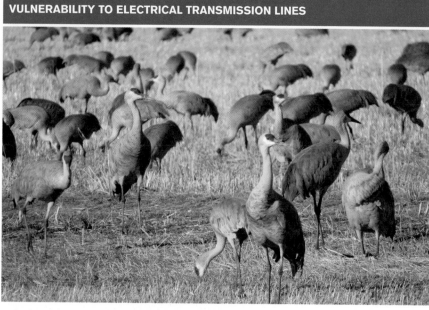

(Manuel Molles)

FIGURE 10.32 Some birds, such as the sandhill crane (*Grus canadenis*), are especially vulnerable to death from collisions with electrical transmission lines.

Although underground transmission lines cost far more than above-ground towers, should underground electrical transmission be phased in gradually to reduce environmental impact?

sandhill cranes (*Grus canadensis*), a species for which transmission lines can be a major source of morality (**Figure 10.32**).

Wind Developments Harm Wildlife Habitat

In the Great Plains, researchers have found that bird diversity and densities are lower in areas where wind-generation facilities have been built, compared with off-site reference areas. In one study, biologists followed the movements of prairie chickens fitted with radio collars and found that they avoid power transmission lines and wind generators. This may be

because they avoid any tall structures where hawks, their natural predators, may roost. Regardless, wind energy development may be subdividing remnant populations of prairie chickens into increasingly isolated habitat fragments. One species of special concern in sagebrush habitats is the sage grouse, *Centrocercus urophasianus*, which abandon their mating areas when tall structures are built near them (**Figure 10.33**). Conservationists fear this could increase their risk of extinction.

⚠ Think About It

1. Considering the relative bird mortality caused by different factors, are concerns about mortality resulting from wind development out of proportion to actual impacts on bird populations? Argue both for and against this position.

2. With house cats currently killing approximately 1,000 times the number of birds that die from collisions with wind turbines, should we consider discouraging outdoor foraging by pet cats?

10.6 Hydroelectric development can have multiple environmental and social impacts

Hydroelectricity comes with a number of environmental costs. We discussed several of these impacts in previous chapters. As we saw in Chapter 6, dams harm aquatic biodiversity by trapping nutrient-rich sediments in reservoirs; in Chapter 8, we reviewed how dams restrict migrations of fish such as salmon. Dams also present social and environmental challenges for people.

How might basic studies of the behavior, ecology, and evolutionary history of sensitive species, such as prairie chickens, help in the design of low-impact renewable energy systems?

PRAIRIE BIRDS IMPACTED BY WIND DEVELOPMENT

(U.S. Fish & Wildlife Service)

Prairie chicken, *Tympanuchus* spp

(Jeannie Stafford/USFWS)

Sage grouse, *Centrocercus urophasianus*

FIGURE 10.33 Prairie chickens and sage grouse inhabit prime areas for wind energy development. Unfortunately, both species can be impacted by such developments because they avoid tall structures, such as wind turbine and transmission line towers, which hawks and other raptors may use as perches from which to spot prey and launch attacks.

they also reduce the need for building high-voltage transmission lines.

Policy Initiatives and Incentives

As with any new technology, there are often practical, legal, and economic barriers to overcome. With regard to developing photovoltaic (PV) systems, obstacles have included difficulty acquiring permits to connect to the electrical grid, gaining approval for financial assistance to install a PV system, and locating trained licensed contractors to install PV systems. In the United States, such hurdles have been partly overcome through the Renewable Portfolio Standard, or RPS, which requires electrical utility companies to obtain a certain percentage of their power from renewable energy.

For example, the RPS for New York State requires that investor-owned utilities produce approximately 30% of their electricity from renewable energy by 2015. Similarly, California and Colorado each have an RPS requiring approximately 30% of electrical generation from renewables by 2020. To facilitate renewable development, the U.S. federal and state governments offer a variety of incentives, including tax credits for businesses and individuals who install renewable energy systems, as well as financial assistance in the form of grants and loans. In addition, in many areas, consumers who install photovoltaic systems tied to the grid are generally paid wholesale rates for any electricity generated in excess of their use.

⚠ Think About It

1. Could water-saving solar technology actually make more water available for natural ecosystems in some circumstances?

2. What does the extensive solar development in Germany, situated in northern Europe, suggest about the potential for solar energy development in a country like the United States or Australia?

3. Some rural electrical cooperatives resist high RPS requirements and do not encourage members to install their own photovoltaic systems. They argue that doing so would reduce demand for centrally generated power and, as a consequence, increase the cost of power to consumers. Argue for and against this position.

10.9 Less damaging wind-generation strategies are under development

Wind energy development impacts more land per unit of energy produced than do many other forms of energy production. The U.S. Department of Energy estimates that

ROOFTOP SOLAR POWER

FIGURE 10.40 Germany, which has generated more than 40% of its electricity using photovoltaic panels, has been particularly successful at installing rooftop photovoltaic generation, as shown by this photo of a solar village in Freiburg, Germany.

50,000 km² (19,300 mi²), an area roughly half the size of Ohio, will have to be developed for wind energy to reach its goal of generating 20% of the nation's electricity by 2030. To avoid damaging wildlife habitats, one proposed strategy would site wind farms on disturbed landscapes, including agricultural lands and areas of oil and gas development. In addition to reducing habitat loss, developments on agricultural land can provide supplemental income to farmers or other landowners (**Figure 10.41**).

Should we wait for costs to decrease before setting aggressive RPS requirements?

WIND TURBINES ON AGRICULTURAL LAND

FIGURE 10.41 Locating wind turbines on agricultural land can increase farm income, while preserving undisturbed natural ecosystems as wildlife habitat.

A recent analysis shows that there are sufficient areas of disturbed land across the United States to meet the Department of Energy goal of 20% wind-generated power by 2030. The area of disturbed lands suitable for wind energy development across the lower 48 states totals 14.5 million km² and could provide about 3,500 gigawatts of energy—more than 14 times the Department of Energy's goal for 2030. Disturbed lands of low conservation value in Kansas alone have the potential to generate over half the Department of Energy's 2030 capacity target.

Reducing Bird Mortality

Changing the way that wind generators are operated and using newer types of wind generators appear to reduce wildlife mortality. For instance, ongoing studies at the Altamont Pass Wind Resource Area (APWRA) have documented that newer wind turbine designs cause significantly lower mortality than older wind turbines operating at the same time (**Figure 10.42**). Researchers have suggested that the older wind turbines attracted birds because the towers supporting them were constructed of a lattice-like framework, which offered perching sites for birds. In contrast, newer wind turbines with a single pole have no perches. Researchers estimate that replacing these older turbines ("repowering") at the APWRA would reduce mortality rates among birds of prey by 54% and among all birds by 65%. These projections are consistent with the results of studies of bird mortality at wind power developments outside of California.

On those wind farms, where wind turbines are mostly of newer designs, researchers have observed rates of mortality among raptors that are approximately 60% lower than on wind farms across California and more than 70% lower than bird mortality rates at the APWRA. In response, utilities are actively replacing older wind turbines with newer ones that are safer for birds *and* that will generate more electrical power in the same area. For instance, 50 newer turbines replaced 235 older turbines installed at Solano Pass, California, which increased the generating capacity of the wind farm more than four-fold, from 25 megawatts to 102 megawatts.

Decreasing Hazards to Bats

While the redesign of wind turbines has benefited birds, bat mortality has increased to the point that in some regions, it exceeds bird mortality by a factor of 10 to 1. The number of bats killed by wind turbines is highest during migration in late summer and fall, and they tend to die from wind turbines spinning at lower speeds. Using this information, researchers performed an experiment at a site in southwestern Alberta, Canada, where high bat mortalities had been documented during migration. In the experiment, wind turbines were reprogrammed so that they only began spinning and generating electricity at higher wind speeds. As a result, bat mortality during migration was reduced by approximately 60%, with little loss in power generation (**Figure 10.43**). Similar changes in wind turbine operations at a wind project in Pennsylvania reduced bat mortalities at the site by 44% to 93%, with losses in annual power generation of less than 1%. However, better solutions are on the horizon. Bladeless wind turbines, now in development, may eventually eliminate the threat of wind turbines to both birds and bats.

Reducing Transmission Line Impacts

One of the simplest ways to reduce bird mortality due to transmission line collisions is to locate new wind-generation facilities as close as possible to existing transmission lines and switching stations. Such siting would not only reduce the amount of new lines, but also reduce development-related costs. Another way to reduce impact is to build transmission lines where existing rights of way can be used. Transmission line routes should avoid roosting and feeding areas of particularly vulnerable birds, such as sandhill cranes and other waterfowl (see Figure 10.32, page 314).

What incentives might federal, state, and local governments create to encourage wind energy development on lands with lower conservation value?

Should siting of transmission lines also take into consideration the potential impacts on human, as well as natural, communities?

REDESIGNING WIND TURBINES FOR REDUCED IMPACT ON BIRD POPULATIONS

Old generation wind turbines

New generation wind turbines

FIGURE 10.42 Research has shown that the design of some of the older wind turbines was responsible for much of the high bird mortality at the Altamont Pass Wind Resource Area. Consequently, the older wind turbines have been largely replaced by newer, taller wind turbines with fewer potential perching sites.

STRATEGIES FOR REDUCING BAT MORTALITY AT WIND FARMS

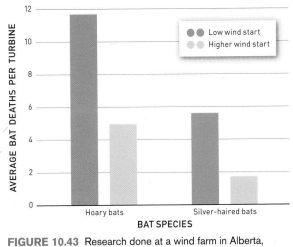

Low wind start

Higher wind start

FIGURE 10.43 Research done at a wind farm in Alberta, Canada, demonstrated a significant reduction in bat mortalities when wind turbines were set to start generating at higher wind speeds. The adjustment resulted in minor losses in total power generated. (Data from Baerwald et al., 2009)

If new transmission lines must be built, marking them with colorful spirals, plates, or spheres has been shown by several independent studies to reduce bird collision mortality by an average of nearly 80% (**Figure 10.44**). However, because the transmission line network is so

extensive, it is not economically feasible to mark all lines. Current research is focused on identifying bird collision "hot spots" so that mitigation efforts will have the greatest likelihood of success.

⚠ Think About It

1. How are concerns over bird and bat mortality leading to innovation in wind-generator designs and operation?

2. How can land classification by conservation value and careful planning help us avoid many impacts of renewable energy development?

10.10 Downsizing can mitigate the impacts of hydroelectric development

Many of the problems with modern hydropower projects stem from their massive size. Focusing on smaller-scale hydroelectric development and building lower dams can mitigate most of these impacts. Lower dams reduce the amount of water stored and the area flooded. Because small dams store little water, they create little change in river flow and have minor effects on the temperature or chemistry of a river. They also displace few or no people.

Run-of-the-River Hydroelectric Systems

One common alternative to erecting large dams on the main channel of a river is to erect smaller power plants. These **run-of-the-river power plants,** which provide little or no water storage in a reservoir, divert a portion of river flow through pipes that pass directly through a turbine (**Figure 10.45**). These systems have their own impacts,

run-of-the-river power plants Hydroelectric systems that provide little or no water storage in a reservoir and divert a portion of river flow through pipes that pass directly through a turbine.

MARKING TRANSMISSION LINES FOR REDUCED BIRD MORTALITY

(Courtesy Melvin Walters)

FIGURE 10.44 Markers like these help birds see transmission lines, substantially reducing the number of collision-related deaths.

RUN-OF-THE-RIVER HYDROELECTRIC SYSTEM

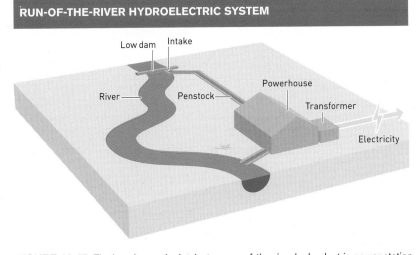

FIGURE 10.45 The low dam at the intake to a run-of-the-river hydroelectric power station impounds only enough water to keep the intake to the penstock (pipe) submerged, therefore avoiding most of the environmental changes caused by large-scale hydroelectric systems. If a small fraction of the total flow on any given season is diverted, the river will continue with something close to its natural flow pattern.

FISH BYPASS AT LOW DAM

(U.S. Fish and Wildlife Service)

FIGURE 10.46 Fish bypass systems are relatively simple to design, where dams are low, as on run-of-the-river hydroelectric systems. Increasingly, fish bypass systems like this one are being designed to blend with the natural landscape.

Should the amount of water diverted by run-of-the-river hydroelectric systems change during droughts and times of abundant flow? If so, what criteria should be used to manage river diversions?

but there are ways to mitigate them. First, water must flow continuously through the main river channel at reasonable water levels to maintain ecosystem health and biodiversity. In addition, fish still need a bypass system to permit them to swim beyond the diversion dam. However, with a low dam, it's much easier to build such a bypass system with relatively natural structures and flows (**Figure 10.46**).

Retrofitting Existing Dams

Many dams were built in the early 20th century for water storage and flood protection—not hydropower production. In such situations, we can add hydropower turbines without building new dams. Researchers at the Oak Ridge National Laboratory point out that there are 54,000 unpowered dams in the United States alone (**Figure 10.47**), and they estimate that the top 100 could add 8 gigawatts of installed hydroelectric capacity to the U.S. electrical grid. That's equivalent to four Hoover Dams (see Figure 10.12, page 304). The greatest potential for such developments is found in the nation's largest rivers, including the Mississippi, Ohio, and Arkansas Rivers—notably, rivers mainly found in regions of the country not particularly rich in wind or solar energy.

Dam-Free Hydropower

It's not always necessary to build a dam to harness the kinetic energy of a river (**Figure 10.48**). In 2009 one such electric turbine was placed directly in the Mississippi River at Hastings, Minnesota. The turbine actually sits beneath the outflow of an existing, conventional hydroelectric power plant. Tests of the turbine installed at Hastings, which spins at 21 revolutions per minute, showed that over 97% of fish were able to pass through it without injury. Similar projects are planned for many of the larger rivers of the eastern half of the United States, with a total potential generating capacity of 500 megawatts.

⚠ Think About It

1. How could hydroelectric power complement solar and wind power in an electrical power system?

2. How are the impacts of hydroelectric systems on fish populations like those of wind generation

What are the potential benefits to spreading generating capacity across 100 sites instead of focusing generation at just a few, Hoover Dam–scale sites?

FIGURE 10.47 There are tens of thousands of existing dams in the United States that could be retrofitted to generate hydroelectricity. Just 100 of the top prospects for retrofitting could add significantly to existing hydroelectric capacity. (Data from Hadjerioua et al., 2012)

UNPOWERED DAMS ACROSS THE UNITED STATES

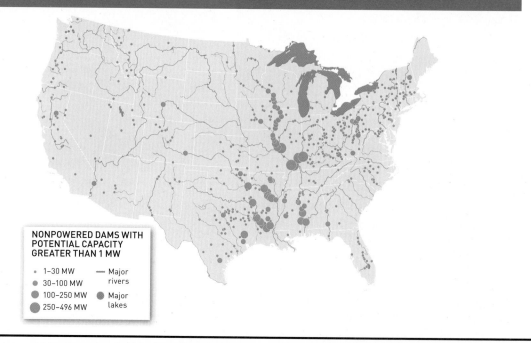

NONPOWERED DAMS WITH POTENTIAL CAPACITY GREATER THAN 1 MW

- 1–30 MW
- 30–100 MW
- 100–250 MW
- 250–496 MW
- — Major rivers
- Major lakes

IN-RIVER HYDROELECTRIC TURBINE

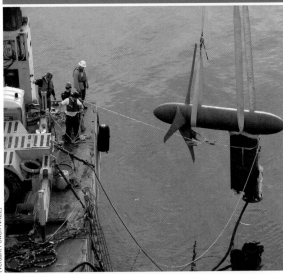

(Verdant Power/NREL)

FIGURE 10.48 Water turbines like this one can be deployed in rivers or tidal channels, where they generate renewable electrical energy but do not endanger fish or other aquatic organisms.

on bird and bat populations? How are they different?

3. How might the cumulative impacts of many run-of-the-river systems on a river equal or exceed the impacts of a few large systems?

10.11 Less damaging, more efficient biofuels are under development as alternatives to oil-based fuels

As we have seen, some forms of biomass fuel development (e.g., corn-based ethanol) come with significant social and environmental costs. Fortunately, a number of processes under development have the potential to avoid most of those impacts. In addition, many are more efficient from an energy perspective. One of those approaches converts whole plants or agricultural and forestry wastes into ethanol.

Reducing Deforestation Due to Biofuels Production

With rising demand for biofuels comes the threat of replacing natural ecosystems with bleak monocultures of oil palms or soybeans. However, Brazil, the second largest producer of biodiesel after Germany, has reduced its rate of deforestation mainly through the creation of indigenous reserves and other protected areas, including some set aside for sustainable use (**Figure 10.49**). The new reserves include over half the Brazilian Amazon

Basin, and indigenous people now control approximately 20% of the basin. Strong enforcement and prosecution of violators, including corrupt government officials, followed the legislation creating these forest reserves.

Moreover, Brazil achieved this land protection while increasing its overall agricultural production, including production of biofuels, especially sugarcane and soybeans. In response to pressure from environmental groups, palm oil buyers and their producers in Southeast Asia have formed a coalition called the Roundtable on Sustainable Palm Oil, which seeks to reduce deforestation and environmental harm from palm oil production. In 2013 Wilmar, the world's biggest palm oil trader, which controls 45% of the market, announced a landmark zero deforestation policy; it will avoid development on areas with high conservation value. Other companies have followed Wilmar's lead, but it remains to be seen whether they will live up to such voluntary commitments.

Algae Can Turn Waste into Fuel

Another way to avoid many of the environmental threats stemming from biofuel production is to turn from terrestrial plants to aquatic algae (**Figure 10.50**). Growing algae, as we noted earlier, will not compete with farm production because they can be grown using brackish water too salty for irrigating crops or for drinking. They also happen to be fabulous environmental remediators. Conveniently, algae need a steady supply of carbon dioxide, along with nutrients such as nitrogen and phosphate. It turns out that agricultural and municipal wastes are rich in nitrogen and phosphorus, which can pollute the water supply by causing harmful algal blooms. Diverting these nutrients to an algae **bioreactor,** which

bioreactor A system designed to cultivate algae; helps filter wastewater before it enters the environment.

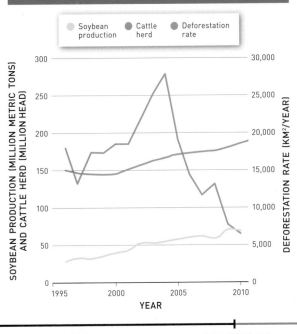

DECREASING DEFORESTATION RATES

● Soybean production ● Cattle herd ● Deforestation rate

SOYBEAN PRODUCTION (MILLION METRIC TONS) AND CATTLE HERD (MILLION HEAD)

DEFORESTATION RATE (KM²/YEAR)

YEAR

FIGURE 10.49 Brazil has reduced the rates of deforestation in the country while increasing agricultural production, including production of soybeans, some of which are used for biofuel. (Union of Concerned Scientists, 2011)

FIGURE 10.50 Algae farms like this one may play an important role in the future production of high-quality biofuels while minimizing collateral damage to the environment.

How might a whole ecosystem approach to biofuel production reduce costs and improve production?

BIOFUELS OF THE FUTURE?

(The Asahi Shimbun/Getty Images)

is a chamber designed to cultivate algae, provides a means to recycle these nutrients and filter the wastewater before it enters the environment. In addition, by diverting the carbon dioxide generated by electrical generating stations into algal cultures, it will be possible to enhance their growth and sequester carbon emissions at the same time.

Cellulosic Ethanol

Corn-based ethanol has an energy return on energy investment (EROEI) that barely exceeds a value of 1.0 (**Figure 10.51**), much lower than gasoline. Although ethanol is a cleaner-burning fuel than gasoline, corn is clearly not going to solve our energy needs. Sugarcane ethanol and soybean biodiesel, by contrast, have EROEIs substantially higher than that of corn ethanol and somewhat higher than gasoline extracted through energy-intensive processes from the Athabasca oil sands. Unfortunately, such fuels are best produced in tropical and subtropical climates, making them poorly suited for most temperate regions. Now look at cellulosic ethanol: It has an EROEI 10 times greater than corn ethanol and approaches that of gasoline from conventional oil. Could cellulosic ethanol be the future? Quite possibly.

In 2013 the cost of cellulosic ethanol was still 30% higher than a liter of corn-based ethanol. However, costs of production are falling rapidly and it should be on a par with corn before 2020. One of the major current costs of cellulosic ethanol is the cost of organic material suitable for production.

Recent reviews identify five promising sources of materials for cellulosic ethanol production (**Figure 10.52**). Developing these resources could improve environmental conditions in several ways. For example, growing perennial plants such as switchgrass (*Panicum virgatum*), native to most of the United States and Canada and that grows to 1 to 1.5 meters (3 to 5 feet)

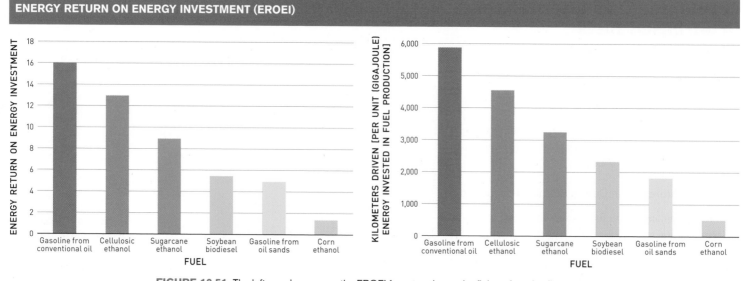

ENERGY RETURN ON ENERGY INVESTMENT (EROEI)

FIGURE 10.51 The left panel compares the EROEI for extraction and refining of crude oil or growing, harvesting, and processing of biomass, for several fuels. The number of kilometers traveled by a vehicle using the amount of fuel produced from investment of 1 gigajoule of energy in production of the fuel is shown in the right panel. A gigajoule is approximately equivalent to the energy content of 27 liters of crude oil. (Data from Schmer et al., 2008; Inman, 2013)

REDUCING THE IMPACTS OF BIOFUEL PRODUCTION

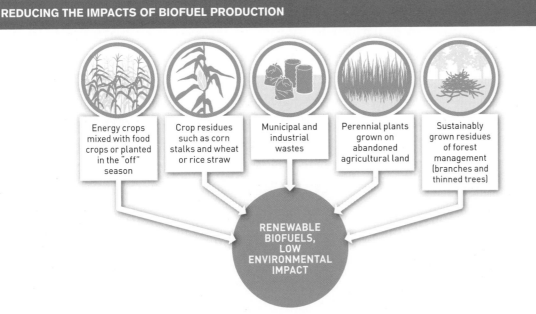

FIGURE 10.52 Potential sources of materials ("feedstocks") for biofuel production that minimize competition with food production and reduce environmental impact.

in height, has the potential to improve soil fertility by adding to soil carbon. In addition, switchgrass can grow well on a variety of soil types, including shallow soils of marginal fertility and low water content.

⚠ Think About It

1. Why is coupling legal protection with strong law enforcement essential to Brazil's efforts to curtail deforestation?

2. How could development of systems for producing biofuels from aquatic algae complement land-based biofuel production?

3. How do improvements in fuel economy complement the development of new technologies for production of biofuels?

How might increasing demand for food from a growing human population influence the relative cost of cellulosic ethanol?

10.8–10.11 Solutions: Summary

A number of strategies and technological developments help address the environmental challenges facing solar energy development. Concentrating solar power plants using molten salt for heat transfer and storage use water-saving hybrid cooling systems that combine water and air-cooling. Photovoltaic generation uses even less water and, when moved to rooftops in population centers, reduces the impacts of solar development on wildlife habitats.

New wind-generator designs have reduced bird mortality by nearly two-thirds but have resulted in *increased* bat mortality; but that, too, can be reduced by programing wind turbines to operate only at higher wind speeds. A key to reducing the impact of hydroelectric

development is to build lower dams, which decreases both the amount of water stored and the area flooded. Smaller dams are also easier to fit with fish bypass structures. Retrofitting existing unpowered dams with hydroelectric generators and installing fish-friendly in-river turbines can also add to hydroelectric generation capacity without further disruption of river ecosystems and biodiversity.

The dependence of the current transportation sector on petroleum-based liquid fuels could be addressed in a sustainable way by developing biofuels of low environmental impact. Cellulosic ethanol converts whole plants or agricultural and forestry wastes into ethanol and has a higher rate of EROEI than corn-based ethanol production. Several methods are being developed to convert algae and terrestrial plant biomass directly into gasoline, diesel, and jet fuels.

Answer the following questions for each chapter section and then answer the Central Question.

Central Question: Can we develop renewable energy resources to help sustain a thriving economy without adversely affecting the environment?

10.1–10.3 Science

- How is solar energy captured and utilized?

- What role do wind, water, and geothermal energy play in the renewable energy mix?

- How is energy in biofuels stored and accessed?

10.4–10.7 Issues

- What are the financial and environmental drawbacks of solar energy?

- How do wind turbines impact wildlife?

- What are the environmental and societal impacts of hydroelectric power?

- How does using biofuel for energy impact food supply and the environment?

Renewable Energy and You

The future of civilization and the biosphere depends on how we manage our remaining fossil fuel resources and transition to renewable energy resources. The speed at which we achieve this energy revolution depends substantially on the actions of governments and industry. Still, individuals can make a difference by staying informed and acting on evolving energy challenges and opportunities.

☐ *Stay informed and engaged in energy issues.*

Keep informed regarding rapid developments in renewable energy, especially with regard to technological advances and growth in electrical generating capacity around the world. Contact your state and federal lawmakers and ask them to support legislation that incentivizes renewable energy and technologies that use it. For a list of incentives, rebates, and other programs, check out www.dsireusa.org.

☐ *Consume energy wisely.*

Cut your own energy consumption: Choose efficient bulbs such as compact fluorescent lights; if you have access and your health permits, set the thermostats in your home and/or workplace to conserve energy—that is, a few degrees lower in winter and higher in summer. In many communities, it is possible to personally invest in renewable energy through your electrical utility by electing to purchase power generated using renewable energy resources.

☐ *Use energy-efficient transportation.*

If you can do so securely, walk or bicycle to school or work. If you can't, take a bus or train. If you must drive, try to carpool. If you purchase a motor vehicle, consider models that are both safe and fuel-efficient. While governmental mandates affect the fuel efficiency of vehicles produced by manufacturers, our choices as consumers can also exert significant influence on manufacturing.

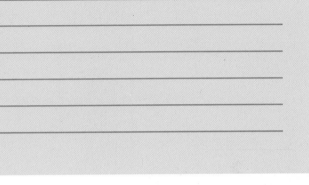

10.8–10.11 Solutions

- **What improvements are being made in the technology of solar energy?**
- **In what ways are the use and technology of wind turbines being improved?**
- **What strategies reduce the impacts of hydroelectric power?**
- **How can we reduce the impact of biofuels?**

Answer the Central Question:

_____

_____

_____

_____

_____

_____

Chapter 10
Review Questions

1. What sources of renewable energy does the Danish island of Samsø use to power its communities?
a. Wind energy c. Solar energy
b. Biomass fuels d. All of the above

2. Which of the following is a major technical problem associated with solar energy?
a. Very little solar energy strikes Earth.
b. The potential for using solar energy is limited to the deserts and tropics.
c. Because of Earth's day–night cycle, solar energy is intermittent.
d. Solar power can be generated only by rooftop photovoltaic systems.

3. Which of the following is not a form of "water power"?
a. Wave energy
b. Geothermal energy
c. Tidal energy
d. Hydroelectric power

4. What is the major challenge to producing cellulosic ethanol?
a. Low supplies of biomass to fuel the process
b. Low energy content of wood, straw, and other forms of biomass
c. Lack of interest by researchers for developing an energy-efficient process
d. The need to break up complex cellulose molecules into simple sugars

5. Which of the following has the highest energy return on energy invested (EROEI)?
a. Hydroelectric power
b. Coal-generated electricity

c. Natural gas–generated electricity
d. Wind-generated electricity

6. Which of these statements about wind turbines and bird mortality is true?
a. Collisions with wind turbines now result in more bird deaths than collisions with buildings.
b. Collisions with wind turbines now result in the deaths of a few hundred birds each year.
c. Domestic cats kill fewer birds than do wind turbines.
d. Bird deaths from collisions with wind turbines are projected to reach 1 million annually in the year 2030.

7. What can be done to reduce the environmental impacts of solar power?
a. Shift from water to air-cooling systems for solar concentrating power plants.
b. Site solar concentrating power plants on disturbed landscapes.
c. Mount industrial-scale photovoltaic systems on the roofs of commercial buildings.
d. All of the above.

8. How do new wind turbine designs mainly help in reducing bird collision deaths?
a. The new turbines spin too fast for birds to hit them.
b. The brighter colors of newer-style wind turbines warn birds to stay away.
c. The single-pole support for the new wind turbines offers no perch sites.
d. The new wind turbines are mounted closer to the ground where few birds fly.

9. Which of the following is not an issue associated with biofuel development?
a. Deforestation
b. Massive ozone depletion
c. Reducing potential human food supplies
d. Threats to endangered species

10. Which of the following statements about renewable energy development is correct?
a. Renewable energy resources do not have environmental impacts.
b. Renewable energy resources have greater impacts on the environment than nonrenewable energy resources.
c. There is no difference in the environmental impacts of renewable and nonrenewable energy resources.
d. Development of renewable energy resources will require addressing a number of associated environmental impacts.

Critical Analysis

1. How might the development to full reliance on renewable energy on Samsø Island in Denmark serve as a model for development of renewable energy resources at larger national or global scales?

2. Discuss the environmental and social issues associated with hydroelectric development and how those issues can be addressed.

3. What contributions can be made by industry, government, and research scientists to facilitate a transition to greater reliance on renewable sources of energy?

4. Even though producing ethanol from corn is simpler, why are major energy companies investing heavily in the development of cellulosic ethanol?

5. Considering the major environmental and human costs associated with corn-based ethanol production and its low EROEI, why do U.S. farmers continue growing large amounts of corn specifically for ethanol production.

Find additional resources and links online at www.macmillanhighered.com/launchpad/molles1e.

Central Question: What is the relationship between the environment and human health and how can we manage that relationship?

Describe the toxic substances and pathogens in the environment and their effects on humans.

SCIENCE

Environmental Health, Risk, and Toxicology

Explain the environmental and human health consequences of exposure to toxic substances and pathogens.

Discuss how we might assess risk and deal with toxic substances and pathogens in our everyday lives.

ISSUES

SOLUTIONS

Workers inspect an area where a toxic chemical that leaked from storage tanks contaminated the Elk River near Charleston, West Virginia, in January 2014. In the aftermath of the spill, which polluted domestic water supplies, residents of Charleston lined up to fill containers with potable water.

Chemical Spill on the Elk River

A toxic leak highlights how little we know about the numerous hazardous substances that affect human health and the environment.

In the mid-morning of January 9, 2014, residents in Charleston, the sleepy capital of West Virginia, noticed a sweet, licorice smell in the air. It wasn't until 5:45 P.M. that the regional water utility began telling customers to avoid drinking, cooking, or washing with the water. One mile upstream of the city, on the banks of the Elk River, a rusty, storage tank owned by a company called Freedom Industries had leaked as much as 7,500 gallons of a chemical used to process coal. Over the next 24 hours, some 700 residents had contacted poison control, complaining of rashes and nausea. Fourteen were hospitalized.

Like many industrial chemicals used widely in the United States, little was publicly known about the toxicity of this chemical, 3-methylcyclohexanemethanol (MCHM). Under the Toxic Substances Control Act, which was passed in 1976, the U.S. Environmental Protection Agency (EPA) can test chemicals that pose a health risk—but only after those chemicals have shown evidence of harm. The 62,000 chemicals that were already on the market at the time the law was passed were exempted from regulation, and, of the 21,000 chemicals registered since then, only 15% have included health-and-safety data. Data on these new chemicals are often hidden from the public because companies argue they contain confidential business information.

On January 13, four days after the spill, officials lifted the water advisory. The little data that existed on MCHM indicated that drinking it could harm the liver and kidneys at high doses, but no one really knew for sure how exposure would affect people in the months or years to come. Nevertheless, the incident revealed how environmental hazards, when unchecked by regulation, can affect our health and well-being.

Much of this book is concerned with the health of the environment—that is, the condition of the environment

"If we are going to live so intimately with these chemicals—eating and drinking them, taking them into the very marrow of our bones—we had better know something about their nature and their power."

Rachel Carson, *Silent Spring*

itself or some part of it—for instance, whether a wetlands ecosystem is clean, productive, and sustainable. In this chapter, we look at the related but distinct issue of **environmental health,** which refers specifically to human health and safety and the way the environment—both natural and human-created aspects—affects them. This brings us to the Central Question of this chapter.

environmental health An area of research and action that assesses and attempts to mitigate the physical, chemical, and biological factors in the environment that impact human health.

Central Question

What is the relationship between the environment and human health and how can we manage that relationship?

(fotog./Getty Images)

11.1–11.2 Science

environmental hazards
Phenomena dangerous to humans, including infectious disease, toxic substances, and pollutants.

pollutant A substance (e.g., oil or pesticides) or condition (e.g., excessive noise) harmful to living organisms that contaminates air, water, or soil.

toxin A poisonous substance produced by a living organism (e.g., a plant, animal, fungus, or bacterium) that can harm human health. See *toxicant.*

Humans frequently encounter **environmental hazards,** natural and human-made phenomena that are dangerous to humans. These hazards range from microscopic organisms to pollutants that spread around the globe (**Figure 11.1**). The World Health Organization (WHO) has estimated that one-quarter of all deaths—13 million people every year—result from exposure to environmental hazards. This chapter explores biological and chemical environmental hazards: toxic substances, infectious disease, and pollutants.

11.1 Chemical hazards include toxic substances and pollutants

Toxic substances can kill you—some of them very quickly, if you are exposed to a sufficiently high concentration

of the substance. A toxic substance that harms living organisms through contamination of the air, water, or soil is considered a **pollutant** (**Figure 11.2**). For instance, an insecticide such as DDT, one focus of Rachel Carson's 1962 book *Silent Spring* (see Chapter 1, page 18), sprayed in the environment is considered a pollutant because its residue contaminates soil, water, and plants and animals. We will discuss pollution and pollutants in detail in Chapter 13. Pollutants can be lethal in high doses, as with the exposure of agricultural workers to various insecticides, and low levels of exposure can damage the embryos of humans and other animals or can make adults ill.

Natural and Manmade Toxic Substances

Toxic substances are generally divided into two categories: toxins and toxicants. As discussed below, a **toxin** is a poisonous substance produced by a living

---

ENVIRONMENTAL HAZARDS HAVE A WIDE RANGE OF SOURCES

(Munshi Ahmed/Bloomberg via Getty Images)

Severe air pollution in Singapore

Foot lesions caused by drinking arsenic-contaminated water

(Majority World/UIG via Getty Images)

(© David Chang/epa/Corbis)

Testing for avian flu in goose flock in Taiwan

FIGURE 11.1 Environmental hazards have both human and natural sources. In Singapore, forest fires in nearby Malaysia have caused severe air pollution. Drinking water from a well contaminated with arsenic has produced lesions on this woman's feet. In January 2015, health workers in Taiwan test a flock of geese for avian flu viruses H5N2 and H5N8, which had been detected on the island.

organism, such as a plant, animal, fungus, or bacterium that can harm human health. For instance, botulin, the cause of the food poisoning botulism, is a type of bacterial toxin. A **toxicant** is a toxic substance produced by humans or as a by-product of human activity. Dioxins, a class of toxicants that are by-products of a number of manufacturing and industrial processes, are among the toxic substances most dangerous to human health.

Some toxic substances are a natural part of the environment (**Figure 11.3**). Arsenic, mercury, and lead are toxic substances called *heavy metals.* Heavy metals are found naturally in the environment and are typically not harmful in small amounts. In most cases, heavy metals become a problem to human health as a result of human actions. Mining and road construction, for example, expose fresh rock surfaces to rain that dissolves heavy metals in the rocks. Rainwater carries the heavy metals into streams and ponds and thus soaks into the ground, contaminating wells.

Humans manufacture a vast number of potentially toxic substances—85,000 industrial chemicals are produced in the United States alone, some of them in enormous quantities. Manufacturing more than 450,000 kilograms (1 million pounds) per year of a single toxic substance is considered a "high production volume," and 2,500 chemicals fall under that category.

Not all chemicals are toxic, of course, but we don't know how many of the 85,000 industrial chemicals created are indeed toxic because most of them have never been tested. Furthermore, 2,000 *new* chemicals are synthesized every year and very few of those are tested for toxicity before being placed in production.

The science that studies the effects of toxic substances on humans (and other organisms) is **toxicology.** Three factors determine the toxic effects of a substance on

POLLUTION, A MAJOR CONCERN OF ENVIRONMENTAL SCIENCE, IS UBIQUITOUS IN THE MODERN WORLD

(Federico Rostagno/Shutterstock)

FIGURE 11.2 Industry, transportation systems, agriculture, and activities in the home are all potential sources of toxic pollutants. Here, an agricultural field is being sprayed with a pesticide, a potential source of pollution.

humans: (1) the duration of exposure, ranging from short to long; (2) the concentration of the substance, ranging from low to high; and (3) the frequency of exposure, which can range from a single exposure to repeated exposures over time (**Figure 11.4**). Humans can be exposed to toxic substances by breathing them in, swallowing them in food or water, or absorbing them through the skin. The goal of toxicology is to define safe levels of exposure to toxic substances.

How Toxic Substances Affect the Human Body

Toxic substances, many examples of which are discussed in this section, can harm the human body in several

What are the pros and cons of requiring that the toxicity of all new chemicals be tested before they are put into production and released into the environment?

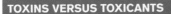

TOXINS VERSUS TOXICANTS

(NPS Photo)

Western diamondback rattlesnake

(Swasdee/Shutterstock)

Open burning of solid waste

toxicant A toxic substance produced by humans or as a by-product of human activity. See *toxin.*

toxicology The science concerned with the effects of toxic substances on humans and other organisms.

?

How do synergistic effects complicate toxicity testing?

neurotoxins Toxic substances that attack nerve cells.

carcinogen A substance that causes cancer by directly damaging the DNA of cells.

teratogen A substance that causes abnormalities during embryonic growth and development, resulting in birth defects.

allergen A substance that activates the immune system, inducing an allergic reaction.

endocrine disruptor A chemical that mimics hormones, including female hormones (estrogen and progesterone), male hormones (testosterone), or thyroid hormones.

antagonistic effect An interaction of two toxic substances wherein the toxicity of one chemical is reduced in the presence of the second chemical, which may be used as an antidote.

additive effect An interaction of two toxic substances wherein their combined toxicity is simply the sum of their individual effects.

synergistic effect An interaction of two toxic substances wherein their combined toxicity is greater than the sum of their individual effects.

pandemic Expansion of a disease affecting a large proportion of a population in a very large geographic area (e.g., across an entire continent).

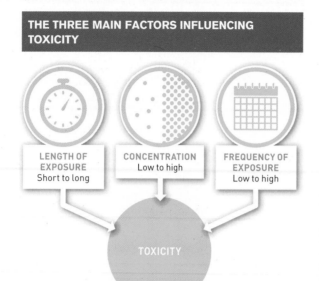

THE THREE MAIN FACTORS INFLUENCING TOXICITY

LENGTH OF EXPOSURE
Short to long

CONCENTRATION
Low to high

FREQUENCY OF EXPOSURE
Low to high

TOXICITY

FIGURE 11.4 The toxicity of a substance to humans, or any other organism, is influenced by a complex interaction among length of exposure, concentration, and frequency of exposure.

different ways. They are generally grouped by the effects they have on the body.

Neurotoxins attack nerve cells. Different types of neurotoxins affect nerves differently: Heavy metals kill nerve cells, whereas chemicals called organophosphates and carbamates (used in insecticides) inhibit signal transmission between nerve cells. Chlorinated hydrocarbons, which are found in some cleaning fluids, disrupt nerve cell membranes.

Carcinogens cause cancer by directly damaging the DNA of cells. Mutations that interfere with the *off* signal of the cell division cycle cause cells to multiply uncontrollably, forming a tumor. Common carcinogens include chemicals such as benzene (which occurs in cigarette smoke), radioactive substances such as radon (a gas that is released by some rocks and can accumulate in basements), and heavy metals including arsenic.

Teratogens cause abnormalities during embryonic growth and development, resulting in birth defects. The brain of a fetus is especially sensitive to pollutants: Exposure to lead and mercury (both heavy metals), alcohol, and insecticides can result in low IQ or neuromuscular defects that interfere with babies' speech and movement.

Allergens activate the immune system, causing an allergic reaction. Allergenic effects can range from mild, such as nasal congestion from pollen, to life-threatening, such as anaphylactic shock from a bee sting.

Endocrine disruptors mimic female hormones (estrogen and progesterone), male hormones (testosterone), or thyroid hormones. The results include damage to or changes in the male or female reproductive

system, behavioral changes, damage to the immune system, neurological problems, and tumors.

How Toxic Substances Interact

Organisms are exposed to many different toxic substances at the same time, and the combined effects of two or more substances may take a variety of forms. Some chemicals have **antagonistic effects.** In this situation, the toxicity of a chemical is reduced in the presence of a second chemical, which may be used as an antidote. In other cases, the toxic substances may have **additive effects.** In this case, the toxicity of the two substances is simply the sum of their individual toxicities, for instance, 2 + 2 = 4. However, toxic substances may have **synergistic effects,** where their combined toxicity is greater than the sum of their individual effects; that is, instead of 2 + 2 = 4, it might be 2 + 2 = 6 or more. For example, both mercury and a toxic compound from detergents called 4-nonylphenol accumulate in the sludge in sewage treatment plants. Each of these chemicals is toxic by itself, but when combined in a laboratory test, each becomes about one-third more toxic to human liver cells than if it were acting alone.

Humans are not the only animals affected by synergistic effects of pollutants. A study of two chemicals (ethylene glycol and methanol) that are used to increase oil well production in the Gulf of Mexico found that exposure to both chemicals reduced the swimming ability of pompano (a popular game fish) much more than did exposure to either of the chemicals alone.

⚠ **Think About It**

1. Why is using sewage sludge as fertilizer on crops and pastures probably not a good idea?

2. What are the several types of toxic substances?

3. How can the antagonistic effects of one toxic substance on another be used as an antidote?

11.2 Bacteria, viruses, and parasites are spread through the environment

The Black Death: The very name evokes horrifying images (**Figure 11.5**). This **pandemic** outbreak of bubonic plague, which peaked in Europe around 1350, killed an estimated one-third of Europe's population, convincing many people at the time that they were witnessing the end of the world. Bubonic plague is only one of the devastating pandemics humans have experienced; smallpox, infant diarrhea, malaria, tuberculosis, and cholera have killed millions of people.

(De Agostini/A. Dagli Orti/Getty Images)

FIGURE 11.5 The bubonic plague swept Europe in the Middle Ages, killing an estimated one-third of the population and leaving the survivors emotionally and psychologically shaken, as suggested by this painting from the period.

At the time, it was thought that the Black Death, and **disease** in general—a condition in which normal biological function is impaired by bacteria, viruses, parasites, improper diet, or pollutants—were caused by foul or noxious air. It was not until 500 years later, in the late 19th century, that one of the key scientific breakthroughs in the fight against infectious disease revealed that infectious bacteria cause many diseases, including bubonic plague.

Bacteria are not the only **pathogens** (organisms that produce illness). Viruses and parasites also produce disease in humans and other organisms. Although we now understand how most diseases are transmitted, infectious disease has remained the leading cause of death due to environmental hazard over the last century. Let's consider the three classes of pathogens—namely, bacteria, viruses, and parasites—beginning with bacteria.

Bacterial Disease

Bacteria (singular *bacterium*) existed billions of years before the first multicellular organisms developed. Most bacteria are single-celled and very small—a fraction of the size of human body cells—with a very simple structure (**Figure 11.6**). Because of their small size, bacteria remained invisible to humans until the invention of the first microscopes in the late 1600s, and it would be more than a century before it was discovered that bacteria cause diseases.

Bacteria live everywhere—from down in the deepest mines to a piece of dust floating in the atmosphere—and are essential to the functioning of the biosphere. Most bacteria are benign—that is, they do not cause disease—and many are, in fact, essential to human health. The body of an adult human includes more than 100 trillion cells, but only about 10 trillion of them are human cells. The remaining 90% of cells in and on the human body, your body, are bacteria; these 90 trillion bacterial cells

How many people would die if a pandemic killed a third of the current global population?

disease A condition in which normal biological function is impaired by bacteria, viruses, parasites, improper diet, or pollutants.

pathogen An organism that produces illness.

bacteria Single-celled organisms (singular *bacterium*) lacking a nucleus or other membrane-bound organelles; the vast majority of bacteria are not pathogens.

Capsule
Cell wall
Plasma membrane
Cytoplasm
DNA
Ribosomes
Pili
Flagellum

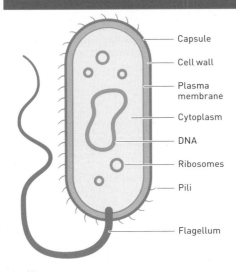

(Hugh Spencer/Science Source/Colorization by Mary Martin)

Nodules on clover roots house nitrogen-fixing bacteria

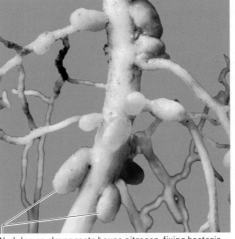

(National Institute of Allergy and Infectious Diseases (NIAID))

Plague bacteria, *Yersinia pestis*

FIGURE 11.6 The simple physical structure shown here belies the metabolic complexity of bacteria. Nitrogen-fixing bacteria are essential to the functioning of all ecosystems. *Yersinia pestis*, the bacterial species that was the source of the Black Plague, is still a hazard in many parts of the world.

make up only a small proportion of the human body's mass, but nearly all of them are beneficial. For instance, bacteria in the intestine help to digest food, and bacteria on the skin and in the nose help prevent harmful bacteria from invading those areas.

It is the harmful bacteria that cause bacterial diseases, but they normally act with an environmental agent. An environmental agent can be as simple as a sharp object that penetrates the skin and allows bacteria to invade the tissues beneath; or it can be as complex as those associated with the cholera bacterium, which adjusts its lethality in response to the pollution of water sources by human waste.

The majority of disease-causing bacteria affect the host organism they invade by producing substances that damage cells in the host's body. Bacteria produce these toxins in two ways. *Exotoxins* are proteins secreted by bacteria into their surrounding environment. Some of these exotoxins, such as botulin, which is responsible for the food poisoning known as botulism, are among the most toxic natural substances known. Cholera, an often fatal disease of the small intestine, is the result of an exotoxin produced by the bacterium *Vibrio cholerae*. By contrast, *endotoxins* are part of the cell membranes of some bacteria, such as *Salmonella,* which is responsible for another type of food poisoning. Endotoxins are released when bacterial cells die and disintegrate.

Other bacterial diseases, such as tuberculosis, are the result not of toxins but of bacteria growing in the tissues of their host and competing with the host's cells for nutrients.

Viral Diseases

Viruses are responsible for numerous health problems, ranging from relatively mild diseases, such as the common cold, to some of the most destructive diseases, including smallpox, rabies, Ebola, and AIDS (acquired immune deficiency syndrome), which is caused by HIV (human immunodeficiency virus).

Viruses, like bacteria, are pervasive in the environment. Although a virus is a microscopic pathogen like a bacterium, the two are quite different biologically. Viruses consist simply of genetic material encased in protein; they do not have the cell structures and biochemical pathways that allow bacteria to carry out all the processes necessary for life (**Figure 11.7a**). A virus can complete its own life cycle only by invading and taking over the cellular systems of animals, plants, or bacteria (**Figure 11.7b**). Viruses essentially highjack the machinery of a cell and use it to produce copies of the virus instead of carrying out the normal functions of the cell. When cells stop their normal processes, the symptoms of viral disease appear.

The damage that some viruses cause can be repaired. Viruses kill the cells they infect, but if healthy cells multiply, they replace the dead cells. Some cells, such as nerve cells, are not actively replaced, however, and

How does the discovery of the numbers and essential services provided by bacteria challenge the concept of human individuality?

Do you think viruses are living organisms? Why or why not?

virus A structurally simple disease-causing agent consisting of DNA or RNA encased in protein; viral diseases include common cold, flu, measles, mumps, chicken pox, smallpox, rabies, herpes, and human immunodeficiency virus (HIV, the virus responsible for AIDS).

VIRUSES: SIMPLIFIED PATHOGENS

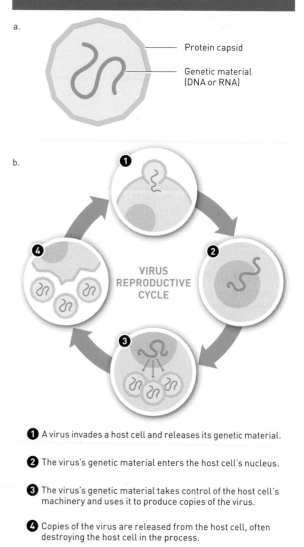

a.

Protein capsid

Genetic material (DNA or RNA)

b.

VIRUS REPRODUCTIVE CYCLE

❶ A virus invades a host cell and releases its genetic material.

❷ The virus's genetic material enters the host cell's nucleus.

❸ The virus's genetic material takes control of the host cell's machinery and uses it to produce copies of the virus.

❹ Copies of the virus are released from the host cell, often destroying the host cell in the process.

FIGURE 11.7 Encased in a protein coat, the genetic material of a virus contains all the instructions necessary for taking over the cells of its host. This simplified reproductive cycle shows how a virus invades a host cell and makes numerous copies of itself, killing the host cell in the process.

the damage to the body can be permanent. Polio is an example of a viral disease that attacks nerve cells with crippling effects. HIV attacks cells of the human immune system, reducing the capacity of the body to resist infectious diseases. The Ebola virus, which is endemic to several African countries, also attacks the immune system. It causes fever and widespread inflammation, which damages the liver, intestine, and blood vessels and potentially leads to bleeding from the eyes and nose.

One of the most common types of viruses that affect humans is the influenza virus. Most of us have had the flu at one time or another and have suffered from its symptoms: fever, cough, congestion of the lungs, sore throat, headache, sore muscles, and fatigue. Though most people recover from the flu within a week or two, the disease still takes a substantial toll on populations, especially among the elderly and the chronically ill. In the United States each year, approximately 100,000 victims of seasonal flu are hospitalized and there are about 20,000 flu-related deaths. Around the world, seasonal flu causes 3 to 5 million cases of severe illness and a quarter to half a million deaths annually.

"It's flu season, be sure to get your shot!"—it's a familiar refrain each year. Annual vaccinations are necessary for protection because viral genetic material, in this case that of the influenza virus, can change. In other words, viruses evolve. One of the most changeable of the viruses is HIV, which makes combating AIDS much more difficult.

Parasitic Diseases

Parasitism is another major source of disease in human populations. A **parasite** is an organism that lives in or on another organism, called the *host* (**Figure 11.8**). Parasitism is a particular type of ecological relationship similar to that between predators and prey (see Chapter 3, page 71), in which the host is harmed by the parasite and the parasite extracts benefits from the host, such as food, protection, dispersal of offspring, and so forth.

Parasitism is a very successful way of life. Because every species of animal, including all insects and other arthropods, is a host to at least one species of parasitic roundworm, and because most animal species serve as hosts for multiple species of roundworms, there are probably more species of roundworms than all other kinds of animals combined. However, they remain understudied, so we don't know for sure.

Humans serve as hosts to at least several hundred species of parasites that live on us and in us. For instance, outbreaks of head lice plague schools every year, and infestations of bed bugs are becoming a problem for travelers, even those who stay in expensive hotels. Both lice and bed bugs are external parasites that suck blood, as are mosquitoes and other biting insects. A variety of worm-shaped organisms are internal parasites of humans. Tapeworms live in the intestine. The largest human parasite, a tapeworm that can reach a length of more than 30 feet and has a life span of 20 years, can infect people eating raw or undercooked fish. Hookworms also live in the intestine, but they enter the body by penetrating the skin on the soles of the feet.

Many parasites do not cause a disease, but a heavy infestation of blood-sucking parasites can produce anemia (an abnormally low content of hemoglobin in the blood) or reduce the levels of important vitamins and minerals in the tissues. Some parasites transmit bacteria or viruses: Some ticks, for example, transmit the bacterium that causes Lyme disease, and mosquitoes transmit disease-causing viruses, such as the West Nile virus and the eastern equine encephalitis virus.

Seven of the top 10 diseases targeted by the World Health Organization's Special Programme for Research and Training in Tropical Diseases are parasitic diseases. Tropical parasitic diseases take a terrible toll on human populations, infecting hundreds of millions of people each year. They are responsible for more than a million deaths annually and, in some cases, are spreading rapidly. The range of some tropical diseases, such as dengue fever, is expanding as global climate change allows the mosquitoes that carry the disease to spread northward.

Transmission of Pathogens

Pathogens are found virtually everywhere in the environment. Bacteria and viruses travel in air, in water, and in the bodies of organisms. They can enter through open wounds, via the lungs, and via the intestine.

parasite An organism that lives in or on another organism, called the host; hosts are harmed by the parasite, while the parasite receives various benefits from the host (e.g., food, protection, dispersal of offspring).

PARASITES HAVE EVOLVED TO EXPLOIT OTHER ORGANISMS

Tapeworm · Mistletoe · *Plasmodium*

FIGURE 11.8 Tapeworms live in the digestive tract of animals, where they siphon off nutrients and energy in the food of their hosts. Mistletoe is a plant that parasitizes other plants such as this poplar tree. *Plasmodium* is the parasite that causes malaria in many animal species, including humans.

THE COMPLEX LIFE CYCLE OF *PLASMODIUM*, THE PARASITE THAT CAUSES MALARIA

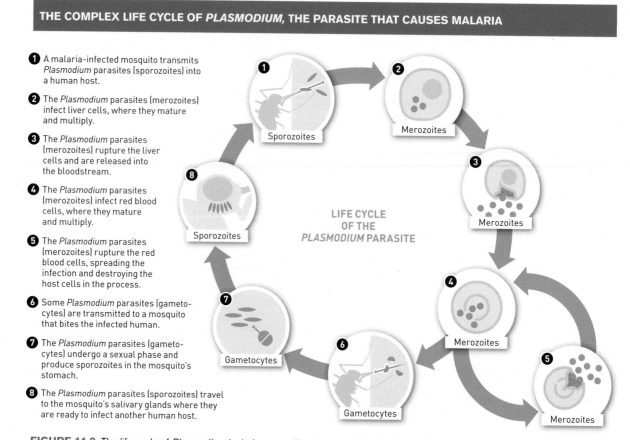

1 A malaria-infected mosquito transmits *Plasmodium* parasites (sporozoites) into a human host.

2 The *Plasmodium* parasites (merozoites) infect liver cells, where they mature and multiply.

3 The *Plasmodium* parasites (merozoites) rupture the liver cells and are released into the bloodstream.

4 The *Plasmodium* parasites (merozoites) infect red blood cells, where they mature and multiply.

5 The *Plasmodium* parasites (merozoites) rupture the red blood cells, spreading the infection and destroying the host cells in the process.

6 Some *Plasmodium* parasites (gametocytes) are transmitted to a mosquito that bites the infected human.

7 The *Plasmodium* parasites (gametocytes) undergo a sexual phase and produce sporozoites in the mosquito's stomach.

8 The *Plasmodium* parasites (sporozoites) travel to the mosquito's salivary glands where they are ready to infect another human host.

FIGURE 11.9 The life cycle of *Plasmodium* includes many life stages, each with unique characteristics and a specialized name. The key element in perpetuating malaria around the world is its insect vector: mosquitoes.

However, pathogens must enter a host's body before they can cause an infection, and hosts are well protected. The dense populations of benign bacteria on the skin and in the nose, mouth, and intestine form the first line of defense against infection by pathogenic bacteria. By occupying every site in the body that is suitable for bacterial growth, these harmless bacteria make it difficult for invading pathogenic bacteria to find a place to live. But, by invading in overwhelming numbers, disease-causing bacteria can break through a host's defensive wall of protective bacteria. Hundreds, thousands, or even millions of bacterial cells are needed to initiate an infection by most pathogenic bacteria.

Parasites are often transmitted indirectly, through secondary organisms called **vectors.** Some of the most significant vectors are biting insects such as mosquitoes and flies. The *Plasmodium* protozoan, which causes malaria, is carried by mosquitoes and is injected into its animal host when a biting mosquito releases saliva to prevent blood from clotting (**Figure 11.9**).

Schistosomiasis is a parasitic disease common in many parts of the world, especially in areas with poor sanitation. The organism that causes schistosomiasis reproduces in humans. Its eggs are excreted with human

urine and feces; when the eggs enter a body of water, they hatch into free-swimming organisms that enter the bodies of snails. There, they pass through more stages and are released into the water as yet another stage, called *cercariae*. The cercariae penetrate the skin of humans who enter the water and travel through the body of the human host, eventually lodging in the intestine, where they mature and produce another generation of eggs.

⚠ Think About It

1. Pathogenic organisms must damage their host to get the nutrients they require, but they die when their host dies. Thus, they must be able to infect a new host before they kill their current host. What relationship would you expect to find between the virulence of a bacterium (i.e., how sick it makes its host) and how it is spread to a new host?

2. What benefit do bacteria gain from releasing exotoxins that rupture the cells of their host?

3. Why would you want to be careful not to kill off a bacterium, which releases endotoxins, too rapidly?

vector An organism that transmits a pathogen or parasite to other organisms (e.g., mosquitoes transmit malaria and other diseases to humans and other species).

11.1–11.2 Science: Summary

Thousands of chemicals are manufactured and released into the environment each year, but a minority have been tested to determine their impact on environmental health. Many of these chemicals persist and accumulate in the environment and have the potential to harm humans and other animals. Toxic substances cause harm through various processes: Neurotoxins damage the nervous system; carcinogens and mutagens damage genetic material; allergens cause mild to severe allergic reactions; teratogens harm embryos; and endocrine disruptors change or damage the hormonal system.

Bacteria exist everywhere on the planet. Most have no effect on humans at all; many bacteria are beneficial, but a few can cause illness or death. Viruses act by invading healthy cells and taking over the machinery of the host's cell, turning it to the production of more viruses. Parasites live in or on other organisms. Many parasites have no measurable effect on the host, but some harm their host directly or carry pathogenic bacteria or viruses.

11.3–11.6 Issues

The release of chemicals into the environment starts a chain of events with profound consequences for the health of humans and other animals, as well as for the environment as a whole. Even pollutants that occur at low levels can have an impact as they become more concentrated as they move through the food chain. Our unfettered release of chemicals leads to situations in which good dietary practices, such as the advice to eat fish that are high in omega fatty acids, must be countermanded by warnings to avoid fish that are high in the food chain because of mercury contamination and other toxic substances.

11.3 Toxic substances move through the environment and can accumulate in large concentrations

Pollution of the environment by humans on a global scale has a long history. The levels of heavy metals in the atmosphere began to increase about 7,000 years ago when humans first started to refine ores to extract metals such as copper and lead. During the height of the Roman Empire, smelters in Italy and Spain were producing as much as 100 million kilograms of lead, 15 million kilograms of copper, and more than 2 million kilograms of mercury each year. Pollution from these smelters spread around the world—elevated levels of lead, copper, and mercury have been detected in the ice of a glacier in Greenland, more than 4,000 kilometers away.

Only a few societies were refining such large quantities of metals 2,000 years ago, and the resulting pollution damaged mainly the environments close to the smelters. Farther away from the smelters, the concentrations of the metals and other toxins had been diluted below levels that could cause damage. In fact, "The solution to pollution is dilution" remained the slogan of many industries through the mid-20th century. But this "solution to pollution" is problematic. First, many toxic substances are *persistent,* that is, they are resistant to degradation and may remain in the environment for a long time, move around the globe, and even increase in toxicity. In addition, as the human population has grown and the standard of living has increased, the quantity of pollutants that are released has increased enormously.

Toxic substances move through the environment, carried by air, water, and even the tissues of plants and animals. The **solubility** of a toxic substance generally determines how and where it moves in the environment or within an organism. Water-soluble compounds, such as alcohols and the metallic form of heavy metals, can move freely and quickly in rivers, lakes, and oceans and within an organism. Oil- or fat-soluble substances, such as methyl mercury (the highly toxic organic form of mercury), are generally carried in body tissues, especially fat deposits, where they may remain for the lifetime of the organism.

Persistent toxic substances that are retained within an organism build up to higher and higher levels over time, a process called **bioaccumulation.** As a consequence, larger fish (e.g., large tuna or swordfish) generally have

Pollution from metal smelting in the Roman Empire had a global reach. What does this suggest about pollution from today's intense industrial activity?

solubility The amount of a substance capable of dissolving in a particular amount of solvent.

bioaccumulation The absorption and increase in concentration of chemicals in organisms, over time, including potentially toxic chemicals. See *biomagnification*.

FIGURE 11.10 As striped bass grow, they accumulate increasing concentrations of mercury in their muscle tissue (left panel). Consequently, the tissues of smaller, younger individuals in the population have lower concentrations of mercury than do larger, older individuals. The concentration of mercury increases from lower to higher trophic levels in this arctic marine food chain (right panel). As a result, ring seals, the top predators, have a much higher concentration of mercury than do the lower trophic levels. (Data from Atwell et al., 1998; Burger and Gochfeld, 2011)

BIOACCUMULATION IN STRIPED BASS AND BIOMAGNIFICATION IN A MARINE FOOD CHAIN

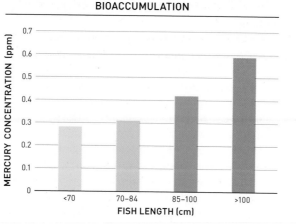

BIOACCUMULATION

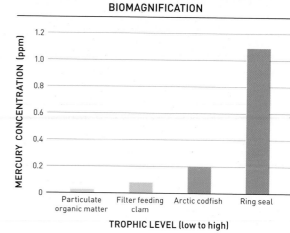

BIOMAGNIFICATION

higher concentrations of toxic substances such as mercury than do smaller, younger individuals in the population (**Figure 11.10**).

The next step beyond bioaccumulation is called **biomagnification** (see also Chapters 3 and 13, pages 81 and 392), a process in which substances increase in concentration at each successive trophic level in a food chain. For instance, mercury released during the burning of coal may be deposited into a body of water with rain or snowfall, where algae, absorb it. Invertebrate animals and small fish retain mercury from the many algae they eat. Larger fish, in turn, eat large numbers of those smaller organisms, and the mercury is transferred to these predators' bodies. As a result, the amount of the toxic substance increases at each successive trophic level—in other words, it is biomagnified (see Figure 11.10).

DDT (dichlorodiphenyltrichloroethane) was first recognized as a contact poison of insects in 1939 and quickly put to use. It was widely used during and following World War II and was initially seen as a shining example of the accomplishments of industrial chemistry. DDT was inexpensive to manufacture, easy to apply, and it killed insects while apparently not harming humans. Photographs from that era show soldiers and civilians being sprayed with DDT to kill lice and other external parasites.

But the dark side of DDT became apparent as it spread into the environment: It is toxic to aquatic animals (fish and amphibians), and it causes cancer in some mammals. DDT also provided the first demonstrations of biomagnification. Its concentration in tissues increases from the bottom to the top of the food chain, and a breakdown product of DDT called DDE interferes with the formation of the eggshells of raptors, such as the California condor and peregrine falcons. The thin eggshells break during incubation, killing the developing embryo (see Chapter 3, page 81). Where the critically endangered condors along the California coast feed on the carcasses of

biomagnification An increase in concentration of a substance (e.g., heavy metals or fat-soluble chemicals) at sequentially higher trophic levels in a food web. See *bioaccumulation*.

sea lions with high concentrations of DDT and DDE, they have a nesting success rate of 20% to 40%. Meanwhile, the Arizona population of condors, which do not have access to the pesticide-laden carcasses, has a nesting success rate of 70% to 80%.

⚠ **Think About It**

1. Why are pregnant women and children advised to limit consumption of certain fish?

2. What is the difference between bioaccumulation and biomagnification?

3. If you're concerned about the potential of consuming unsafe amounts of a known toxic substance, such as mercury, what should be the trophic position of the seafood you eat?

11.4 **Exposure to endocrine disruptors can affect the health of humans and other organisms**

Many environmental health issues are worsened, or even caused, by our failure to apply basic biological information. For example, we know that predatory insects control populations of plant-eating insects; yet we spray crops with pesticides that kill harmful and beneficial insects indiscriminately and then find their way into bodies of water, where they harm aquatic organisms. We also know that insects can evolve resistance to pesticides rapidly because they have short generation times, yet we continue to use those chemicals in ways that almost guarantee the evolution of resistance (see Chapter 7, page 208). In addition, we understand the effects of hormones on humans and other animals, yet we release endocrine disruptors, chemicals that have hormone-like effects, into the water that we use for drinking, cooking, and bathing.

Endocrine Disruptors in the Environment

Hormones control many elements of the physiology and development of organisms, including embryonic growth and the storage and release of glucose for cellular metabolism. Certain chemicals called *endocrine disruptors* interfere with these processes (**Figure 11.11**). They can turn on the signals that hormones carry to cells, turn them off, or change the meaning of the message a hormone is carrying. All these actions can have drastic effects on tissue function. Interference with the development and function of sex organs is one of the most conspicuous effects of endocrine disruptors.

Many endocrine disruptors are by-products of manufacturing processes. For example, a chemical called *androstenedione* is an anabolic steroid, one of the hormones with masculinizing effects that gained notoriety because many athletes misused it in hopes of enhancing their performance. Problems have arisen where this hormone is released into the environment. For example, androstenedione is released as a manufacturing by-product in the waste from a pulp paper mill on the Fenholloway River in the panhandle of Florida. The mosquitofish upstream from the mill have the expected 50:50 ratio of males to females, but downstream from the mill 90% of the females are partially or fully masculinized.

Androstenedione is unusual because it has a masculinizing effect; most endocrine disruptors are estrogen mimics that produce female characteristics. On a daily basis, we are surrounded by estrogen mimics: PCBs (polychlorinated biphenyls; see Chapter 13, page 392) are in the lubricants in our garages or basements, alkylphenols are in the detergents in our laundry rooms and beneath

our kitchen sinks, styrenes are in cigarettes and automobile exhaust, and parabens are in cosmetics. Two estrogen mimics are especially pervasive—bisphenol A, which is in some plastics, and the synthetic form of estrogen that is in birth control pills.

Bisphenol A (BPA) keeps polycarbonate plastics flexible and is used in epoxy resins. Infant bottles and plastic water bottles contain BPA, and BPA is present in the epoxy resins that coat the insides of food cans. Most human exposure to BPA comes from these sources. In 2012 the Food and Drug Administration (FDA) did not ban BPA from food containers because their studies showed that the human liver detoxifies BPA, thus concluding that humans have very low levels of BPA in their blood. The same studies also showed, however, that other animals are *not* able to detoxify BPA, so the FDA's decision will allow the environmental effects of BPA to continue. Regardless of the decision, great demand among consumers for BPA-free plastic products may lead to a phasing out of its use in plastics manufacturing.

The natural estrogen that women produce is excreted in urine, and the synthetic estrogen found in birth control pills is not completely broken down in sewage treatment plants. Consequently, the sewage effluent that is released into lakes, rivers, or the sea contains mixtures of these estrogens, which have a feminizing effect on aquatic organisms. Feminized male fish have been found in rivers even several kilometers downstream from the discharge pipes of sewage treatment plants.

The list of species that have been feminized by environmental estrogens is long. It includes aquatic species (fish, frogs, salamanders, turtles, and alligators); terrestrial birds; mammals, invertebrates as well as vertebrates; and marine species (harbor seals and harp seals), including freshwater forms. Human males are not immune to the effects of environmental estrogens; the average sperm count of men in Europe and the United States has decreased by about 50% since 1950. Because

Why do humans continue to cause harm to the environment when we know in advance the consequences of our actions?

How might the mosquitofish in the Fenholloway River and other species in other environments function as canaries once did in coal mines?

What might be some long-term effects of endocrine disruptors on biodiversity?

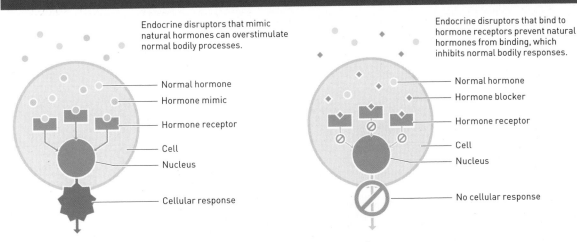

HOW ENDOCRINE DISRUPTORS WORK

Endocrine disruptors that mimic natural hormones can overstimulate normal bodily processes.

- Normal hormone
- Hormone mimic
- Hormone receptor
- Cell
- Nucleus
- Cellular response

Endocrine disruptors that bind to hormone receptors prevent natural hormones from binding, which inhibits normal bodily responses.

- Normal hormone
- Hormone blocker
- Hormone receptor
- Cell
- Nucleus
- No cellular response

FIGURE 11.11 We are surrounded by known and suspected endocrine disruptors to which we are exposed through water, food, and air. These are two known modes of action by endocrine disruptors. (Data from NIH, 2010)

these environmental estrogens affect so many species, including humans, it's clear that the hormones and receptors of these organisms must have similar structures.

⚠ Think About It

1. Where are endocrine disruptors found and how are people exposed to them?

2. What effects do endocrine disruptors have on organisms?

3. If endocrine disruptors are known to have such wide-ranging negative impacts, why do we continue to release them into the environment?

11.5 Misuse and overuse have promoted resistance to antibiotics and insecticides

In March 2013 the director of the U.S. Centers for Disease Control (CDC), Tom Frieden, held a press conference to discuss a "nightmare bacteria" killing patients in long-term medical care facilities. The bacteria, carbapenem-resistant enterobacteriaceae (CRE), are resistant to nearly all **antibiotics,** cause high mortality rates, and transfer their resistance to other bacteria. Resistance to antibiotics, substances that suppress bacterial growth or attack, makes this bacterial pathogen very dangerous because antibiotics are one of modern medicine's main tools in the treatment of bacterial diseases.

Furthermore, it appears that CRE are spreading. In the first half of 2012 alone, nearly 200 medical facilities treated at least one patient who was infected with these bacteria. Scientists at the CDC have tracked CRE from a single health-care facility in one state in 2001 to health-care facilities in 48 states in 2015. That's a very troubling

increase, given that CRE can spread the genes that destroy the effectiveness of antibiotics to other bacteria. When this happens, people can get severe infections, for which few effective treatments exist.

Why is the spread of CRE an environmental issue? If it were a single deadly case, it would be a tragedy for the patient and the patient's family, but it is an environmental issue because antibiotic resistant bacteria do not remain isolated. For example, penicillin, the first antibiotic to be discovered, became available for clinical use in 1944 and in 1945 the first penicillin-resistant strains of golden staph (*Staphylococcus aureus*) were reported. By 1959 half of the infections caused by golden staph were resistant to penicillin, so a new drug, methicillin, was introduced. Initially, methicillin killed penicillin-resistant golden staph, but within a year methicillin-resistant strains of golden staph appeared. These strains, known as **MRSA** (methicillin-resistant *Staphylococcus aureus*), originated in hospitals and then spread to the community.

Though drug resistance evolves as bacteria adapt through **mutation** and natural selection, human behavior is also implicated. Failure to take antibiotics as prescribed, as well as overuse of antibiotics, leads to increased prevalence of antibiotic resistance in bacterial populations. Health providers sometimes prescribe antibiotics for patients who ask for them, even when there is no medical necessity. Antibiotics will not cure a cold, for instance, because viruses cause colds and antibiotics have no effect on viruses. Regardless, many people demand them from their doctors or get them from friends in the belief that antibiotics can cure anything.

The overuse of these drugs can speed up the natural selection process for antibiotic resistance. This is because one mode of competition among microorganisms is the production and release of antibiotics. Thus, bacteria were coping with naturally occurring antibiotics billions of years before humans started to use antibiotics to

antibiotics Substances that suppress bacterial growth or attack and that are used in modern medicine in the treatment of bacterial diseases.

MRSA (methicillin-resistant *Staphylococcus aureus*) A pathogenic bacterium resistant to the antibiotic methicillin; MRSA originated in hospitals and then spread to the broader community.

mutation A change in the structure of an organism's DNA, i.e., in its genes.

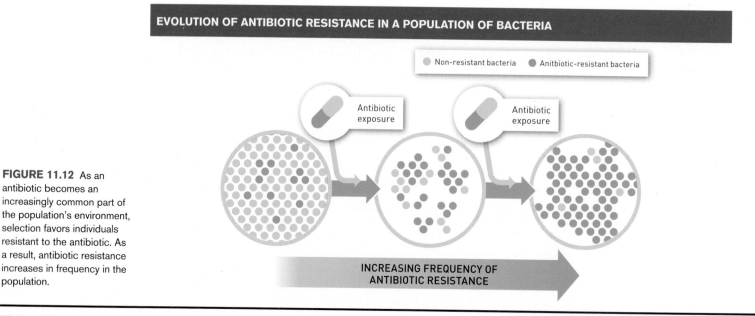

EVOLUTION OF ANTIBIOTIC RESISTANCE IN A POPULATION OF BACTERIA

● Non-resistant bacteria   ● Anitbiotic-resistant bacteria

Antibiotic exposure

Antibiotic exposure

INCREASING FREQUENCY OF ANTIBIOTIC RESISTANCE

FIGURE 11.12 As an antibiotic becomes an increasingly common part of the population's environment, selection favors individuals resistant to the antibiotic. As a result, antibiotic resistance increases in frequency in the population.

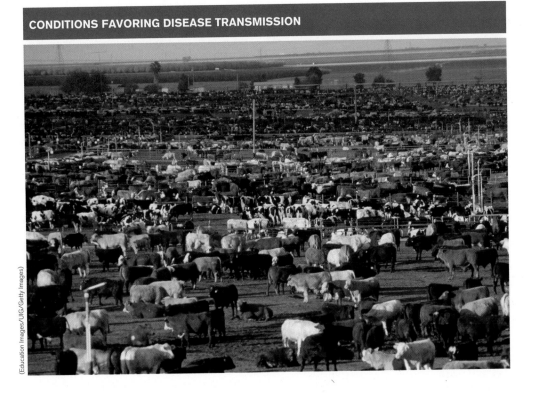

CONDITIONS FAVORING DISEASE TRANSMISSION

(Education Images/UIG/Getty Images)

FIGURE 11.13 Because infectious disease is more readily transmitted in dense populations, high population density was one of the factors that led to the heavy use of antibiotics as growth promoters in industrial-scale livestock-raising operations.

treat bacterial infections. Penicillin is produced by the mold *Penicillium crysogenum,* and most antibiotics used in medicine are still derived from microorganisms. Thus, exposure to antibiotics and the evolution of antibiotic resistance have been part of bacterial biology throughout their evolutionary history. What is new is the use of antibiotics to treat infections, the high frequency of antibiotic resistance found in bacteria in some settings, and the increasing occurrence of bacteria that are resistant to multiple antibiotics (**Figure 11.12**). The consequences of these new factors have enormous significance for human health.

Industrial Meat Production and Antibiotic Resistance

Medical misuse and overuse of antibiotics are important issues, but the use of antibiotics as growth promoters in agriculture overshadows them. Antibiotics—in many cases, the same ones prescribed to humans—are administered in food and water to cattle, hogs, sheep, and poultry to reduce the incidence of bacterial disease when animals are kept in crowded conditions. The crowding reduces both the amount of feed and the time required to bring an animal to slaughter weight. However, crowding also creates ideal conditions for the evolution and spread of antibiotic-resistant bacteria (**Figure 11.13**).

Meat producers in the United States administer more antibiotics per kilogram of meat produced than in any other country. A study by the Pew Charitable Trusts estimated that in 2011, 13.6 million kilograms (30 million

pounds) of antibiotics were used in the production of food-producing animals in the United States, mainly as additives to feed. This accounts for approximately 80% of the total use of antibiotics in the United States (**Figure 11.14** on the next page).

It is hard to imagine a more effective method of selecting for antibiotic resistance than to expose bacteria in the guts of livestock to constant low levels of antibiotics. Resistant bacteria that originated in livestock have been found in soil, ponds, and groundwater near sites where large numbers of animals are being reared. Furthermore, the transfer of antibiotic-resistant bacteria from animal to animal and from animals to farm workers has been documented.

Antibiotic-resistant intestinal bacteria have been isolated from wild animals as well—from field mice and bank voles in England, from magpies and rabbits in Wales, from forest birds in Brazil, from Canada geese in New Jersey, and from wild frogs in New York. None of these species has direct contact with humans, and the antibiotic resistance of their bacteria probably originated in farm animals.

⚠ **Think About It**

1. What drives antibiotic resistance?

2. How do the large population sizes and short generation time of pathogenic bacteria add to the challenge of controlling them?

3. Is antibiotic resistance just about the misuse of antibiotics?

Do individuals have an ethical obligation to the larger community to avoid misusing antibiotics?

Might livestock farmers be able to develop a profitable market for meat produced without the overuse of antibiotics?

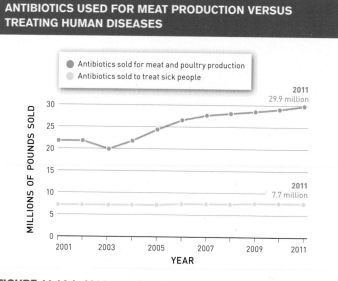

ANTIBIOTICS USED FOR MEAT PRODUCTION VERSUS TREATING HUMAN DISEASES

FIGURE 11.14 In 2011 approximately 80% of the antibiotics sold in the United States were used for the production of beef, pork, and poultry. About 90% of that total was administered to promote growth of the livestock. (Data from Pew Charitable Trusts, 2013)

11.6 Infectious diseases spill over from wild species and continue evolving to evade our defenses

How does a parasitic disease like malaria show the complex relations between environment and infectious disease?

zoonotic disease Any infectious disease that can spread from animals to humans.

bushmeat The butchered meat of wild animals, most commonly from African forests.

malaria A disease transmitted by mosquitoes that results from infection by a protozoan parasite of the genus *Plasmodium;* its life cycle uses hosts: mosquitoes and humans.

The infectious diseases that plague human societies emerged from natural ecosystems. Disease epidemics can often be linked to environmental problems and, sometimes, our misguided attempts to solve them.

Ebola, Deforestation, and Bushmeat

On December 26, 2013, an 18-month-old boy in the village of Meliandou, Guinea, came down with a fever, began vomiting, and had bloody stool. Two days later, he was dead. Within a couple of weeks, several members of his family came down with a similar illness, which was later identified as Ebola. By then, the epidemic was in full swing, spreading to the neighboring countries of Sierra Leone and Liberia. As of July 2015, there were 27,678 suspected cases of Ebola and 11,678 deaths, according to the World Health Organization (WHO).

Ebola is a **zoonotic disease,** which is any infectious disease that can spread from animals to humans. According to the U.S. CDC, 6 out of 10 infectious diseases are zoonotic diseases. When humans come in close contact with wild animals through the pet trade or when they kill and butcher wild animals for so-called **bushmeat,** they are at risk of catching new and emerging zoonotic diseases.

Deforestation in tropical countries also puts humans in closer contact with wild animals. Scientists have found Ebola in chimpanzees, gorillas, antelopes, and porcupines; but, like many zoonotic viruses, they believe the natural reservoir is fruit bats. Although it's not known exactly how the child in Meliandou caught Ebola in the first place, before he fell ill, he was seen playing near a hollow tree where bats roost. Another human-infecting virus harbored by fruit bats is severe acute respiratory syndrome, or SARS, which resulted in an outbreak in China in 2003 after it spread to a farm where wild civets were being raised for meat. Researchers also suspect that HIV was initially transferred to humans who came in contact with meat from chimpanzees or gorillas.

Evolution Challenges Efforts to Control Malaria

Diseases have evolved over millions of years to persist in species and ecosystems, and our efforts to eliminate them are often foiled by natural selection. One such disease is **malaria,** which is caused by a protozoan parasite carried by mosquitoes. Scientists have identified more than 100 species of malaria that infect birds, reptiles, and mammals. Six of these species are known to infect humans, primarily in tropical countries where malaria-carrying mosquitoes live. Each year, 300 to 500 million people become infected, and approximately 700,000 people die from the disease. Ninety percent of these cases occur in sub-Saharan Africa, where a child under 5 years of age dies from malaria every minute (**Figure 11.15**).

The best weapons we have against malaria after a person has already been infected are antimalarial drugs and pesticides that can reduce infection rates by reducing mosquito populations. However, both malaria parasites and mosquitoes are living organisms, and purely by chance some individuals within a population will be resistant to any new pesticide or antimalarial drug. When a pesticide or antimalarial drug is used, some of the resistant individuals will survive the treatment; those individuals will give rise to a new generation with greater resistance than their parents' generation. This process repeats itself generation after generation, ultimately rendering the pesticide or antimalarial drug ineffective. Both these evolutionary processes have frustrated efforts to control malaria.

Quinine was an effective antimalarial drug during the 18th and 19th centuries, as European nations established colonies in tropical regions around the world. But by the mid-20th century, resistant strains of malaria in many areas forced quinine to be replaced by chloroquine. Resistance to chloroquine has now become common in sub-Saharan Africa, Asia, and South America, and attention has turned to a compound called artemisinin, which is extracted from a plant called sweet sage. The first appearance of reduced susceptibility of the malaria parasite to artemisinin was reported in 2011. The team reviewing the international campaign to eradicate malaria wrote: "Losing . . . [artemisinin] to resistance

GEOGRAPHIC DISTRIBUTION OF MALARIA IN 2011

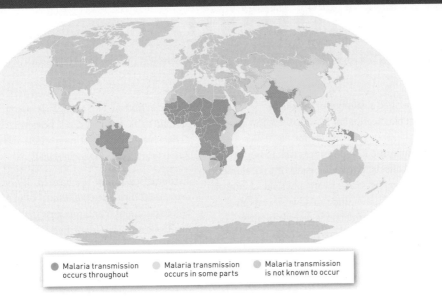

● Malaria transmission occurs throughout ○ Malaria transmission occurs in some parts ○ Malaria transmission is not known to occur

FIGURE 11.15 Although once common in temperate parts of the world such as southern Europe and the southern United States, malaria is now found mainly in tropical and subtropical parts of the world. It is especially prevalent in Africa. (After CDC, 2012)

would be a disaster for the control of malaria and would bring eradication efforts to a standstill."

Reports in a 2012 issue of the medical journal *The Lancet* further emphasized the point: "Antimalarial control efforts are vitally dependent on artemisinin combination treatments. Should these regimens fail, no other drugs are ready for deployment, and drug development efforts are not expected to yield new antimalarials until the end of this decade."

If the parasite can't be eliminated, is it possible to eliminate the mosquitoes that transmit the parasite? In the 1950s and 1960s, the WHO conducted an anti-malarial campaign that employed widespread application of DDT, often from trucks that cruised through neighborhoods every evening spraying DDT. Not surprisingly, mosquitoes responded by evolving resistance to DDT, and within a few years the effectiveness of DDT had declined in many parts of the world. Other insecticides replaced DDT, but, in a repetition of the earlier events, targeted mosquito populations are rapidly evolving resistance to the new chemicals.

Think About It

1. How does a lack of education hinder efforts to stop the spread of zoonotic diseases such as Ebola?

2. How might malaria control programs be managed to reduce the likelihood that the parasite or vector will evolve resistance?

3. Why are environmental health scientists concerned about how global warming may affect efforts to control malaria?

11.3–11.6 Issues: Summary

Movement of toxic substances through the environment, persistence, bioaccumulation, and biomagnification add to the potential for harm to humans and other organisms. The use of chemicals such as endocrine disruptors, antibiotics, and pesticides can cause significant harm to the environment and to human health. Endocrine disruptors interfere with a variety of essential metabolic and developmental processes. Alterations of the normal development and function of sex organs are among their more conspicuous effects.

Human factors such as misuse and overuse of antibiotics can increase the spread of antibiotic-resistant bacteria. The use of antibiotics in agriculture, which now far exceeds their use for treating human disease, appears to be a major contributor to the evolution of antibiotic resistance among pathogen populations. When humans come in close contact with wild animals, they are at risk of catching diseases that can spread from animals to people. Studies during the last half of the 20th century have demonstrated repeatedly that pathogens can rapidly evolve resistance to pesticides, antibiotics, and other drugs. As a consequence of our ignoring the evolutionary potential of pathogens, many serious diseases are now resistant to one or several antibiotics.

11.7–11.9 Solutions

We have described a number of threats to environmental health, but how do we know how much of a risk each represents? While the word *risk* has many definitions, the EPA considers **risk** to be "the chance of harmful effects to human health or to ecological systems resulting from exposure to any physical, chemical, or biological agent that can induce an adverse response." The practice of controlling infectious disease is fairly straightforward because the underlying causes of disease and antibiotic resistance are well understood. Solutions to problems related to toxic matter, in contrast, require a hard look at how much risk we are willing to tolerate.

11.7 We assess risk both qualitatively and quantitatively

Generally speaking, a risk is anything with the potential to cause us harm or loss or to put us in danger. How we assess risk in any given situation, however, will be influenced by our circumstances and the kinds of possible hazards we are considering. We evaluate a possible risk both qualitatively—how dangerous it seems to us personally, for instance—and quantitatively—its statistical likelihood of harming us or its economic cost. Environmental scientists, economists, and epidemiologists have developed a number of tools for assessing risk.

Risk Assessment Practices

In the 1970s, vinyl chloride gas—the building block for PVC pipes, automobile interiors, and dishes—became a ubiquitous industrial chemical. The United States was its number one producer. But researchers began to notice that the chemical could cause liver and bone damage in animals at very low doses. The pivotal moment came on January 23, 1974, when the B. F. Goodrich Company announced that it was investigating the cause of cancer deaths among three of its workers. In response, the EPA published its first-ever assessment of environmental risk for a chemical.

The process of risk assessment became formalized with the National Academy of Science's (NAS) groundbreaking report *Risk Assessment in the Federal Government: Managing the Process.* Since its publication in 1983, the EPA has integrated the principles of risk assessment into its practices and outlines four basic steps in the assessment of a potentially toxic substance: Identification of the hazard, dose–response assessment, exposure assessment, and risk

characterization. Not every risk assessment encompasses all four steps. In most cases, however, risk assessment is an expensive process, costing from $1 to $2 million and taking 3 to 5 years for a single chemical.

1. Identification of the hazard. *What health problems are caused by a given pollutant?* During the first step of risk assessment, contaminants that are suspected to pose health hazards are identified. They are then tested to determine whether exposure may cause specific health problems (e.g., cancer, chronic disease) and whether the adverse health effect is likely to occur in humans. To obtain information for this step, existing scientific data for a specific chemical are evaluated. In addition, researchers may study populations that have been exposed to the chemical, or they may experimentally test the effects of the chemical on an animal (e.g., rats, mice, or monkeys).

2. Dose–response assessment. *What are the health problems associated with different exposures?* The likelihood and severity of adverse health effects (response) are related to the amount of exposure to an agent (dose). This relationship is described in a **dose–response assessment.** Typically, as the dose increases, the measured response also increases. At low doses there may be no response until the exposure reaches a **threshold dose,** which is the lowest dose (concentration) of a toxic substance that induces a toxicity response in an organism.

The most common type of dose–response testing is done on animals such as mice or rats; it involves exposing the subjects to varying amounts of the toxic substance. Initial toxicity testing is intended to find acute or short-term toxicity effects. This normally requires an LD_{50} test, which determines the amount and exposure of the toxic substance that is a lethal dose to 50% of the animals in the test at the end of 14 days (**Figure 11.16**). Then researchers examine the animals further to discover which organs are affected, determine the reversibility of the toxic response, and develop dosages for continuing experiments. Researchers can then follow with more toxicity tests, such as chronic (long-term) exposure to the toxic substance at low levels of exposure.

3. Exposure assessment. *How much of the pollutant are people exposed to over a given period of time? How many people are exposed?* **Exposure assessment** defines the population that might be exposed to the agent of concern and identifies the routes through which exposure can occur. Exposure assessment also estimates the amount,

Can risk be entirely eliminated from our lives?

risk The chance of harmful effects to human health or to ecological systems resulting from exposure to any physical, chemical, or biological agent.

dose–response assessment A test of the response of an organism to a range in the dose, or concentration, of a potentially toxic substance.

threshold dose The lowest dose (concentration) of a toxic substance that induces a toxicity response in an organism.

exposure assessment An assessment of the population that might be exposed to an agent of concern and of potential routes of exposure.

DOSE–RESPONSE CURVES

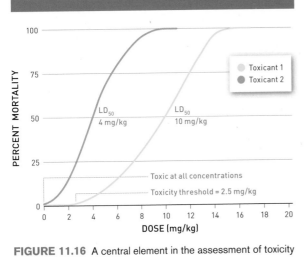

FIGURE 11.16 A central element in the assessment of toxicity is determining the concentration of a substance (dose) under study that causes 50% mortality of test animals (LD_{50}), such as mice or fish, during a predetermined time of exposure (e.g., 14 days). In this hypothetical example, toxicants 1 and 2 differ significantly in toxicity.

duration, and frequency of the doses that people might receive as a result of their exposure.

4. Risk characterization. *What is the extra risk of health problems in the exposed population?* This is the step in which risk assessment results are articulated. It includes the analysis of information from the first three steps to develop a qualitative or quantitative estimate of the likelihood that any of the hazards associated with the agent of concern will occur in exposed people.

Assessment in Action

The EPA conducts risk assessments for a variety of agents, such as diesel exhaust, mercury, secondhand smoke, and ozone. The toxic metal lead is one such agent. Known to damage the nervous system, kidneys, and other internal organs, ingestion of lead in children can cause developmental delays or mental retardation. Since the 1980s, the EPA has phased out lead in gasoline and has banned or limited lead used in consumer products like residential paint. As a result, the levels of lead in the air have decreased by 94% between 1980 and 1999. The amount of lead in people's blood has also decreased significantly in recent years. Of particular importance is the decrease in blood lead concentrations among U.S. children (**Figure 11.17**).

Furthermore, the EPA's National Center for Environmental Assessment periodically evaluates the latest research concerning the public health and welfare effects of lead and publishes the most up-to-date findings. The data are used for the establishment of the most current national air-quality standards for lead.

Precautionary Approach Revisited

Health and environmental regulations today are designed to keep the amount of a given contaminant released into the environment at a "safe" level, or to clean it up after it's already entered the environment. New products and chemicals are often subjected to limited testing and assumed to be "innocent until proven guilty"—that is, until scientific evidence demonstrates them to be harmful. With this approach, it is possible for toxic chemicals to be released into the environment or into our bodies until sufficient evidence suggests that harm is being done. Instead of the assumption of safety, the **precautionary principle** (see Chapter 1, page 12) offers protection *before* harm is done—an approach that can be characterized by the phrase "Better safe than sorry" (**Figure 11.18**). Invoking the precautionary principle, Canada banned the use of BPA in baby bottles. The European Union has also integrated the precautionary principle into its process for making decisions on environmental and health-related issues.

In the 1992 Rio Declaration on Environment and Development of the Earth Summit, the precautionary principle was proposed in the context of protecting the environment. Principle 15 of the Rio Declaration states that "in order to protect the environment, where there are threats of serious or irreversible damage, lack of full scientific certainty shall not be used as a reason for postponing cost-effective measures to prevent environmental degradation." The basic tenets of the precautionary principle involve taking preventative action before scientific certainty of cause and effect, seeking out and evaluating alternative products or services, and disclosing the potential impact on human health and environment associated with the selection of products or services.

What levels of risk should be too high to allow the release or use of a toxic chemical?

CHANGE IN PERCENTAGE OF U.S. CHILDREN WITH ELEVATED LEVELS OF LEAD

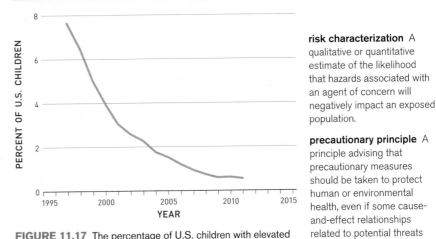

FIGURE 11.17 The percentage of U.S. children with elevated concentrations of lead in their blood fell from 7.6% to 0.56% between 1997 and 2011. (Data from CDC, 2013)

risk characterization A qualitative or quantitative estimate of the likelihood that hazards associated with an agent of concern will negatively impact an exposed population.

precautionary principle A principle advising that precautionary measures should be taken to protect human or environmental health, even if some cause-and-effect relationships related to potential threats are not fully understood scientifically.

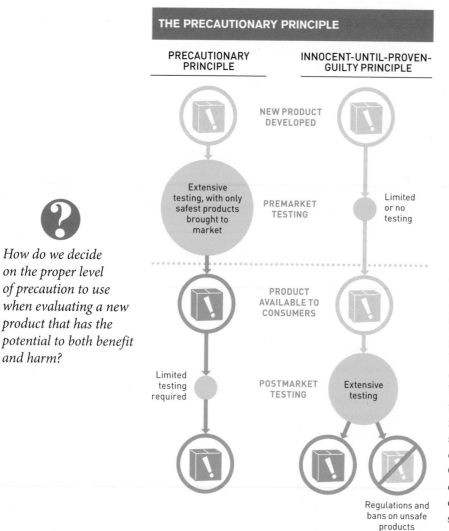

THE PRECAUTIONARY PRINCIPLE

PRECAUTIONARY PRINCIPLE

INNOCENT-UNTIL-PROVEN-GUILTY PRINCIPLE

NEW PRODUCT DEVELOPED

Extensive testing, with only safest products brought to market

PREMARKET TESTING

Limited or no testing

PRODUCT AVAILABLE TO CONSUMERS

Limited testing required

POSTMARKET TESTING

Extensive testing

Regulations and bans on unsafe products

?

How do we decide on the proper level of precaution to use when evaluating a new product that has the potential to both benefit and harm?

FIGURE 11.18 The precautionary principle emphasizes investing heavily in testing before a product is released to the public to ensure that only safe products are marketed. In contrast, the "innocent-until-proven-guilty" approach can involve more testing after a product is released and as problems arise.

Some cities have successfully used the precautionary principle to guide policy. In 2005 San Francisco passed a purchasing ordinance that requires the city to use safer alternatives when purchasing commodities for the city, such as cleaning products or electronics. The idea is to minimize harm by using the best available science to identify safer, cost-effective alternatives. The logistics of that implementation requires many different groups to participate. City employees with scientific backgrounds read and interpret academic research and then identify alternative products with less toxicity. Purchasers accustomed to prioritizing cost over other factors now think more holistically about purchasing decisions, and janitors and other maintenance workers find the best way to use new cleaning materials. Following implementation in San Francisco, the ideas of the precautionary principle have been applied in other U.S. cities, including Portland, Oregon, and Berkeley, California.

The precautionary principle is one useful tool to apply to environmental analysis and decision making, but it is not without its critics. Some say that regulation of products and chemicals could deprive society of significant benefits. In the United States, for example, the FDA requires all new drugs to be tested before they are put on the market. In other words, the FDA requires testing as a precaution to prevent harm to human health. Careful safety testing protects people from dangerous side effects, but it slows the introduction of new drugs. In the interest of precaution, the sickest of the sick may be prevented from receiving beneficial new medications.

The Exposome

One promising approach to studying how the environment impacts your personal health has been dubbed "the exposome." The exposome consists of all the environmental exposures in an individual's life and how those exposures affect that individual's health. Because everyone is exposed to a wide range of pollutants, stresses, and other environmental variables, it has often been difficult to link a specific disease, say, a type of liver cancer, to a specific cause. Recently, researchers have begun measuring the exposome using a variety of high-tech devices alongside chemical and genetic tools. For instance, a scientist at Columbia University has created special backpacks that children can wear to continuously collect air samples while they are at home and at school. Other researchers are analyzing exposures by identifying chemicals in a person's blood or looking for the fingerprint of chemical exposure on a person's genome. Such work is still in the early stages, but it provides more data to give us greater confidence in our risk assessments.

⚠ Think About It

1. Would it be possible for your school to implement the precautionary principle? If so, where could it do this?

2. What can individuals do to protect themselves from environmental contaminants?

3. What are the pros and cons of strict application of the precautionary principle in the development of new medications or other products?

11.8 Risk management involves reducing environmental hazards and controlling disease

Modern approaches to disease control include cooperation among nongovernmental organizations and governments, education, and awareness of culture and values, as well as medicines and vaccines.

Diarrheal Diseases

Diarrheal diseases are caused by a variety of bacteria, viruses, and parasites transmitted in water that is contaminated by feces. Thus, diarrheal diseases, which cause 2.5 million deaths per year, are most prevalent in countries with poor sanitation systems. According to the WHO, 88% of diarrheal diseases can be directly attributed to poor sanitation and hygiene. Young children are especially susceptible to diarrheal diseases, and half the victims of diarrheal diseases are children younger than 2 years old.

Reducing or eliminating diarrheal diseases hinges on better conservation of drinking water and management of waste. Improved water supplies can reduce diarrhea cases by 21%, while better sanitation can reduce them by 37.5%. Simply providing access to a latrine reduced incidence of diarrhea by 24% among children under 5 years of age in Lesotho, South Africa.

Ending the Bushmeat Trade

In remote villages and logging camps in Africa and elsewhere, locals continue to hunt and eat wildlife, including fruit bats, gorillas, and porcupines, that may harbor deadly viruses such as Ebola. Although many people turn to bushmeat and wild game because it is a cheap and readily available source of protein, it also plays a cultural role in some societies; this has led to a rampant national, and sometimes international, trade in bushmeat.

At John F. Kennedy airport in New York, U.S. customs agents frequently confiscate bushmeat that African nationals have brought into the country for friends and relatives. Some of these wild species are endangered or already protected, thanks to local laws and international treaties; but one of the best ways to reduce the probability of zoonotic outbreaks is through public health campaigns that emphasize the dangers of harvesting certain animals. In addition, ensuring food security among rural populations will also go a long way toward reducing reliance on bushmeat.

Malaria Control

Most attempts to control malaria, primarily by eradicating mosquitoes, have not been sustainable. Health workers have long recognized that environmental changes can impact mosquito populations and the spread of malaria. Dams and irrigation projects provide breeding grounds for mosquitoes and reducing standing water is critical to limiting their populations. Attacking the disease requires a two-pronged approach, focusing on both the environmental factors that allow mosquitoes and malaria to thrive, and a way to prevent it from being transmitted to humans.

One of the most promising approaches to reducing rates of malaria infection, especially of young children and pregnant women, combines insecticide-treated

mosquito nets with indoor spraying. Mosquito nets treated with pyrethroid insecticides, which are natural plant products and biodegradable, are hung over beds to protect people while they sleep (**Figure 11.19a**). Spraying the interior of houses with one of several insecticides, including pyrethroids, further reduces the rate of malaria infection (**Figure 11.19b**). As shown in **Figure 11.19c**, combining indoor spraying with insecticide-treated nets reduced rates of malaria infection by over 50% in areas with a medium risk of transmission and over 30% in high-risk transmission areas.

The research on the effectiveness of insecticide-treated nets is a cooperative project between the London School

MALARIA CONTROL

a. Family under treated mosquito netting, São Tomé, Africa

b. Indoor spraying for mosquitoes, Zambia, Africa

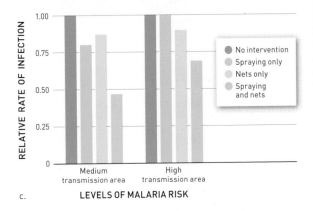

c. **LEVELS OF MALARIA RISK**

(Legend: No intervention, Spraying only, Nets only, Spraying and nets)

Y-axis: RELATIVE RATE OF INFECTION (0, 0.25, 0.50, 0.75, 1.00)
X-axis: Medium transmission area, High transmission area

How would improving living conditions around the world reduce mortality from infectious diseases?

Both historic attempts to control malaria by widespread spraying of mosquito populations and using protective mosquito nets have employed pesticides. What are the differences between the two approaches?

FIGURE 11.19 Two of the most effective ways to reduce the rate of malaria infection have been sleeping under insect nets treated with long-lasting insecticide proven safe for humans and spraying the interior of the home with other long-lasting insecticides. Studies in 17 countries in sub-Saharan Africa showed that combining the use of insecticide-treated mosquito nets with the spraying of the houses' interiors reduced rates of malaria infection in children more than did either using the nets alone or spraying alone. (Data from Fullman et al., 2013)

CHANGING MALARIA INCIDENCE AND DEATHS IN AFRICA

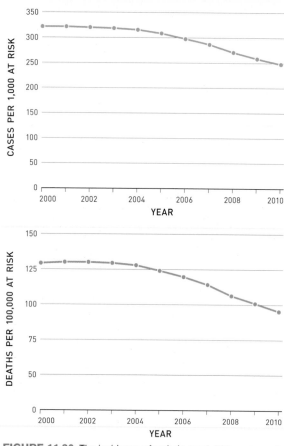

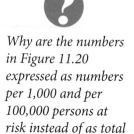

Why are the numbers in Figure 11.20 expressed as numbers per 1,000 and per 100,000 persons at risk instead of as total numbers?

FIGURE 11.20 The incidence of malaria per 1,000 persons at risk of infection decreased by 23% between 2000 and 2010. Over the same interval, deaths from malaria decreased by 33%. (Data from WHO, 2012)

of Hygiene and Tropical Medicine (LSHTM), a leader in research on tropical diseases for over a century, and the WHO. This project and others developed by LSHTM have received extensive funding by the Bill & Melinda Gates Foundation, a leading donor to malaria research and response. Such cooperation is a crucial part of finding a solution to any complex, large-scale environmental health issue. In the case of malaria in Africa, the results of such cooperation have been significant. As insecticide-treated nets and indoor spraying techniques have been widely applied, the number of malaria cases and deaths from the disease has decreased substantially across sub-Saharan Africa (**Figure 11.20**).

Tuberculosis

Tuberculosis (TB) is a bacterial infection that usually attacks the lungs. Cases of TB occur throughout the world, but it is most prevalent in sub-Saharan Africa and Central, South, and Southeast Asia. Controlling the disease has become a major challenge to public health systems worldwide. People at greatest risk of contracting TB are the elderly, infants, and those with weakened immune systems. The risk of contracting TB increases when individuals are in frequent contact with someone already infected, have poor nutrition, or live in crowded or unsanitary conditions (**Figure 11.21**). This is part of what we mean when we talk about an individual's cultural and social environment. Stamping out TB requires changing that environment.

One program receiving one of the widest applications is the directly observed treatment system, or DOTS, of the WHO. The centerpiece of the DOTS therapy

SANITATION: A KEY TO PREVENTING MANY DISEASES

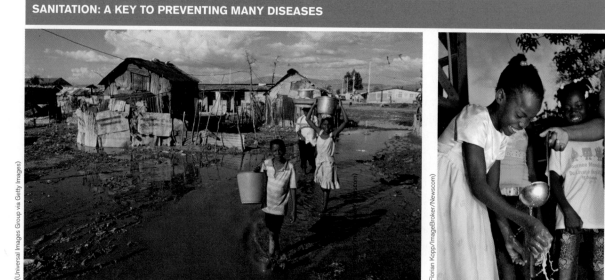

FIGURE 11.21 One of the most cost effective ways to prevent disease, including tuberculosis, is for a community to sustain sanitary living conditions for people, including good air quality, proper sewage treatment, and safe drinking water.

Children walking barefoot through a flooded dirt lane, Haiti

Hygiene education, Haiti

is that patients diagnosed with TB take the treatment while being directly observed by trained personnel over the 6 to 8 months generally required for effective treatment. Direct observation ensures that all treatments are administered, thereby maximizing the chance for a successful cure while minimizing the risk of incomplete treatment, which runs the risk of producing antibiotic-resistant bacteria.

The DOTS system of treating TB combines the advantage of low expense with effectiveness. As a result, estimates by the WHO indicate that DOTS treatment prevented approximately 20 million deaths between 1995 and 2013. However, the DOTS treatment is effective only against strains of TB that are not drug resistant. Treating cases of multi-drug-resistant TB (MDR TB) requires up to 2 years and is much more expensive. Treating rare cases of extensively drug-resistant TB (XDR TB) are even more difficult and costly.

A broad partnership of organizations has been formed to combat these much more difficult forms of TB on a global scale. The partnership involves several governmental and nongovernmental organizations, including (again) the Bill & Melinda Gates Foundation, which has been one of the largest nongovernmental contributors to programs aimed at combating TB. For example, the foundation donated over $112 million in 2011 to support development of better tools for addressing the global TB epidemic, including shorter and simpler treatments, new and improved vaccines, and better diagnostic tools. The foundation works as an advocate for wider access and reduced TB treatment costs around the world.

⚠️ Think About It

1. How might cooperation between private and governmental organizations make work on disease prevention and control logistically easier?

2. What are some factors that make addressing a biological hazard, such as malaria, fundamentally different from addressing a chemical hazard, such as lead?

3. How might focusing on houses and sleeping areas for mosquito control reduce selection on mosquito populations for pesticide resistance?

11.9 Evolutionary biology can help manage antibiotic and insecticide resistance

The pathogens and insect disease vectors, such as mosquitoes, that we are trying to control evolve in response to selection we create by using antibiotics or insecticides. Adopting common-sense measures to limit infections and infestations, and to reduce the chances of resistance genes increasing in frequency offers the best hope of a long-term solution.

Prevention and Appropriate Treatment

Because the likelihood increases that bacteria will evolve resistance to antibiotics each time we use such treatment, it pays to try to prevent infections in the first place. For example, good hygiene in medical care facilities and in the home can reduce the spread of infectious disease. Reducing the rate of infection decreases the need for treating patients. Having to treat fewer patients lowers selection pressure on pathogenic bacteria for antibiotic resistance. When treatment with antibiotics is necessary, appropriate treatment is essential. Antibiotics should be used only when they can benefit the patient and should target a specific pathogen. In addition, the antibiotic must be taken in the appropriate dosage and for the full length of time prescribed for effective treatment.

Lower Antibiotic Use in Livestock Production

Reducing the use of antibiotics in livestock production can also lower the prevalence of antibiotic resistance among bacteria in meat products and animal waste without compromising the welfare of the industry. For example, the amount of antibiotics given during the production of a kilogram of meat in Denmark, the world's leading exporter of pork, is one-sixth that given in the United States. Concerned about the growing problem of antibiotic resistance, Denmark banned the use of antibiotics as growth promoters for livestock in 1998. Since then, the antibiotics used in Denmark's highly industrialized livestock production systems can only be administered to treat illness and upon prescription by a licensed veterinarian. As a result, the amount of antibiotics administered to livestock in Denmark has been cut in half.

To combat disease, Danish livestock producers now rely less on antibiotics and more on preventative hygiene—for example, more frequent cleaning of pens, better ventilation, and reduced crowding. Adjustments such as these have increased the costs of production by less than 1% and the meat-producing industry in Denmark continues to grow. Meanwhile, the occurrence of antibiotic-resistant bacteria in livestock and meat products in Denmark has decreased dramatically (**Figure 11.22** on the next page).

Surveillance Is Essential for Managing Resistance

Surveillance, or monitoring, of pest and pathogen populations aimed at detecting the appearance and prevalence of resistance is essential to any program aimed

?

Why might some agricultural interests oppose changing the use of antibiotics in meat production in the United States?

RESISTANCE DECLINE IN BACTERIA FROM PIGS

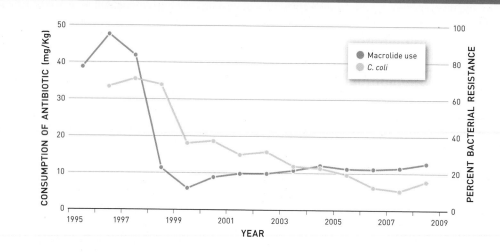
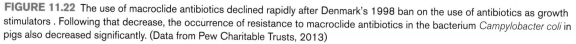

FIGURE 11.22 The use of macroclide antibiotics declined rapidly after Denmark's 1998 ban on the use of antibiotics as growth stimulators . Following that decrease, the occurrence of resistance to macroclide antibiotics in the bacterium *Campylobacter coli* in pigs also decreased significantly. (Data from Pew Charitable Trusts, 2013)

at limiting antibiotic or insecticide resistance. As we saw in Chapter 7, monitoring of pest populations is critical to effective Integrated Pest Management in agricultural settings (see page 221), and it is just as important to controlling disease vectors. A surveillance program that detects the appearance of resistance early permits a timely and effective response. An early response, such as switching to an alternative antibiotic or pesticide, is essential to prevent resistance from becoming too frequent in pest and pathogen populations.

Diversifying Chemical Treatments

Developing a diversity of antibiotics and insecticides with different modes of action is crucial for combating resistance. In their attempts to control malaria vectors, researchers have identified several new insecticides that are highly effective against populations of malaria-transmitting mosquitoes. Of particular interest are those new insecticides that are effective at controlling mosquito populations now resistant to pyrethroids and DDT. For example, controlled studies in Benin, Africa, conducted by researchers from the London School of Hygiene and Tropical Medicine and local scientists have demonstrated the effectiveness of an inexpensive, long-acting substitute for DDT and pyrethroid insecticides for indoor spraying (**Figure 11.23**). This insecticide, chlorpyrifos methyl, which was developed by Dow AgroSciences, also has

What role can the business community play in efforts to control malaria in Africa and elsewhere?

low toxicity to mammals and is rated as safe for indoor spraying. As we continue our quest for sustainable solutions, partnerships between research scientists, environmental health specialists, and business will be essential.

MORTALITY OF MOSQUITOES EXPOSED TO DIFFERENT HUT-SPRAYING TREATMENTS

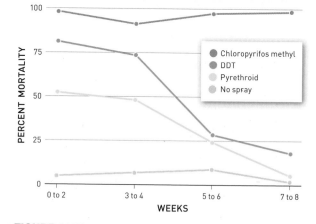

FIGURE 11.23 A new insecticide, Chlorpyrifos methyl, shows promise for reducing malaria transmission by a population of the mosquito *Anopheles gambiae* known to be resistant to pyrethroid insecticides and DDT. (Data from N'Guessan et al., 2010)

⚠ Think About It

1. Why should there be an evolutionary cost to antibiotic or insecticide resistance?

2. Is there likely to be any chemical defense against pathogenic bacteria or insect vectors of disease that will not eventually lead to resistance in the target population? Why or why not?

3. There are significant benefits to using large amounts of antibiotics in livestock production. What are the risks to people and the environment?

11.7–11.9 Solutions: Summary

A risk is anything with the potential to cause us harm or loss or put us in danger. Risk management practices include four basic steps: identification of the hazard, dose–response assessment, exposure assessment, and risk characterization. Based on risk assessment, the EPA phased out lead in gasoline and banned or limited lead used in consumer products such as residential paint.

The precautionary principle can be used as a tool for making better health and environmental decisions. It aims to prevent harm from the outset rather than manage it after the fact. Modern approaches to disease control go beyond vaccines and medicines to include cooperation among health organizations and governments, innovative research, education, and awareness of culture and values. The health risks associated with the bushmeat trade may be reduced by public health campaigns that emphasize the dangers of harvesting certain animals. Insecticide-treated mosquito nets and indoor spraying are being used to reduce the rates of malaria infection, especially among children and pregnant women. The principles of evolutionary biology can contribute to more sustainable control of pathogens and pests.

Answer the following questions for each chapter section and then answer the Central Question.

Central Question: What is the relationship between the environment and human health and how can we manage that relationship?

11.1–11.2 Science

- What chemical hazards are present in the environment and what are their effects?

- How are pathogenic bacteria, viruses, and parasites spread through the environment?

11.3–11.6 Issues

- How do bioaccumulation and biomagnification affect the concentration of toxic substances?

- How do endocrine disruptors affect the health of humans and other organisms?

- How has resistance to antibiotics and insecticides developed?

- What are the challenges to controlling malaria?

Environmental Health, Risk, Toxicology, and You

Environmental hazards can seem overwhelming because they come from all directions, and they can change or simply overwhelm us. In the face of such a challenge, what can you do to make a difference?

☐ *Keep up to date.*

Stay current with developing issues by following news on the WHO website and other global online sources. Keep abreast of local and federal legislation that addresses health issues, for example, the FDA's requirements for the testing of bottled water.

☐ *Apply your knowledge.*

Practice good hygiene—20 seconds of hand washing with soap is more effective at removing infectious bacteria and viruses from your skin than sanitizers. Sustain healthy eating habits and take other preventative measures to avoid contracting infectious disease. Get an annual influenza vaccination. Take any antibiotics that might be prescribed by your physician for bacterial infections responsibly; finish the entire prescription and do not share your pills.

☐ *Stay involved.*

Consider supporting or working with organizations that are concerned about the most pressing environmental health issues around the world. Organize neighborhood water clean-ups to remove any containers or clogged gutters that might harbor mosquito populations.

11.7–11.9 Solutions

- How do we assess risk?

- What does risk management involve?

- How can evolutionary biology help manage antibiotic and insecticide resistance?

Answer the Central Question:

_____

_____

_____

_____

_____

_____

Chapter 11

Review Questions

1. What is environmental health?
a. Environmental health refers to the level of biodiversity in an ecosystem.
b. Environmental health is determined by how close the actual primary production in an ecosystem is to its potential.
c. Environmental health is a field of study concerned with how environmental factors affect human health and safety.
d. Environmental health refers to how comfortable people feel in an environment.

2. Which of the following interactions between two toxic substances results in a level of toxicity much greater than the sum of their individual toxicities?
a. Antagonistic effects   c. Additive effects
b. Synergystic effects   d. Hyper effects

3. Approximately what percentage of the cells in your body are human?
a. 100%   c. 50%
b. 75%   d. 10%

4. Which of the following classes of toxic substances are most likely to show bioaccumulation?
a. Water-soluble toxic substances
b. The various types of alcohols
c. The metallic form of mercury
d. Methyl mercury, an organic form of mercury

5. Why are endocrine disruptors an especially serious class of toxic substances?
a. They affect a wide range of organisms.
b. They occur in so many common products.
c. They can alter sexual development.
d. All of the above

6. Which of the following conditions contributes to the evolution of antibiotic resistance in bacteria populations?
a. Small population size where mutations are more likely to occur
b. Infrequent exposure to antibiotics, which allows resistance to build up in the population
c. Patients always completing a full course of treatment with antibiotics
d. Frequent exposure to antibiotics, which exert strong natural selection favoring resistance

7. What is the meaning of an LD_{50} in a dose–response test?
a. An LD_{50} is half the lethal dose of a toxic substance.
b. An LD_{50} is the dose that is lethal to the population after 50 hours.
c. An LD_{50} is the dose that kills half the test population.
d. An LD_{50} is the percentage of the population killed by a lethal dose.

8. Why are access to sanitation and clean water serious environmental issues?
a. Sanitation and clean water are engineering problems, not environmental health issues.
b. Access to sanitation and clean water are social problems, not environmental health issues.
c. Access to sanitation and clean water are not environmental health issues because those basic necessities were addressed long ago.
d. Millions of people die each year because their environment does not provide access to sanitation and clean water.

9. Why is treating and controlling the spread of tuberculosis (TB) as much a social problem as it is a medical problem?
a. Poor nutrition and crowded living conditions increase the likelihood of contracting TB.
b. Ensuring completion of treatment often requires careful guidance by caregivers.

c. Contact with individuals infected with TB is a common way to contract the disease.
d. All of the above

10. How much has decreased use of antibiotics and more attention to prevention of disease increased the cost of livestock production in Denmark?
a. 1%
b. 5%
c. 15%
d. 25%

Critical Analysis

1. What basis is there for considering individual humans as ecosystems or complex ecological communities? How might these ecological concepts contribute to managing human health?

2. How are DDT and mercury pollution similar? How are they different?

3. How should the precautionary principle be applied to the release of chemicals into the environment? Include detailed conditions for invoking the precautionary principle.

4. Outline and explain in detail the elements of a chemical risk assessment.

5. Discuss how the health of individuals is largely determined by the interaction between their genome, their genetic makeup, and their exposome.

Find additional resources and links online at www. macmillanhighered.com/launchpad/molles1e.

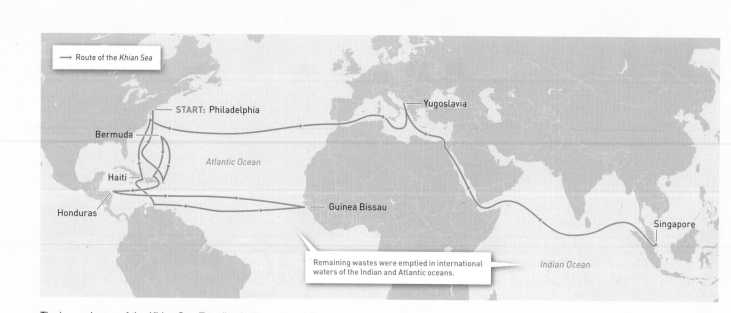

The long odyssey of the *Khian Sea*. Traveling halfway around the world looking for a place to dump its cargo of waste, the ship left a trail of deception and refuse that spanned three seas and two oceans.

Philadelphia's Traveling Trash

One city's refuse changed the way the world deals with garbage.

In the 1970s, Philadelphia had a serious trash problem. Its dumps were filled to maximum capacity and the city started sending its paint cans, car tires, watermelon rinds, and disposable diapers to New Jersey and other neighboring states. After New Jersey stopped accepting such waste in 1984, city officials were paying to have their trash hauled to places as far away as Houston, Texas. To reduce soaring costs, Philadelphia started burning its waste. Though incineration decreases waste volume by 70% or more, it leaves behind ash laden with toxic agents that must be safely stored somewhere. So Philadelphia came up with a plan: In 1986 a barge named the *Khian Sea* was piled with 15,000 tons of Philadelphia's incinerated trash. *Khian Sea* would carry the ash to the Bahamas, where it would be dumped on a man-made island. When the Bahaman government got wind of the plan, however, it turned the ship away.

Next, the *Khian Sea* tried the Dominican Republic. No luck there, either. The ship was also turned away by Honduras, Panama, Bermuda, Guinea Bissau, and the Dutch Antilles. In Haiti, the crew unloaded some 4,000 tons of the ash, claiming it was "topsoil fertilizer." By the time the Haitian government was alerted to the true nature of the material, the barge had departed.

The *Khian Sea* sailed onward, crossing the Atlantic, passing through the Suez Canal, and drifting into Southeast Asia. Despite changing the name and registration of the ship twice in order to conceal its identity, no country would allow the foul ash to be unloaded onto its shores. After a final attempt at Singapore, the barge reversed course, crossed the Indian Ocean, and dumped its toxic load into international waters—16 months after first setting sail from Philadelphia. Meanwhile, the ash left in Haiti festered there for a dozen years before finally being shipped to Florida, where it sat on a barge for two more years until at long last, it was tested for carcinogens. The waste was deemed safe and then sent back to Pennsylvania for burial.

> "There is no 'waste' in nature and no 'away' to which things can be thrown."
>
> Barry Commoner, second law of ecology

The global reaction to the *Khian Sea* incident and others would contribute to the development of an international treaty known as the Basel Convention, discussed later in this chapter. The bottom line is that the problem of how to safely handle refuse touches every person and every community on Earth. Shrinking landfill space, rising waste disposal costs, incinerating waste, and exporting hazardous waste are challenges we still face today.

Prior to the environmental movement of the 1970s, relatively little regulation existed regarding the disposal and treatment of solid waste. Dump sites created before environmental legislation might be undocumented and contain hazardous waste, which might leach into the soil. Moving into the future, we need to recycle as much as we can and limit the amount of nonrecyclable waste produced, while ensuring that it is treated and contained safely. These goals lead to the central question of this chapter.

Central Question

How can we reduce the environmental impact of solid waste and dispose of hazardous waste safely?

(USFWS photo by Susan White)

12.1–12.2 Science

How do modern garbage heaps differ from those left by ancient people, such as that at Emeryville?

Just north of Oakland, California, lies the Emeryville Shellmound, a pile of discarded clam, mussel, and oyster shells that was once over 60 feet high and 350 feet wide. These are the remnants of more than 2,000 years of habitation by Native Americans. Around the world, it was common to toss out waste in open dumps, and everywhere that humans lived, their trash accumulated. However, as the human population grew and the intensity of economic activity increased, we littered Earth with more diverse types of waste, including toxic waste. In

some places, where open dumps are still in use, waste deposits have reached the size of small mountains (**Figure 12.1**). These growing accumulations on land and sea loom as one of the great environmental challenges. By understanding the way natural ecosystems generate and recycle their own waste, we can find models for how societies might address this challenge.

12.1 The "waste" generated by economic systems does not occur in ecosystems

In the early days of human societies, it didn't matter much if people just tossed waste out their back windows. They produced little of it and their settlements were small. Critically, most of the waste consisted of natural materials—which simply decomposed over time due to the activities of scavenging animals, plants, and microorganisms. The chemical elements contained in those wastes were eventually recycled by the ecosystem. If one patch of land became too polluted, communities could always just pick up and move.

With explosive human population growth and development of the modern industrial society, however, the quantity, types, and sources of wastes produced by humans have changed radically (**Figure 12.2**). Today, much of human waste no longer cycles freely and rapidly but too often meets a dead end—buried in a landfill where it cannot be decomposed easily, preventing the elements it contains from re-entering the biosphere. Many human-made chemicals, such as plastics, have no natural cycling pathway, or are manufactured in such abundance that those pathways cannot degrade them quickly enough.

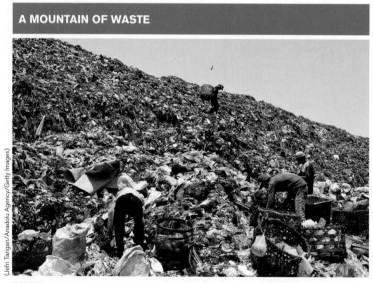

A MOUNTAIN OF WASTE

(Jefri Tarigan/Anadolu Agency/Getty Images)

FIGURE 12.1 The amount of waste generated by the billions of humans living today is astounding. Solid waste from cities, called *municipal solid waste,* is especially apparent because vast quantities can accumulate in waste dumps. In several developing countries, waste dumps attract many people searching for usable resources.

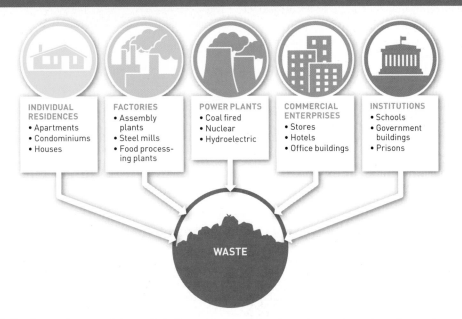

INDIVIDUAL RESIDENCES
• Apartments
• Condominiums
• Houses

FACTORIES
• Assembly plants
• Steel mills
• Food processing plants

POWER PLANTS
• Coal fired
• Nuclear
• Hydroelectric

COMMERCIAL ENTERPRISES
• Stores
• Hotels
• Office buildings

INSTITUTIONS
• Schools
• Government buildings
• Prisons

WASTE

FIGURE 12.2 Nearly every human activity results in the production of some type of waste. In modern urban environments, the numerous sources of waste range from individual residences to factories, power plants, commercial enterprises (e.g., stores and hotels), as well as institutions such as schools, hospitals, and nursing homes.

A first step toward efficiently recycling human-generated waste is to understand the cycling of materials in nature. The law of conservation of matter (see Chapter 2, page 42) states that matter in a closed system cannot be created or destroyed, only changed in form. As a result, the chemical elements of which all matter is composed can cycle indefinitely in ecosystems. For example, the carbon, nitrogen, phosphorus, and sulfur cycles (described in Chapters 2, 7, 8, and 13) trace the movement of these elements through ecosystems. Similar cycling occurs for all the other elements, like calcium and iron, found in living things.

Every living organism can be reduced to its building blocks, making them available for other organisms. Every resource made by nature returns to nature—decomposed by microbes, plants, and animals. Even crude oil will degrade under the right conditions. And, at the very largest scale, the Earth system recycles the minerals that make up rocks, the most basic geologic component of the planet (see Appendix B).

⚠ **Think About It**

1. What is the eventual fate of elements found within dead plants, animal waste, and other matter?

2. Earth is a closed system to nearly all forms of matter. How, then, has the biosphere been able to support countless life forms across millions of years?

3. What basic steps might be taken to make our wastes better able to enter Earth's natural cycles? (Hint: Consider the carbon cycle in Figure 2.13, page 43, as a model.)

12.2 Waste has diverse sources and properties and varies with level of economic development

Archeologists can infer a great deal about past societies by studying waste dumps, such as the Emeryville Shellmound. In the modern world, waste comes from many sources and is made up of countless forms of discarded material. This mixture is called the **waste stream.** What would some future archeologist conclude about our contemporary world from studying our waste stream? Two conclusions would be inescapable: Rich countries produce a lot more waste than poor countries, and the types of things they each throw out differs greatly.

Municipal Solid Waste

While people everywhere generate solid waste, the amount of waste generated is particularly large in cities because urban populations are particularly concentrated. **Municipal solid waste (MSW)** includes all solid waste from institutions, households, and businesses, including

waste stream The flow of discarded materials, especially municipal solid waste, from institutions, homes, and businesses.

municipal solid waste (MSW) Solid waste from institutions, households, and businesses, including paper, packaging, food scraps, glass, metal, textiles and other solid discards.

paper, packaging, food scraps, glass, metal, textiles, and all other solid discards. Consequently, waste management is one of the most critical services that cities provide. In low-income regions, waste management is the most expensive item in municipal budgets. It's so fundamental to life in the city that most people don't even think about it—at least, not until the system fails. For example, in 2011, garbage collectors in Athens, Greece, went on strike to protest potential tax hikes and salary cuts, allowing trash to pile up on curbsides for 17 days. Strikes by garbage collectors have had similar consequences in many places around the world (**Figure 12.3**).

Because municipal solid waste is a product of economic activity, income level is correlated with the amount and composition of waste generated by a population. As shown in **Figure 12.4**, the per capita production of MSW in the world's richest countries is triple that of the poorest countries.

There are many different types of solid waste, which must be handled differently. The World Bank classifies MSW into six categories: organic, paper, plastic, glass, metal, and other (**Table 12.1**).

Under the World Bank's scheme, organic waste includes food scraps, yard clippings, and wood. Paper and plastics are also chemically organic, but paper is highly processed organic material and plastics are mostly synthesized from petroleum. The "other" category includes processed organic material as well, including leather and rubber, but also appliances, electronic waste, and ash.

hazardous waste A flammable, reactive, corrosive, or toxic waste capable of causing illness, death, or other harm to humans and other organisms.

flammable Easily ignited; a flammable substance can ignite and burn easily (e.g., from friction, absorption of moisture, or contact with other waste materials).

reactive Chemically responsive; a reactive substance will readily undergo a violent chemical change when in contact with other substances.

corrosive Capable of causing permanent damage to a variety of surfaces, including living tissue; corrosive substances include strong acids (pH of 2 or less) or strong bases (pH of 12 or greater).

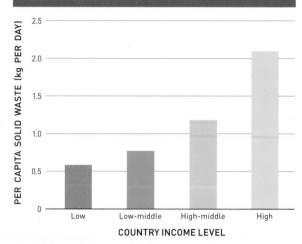

MUNICIPAL SOLID WASTE GENERATION AND INCOME LEVEL

FIGURE 12.4 On average, municipal populations in high-income countries produce more than 3 times the amount of solid waste per capita than do individuals in low-income countries. (Data from Hoornweg and Bhada-Tata, 2012)

In poor countries, organic wastes such as food, wood, and garden waste make up a full 64% of municipal solid wastes. By contrast, such items make up only 30% of waste in cities of wealthy countries. Meanwhile, the proportions of paper, glass, and metal in solid waste increase with income level (**Figure 12.5**). We can assume that as poor nations develop economically, the generation of these types of wastes will increase.

Properties of Hazardous Waste

Some wastes are so dangerous and toxic that they cannot be disposed of with conventional means. The U.S. Environmental Protection Agency (EPA) defines **hazardous waste** as having a "chemical composition or other property that makes it capable of causing illness, death, or some other harm to humans and other life forms when mismanaged or released into the environment." More specifically, waste material that has any of the following four properties is considered hazardous:

1. Flammable: Substances that ignite and burn quickly and easily. These can spontaneously ignite from friction, absorption of moisture, or contact with other waste materials.

2. Reactive: Unstable substances that readily undergo a violent chemical change when in contact with other substances, especially with water.

3. Corrosive: Any strong acids (pH of 2 or less) or strong bases (pH of 12 or greater). These substances can permanently damage a variety of surfaces, including living tissue.

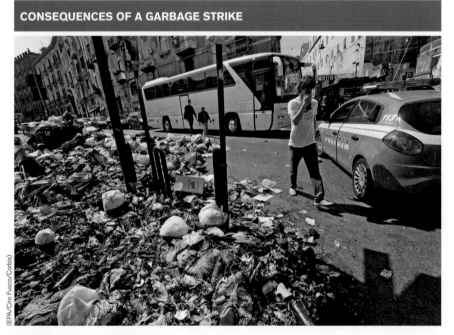

CONSEQUENCES OF A GARBAGE STRIKE

(EPA/Ciro Fusco/Corbis)

FIGURE 12.3 The vast amounts of waste produced in a large city and the essential service represented by waste management are never more evident than during a garbage strike. When negotiations between Naples, Italy, and waste collectors broke down in 2011, solid waste began to accumulate in the streets, creating a health hazard, filling the city with noxious smells, and partially blocking foot and automobile traffic.

TABLE 12.1
CATEGORIES OF MUNICIPAL SOLID WASTE

The World Bank has made a thorough inventory of the composition and amount of solid waste generated by populations around the world.

Waste category	Examples
Organic	Food waste, garden and yard waste, wood
Paper	Office paper, cardboard, newspaper
Plastic	Packaging, bottles, food containers, bags
Glass	Bottles, colored glass, glassware
Metal	Cans, household appliances, foil
Other	Leather, rubber, ash, electronic waste

Information from Hoorweg and Bhada-Tata, 2012.

4. Toxic: Relatively low amounts of these substances are harmful to living organisms.

Hazardous wastes cannot be handled, treated, or stored in the same way as other categories of waste. The risk of an explosive chemical reaction, harm to workers, or damage to lining of waste containers or landfills is simply too great. (Hazardous chemicals are also discussed in Chapter 11, as an aspect of environmental health, and in Chapter 13, as pollutants.)

Sources of Hazardous Waste

In the United States, the "basic chemical" industry, which produces everything from nail polish remover to chlorine for swimming pools, is responsible for over half of all hazardous waste (**Figure 12.6**). In second place is the production of petroleum- and coal-based products, such as gasoline, plastics, and lubricants. Taken together, five industry sectors accounted for 94% of the approximately 31 million metric tons of hazardous wastes produced in the United States in 2011. The remaining 6% of hazardous wastes produced in the United States comes from economic activities by 45 commercial and industrial sources, ranging from paint manufacturing to sawmills.

While most household wastes aren't dangerous, there are some hazardous items commonly found in the home that individuals should never simply place in a trash can or pour down a drain, on the ground, or into storm sewers. For example, the following household wastes meet at least one of the EPA's hazardous waste criteria:

- Medical waste—expired prescriptions, discarded needles, bandages (toxic)
- Chemical drain openers (corrosive)
- Paint thinners (flammable)
- Pesticides (toxic)

INCOME LEVEL AND MUNICIPAL SOLID WASTE COMPOSITION

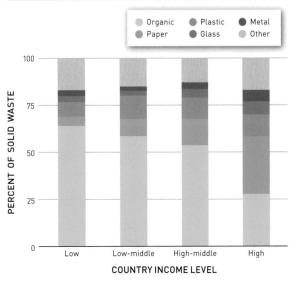

FIGURE 12.5 Relative decreases in organic waste and increases in paper in the waste stream appear to be the main changes in waste composition from low-income to high-income populations. (Data from Hoorweg and Bhada-Tata, 2012)

- Used motor oil (toxic, corrosive, flammable)
- Mercury-containing products—old thermometers, fluorescent light bulbs (toxic)
- Antifreeze (toxic)
- Batteries (reactive, flammable)

Because wealthier countries tend to have greater capacity for manufacturing chemicals than do poorer countries, they tend to produce more hazardous

SOURCES OF HAZARDOUS WASTE IN THE UNITED STATES

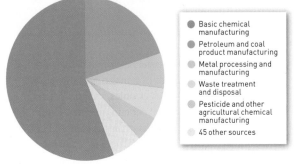

FIGURE 12.6 Five economic activities account for 94% of the hazardous waste generated in the United States. Just two of these sectors are responsible for over three-quarters of the hazardous waste produced in the country. (Data from Environmental Protection Agency [EPA], 2011c)

Does an increase in disposable income lead inevitably to increased per capita waste production?

toxic Poisonous; a toxic substance is harmful to living organisms in relatively low amounts.

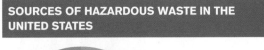

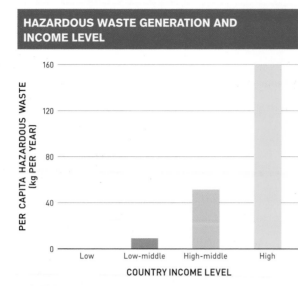

FIGURE 12.7 Hazardous waste generation is much higher in high-income countries than in high-middle-income and low-middle-income countries. Insufficient data were available for low-income countries. (Data for 2006 from Wielenga, 2010)

Does disposal of household wastes have potential ethical implications?

waste. As shown in Figure 12.7, the average per capita production of hazardous waste in high-income countries is approximately 3 times that in high middle-income countries and over 16 times the rate in low middle-income countries.

⚠ Think About It

1. What might future archeologists conclude about today's human societies by studying our MSW?

2. How might income level influence the composition of waste in different countries? How do differences in income level among cities affect waste management issues?

3. While federal law requires industry to safely dispose of hazardous waste, individuals and households are not held to the same standards. Do you believe the EPA should extend these laws to include all waste producers? Explain why or why not.

12.1–12.2 Science: Summary

As long as human populations were small, the volume of waste they produced was low, and the waste stream consisted of natural materials. The situation changed with population growth and the development of intensive industrial economies. Municipal solid waste includes all solid waste from institutions, households, and businesses, making solid waste management one of the most critical and expensive public services. National income is correlated with the amount and composition of waste. The EPA considers waste that is flammable, reactive, corrosive, or toxic as hazardous. In the United States, the largest sources of hazardous waste are basic chemical manufacturing and the manufacturing of products from petroleum and coal, such as gasoline, plastics, and lubricants. Significant amounts of hazardous waste can also be generated in households.

12.3–12.6 Issues

In the Chinese town of Guiyu, hazardous waste piles up on the streets as hundreds of thousands of untrained workers dismantle cast-off electronics, including flat-screen television displays that contain mercury. Out in the Pacific Ocean, so many tiny bits of plastic bags and bottles have been trapped in a circular ocean current that the water appears cloudy. At nuclear power plants around the United States, spent fuel rods pile up with no permanent repository open to store them. Every type of waste brings its own challenges, and as the population grows, these issues grow more severe with each passing year.

12.3 Municipal solid waste management is a growing problem

No one likes to take out the trash from his or her own house. But that's just the beginning of the journey for our waste. Consider, for a moment, the epic task of gathering trash from each and every household in a city of several million and finding a place to put it. Waste collection is a critically important task because uncollected solid waste can block drainage systems and cause air and water

INCREASING GENERATION OF MUNICIPAL SOLID WASTE

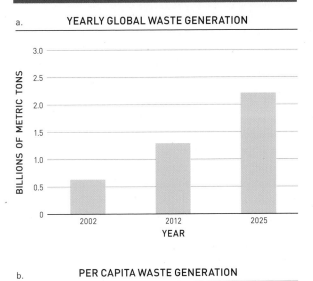

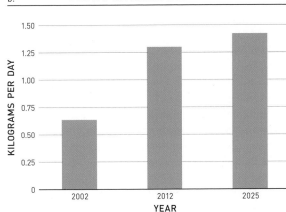

FIGURE 12.8 (a) As a result of increasing per capita waste generation and growth in the urban population, the total amount of MSW will likely triple between 2002 and 2025. (b) The per capita generation of MSW around the world approximately doubled between 2002 and 2012 and is predicted to increase another 20% by 2025. (Data from Hoornweg and Bhada-Tata, 2012)

pollution. Waste disposal is yet another issue. What happens in the long term to all that buried municipal waste? Does it ever decompose? Does any of the waste material find its way back into the soil and water systems of the nearby environment? And if so, is it toxic and therefore hazardous?

Municipal Solid Waste

As the odyssey of the *Khian Sea* showed, getting rid of waste can be a fraught process. Around the globe, the amount of municipal solid waste generated is soaring. As of 2002, just over 600 million metric tons of MSW was being generated per year; and in the following 10 years, that number doubled to 1.3 billion metric tons.

By 2025 that total is expected to increase even further to approximately 2.2 billion metric tons of MSW per year (**Figure 12.8a**).

These increases represent not only the growth of urban populations around the world, but also increased economic development. As disposable incomes and living standards increase, consumption correspondingly increases. More consumption equals more trash. Between 2002 and 2012, the amount of waste an individual produced doubled from approximately 0.6 kilogram to 1.2 kilograms per day. These numbers are predicted to rise by nearly 20% by 2025 (**Figure 12.8b**). When you consider the increase in per capita waste production in combination with urban population growth across the globe, solid waste will have tripled between 2002 and 2025.

In the United States, however, we are beginning to see a slight decline, suggesting we've reached "peak trash" (**Figure 12.9**). Indeed, people today are throwing away less trash than they did 15 years ago. Let's take a closer look at the numbers. In 1960 the United States generated

41 YEARS OF GENERATING MUNICIPAL SOLID WASTE IN THE UNITED STATES

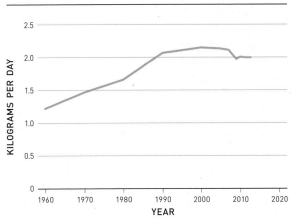

FIGURE 12.9 Total annual waste generation increased from 1960 until 2007, then decreased about 2.5% by 2011 (top). Per capita generation of waste (bottom) peaked in 2000, decreasing a full 7.5% by 2011. (Data from EPA, 2013c)

80 million metric tons of MSW; that number increased to 233 million metric tons by 2007. After 2007 solid waste generation decreased by 1.3% to 230 million metric tons by 2013. What accounted for that decline? From 1960 to 2000, people were becoming more wasteful: The per capita generation rate went from 1.22 kilograms per person per day to 2.15 kilograms per person per day. However, the per capita rate then declined 7.5% over the next decade, down to 2 kilograms per person per day in 2013.

According to World Bank statistics, these per capita rates of solid waste generation are comparable to those of other developed countries—such as Australia, Canada, Norway, and Denmark—but are higher than those of Japan, the United Kingdom, and Sweden. All these nations have a much higher rate of waste generation than the low-income nations (see Figure 12.4, page 362). This decrease may indicate a trend toward waste reduction among companies and consumers, a strategy that we will discuss further in the solutions section of this chapter.

Food Waste

According to the EPA, Americans threw out approximately 35 million tons of food in 2012. That's more than 3 times the amount we were throwing out in 1980, and it represents more than 20% of our waste stream. In fact, we throw away more food than plastic, paper, or metal. The Natural Resources Defense Council, an environmental advocacy organization, has estimated that 40% of our food goes to waste.

You may think that it's no big deal because food and other organic waste are **biodegradable,** meaning complex molecules can be broken down into their simpler elements and compounds. Even though such

waste will decompose in the long term, it is filling up our country's landfills; and as it decomposes due to the actions of bacteria, fungi, and some species of insects, it releases climate-warming methane into the atmosphere. In addition, finite resources, such as the phosphate in fertilizer used to grow the food, are removed from Earth's systems and locked away in landfills.

Plastic Waste

While some of our wastes, such as paper and food scraps, are biodegradable, others are not. All of our products begin as naturally occurring substances, including crude oil, but through manufacturing they undergo processes of heating, molding, coloring, or are chemically changed. The end products, in many cases, are **non-biodegradable,** meaning living organisms cannot break them down. As a consequence, the elemental components (mainly carbon and hydrogen) of these substances will remain locked in this new form indefinitely.

Because plastics are not biodegradable, they accumulate in waste dumps and, more significantly for the environment, in natural ecosystems. The Great Pacific Garbage Patch (**Figure 12.10**) consists of particles of plastic that have concentrated in the central Pacific Ocean due to an oceanic current known as the North Pacific Gyre (see Figure 8.2, page 232). Although the area with a significant amount of plastic waste in the Pacific Ocean is undoubtedly large, according to the U.S. National Oceanographic and Atmospheric Administration (NOAA), there is currently no scientifically sound estimate of the area of this accumulation. Garbage patches have also been found in both the Atlantic and Indian Oceans.

biodegradable A substance that can be decomposed to its chemical constituents by biological processes.

non-biodegradable A substance that cannot be decomposed to its chemical constituents by biological processes.

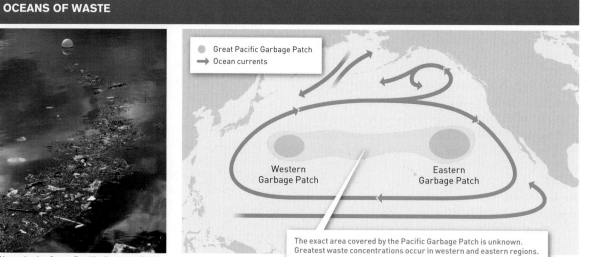

OCEANS OF WASTE

● Great Pacific Garbage Patch
→ Ocean currents

Western Garbage Patch

Eastern Garbage Patch

The exact area covered by the Pacific Garbage Patch is unknown. Greatest waste concentrations occur in western and eastern regions.

Waste in the Great Pacific Garbage Patch

(© Andrew Payne/Alamy)

FIGURE 12.10 The oceans have been accumulating vast quantities of plastic waste that are concentrated by large circular currents called *gyres*. The first of these garbage patches to be discovered, a bit of which is shown here, was the Great Pacific Garbage Patch.

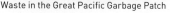

PLASTIC WASTE: DEADLY TO SEA LIFE

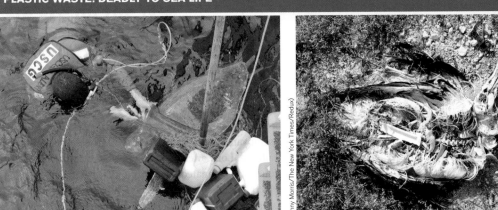

Sea turtle entangled in plastic waste

Albatross remains, found on the Hawaiian Islands, filled with ingested plastic debris

(Coast Guard photo by Petty Officer 3rd Class Nicole Pickio)

(Linny Morris/The New York Times/Redux)

FIGURE 12.11 Larger pieces of plastic waste in the oceans regularly ensnare unsuspecting sea life. This sea turtle (left) was trapped in a bundle of fishing gear off the coast of Central America and would almost certainly have died if not released by this Seaman of the U.S. Coast Guard. The Laysan albatross chick (right), like thousands of others, was fed bits of plastic that its parents mistakenly ingested as they skimmed the ocean surface in search of squid or fish eggs. Unlike the parents, however, the chick cannot regurgitate the plastic; thus, its digestive tract gradually filled with the indigestible material and the chick eventually died.

The dominant waste in all these patches is plastic, ranging in size from large chunks to microscopic particles. Because they are not biodegradable, these plastics do not decompose but rather are broken into smaller and smaller pieces by mechanical action and by sunlight. The larger pieces are hazardous to sea life, including seabirds and turtles that can become entangled and drown. Many marine animals also mistake plastic particles for their natural food, with potentially fatal consequences. A particularly hard-hit population is that of the Laysan albatross of Midway Island, where many chicks die as a consequence of being fed large amounts of plastic debris by their parents (**Figure 12.11**). The U.S. Fish and Wildlife Service estimates that adult Laysan albatross unwittingly feed approximately 5 tons of plastic debris to their chicks each year.

The impact of this plastic waste goes beyond its physical effects. Plastic debris in the oceans allows persistent organic pollutants such as PCBs, polychlorinated biphenyls (see Chapter 13, page 392), to enter the marine food chain. While these pollutants are not very soluble in water, they adsorb on plastic particles and can be released into the tissues of animals ingesting them. Because these compounds biomagnify (see Chapter 11, page 340), their concentrations can be substantial in a variety of popular seafoods.

⚠ **Think About It**

1. Can total waste generation by a population increase over time, while per capita waste generation declines significantly? Explain.

2. Why can we expect the amount of municipal waste generated around the world to continue to increase for the foreseeable future?

3. What does the impact of plastics on Laysan albatrosses tell us about the municipal solid waste stream?

12.4 **Hazardous waste generation is increasing and is often handled unsafely**

Uncontrolled dumping of hazardous waste was once the norm—and it was also completely legal. Today, we must deal with contamination left behind by companies operating during this time. In 1980 the U.S. Congress passed the **Comprehensive Environmental Response, Compensation, and Liability Act (CERCLA),** or Superfund Law, to address this problem. One of the goals of the Superfund program is to identify sites where the contamination of air, soil, and water by hazardous substances has been sufficient to threaten human health or harm the environment. As of 2011, there were more than 1,350 Superfund sites, which the EPA continues to actively manage.

Love Canal

In the spring of 1962, following an extremely wet year, residents living in a new development in Niagara Falls, New York, reported noxious fumes and colored liquids oozing from the ground and flooding roadways and

Whose responsibility is it to clean up the Great Pacific Garbage Patch?

Comprehensive Environmental Response, Compensation, and Liability Act (CERCLA) Superfund Law enacted in 1980 to regulate hazardous waste and require companies to dispose of it safely.

INFORMAL E-WASTE PROCESSING IN GUIYU, CHINA

(Norman Ng/MCT via Getty Images)

FIGURE 12.15 The methods for recovering valuable components from e-waste used by the informal sector in Guiyu, China, and elsewhere create conditions harmful to the health of workers and contaminate the environment with hazardous waste.

How is poverty connected to unsafe recycling of e-wastes?

e-waste business involves labor-intensive and often hazardous manual dismantling of equipment, using simple tools like hammers and screwdrivers and salvaging and selling reusable component parts from circuit boards, compressors, appliances, and other electronic devices. The most lucrative practice is to separate out precious metals and then refine them. However, these cottage e-waste workers lack the technology, equipment, and training to safely manage the process.

Substandard informal recycling practices include open burning or melting of plastics, toner sweeping, dumping of lead-containing CRTs, acid stripping of printed wiring boards, and de-soldering of chips, as well as dumping waste chemicals onto the soil or into water sources. These common practices pose direct risks to the health of workers and to the local environment.

The global center for unsafe recycling practices has been Guiyu, China, a coastal city that imports electronic waste from all over the world (**Figure 12.15**). People from rural areas migrate to Guiyu, seeking jobs disassembling, burning, and melting the precious metals from circuit boards. With a population of 150,000, including 100,000 migrants, Guiyu is home to more than 300 companies and 3,000 individual workshops that are engaged in e-waste recycling. These e-waste workers, many of whom are women and children, earn an average wage equivalent to USD$1.50 per day.

According to the Basel Action Network, the majority (80%) of children in Guiyu have elevated levels of lead in their blood. Workers at these recycling sites face many health hazards. For instance, women working in these conditions run the risk of having children with birth defects, and the child workers are likely to suffer impaired neurological development. Although the Chinese government has banned these informal e-waste recycling practices, the environmental damage they cause will persist for many years and will require substantial effort and resources to mitigate.

⚠ **Think About It**

1. Many electronic items have a label of a trash bin with a line crossing through it. How might disposing of electronic waste with the rest of a household's waste have far-reaching environmental consequences?

2. Why does the problem of e-waste receive a great deal of media attention, while the much larger amount of other hazardous wastes generated goes largely unreported?

12.6 Safe nuclear waste disposal requires long-term security

Nuclear power plants produce two types of radioactive waste that must be disposed of (see Chapter 9). **Low-level nuclear waste** includes any item that has become

low-level nuclear waste
Radioactive waste, including any item that has become contaminated with small amounts of radioactive particles, including instruments, protective suits, or clothing from nuclear facilities.

contaminated with small amounts of radioactive particles, including instruments, protective suits, or clothing from nuclear facilities. **High-level nuclear waste** is radioactive waste consisting primarily of nuclear fuel rods that have been depleted to the point that they can no longer contribute to efficient production of electricity. These fuel rods still contain uranium in addition to several other waste products of nuclear fission.

Over very long periods of time, radioactive waste will completely decay and become harmless. However, most radioactive isotopes take so long to decompose that the timeframe is essentially meaningless in terms of human timescales. Meanwhile, tiny particles and rays of energy are released during active decay, which can damage living tissue, including the DNA blueprint found within the nucleus of all organisms. Safety requires that nuclear wastes be disposed of properly so that they do not contaminate the environment and endanger humans and other organisms.

Technical Challenges

Developing a permanent repository for high-level nuclear waste requires addressing complex technical issues. To prevent radiation from escaping, nuclear waste must be securely stored within several feet of steel and concrete. In the short term, this is a relatively simple task. What makes storing nuclear waste so difficult is the great length of time needed before some unstable elements decay into harmless ones.

Radioactive isotopes all decay, but at different rates. The time needed for half of a given amount of a radioactive isotope to decay is known as its **half-life.** While some nuclear fission waste products decay quickly, in a matter of hours or days, others require thousands of years. The half-life of plutonium-239, for example, is approximately 24,000 years. The durability of even the oldest known human structures, such as the 4,500-year-old pyramids at Giza, pales in comparison. Finding a suitable storage structure and location to accommodate these slow rates of radioactive decay presents a tremendous technical challenge. For the time being, most of the high-level nuclear waste in the United States is temporarily stored in concrete and steel towers called **dry casks** on the site of the reactors themselves, as the country searches for a long-term solution. These casks, while stable, cannot realistically be expected to last for the full length of time needed for the radioactive isotopes to decay to safe levels (**Figure 12.16**).

Social Resistance

One of the biggest challenges to creating a long-term solution for nuclear waste is that no community wants it in their backyard, due to fears of radioactive contamination. The United States has been successful in building and operating a repository for defense-related radioactive wastes at the Waste Isolation Pilot Plant in southern New Mexico. The proposed facility for storing high-level waste from civilian nuclear reactors is located at Yucca Mountain, Nevada, where nuclear wastes would be stored approximately 300 meters below ground and a similar distance above the level of groundwater (**Figure 12.17**). However, plans for the facility were halted in 2009, a year before it was to begin storing waste (see Chapter 9). The main reason for the cancellation was political opposition by local communities, who didn't want nuclear waste transported through their towns.

There were also questions concerning the suitability of the Yucca Mountain geology. Fractures in the geologic formation could provide an avenue for radioactive

ONSITE MANAGEMENT AND STORAGE OF HIGH-LEVEL NUCLEAR WASTE

(Guillaume Souvant/AFP/Getty Images)

(Nuclear Regulatory Commission)

20-meter-deep cooling pool for spent fuel rods at a nuclear power plant

On-site storage casks for spent nuclear fuel at a nuclear power plant

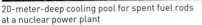

FIGURE 12.16 Following a period of cooling in heavily reinforced and carefully monitored pools of water called *spent fuel pools,* high-level nuclear waste can be moved to onsite dry storage casks constructed of steel and concrete until a permanent repository is available.

high-level nuclear waste Radioactive waste, primarily nuclear fuel rods that have been depleted to the point that they can no longer contribute to the efficient production of electricity.

half-life The time needed for half of a given amount of a radioactive isotope to decay.

dry casks Steel and concrete structures used for temporary storage of nuclear waste.

FIGURE 12.17 After investments of billions of dollars in geologic and engineering studies, construction, and development of detailed handling procedures for high-level nuclear waste, plans for the Yucca Mountain Repository were abandoned in 2009. Discussions concerning opening the site were reopened in the U.S. Congress in 2015, which added uncertainty to the proposed repository. (From U.S. Nuclear Regulatory Commission)

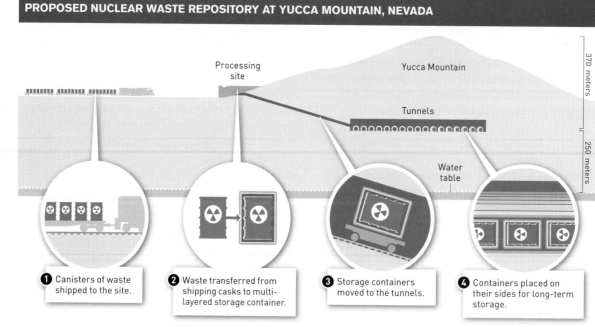

PROPOSED NUCLEAR WASTE REPOSITORY AT YUCCA MOUNTAIN, NEVADA

Processing site

Yucca Mountain

Tunnels

Water table

370 meters

250 meters

1 Canisters of waste shipped to the site.

2 Waste transferred from shipping casks to multi-layered storage container.

3 Storage containers moved to the tunnels.

4 Containers placed on their sides for long-term storage.

What roles should science, economics, and politics play in choosing a repository for high-level nuclear waste?

materials to contaminate groundwater below the storage facility. This would be particularly likely if the regional climate became wetter, which is possible during the million-year time frame that the facility is supposed to remain secure. Despite these concerns, in 2015, the U.S. Congress renewed efforts to open the Yucca Mountain repository, but whether the proposal will be approved remains uncertain. Meanwhile, the country remains without a site for disposing of high-level nuclear waste.

⚠ Think About It

1. Explain what physical properties of nuclear waste preclude its disposal using conventional means.

2. With the many hazardous wastes to which we may be exposed, why does nuclear waste appear to generate an especially high level of concern among the public?

12.3–12.6 Issues: Summary

Around the world, the amount of municipal solid waste generated is increasing with urban population growth and economic development. Food waste and other biodegradable waste take up valuable landfill space and represent an unnecessary and avoidable problem. Non-biodegradable waste also creates significant environmental problems. For example, plastics accumulate in natural ecosystems, such as the Great Pacific Garbage Patch, where they threaten sea life.

Hazardous waste represents a great threat to the environment—contaminating land and water the world over. This type of contamination produced brownfields, abandoned industrial sites contaminated with hazardous waste that are unusable without remediation. The increase in the use of electronic devices has created a new source of waste—e-waste. Finally, radioactive waste poses unique challenges in that it must be sequestered in structures sufficient to shield the environment from harmful radiation. Nuclear waste remains in temporary storage at nuclear power plant sites across the United States and around the world.

12.7–12.10 Solutions

The world was outraged by the Khian Sea saga. In 1993, as the environmental organization Greenpeace held a protest outside the Philadelphia City Hall, the ship's operators were convicted of perjury for lying to a federal grand jury about the dumping. The imbroglio catalyzed the development of an international treaty known as the Basel Convention, which was drawn up in 1989. Known formally as the Basel Convention on the Transboundary Movements of Hazardous Wastes and their Disposal, the treaty's intent was to limit the exportation of hazardous waste, including electronic waste, from developed to developing countries and to encourage the safe treatment and disposal of hazardous wastes within the countries where they were generated. The Basel Convention, which came into force in 1992, now includes 179 countries and the European Union as parties. The United States signed the treaty in 1990 but has not yet ratified it.

Laws and treaties are just one piece of the puzzle needed to find solutions to our growing garbage problem. As global economic development continues, waste generation will continue to increase unless individuals embrace the four R's: Refuse, Reduce, Reuse, and Recycle. Refuse means to refuse unnecessary products and disposable items like plastic bags—and even to refuse the latest technological gadgets. Reduce means to buy less and use less. Reuse means to use things you already own and opt for reusable items rather than disposable ones. Recycle means to sort your waste stream and recycle objects made of paper, plastic, metal, glass, and organic matter for composting. In this section, we discuss how such a simple strategy is being implemented.

12.7 Modern waste management emphasizes reduced disposal

The first step to solving our waste problem is reducing the amount of trash being generated in the first place. The main federal legislation governing the disposal of solid and hazardous wastes in the United States is the **Resource Conservation and Recovery Act,** or **RCRA,** which was passed by the Congress in October 1976. RCRA banned open dumping of wastes and set standards for solid waste landfills. The act recommended reducing the amount of waste and recycling. Under the auspices of RCRA, the EPA developed a management framework called **integrated waste management,** an approach that aims ultimately to minimize the amount of waste

that finds its way to a disposal site. The centerpiece of integrated waste management is a hierarchy of options in which reducing waste and reusing materials are given highest priority, followed by recycling and composting and energy recovery, with disposal as the least preferred option (**Figure 12.18**).

Reducing the Waste Stream

Reducing the amount of material that enters the waste stream is referred to as **source reduction.** As consumers, we can contribute by reusing, repairing, borrowing, or renting goods, which, for many, means adopting a new approach to living and consuming. Consider the purchases by a typical family. Boxes, cans, plastic wrappers, bags, and other packaging material surround nearly every type of food available at the grocery store. Plastic shopping bags, disposable diapers, bottled water, and other disposable products represent conveniences to consumers. The short-term benefit of these goods to the consumer is offset by the long-term consequences to the environment. In addition, nearly all the packaging for these items is non-biodegradable plastic.

An effective way to encourage consumers to make a different choice is to use incentives or penalties. Incentives include offering discounts for bringing your own beverage container or shopping bag. Several high-profile

INTEGRATED WASTE MANAGEMENT REDUCES THE WASTE STREAM

WASTE MANAGEMENT PRIORITIES

REUSE MATERIALS AND REDUCE WASTE

COMPOST ORGANIC WASTE AND RECYCLE

USE AS ENERGY RESOURCE

TREAT AND DISPOSE

Highest · PREFERENCE · Lowest

FIGURE 12.18 The U.S. EPA takes a prioritized, hierarchical approach to waste management, with the highest priority given to waste reduction and reuse. Of the amount of waste eventually discarded, as much as possible is recycled, composted (organic wastes), or used as an energy source. The remainder is stored in sanitary landfill disposal sites constructed to avoid environmental contamination. (After EPA, 2013c)

Resource Conservation and Recovery Act (RCRA) A law passed by the U.S. Congress that banned open dumping of wastes and set standards for solid waste landfills.

integrated waste management A management strategy that minimizes waste disposal by stressing the importance of reducing waste, reusing materials, recycling, composting, and recovering energy from waste materials.

source reduction A waste management tactic aimed at reducing the amount of material that enters the waste stream.

businesses offer such incentives. Penalties are imposed when customers are asked to pay extra for these items. Governments can also intervene. In California, 138 cities and counties have banned single-use plastic bags, and in September 2014, a statewide ban went into effect, forcing retailers to charge 10 cents or more for recycled paper bags in order to encourage customers to bring in reusable bags. The bill did not pass without controversy, however. Within days of it being signed into law, plastic bag manufacturers launched a drive to put a referendum on the issue before the voters, arguing that banning single-use plastic bags would eliminate jobs. As a result, the fate of the law will be decided in a statewide vote in November 2016. Other states and municipalities are also considering such laws.

The business community can make fundamental contributions to source reduction by redesigning products and packaging. Simply reducing the weights of packaging or using lighter-gauge metals in appliances and other products, without impairing function, could reduce the amount of solid waste. For example, manufacturers have reduced the weights of aluminum cans by over 15% in recent years. Another example of the business community reducing waste is through the reuse of shipping pallets. Millions of the wooden pallets used for shipping heavy goods are now being reused instead of being discarded after one use. The online retailer Amazon.com has introduced a frustration-free packaging program, which requires manufacturers to offer products for shipment in easy-to-open, recyclable packages and to avoid using plastic clamshell casings. Tallying up the total environmental impact of a product from the cradle to the grave—that is, from raw material extraction to recycling and disposal—is known as **life cycle assessment**; it has the potential to help consumers make informed choices about their buying decisions (see Chapter 14, page 453).

Hazardous Waste Reduction

Because of the threat they pose to humans and other organisms, hazardous wastes are particularly problematic. As individuals, we need to take care to properly dispose of any hazardous household wastes. Fortunately, many local communities have special collection sites available or dedicated days scheduled to dispose of this waste safely. The activities of significant industrial generators of hazardous waste are subject to a variety of local, state, federal, and international regulations. One of the most notable provisions in the RCRA requires manufacturers to keep records on the generation, transport, treatment, and eventual disposal of hazardous wastes. Had such a law been enacted at an earlier time in U.S. history, it would likely have prevented the creation of the hundreds of thousands of brownfields that dot the country today.

As with municipal solid waste, reducing the amount of hazardous waste is critical. Preventing the release of hazardous waste is not only good for the environment but can also be economically beneficial by reducing the

expense of hazardous waste treatment, transport, and disposal. For example, Siemens Water Technologies, a leader in hazardous waste treatment and source reduction, helped Marathon Norco Aerospace in Waco, Texas, reduce its generation of hazardous waste by 70%. Marathon Norco Aerospace manufactures high-quality nickel cadmium batteries for the aerospace industry, but in the process generates waste containing hazardous concentrations of heavy metals, especially cadmium and chromium. The volume of these wastes and the concentrations of metals in them add to manufacturing expenses due to the high costs of transport, disposal, and recordkeeping associated with hazardous waste management.

By installing a modern Siemens wastewater treatment system, Marathon Norco Aerospace reduced the concentrations of cadmium and chromium in effluent at the plant to below permitted levels. The new equipment also lowered costs by decreasing the volume of waste sent for disposal and eliminating the high labor costs associated with the outdated equipment that the Siemens' system replaced. In most settings, waste reduction has not entirely eliminated the need for hazardous waste disposal.

⚠ Think About It

1. Explain how the term "cradle to grave" applies to the RCRA. How might requirements such as these have reduced the number of brownfields across the United States?

2. Reducing the amount of material used in manufacturing goods seems like it will reduce waste. Can you think about how such efforts might cause problems?

12.8 Food waste and other biodegradable trash can be reduced and repurposed

According to the EPA, in 2013, Americans threw away 35 million tons of food. That's over 100 kilograms (220 pounds) of food for every person in the country. Therefore, it is imperative that we find alternative uses for nutrient-rich food waste; keeping it from taking up valuable landfill space represents an important environmental challenge. The nutrients this waste contains could be cycled back into the environment, making it available for use in natural and human-dominated ecosystems.

Reducing Food Waste

Several non-profit organizations, such as Waste No Food in California and the industry-backed Food Waste Reduction Alliance, are working to raise awareness of the problem and create a roadmap for industry and consumers to reduce food waste. It's important to note that problems of food waste differ in the developed and the developing

How might banning single-use plastic bags create economic opportunities?

Could a chemical company that developed ways to reduce its hazardous waste production become more profitable as a result?

How long could 220 pounds of food feed you personally?

life cycle assessment An estimate of the total environmental impact of a product or technology as a result of activities such as extraction of an energy source (e.g., coal), transport, processing of raw materials, construction, maintenance, dismantling, removal, and recycling or disposal of structures.

world. In the latter, about one-third of crops are spoiled or wasted before they even reach consumers. A lack of refrigeration and poor transportation systems in these parts of the world make food spoilage a fact of life; food waste can only be reduced through the development of mobile processing technologies and improved methods for getting food to where it's needed most.

In the developed world, however, food waste occurs primarily among end users, restaurants, and residences. For instance, a restaurant may stock too much of a perishable ingredient, which can't be used in its entirety before spoiling. Preventing such problems requires making a periodic "dumpster dive" and seeing what ingredients are consistently getting thrown out. Of course, it's not always possible to run a business—or a family kitchen—and be certain which items will get used and which will not. Many restaurants have partnered with humanitarian organizations to make daily or weekly donations of surplus food they no longer anticipate using. A significant amount of food waste comes from unused trimmings and leftovers, some of which can be used in animal feed. Finally, used cooking oil and grease can be sold to biofuel manufacturers (see Chapter 10, page 309).

Composting

Rather than dumping food scraps and yard waste in landfills, such material can be composted. **Composting** is a method for returning the nutrients found in organic matter back to the environment, using processes that mimic the natural biogeochemical cycling of matter. The process involves naturally occurring bacteria, fungi, and invertebrates that gradually break down organic wastes into "compost," a dark-colored substance resembling soil. Compost, which adds both nutrients and organic matter to soils, makes a great fertilizer for growing crops.

An example of an effective composting program is found in the sanitation system of San Francisco, California. This city has a food scrap collection program in areas with high numbers of restaurants. The food waste is collected in special bins every week by garbage trucks in a separate run. The waste is delivered to a facility that grinds and mixes it, then places it in huge black plastic bags for about 2 months. Once the composting process is complete, the "black gold" is delivered to farms and wineries throughout the state, where it is used as organic fertilizer. San Francisco's composting system mimics the nutrient cycles found within nature and does an effective job of preventing this biodegradable waste from entering landfills. Home gardeners can also compost in their own backyards, layering food waste from the kitchen with leaves and sawdust and turning it periodically.

Turning Decomposing Waste into Energy

When organic material such as food waste ends up in landfills, naturally occurring bacteria gradually break it down. However, landfills are largely **anaerobic,** meaning very little oxygen is available. Therefore, the bulk of decomposer organisms found in landfills are anaerobic bacteria. In this anaerobic environment, bacteria produce methane (see Chapter 2, page 36) as a waste product—methane is a potent greenhouse gas (see Chapter 14, page 437). This biologically generated methane is chemically identical to the main constituent of the natural gas used for household heat and cooking, as well as for industrial purposes (see Chapter 9, page 268). However, so-called landfill gas is usually a 50-50 mixture of methane and carbon dioxide. Nonetheless, in the United States, landfills are the third largest source of methane emissions to the atmosphere, after industry and agriculture.

Increasingly, landfills have systems that collect emerging gas to reduce their emissions of methane, which can be used as an energy source. The level of treatment needed for landfill gas depends on the intended use. Direct use of landfill gas to heat boilers and kilns generally requires only primary treatment, which removes moisture, particulates, and the trace amounts of sulfur dioxide generally present (**Figure 12.19**). Because primary-treated landfill gas still consists of roughly half carbon dioxide, it has lower energy content than does natural gas. For use in applications where higher energy content is required, such as heating

How could you reduce food waste in your home?

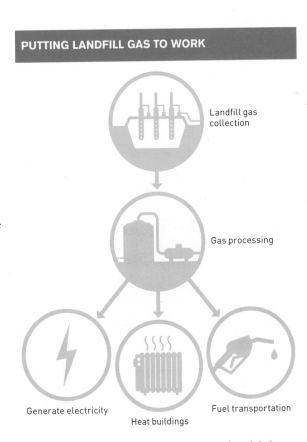

PUTTING LANDFILL GAS TO WORK

Landfill gas collection

Gas processing

Generate electricity

Heat buildings

Fuel transportation

FIGURE 12.19 The energy-rich methane produced during decomposition of organic matter under anaerobic conditions in landfills is widely collected and burned to run electrical generators, heat buildings, fuel industrial processes, and power transportation networks.

composting A process involving aerobic decomposition of organic material used to recycle garden waste and organic components of municipal solid waste.

anaerobic An environment without molecular oxygen (O_2).

homes and running natural gas vehicles, secondary treatment is required. The additional treatment involves removal of carbon dioxide, nitrogen, oxygen, and other trace gases. The result is "pipeline quality gas," which is nearly pure methane and approximately equivalent in energy content to natural gas.

⚠ Think About It

1. What are some other beneficial uses for food waste?

2. Why is the methane emitted by landfills now considered a valuable resource?

12.9 Recycling and demanufacturing are critical to reducing waste

No matter how much we reduce our waste, it's a fact of life that many consumer products must be purchased with a disposable container. However, these materials can be kept out of the waste stream by promoting **recycling,** the process of returning the raw materials present in a form of waste back to the manufacturer to be used again. Commonly recycled materials include glass, plastics, metal, paper, and cardboard (**Figure 12.20**).

Recycling has a number of benefits. When a disposable product, such as an aluminum can, is recycled, both raw material and energy are conserved. Less source material, bauxite ore in this case, must be mined and processed. The amount of energy needed to melt and mold recycled material is much less than when starting with raw ore.

Finally, the product is kept out of the waste stream for at least one more round of use.

Financial Incentives for Recycling

One way to increase recycling rates is to require deposits on bottles and cans. The state of Michigan, for example, imposes a 10-cent deposit on all its recyclable beverage containers, the highest of any state. This deposit is returned to the customer when the container is recycled. In response to this incentive, the recycling rate for these containers in Michigan from 1990 to 2012 was 97%—more than double the national average. Over this period more than $9 billion in deposits were paid by consumers and refunded. Of the approximately $300 million in unclaimed deposits, 75% was returned to the state government to fund environmental programs, and beverage retailers retained 25%.

Take Back Laws and Demanufacturing

One recent development is **take back laws,** which typically require manufacturers of televisions, computers, and other electronics to pay for e-waste recycling programs. Such laws have been passed in 24 states, including New York and Texas. California's Electronic Waste Recycling Act is similar, but it requires consumers to foot the bill for recycling, paying a special fee to retailers when they purchase certain devices. Utah lacks a take back law, but it does require manufacturers to participate in e-waste recycling programs and to inform consumers.

Dismantling electronic equipment into constituent components and scrap metals is called **demanufacturing.** Although the process can be dangerous when conducted

recycling The process of returning raw materials in waste (e.g., glass, plastics, metal, paper) to the manufacturer for reuse.

take back laws State regulations that require manufacturers of various electronics to pay for e-waste recycling programs.

demanufacturing The dismantling of equipment, especially electronics, into constituent components and scrap metals that can be reused or recycled.

RECYCLING: AN ESSENTIAL AND GROWING PROCESS

Taking out materials for curbside recycling

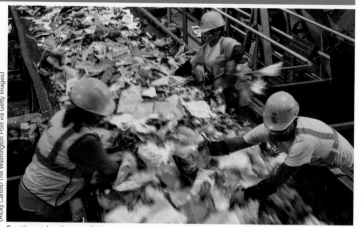
Sorting mixed materials at a recycling center

FIGURE 12.20 Recycling is essential to sustaining economic productivity over the long term. One way to encourage recycling by consumers is to provide collection centers in convenient places. From such locations, materials to be recycled are generally transported to a sorting center.

CHANGES IN THE MANAGEMENT OF MUNICIPAL SOLID WASTE IN THE UNITED STATES

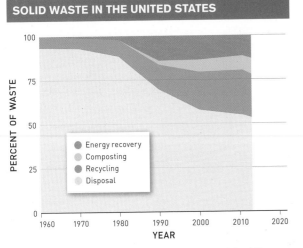

FIGURE 12.21 In 1960 nearly all MSW in the United States was disposed of in a landfill or was incinerated. By 2013 nearly half was recycled, composted, or used as an energy source. (Data from EPA, 2015)

informally, without the oversight of environmental or health regulators, it has the potential to separate hazardous waste from valuable materials that can be recycled. Several electronics recycling centers have emerged in urban centers, supporting the view that demanufacturing could be a vital source of jobs and income.

Progress in Municipal Waste Management

The problem of waste disposal in Philadelphia, now managed in a more integrated fashion, is much less pressing than during the time of the Khian Sea. City ordinances were passed to support a goal of a 50% recycling rate. Although this goal has not yet been reached, recycling rates have steadily improved, as they have across the United States. From 1960 to 2013, the amount of MSW recycling increased from just 6% to 25%. Over this same period, the amount of waste composted increased to 9% and the amount burned for energy recovery rose to 13%. As a consequence, the amount of MSW ending up in landfills across the country fell from 94% to 53% between 1960 and 2013 (**Figure 12.21**).

While the progress made in reducing disposal rates across the United States is encouraging, some cities are doing better than others. For example, San Francisco has succeeded in reducing the amount of solid waste ending up in landfills to just 20% of the total generated. The city has achieved this low level of waste disposal by emphasizing the key elements of integrated waste management: reducing and reusing, recycling, and composting. To facilitate these management goals, residents have access to three refuse bins, one for recyclable materials, one for compostable materials

(e.g., food and garden wastes), and one for solid waste destined for the landfill (**Figure 12.22**).

Based on San Francisco's waste handling system and other measures, an independent study has ranked it as the "greenest city" in North America. However, the city's ultimate goal is zero waste. While no municipality has yet attained this goal, Europe's greenest city, Copenhagen, Denmark, with a disposal rate of only 2% of the waste generated, comes close. Like San Francisco, Copenhagen stresses waste reduction and reuse, recycling, and composting. In addition, any waste that cannot be recycled or composted is burned to generate energy in electrical power plants.

Incineration and Energy Recovery

After removing everything that can be recycled or composted, the remaining trash can be burned to reduce the waste volume by up to 90%. Less waste volume means less landfill space is needed. However, incineration is economically viable only where space for landfills is highly restricted. Incineration facilities are expensive to build and pollution control systems are essential for reducing hazardous emissions. One potential source of environmental impact is the ash produced by incinerators. There are two forms of ash produced by incinerators. **Fly ash** is made of lighter, noncombustible material that is normally sent airborne during incineration. **Bottom ash** is the heavier, noncombustible material left over after incineration. Fly ash has a high concentration of toxins, and incinerators are required to have filters and electrostatic precipitators, similar to those

How can the recovery of valuable resources from e-waste be transformed from a process that negatively impacts both human and environmental health to a positive for both the economy and the environment?

fly ash Particles formed during combustion that are light enough to become airborne and exit a combustion chamber with exhaust gases, including soot and dust.

bottom ash The ash that accumulates at the bottom of an incinerator during the combustion of solid waste.

MAKING RECYCLING EASY IN SAN FRANCISCO

FIGURE 12.22 Providing all residents with color-coded bins to dispose of materials for recycling, composting, and disposal has helped the city of San Francisco achieve exceptional reductions in its rates of waste disposal.

CONVERTING MUNICIPAL SOLID WASTE TO ELECTRICAL ENERGY

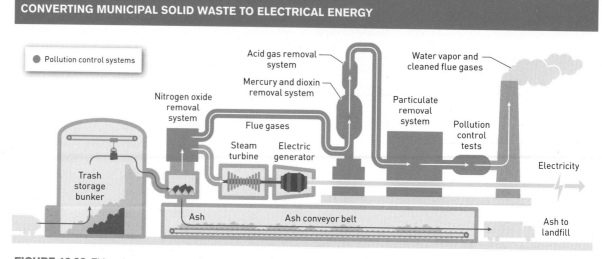

FIGURE 12.23 This schematic shows that using solid waste as a fuel source to generate electricity via a steam-powered turbine is similar to using the other fuels reviewed in Chapter 9 (page 270). However, because of the many potentially hazardous contaminants in solid waste, the U.S. EPA requires that such plants be equipped with advanced pollution control systems.

What opportunities are there for turning "wastes" into valuable resources?

in coal-fired power plants, to collect fly ash and prevent it from being released into the air. Many substances that will not burn (e.g., heavy metals and dioxins) will be left behind in bottom ash.

Once recyclable metals have been recovered from incinerator ash, the residual is sent to a disposal site. The EPA requires regular testing of ash from incinerator waste for the presence of hazardous waste. When the presence of hazardous substances exceeds certain thresholds, the ash must be treated as hazardous waste and disposed of in specially designed hazardous waste disposal sites. Otherwise, in the United States, ash in these plants is disposed in sanitary landfills designed to accept standard MSW. Several European countries use nonhazardous incinerator ash for highway building and other construction.

While properly built and managed incinerators are costly, they can be used to recapture some of the chemical energy that remains in MSW by including a steam-powered turbine in the system (**Figure 12.23**). Increasingly, MSW is being recognized as a valuable energy resource. According to the EPA, in 2013 there were 86 "waste-to-energy" power plants in the United States using MSW to generate electricity. The agency estimates these plants burned 32.7 million tons of MSW—approximately 13% of the total municipal solid waste stream during that year.

These power plants must have the latest pollution control systems to avoid releasing hazardous wastes into the air. Similar plants are in operation in Europe. For example, two waste-to-energy power plants in Oslo, Norway (**Figure 12.24**), which are paid to accept solid waste from the United Kingdom and elsewhere, make additional income by selling electrical power

and heat to consumers. The two plants have a combined capacity to process 410,000 metric tons of waste annually, with an energy equivalent of over 100,000 metric tons of oil.

⚠ Think About It

1. San Francisco's drive toward a zero waste future is accompanied by strict ordinances that require the sorting of waste into recyclable, compostable, and other fractions by residents. Penalties for not doing so include stiff fines. Could the achievements in waste reduction in the city have been attained without such laws and penalties? Explain.

2. Compare and contrast the typical human system of waste management with an ecological cycle. Why is the human system considered unsustainable?

3. What contributes to the fact that incinerating waste to generate electricity is a more common part of waste management in Europe than in North America?

12.10 Safe and secure long-term disposal is the last resort

Today, the brownfield site of Herr's Island in Pennsylvania has been renamed Washington's Landing and is home to a marina, office buildings, and an exclusive townhouse development. In order to achieve this transformation, the soil contaminated with hazardous waste was encapsulated

TURNING WASTE INTO ENERGY

(Klemetsrudanlegget AS)

Klemetsrud waste-to-energy electrical power plant in Norway

(Klemetsrudanlegget AS)

Waste for incineration at Klemetsrud power plant

FIGURE 12.24 Klemetsrud is one of two waste-to-energy power plants in Oslo, Norway, that, combined, have the capacity to process 410,000 metric tons of solid waste per year. Municipal solid waste generated locally and transported to Norway from other European locations provides enough fuel for the two plants to generate sufficient electricity and heat to meet the energy needs of 84,000 homes.

with an impermeable barrier and buried under tennis courts. Even as communities around the world push toward a zero waste future, we are still going to need a safe way to dispose of waste that cannot be reused, recycled, or composted. This includes municipal waste, hazardous waste, and nuclear waste.

Municipal Landfills

The most common and often most economical choice for municipal solid waste disposal is the **sanitary landfill,** which consists of a lined pit constructed and managed in ways to minimize environmental impacts (**Figure 12.25**). A bottom liner provides a barrier between the waste in the landfill and soils and groundwater. In the modern landfill, this liner is made of thick, tough plastic placed over a layer of packed clay, which acts as an additional barrier between the waste and local soil and groundwater. Immediately above the liner is a layer of porous gravel or sand, through which water that has seeped down through the waste in the landfill passes easily until it reaches the plastic landfill liner. This seepage, called **leachate,** flows along the top of the liner to a low point called a sump, where leachate can be pumped out and treated. Because leachate may contain many hazardous substances, groundwater in the vicinity of the landfill should be tested on a regular basis, to ensure that the bottom liner has maintained its

BASIC STRUCTURE OF LANDFILL DESIGNED FOR DISPOSAL OF MUNICIPAL SOLID WASTE

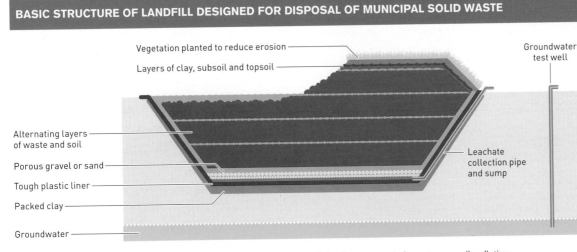

Vegetation planted to reduce erosion

Layers of clay, subsoil and topsoil

Groundwater test well

Alternating layers of waste and soil

Porous gravel or sand

Tough plastic liner

Packed clay

Groundwater

Leachate collection pipe and sump

sanitary landfill A solid waste disposal site consisting of a lined pit constructed and managed in ways to minimize environmental impacts.

leachate Water that has seeped down through the waste in a landfill; flows to a sump in a modern landfill, where it can be pumped out and treated.

FIGURE 12.25 The modern landfill is designed to contain MSW sufficiently to prevent air, water, or soil pollution.

structural integrity and is not contaminating local water supplies.

Once the basic structure of the landfill is in place, filling it becomes a relatively simple task. Garbage trucks unload their contents into an open part of the landfill, where bulldozers compress and spread the material as much as possible. At the end of each day, a layer of soil is placed over the day's trash collection, which prevents the wind from blowing lighter materials away and protects it from scavenging birds and other animals. The layers of soil also help to dampen unpleasant odors emanating from the waste. Daily additions to the landfill continue until the pit reaches its capacity. At that point, the landfill is capped, usually with successive layers of clay, subsoil, and topsoil, which will be planted with vegetation chosen to reduce erosion.

Even after sanitary landfills are filled and capped with layers of clay and soil, they are not waterproof. Rainwater and water from melting snow can continue to percolate through the soil and through the buried waste, emerging as leachate, which will need to be collected and treated.

Treatment and Disposal of Hazardous Waste

The RCRA requires that hazardous wastes be treated before disposal to reduce the danger they pose to the environment. Predisposal treatments involve a variety of physical, chemical, or biological processes to reduce the threat hazardous wastes pose to the environment.

Incineration

Burning, or incinerating, some types of hazardous waste can reduce the dangers of certain kinds of waste and cuts back on its volume. For example, burning medical wastes will kill any biological pathogens that may be contaminating the waste. Also, the toxicity of some chemical wastes can be reduced by incineration at very high temperatures, up to 1,200°C (2,200°F). As in the case of MSW, however, the by-product of incineration, ash, will require testing and disposal.

Hazardous Waste Landfills

Landfills intended for disposal of hazardous wastes must be secure and therefore must meet much stricter design and management standards than municipal landfills. Under RCRA, the EPA requires a double liner for hazardous waste landfills and a double leachate collection and removal system (**Figure 12.26a**). Such landfills must also include a leak detection system and ways to prevent the run on and runoff of storm water. Once a hazardous waste landfill is filled and covered, removal of leachate must continue until it is no longer produced. Ongoing monitoring for leaks and for groundwater contamination is also required. Approximately 10% of hazardous waste disposed in the United States is stored in secure landfills. RCRA prohibits the storage of liquid hazardous wastes in landfills.

Surface Impoundments

Natural or excavated depressions can be used as surface impoundments for temporarily storing or treating liquid hazardous wastes (**Figure 12.26b**). These structures must have double liners, a system for collecting and removing leachate, and a leak detection system. They must be regularly monitored, inspected, and eventually sealed.

Why is a reduction in hazardous waste production especially important not just from a health perspective but also from an economic perspective?

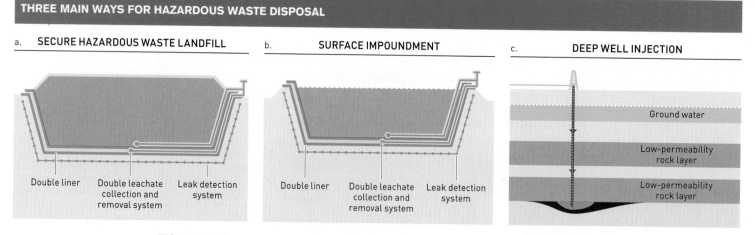

THREE MAIN WAYS FOR HAZARDOUS WASTE DISPOSAL

a.   SECURE HAZARDOUS WASTE LANDFILL

Double liner   Double leachate collection and removal system   Leak detection system

b.   SURFACE IMPOUNDMENT

Double liner   Double leachate collection and removal system   Leak detection system

c.   DEEP WELL INJECTION

Ground water

Low-permeability rock layer

Low-permeability rock layer

FIGURE 12.26 Landfills dedicated to the disposal of hazardous waste are subject to much more stringent standards of construction and management than landfills for the disposal of MSW. Surface impoundments used for temporary storage of liquid hazardous waste are carefully constructed to avoid leakage. Deep well injection of liquid hazardous waste is the means by which approximately 90% of hazardous waste is disposed of in the United States. (Information from EPA, 2013a)

Deep Well Injection

Of the liquid hazardous waste disposed of in the United States, approximately 90% is injected into wells drilled into deep rock formations (**Figure 12.26c**). Deep well injection is regulated by RCRA and the Safe Drinking Water Act of 1974. The average depth of deep injection wells is 1,200 meters (4,000 feet), far below the level of groundwater supplies. The EPA requires that deep injection wells be located in areas with stable geology and without fractures that might allow injected waste to migrate upward and contaminate drinking water supplies. The drilling and casing of deep injection wells also include multiple safety features to minimize the chance of groundwater contamination.

Nuclear Waste Disposal: An Unresolved Hazardous Waste Problem

Nuclear reactors around the world continue to produce waste, yet the problem of waste disposal remains unresolved. Several countries at the forefront of nuclear energy technology favor deep geological storage of intermediate- and high-level nuclear waste. Despite the consensus among these nations, there is only one deep geological disposal site that is currently licensed and operating, the Waste Isolation Pilot Plant (WIPP) located in the United States near Carlsbad, New Mexico.

The WIPP facility, which stores defense-related, intermediate-level nuclear waste in thick salt deposits, has been in operation since 1999. However, given the uncertainty associated with the Yucca Mountain nuclear repository (see page 372), the United States is without a permanent disposal site for high-level nuclear wastes generated by civilian power plants.

While the United States continues its search for alternatives to Yucca Mountain, Finland and Sweden are going forward with the development of deep geological disposal in bedrock. Disposal at a site on an island in southwest Finland is planned to begin in 2020. A different type of geology is being developed for nuclear waste disposal in France, a deep clay formation east of Paris, which geologists estimate has been stable for millions of years. The French repository is planned for opening in 2025. However, the incident at Fukushima in Japan (see Chapter 9, page 283) reminds us that much can happen to influence such plans.

In the aftermath of that accident, Germany decided to eliminate plans for developing nuclear power and to phase out its existing reactors. While such phase-outs, if they occur, will not eliminate the need for nuclear waste disposal, they will reduce the long-term need for disposal space. In the meantime, as geologists, politicians, lawyers, judges, and environmental activists debate and study the issue, the amount of radioactive waste in "temporary" storage continues to accumulate, like all those shells that Native Americans piled up near Oakland, California.

⚠ Think About It

1. Within a hierarchy of preferred options in an integrated waste management system, why would landfills be at the very bottom, even below incineration?

2. The intent of sanitary landfills is to safely contain waste indefinitely. Is this realistic? Explain why or why not.

3. Describe in detail the approach you would use to solve the problem of nuclear waste disposal, if you were given the authority to do so.

What factors make nuclear waste disposal so controversial to the general public?

12.7–12.10 Solutions: Summary

Reducing the waste stream involves reducing packaging, recycling materials when possible, and composting biodegradable food and yard scraps. Using these approaches can substantially reduce the volume of waste incinerated or disposed in a landfill. The modern sanitary landfill is a complex structure designed to reduce water, air, and soil contamination. Decomposer bacteria in landfills produce landfill gas, which is approximately half methane and half carbon dioxide and is increasingly used as an energy resource. As with municipal solid waste, reducing the amount of hazardous waste produced is generally given highest management priority because it reduces the threat to the environment and the expense of hazardous waste treatment, transport, and disposal.

The Resource Conservation and Recovery Act (RCRA) requires industries that produce hazardous wastes to track, treat, and eventually dispose of these wastes. The goal of the Basel Convention treaty is to limit the export of hazardous waste from developed to developing countries and to encourage the safe treatment and disposal of hazardous wastes within the countries where they were generated. Nuclear reactors around the world continue producing high-level nuclear waste, yet the problem of permanent waste disposal remains unresolved.

Answer the following questions for each chapter section and then answer the Central Question.

Central Question: How can we reduce the environmental impact of solid waste and dispose of hazardous waste safely?

12.1–12.2 Science

- How does "waste" generated by economic systems compare to waste in natural ecosystems?

- What are the sources and properties of waste?

12.3–12.6 Issues

- What problems arise in managing municipal solid waste?

- How is hazardous waste generated?

- What are some new forms of hazardous waste?

- What are the issues surrounding the disposal of nuclear waste?

Waste Management and You

While many environmental issues can seem only indirectly related to your everyday life, the problem of waste is definitely not. We all make decisions that affect the waste stream in our local community and beyond, and can therefore make choices that reduce the amount of material entering the waste stream.

☐ *Stay informed.*

There is a wealth of information available from local, state, and federal environmental agencies to help individuals reduce, recycle, and compost waste. The U.S. Environmental Protection Agency, for example, provides detailed information on how to make constructive contributions to solving waste management issues, along with reasons for making such contributions.

☐ *Use less.*

Use the products you own (e.g., appliances, clothing, vehicles) for as long as they continue to function properly. Purchase reusable, rather than disposable, products like reusable shopping bags and glass food storage containers. Consider buying used products in good condition. Experiment for a week by carrying with you all the trash you create in one day—try to decrease your trash each successive day.

☐ *Recycle.*

Consider donating to charity any functional household appliances, electronics, books, or clothing that you no longer need or cannot use instead of disposing of them. Contribute to recycling in your community by saving and recycling glass, paper, cardboard, and recyclable plastics. Most communities have recycling centers or, in many larger communities, programs for collecting recyclable materials at residences. Purchase products made from recycled materials, which helps sustain recycling by making it profitable. Collect hard-to-recycle items like water filters and print cartridges from friends and family and check earth911.com to make a mass dropoff at a nearby organization that will accept those items.

☐ *Safely dispose of hazardous wastes.*

Take particular care to recycle certain items like electronic equipment and batteries to avoid adding to the hazardous waste stream. Because they contain valuable materials like copper, silver, and gold, many manufacturers and retail business will accept electronic devices for recycling.

12.7–12.10 Solutions

- **What is the focus of modern waste management?**

- **How can we reduce and repurpose food waste and other biodegradable trash?**

- **What are critical steps to reducing waste?**

- **How do we manage hazardous waste?**

- **How do we handle nuclear waste?**

Answer the Central Question:

_____

_____

_____

_____

_____

_____

_____

Chapter 12

Review Questions

1. How is the law of conservation of matter related to waste management?
a. The two are unrelated; one is a law of theoretical physics, while the other is a practical problem.
b. This law of physics tells us how to permanently dispose of waste.
c. This law of physics reminds us that waste matter may change form but does not go away.
d. This law of physics provides a guide for transforming one type of chemical element into another.

2. Which of the following is "organic" waste, according to the World Bank?
a. Garden waste
b. Cardboard
c. Plastic water bottles
d. Newspaper

3. Which is the largest source of hazardous waste in the United States?
a. Pesticide manufacturing
b. Petroleum product manufacturing
c. Metal processing
d. Basic chemical manufacturing

4. Even though growth in global per capita waste production is slowing, why has growth in total waste production not slowed?
a. A few countries continue to produce waste at a high rate.
b. The global population is growing.
c. Development is discouraging recycling.
d. Increases in global per capita waste production are actually accelerating.

5. Plastic waste in the Pacific Ocean threatens marine life. How does it potentially impact humans?
a. It doesn't because ocean shipping is unaffected.
b. The accumulated plastics are already catching fire.
c. The toxic chemicals leaching from the plastics have contaminated seafood.
d. The plastic waste has caused a loss of profits to the cruise ship business.

6. Which was not an impediment to developing the nuclear waste repository at Yucca Mountain?
a. Local political resistance
b. Concerns about safety of nuclear waste transport
c. The possibility of groundwater contamination
d. International pressures

7. What was the main purpose of the Resource Conservation and Recovery Act (RCRA)?
a. To set rules for endangered species conservation
b. To conserve forests in the United States
c. To set standards for solid waste landfills
d. To begin cleanup of the Great Pacific Garbage Patch

8. How does landfill gas differ from the natural gas extracted from deep wells in gas and oil fields?
a. Landfill gas is composed of entirely different gases.
b. Unrefined landfill gas has lower energy content.
c. Landfill gas cannot be burned without refining.
d. There is no market for landfill gas.

9. What mechanisms have been used to increase recycling rates?
a. Deposits on beverage containers
b. Laws requiring recycling

c. Access to recycling bins
d. All of the above

10. What is the main difference between landfills for municipal solid waste and hazardous waste landfills?
a. Hazardous waste landfills must have double liners; a single liner is required for municipal landfills.
b. In contrast to municipal landfills, leachate from hazardous waste landfills must be collected.
c. To provide convenient access to stored materials, hazardous waste landfills cannot be capped.
d. Municipal landfills are required to have a leak detection system.

Critical Analysis

1. Discuss how the Great Pacific Garbage Patch represents a Tragedy of the Commons (Chapter 2, page 49)?

2. What would be the advantages and disadvantages of an economic system in which all products are recyclable?

3. What models related to waste might lower-income countries offer to high-income countries?

4. Outline a management plan for brownfields that would turn them into economic and environmental assets to communities where they occur.

5. How would engineers and others ensure that a nuclear waste repository is safe from not only physical disturbance but also human intrusion for the next million years?

Find additional resources and links online at www.macmillanhighered.com/launchpad/molles1e.

Central Question: How can we control and reduce environmental pollution?

Explain the sources of pollution and how they move around the biosphere.

SCIENCE

CHAPTER 13

Air, Water, and Soil Pollution

Describe how air, water, and soil pollution impact biodiversity, ecosystems, and human health.

Analyze the effectiveness of pollution regulation and other tactics to treat polluted environments.

ISSUES

SOLUTIONS

(Geoff Liesik/The Deseret News via AP)

13.1–13.3 Science

Is there a connection between population growth and pollution? Explain.

Pollution is defined as contamination or physical alteration of the environment at levels harmful to living organisms. The factors causing pollution, **pollutants,** include chemical substances, such as oil or pesticides, or altered physical conditions, such as excessive noise or light (**Figure 13.1**). Natural events, such as wildfires or volcanic eruptions, can cause pollution, but at this point in the history of Earth, pollution mainly results from human activity.

13.1 Industry releases pollutants

Concerns about environmental contamination began long ago. As we saw in Chapter 1, Benjamin Franklin

and other residents of 18th-century Philadelphia tried to reduce water pollution by tanneries (see page 15). However, the sources of environmental pollution have grown far beyond what occurred in centuries past. While each form of pollution has some unique characteristics and impacts human health or the environment, pollutants can be categorized according to a few overlapping features, such as how they are released into the environment.

Point Versus Nonpoint Sources of Pollution

Pollutants are released into the environment through one of two avenues: from point sources or from nonpoint sources (**Figure 13.2**). The smokestack of a factory or a

pollution Contamination of the environment, generally of air, water, or soil, by substances or conditions (e.g., noise, light) at levels harmful to living organisms; generally a result of human activity but may result from natural processes (e.g., wildfires, volcanic eruptions).

pollutant A substance (e.g., oil, pesticides) or physical condition (e.g., excessive noise) harmful to living organisms that contaminates air, water, or soil.

PHYSICAL SOURCES OF POLLUTION

(Vicent de los Angeles/Getty Images)

Lights along Gandia Beach, Costa Brava, Spain

(pbombaert/Shutterstock)

Jet airliner flying low over city

FIGURE 13.1 Artificial light can be a source of environmental pollution, that is, "light pollution," in certain contexts. For example, lights near the nesting beaches of sea turtles can stimulate newly emerged hatchling sea turtles to move inland, to their deaths, rather than toward the reflection of the sea. Similarly, disturbing sounds may be a source of "noise pollution" that can cause stress in humans and interfere with wildlife, from birds to whales.

POINT AND NONPOINT SOURCES OF POLLUTION

(Brent Lewis/Getty Images)

Polluted water from the Gold King Mine, Colorado

(Keith Getter/Getty Images)

Genesee River, New York, contaminated by melting snow

FIGURE 13.2 The acidic and heavy metal–laden water from the breached Gold King Mine of southern Colorado (see chapter-opening photo) gushing from this culvert (left) poured millions of gallons of contaminated water into the Animas River. This is a clear example of a point source of pollution. In contrast, the Genesee River near Rochester, New York (right), has been polluted by runoff during spring snowmelt in an urban landscape. Consequently, this contamination cannot be traced to a particular point in the landscape.

Why might secondary pollutants present greater challenges to management than primary pollutants?

sewer pipe discharging pollutants into the environment from a clearly definable place is called a **point source of pollution. Nonpoint sources of pollution** introduce pollutants from scattered locations or may move around in the environment. Examples of nonpoint sources include automobile exhaust or polluted water draining from city streets or pesticide-laden farmland. In general, it is much easier to identify and manage pollution from point sources than from nonpoint sources.

Pollutant Persistence

Some pollutants break down rapidly when released into the environment, whereas others persist for months, years, centuries, or millennia. Bacteria and fungi will quickly metabolize simple organic compounds, such as sugars and amino acids. More complex organic compounds, such as the cellulose component of wood, though biodegradable (see Chapter 12, page 366), will decompose much more slowly. Decomposition of such compounds will occur over a period of months or years, depending on environmental conditions, such as oxygen concentration or temperature. In contrast, atmospheric scientists estimate that some fraction of the carbon dioxide released by fossil fuel combustion will remain in the atmosphere for thousands of years before being taken up by natural processes, such as photosynthesis. Meanwhile, heavy metal pollutants do not break down; they simply persist.

Primary and Secondary Pollutants

Substances that are harmful when released into the environment are called **primary pollutants.** Primary pollutants in air include carbon monoxide, lead, nitrogen oxides, particulate matter, and sulfur dioxide. Meanwhile, some primary air pollutants react chemically to form **secondary pollutants.** For example, ozone (O_3) is formed through the reaction of oxides of nitrogen (NO_x) with volatile organic compounds, or VOCs—such as those in the fumes from gasoline, paints, and other substances rich in organic chemicals. Note that while ozone in the upper atmosphere acts as a protective shield for the biosphere, ozone in the lower atmosphere is a serious pollutant that can damage both animal and plant tissues.

Another notable example of the formation of a secondary pollutant occurs when SO_2 and NO_x undergo reactions in the atmosphere that produce strong acids: SO_2 reacts to form sulfuric acid and NO_x reacts to form nitric acid (**Figure 13.3**). An **acid** is a substance that releases hydrogen ions (H^+) upon dissociation when dissolved in water, resulting in reduced **pH,** an indicator of relative hydrogen ion concentration, of the solution. A pH of 7 indicates a neutral solution, while a pH of less than 7 is acidic (elevated hydrogen ion concentration), and a pH greater than 7 is alkaline (reduced hydrogen ion concentration). An important feature of the pH scale is that the units are base-10 logarithms, which means that a pH difference of 1, for example, pH = 6 versus pH = 7, indicates a 10-fold difference in hydrogen ion concentration. One of the major environmental consequences of the formation of these secondary pollutants is acid rain, which can have major impacts on soils and aquatic ecosystems (discussed later).

⚠ Think About It

1. How might degree of persistence affect management of different pollutants?

point sources of pollution Clear-cut, generally stationary, sources of pollution (e.g., power plants, factories, or sewage outfalls of cities) that are easier to identify, monitor, and regulate.

nonpoint sources of pollution Diffuse, and sometimes mobile, sources of pollution (e.g., runoff from an industrial, municipal, or agricultural landscape or the exhaust from automobiles).

primary pollutant A substance that is harmful when released into the environment (e.g., carbon monoxide, crude oil).

secondary pollutant A pollutant formed from the chemical reactions between other pollutants (e.g., ozone in the lower atmosphere).

acid A substance that releases hydrogen ions upon dissociation when dissolved in water, resulting in reduced pH of the solution; acids neutralize bases.

pH An indicator of the relative hydrogen ion concentration of a solution. A pH of 7 indicates a neutral solution; a pH of less than 7 is acidic (elevated hydrogen ion concentration); a pH greater than 7 is basic (reduced hydrogen ion concentration).

EXAMPLES OF ATMOSPHERIC TRANSFORMATIONS OF PRIMARY POLLUTANTS TO SECONDARY POLLUTANTS

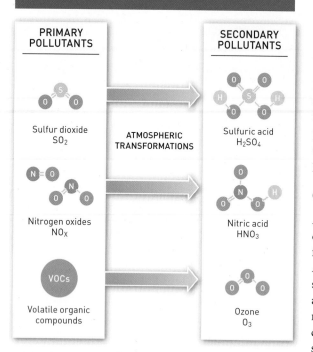

FIGURE 13.3 The primary pollutants—sulfur dioxide (SO_2), nitrogen oxides (NO_x), and volatile organic compounds (VOCs)—undergo chemical reactions in the atmosphere, resulting in the secondary pollutants, sulfuric acid (H_2SO_4), nitric acid (HNO_3), and ozone (O_3).

2. Why is a reduction in the pH of lake water by just 2 units, from pH = 7 to pH = 5, a significant change in the environment?

13.2 Humans produce a wide variety of pollutants

Pollution comes in many forms, some of which we've discussed already. In Chapters 7 and 8, we introduced issues associated with adding excessive organic matter, plant nutrients, and sediments to aquatic environments, and Chapter 9 covered oil spills and nuclear accidents. In addition, Chapter 11 examined pollutants of soil and water within the context of environmental health, including endocrine disruptors, heavy metals, pathogens, pesticides, and pharmaceuticals. Here, we expand those earlier discussions for some of the more problematic pollutants of water and soil (Table 13.1).

Criteria Air Pollutants

Although there are many substances that can contaminate air, some pollutants have been singled out for careful regulation. The U.S. Environmental Protection Agency (EPA) has established air-quality standards for six major air pollutants, which are generally referred to as **criteria pollutants** and include ozone and particulate matter (Table 13.2). These particular pollutants were chosen for regulation because they are very common sources of air pollution and, as indicated in Table 13.2, they are hazardous to human health and to the health of the environment. For example, exposure to sulfur dioxide can aggravate asthma symptoms and damage plant tissues.

Sulfur and Acid Rain

The element sulfur, as a component of proteins, many vitamins, and antioxidants, is essential to life. Like other elements, sulfur moves through the biosphere in a

criteria pollutants Very common sources of air pollution (e.g., sulfur dioxide) chosen by the EPA to be regulated because they are hazardous to human health and the environment.

TABLE 13.1

MAJOR SOIL AND WATER POLLUTANTS

This list of major environmental pollutants of soil and water includes some that also contaminate air, notably heavy metals and pesticides.

Pollutants and Examples	Main Sources	Persistence and Transformations	Health and Environmental Impacts
Heavy metals; lead, mercury, arsenic, cadmium, chromium, copper, nickel	Coal burning, metal smelting, oil, sewage	Persistent; biomagnify in food chains	Interfere with central nervous system function and development; toxic
Nutrients promoting growth of plants and algae; mainly phosphorus and nitrogen	Runoff from agricultural lands, urban areas; treated sewage, aquaculture	Incorporated into biomass through algal and plant growth	Algal blooms in water followed by oxygen depletion as biomass decomposes; in soils, leads to reduced plant and fungal diversity
Organic matter	Sewage, agricultural wastes, aquaculture	Gradually decomposes	Reduces oxygen concentrations, leading to mortality among organisms requiring higher oxygen levels; can shift composition of aquatic community to pollution-tolerant species
Persistent organic pollutants; dioxin, PCBs, DDT	Manufacturing, agriculture	Many highly persistent in environment; biomagnify in food chain	Developmental and reproductive abnormalities, reduced immunity; some cause birth defects and cancer

Data from multiple sources.

TABLE 13.2
EPA CRITERIA POLLUTANTS AND VOLATILE ORGANIC COMPOUNDS

The first six pollutants listed are common air pollutants, deemed hazardous to human health and the environment, which the U.S. EPA is required to monitor. VOCs are also included in the list because they are involved in the formation of ozone in the lower atmosphere, one of the criteria pollutants.

Pollutants and Examples	Main Sources	Persistence and Transformations	Health and Environmental Impacts
Carbon monoxide (CO)	Incomplete combustion of fossil fuels or biomass	Residency time in atmosphere averages approximately 2 months	Binds irreversibly to hemoglobin, reducing blood's capacity to carry oxygen; induces dizziness and nausea at low concentrations; causes death at high concentrations
Lead	Coal burning, metal smelting, oil, sewage	Persistent; biomagnifies in food chains	Interferes with central nervous system function and development; exposure in children can result in lowered IQ and learning difficulties
Nitrogen oxides (NO_x, NO_2)	Burning of coal, oil, gasoline, and biomass; soil bacteria	Converts to nitric acid (HNO_3) via a series of reactions in the atmosphere	Irritates eyes, nasal passages, and lungs; increases susceptibility to lung infections; reduces plant growth; can kill plants; creates brownish haze, reducing visibility
Ozone (O_3)	Forms in lower atmosphere (troposphere) via a series of reactions, in the presence of sunlight, between oxygen, water, NO_x, carbon monoxide, and volatile organic carbon compounds	Residence time in lower atmosphere averages a few months	Irritates respiratory passages; damages lungs and causes respiratory difficulties; damages plant foliage and reduces plant growth; degrades plastics and rubber
Particulate matter	Fossil fuel and biomass burning; wind and water erosion associated with agricultural and construction activity	Removal of suspended particulates from either air or water related to size, with smallest particulates having the longest residence time in the medium	Causes respiratory difficulties; reduces visibility; deposition inhibits respiration and photosynthesis in terrestrial plants; can alter soil pH and damage aquatic habitat by sedimentation
Sulfur dioxide (SO_2)	Burning of coal, oil, and gasoline; metal smelting; volcanic eruptions	Converts to sulfuric acid (H_2SO_4) via a series of reactions in the atmosphere	Creates breathing difficulties, particularly among asthma sufferers; causes lesions and yellowing of plant leaves; can cause atmospheric haze and reduce visibility
VOCs, acetone, benzene, formaldehyde, toluene	Evaporation of solvents, paints, gasoline, and other fuels; leaking underground storage tanks; improper disposal of household products; degassing of particle board	Reacts with NO_x to produce ozone	Eye, nose, throat irritation; liver, kidney, central nervous system damage; many are carcinogens, toxic to a variety of aquatic and soil organisms

Data from multiple sources.

global biogeochemical cycle (**Figure 13.4**). Most sulfur on Earth is tied up in rocks and is released into the biosphere as these geologic materials are exposed by the rock cycle and weathered (see Appendix B). Seawater holds another major portion of Earth's store of sulfur, which is actively exchanged with the atmosphere both as sea salt blown from the ocean's surface and as a variety of gaseous, sulfur-containing compounds produced mainly by marine algae and bacteria. Sulfur also becomes incorporated into ocean sediments through physical and biological processes. Volcanic eruptions, both on land and in the oceans, emit a significant amount of sulfur as well, mainly as sulfur dioxide, SO_2.

The ocean floor is also a site of significant release of sulfur into the biosphere at hydrothermal vents, where hot water rich in hydrogen sulfide (H_2S) emerges and is quickly transformed to SO_2 by sulfur-oxidizing bacteria.

In complete darkness, these bacteria provide the primary production supporting a rich ocean floor ecosystem (see Figure 4.4, page 100). Human economic activity has also become a major contributor to the cycling of sulfur, through our mining and use of fossil fuels, smelting of metal ores, and, to a lesser extent, mining of elemental sulfur. In fact, for decades, burning of fossil fuels has been a much larger emitter of SO_2 than volcanic eruptions and, as we will see, is a major source of pollution.

The chemist Robert Angus Smith appears to have been the first person to recognize and systematically study the phenomenon of **acid rain.** In 1852 he noticed that the acidity of rain increased as you moved from the countryside to cities throughout the British Isles, and understood that it had the potential to affect ecosystems. Acid rain forms when sulfur dioxide, SO_2, undergoes chemical reactions in the atmosphere, forming sulfuric

Should any measureable amount of sulfur dioxide in the atmosphere be considered a pollutant?

acid rain Acidified rainfall. See *acid deposition.*

THE GLOBAL SULFUR CYCLE

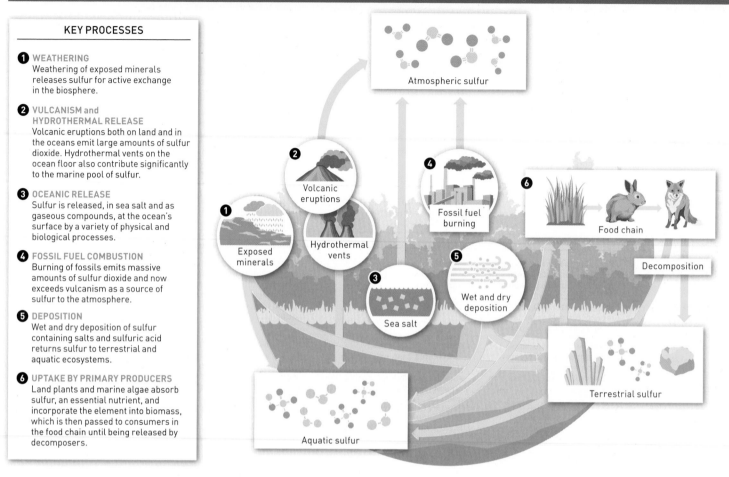

KEY PROCESSES

① WEATHERING
Weathering of exposed minerals releases sulfur for active exchange in the biosphere.

② VULCANISM and HYDROTHERMAL RELEASE
Volcanic eruptions both on land and in the oceans emit large amounts of sulfur dioxide. Hydrothermal vents on the ocean floor also contribute significantly to the marine pool of sulfur.

③ OCEANIC RELEASE
Sulfur is released, in sea salt and as gaseous compounds, at the ocean's surface by a variety of physical and biological processes.

④ FOSSIL FUEL COMBUSTION
Burning of fossils emits massive amounts of sulfur dioxide and now exceeds vulcanism as a source of sulfur to the atmosphere.

⑤ DEPOSITION
Wet and dry deposition of sulfur containing salts and sulfuric acid returns sulfur to terrestrial and aquatic ecosystems.

⑥ UPTAKE BY PRIMARY PRODUCERS
Land plants and marine algae absorb sulfur, an essential nutrient, and incorporate the element into biomass, which is then passed to consumers in the food chain until being released by decomposers.

FIGURE 13.4 The global sulfur cycle involves active terrestrial, oceanic, and atmospheric exchanges of sulfur-containing compounds. Sulfur dioxide is a major component of the atmospheric pool of sulfur, to which burning of fossil fuels is a major contributor. (After Schlesinger, 1991)

How might the fact that the pH scale is logarithmic affect public perception of the significance associated with a pH shift of half a pH unit, for example, from 5.6 to 5.1?

acid deposition An inclusive term that includes both wet and dry deposition of acids.

persistent organic pollutants (POPs) Organic chemicals (e.g., PCBs) that remain in the environment indefinitely; can biomagnify through the food web and pose a threat to human health and the environment.

acid, H_2SO_4, a major source of acid rain, along with nitric acid, HNO_3 (see Figure 13.3, page 390). Although we commonly use the term "acid rain" to refer to this phenomenon, it is more precise to call it **acid deposition** because it can occur as dry deposition, which happens when gases or tiny particles suspended in air are deposited directly onto a surface such as a plant leaf or lake surface, as well as in the form of snow and sleet. Even without pollution, rainfall is slightly acidic with a pH of around 5.6 (recall that a pH of 7.0 is neutral). Generally, rainfall with a pH of less than 5.3 is considered to be acid rain.

Persistent Organic Pollutants

Persistent organic pollutants, or **POPs,** are chemicals that remain in the environment indefinitely, can biomagnify through the food web, and pose a threat to human health as well as to the environment. Because they break down only very slowly and generally under a narrow range of environmental circumstance, POPs can be transported over long distances, as air or water pollutants, across international and regional boundaries. These pollutants ultimately find their way to all corners of the planet, where they accumulate in the fatty tissues of consumers, including those of humans. Some of the POPs considered by the EPA to be particularly problematic are listed in **Table 13.3.**

Most of the chemicals listed in the table are manufactured and most are pesticides, although not all. For example, dioxins and furans are not manufactured but are produced unintentionally as a by-product of waste burning. Also, polychlorinated biphenyls (PCBs) are not pesticides but are manufactured for a wide variety of uses in industry. While all of these chemicals are now banned from being produced, imported, or exported from the United States or are heavily regulated as hazardous toxins or pollutants, some are still used elsewhere, particularly in developing countries.

TABLE 13.3
PERSISTENT ORGANIC POLLUTANTS

The U.S. EPA refers to the 12 persistent organic pollutants listed here as the "Dirty Dozen." Because they are all hazardous and persistent, their production, import, or export have been banned in the United States, or they are regulated as hazardous toxins or pollutants.

Industrial Chemical	Source	Historical Uses
PCBs (polychlorinated biphenyls)	Manufactured; unintentional production during burning	Used in electrical transformers and capacitors, as heat exchange fluids, as paint additives, in carbonless copy paper, and in plastics
Biocides		
Aldrin	Manufactured	Insect control on crops (e.g., corn, cotton) and for termite control
Chlordane	Manufactured	Insect control on crops, including vegetables, small grains, potatoes, sugarcane, sugar beets, fruits, nuts, citrus, and cotton; also used on home lawn and for garden pest and termite control
DDT	Manufactured	Used primarily to control insects on cotton; also used to control insects that carry diseases (e.g., malaria, typhus)
Dieldrin	Manufactured	Insect control on crops (e.g., corn, cotton) and for termite control
Heptachlor	Manufactured	Used to control soil insects, termites, some crop pests, and malarial mosquitoes
Hexachlorobenzene	Manufactured; also produced unintentionally during burning and the manufacture of certain chemicals; an impurity in certain pesticides	Fungicide used to treat seeds (e.g., wheat seeds) to reduce fungal infections
Mirex	Manufactured	Used to control fire ants and yellowjacket wasps; also as an additive to fire retardants
Toxaphene	Manufactured	Used to control insects on crops (e.g., cotton, pineapple, bananas); also to control insect pests of livestock and poultry
Unintentional POP		
Dioxins (polychlorinated dibenzo-p-dioxins) Furans (polychlorinated dibenzofurans)	Produced unintentionally during most forms of combustion, including burning of municipal and medical wastes, backyard burning of trash, and industrial processes; also found as trace contaminants in certain herbicides, wood preservatives, and in PCB mixtures	Waste product, no commercial uses

Information from the U.S. EPA, 2015b.

However, there are many POPs (e.g., flame retardants) still being manufactured and in widespread use, which may merit future regulation.

Heavy Metals

Heavy metals are metallic chemical elements with high atomic weights. Although there are many heavy metals, those of environmental concern are toxic to humans, animals, and plants (Table 13.4, page 395). Humans, plants, and animals require low doses of some heavy metals, such as copper and zinc, but at high concentrations the metals become toxic. Other metals, including mercury and lead, are toxic even at low concentrations. Arsenic and selenium are not technically metals but are generally included in lists of heavy metals because of their similar toxic effects and behavior in the environment. One of the most common sources of heavy metals is the burning of coal. Because coal represents the fossil remains of living organisms, it contains all of the elements in living systems in a concentrated form, including a wide array of heavy metals.

Organic Pollution

Excessive organic matter can stress aquatic ecosystems in particular. Organic pollution has several potential sources, including domestic sewage, aquaculture, and agriculture, especially concentrated animal feeding operations, or CAFOs. All three are potential sources of massive inputs of organic matter into aquatic ecosystems—generally rivers, estuaries, or coastal waters (Figure 13.5). The discharge of organic matter into an aquatic ecosystem increases **biochemical oxygen demand,** or **BOD,** an indicator of the amount of organic matter in water. Biochemical oxygen demand is measured as the quantity of oxygen consumed

Does all natural water have some BOD?

biochemical oxygen demand (BOD) An indicator of the amount of organic matter in water, measured as the quantity of oxygen consumed by microorganisms as they break down the organic matter in a sample of water.

FIGURE 13.5 (a) Untreated or incompletely treated domestic waste can be a significant source of organic pollution. Here, a broken sewer line releases 6 tons of raw sewage per second into the ocean off Ipanema Beach in Rio de Janeiro, Brazil. (b) Discharge from land-based aquaculture facilities or waste feed from water-based systems often contains substantial amounts of organic matter. (c) Meanwhile, concentrated animal feeding operations, including cattle feedlots, pig farms, and poultry farms, are a well-known source of organic pollution.

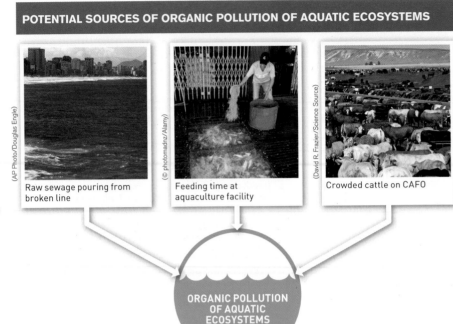

POTENTIAL SOURCES OF ORGANIC POLLUTION OF AQUATIC ECOSYSTEMS

(AP Photo/Douglas Engle)
Raw sewage pouring from broken line

(© photomadnz/Alamy)
Feeding time at aquaculture facility

(David R. Frazier/Science Source)
Crowded cattle on CAFO

ORGANIC POLLUTION OF AQUATIC ECOSYSTEMS

eutrophication A natural process by which nutrients, especially those that limit primary production, build up in an ecosystem.

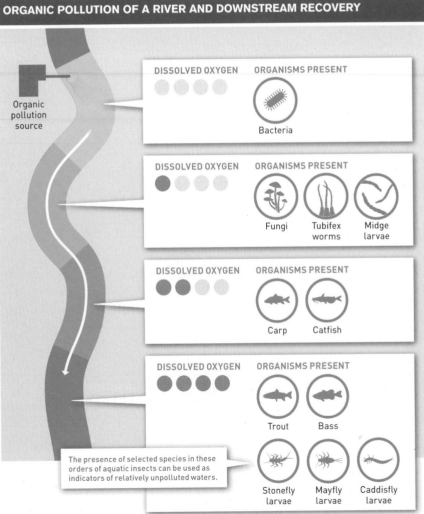

ORGANIC POLLUTION OF A RIVER AND DOWNSTREAM RECOVERY

Organic pollution source

DISSOLVED OXYGEN   ORGANISMS PRESENT
Bacteria

DISSOLVED OXYGEN   ORGANISMS PRESENT
Fungi   Tubifex worms   Midge larvae

DISSOLVED OXYGEN   ORGANISMS PRESENT
Carp   Catfish

DISSOLVED OXYGEN   ORGANISMS PRESENT
Trout   Bass
Stonefly larvae   Mayfly larvae   Caddisfly larvae

The presence of selected species in these orders of aquatic insects can be used as indicators of relatively unpolluted waters.

by microorganisms as they break down the organic matter in a sample of water.

Where additions of organic matter are large and, consequently, BOD is high, respiration by the bacteria and fungi consuming the added organic matter can deplete dissolved oxygen to levels low enough that all organisms are eliminated except bacteria capable of living in the absence of oxygen (**Figure 13.6**). Downstream from this "anaerobic" zone, where some dissolved oxygen is present, you can commonly find dense growths of fungi feeding on the abundant organic matter, along with aquatic invertebrate animals, such as tubifex worms and midge larvae, which are tolerant of low oxygen conditions. Farther downstream, where oxygen levels are still low but somewhat higher, the fish community will be limited to species such as carp and catfish, which are tolerant of lower oxygen concentrations.

Cultural Eutrophication

In natural ecosystems, nutrients gradually build up through a process called **eutrophication.** Most frequently studied in lakes, eutrophication includes increases in nutrient availability and sediments, decreases in lake depth, shifts in the makeup of the biological community, and increases in primary production (**Figure 13.7**). It can also occur on land and involve the accumulation of organic matter and inorganic nutrients in soils. Humans

FIGURE 13.6 Organic pollution of a river or stream from a point source commonly creates a series of predictable physical and chemical environments downstream, which are in turn associated with predictable changes in the biological community.

TABLE 13.4
HEAVY METALS OF CONCERN FOR SOIL POLLUTION

Heavy Metal	Symbol	Sources	Health Effects
Arsenic	As	Erosion of natural deposits; agricultural and industrial practices; coal burning	Thickening and discoloration of the skin; stomach pain, nausea, vomiting; diarrhea; numbness in hands and feet; partial paralysis; blindness; linked to cancer of the bladder, lungs, skin, kidney, nasal passages, liver, and prostate
Cadmium	Cd	Coal and municipal waste burning; emissions from metal	Animal studies indicate potential damage to kidney, liver, lung, bone, immune system, blood, and nervous systems; interferes with neurological development; probable carcinogen
Chromium	Cr	Refining of metal ores and several industrial processes; coal burning	Irritation of the lungs; shortness of breath; decreased lung function; lung cancer
Copper	Cu	Occurs in natural ores with other metals; emissions from metal smelters and refineries; coal burning	Irritation of nose, mouth, and eyes; stomach distress, vomiting; kidney and liver damage
Lead	Pb	Paints in older buildings and furniture; plumbing; metal smelters; coal burning	Developmental disorders, especially of the central nervous system; hearing problems; headaches; slowed growth in children; reproductive problems; high blood pressure; nerve disorders; joint and muscle pain in adults
Mercury	Hg	Coal burning; industrial processes; consumption of contaminated fish and shellfish	Impaired neural development in developing infants and children; impairment of speech, hearing, walking; muscle weakness in people of all ages; kidney damage, respiratory failure, and death at higher exposure
Molybdenum	Mo	Coal burning; metal smelters and refineries; mines; dust from contaminated soils; sewage sludge; contaminated food	Weakness; fatigue; headaches; liver damage; mental disorientation; swollen and painful joints
Nickel	Ni	Oil and coal burning; metal smelters; nickel metal refining; sewage sludge incineration; industrial processes; contaminated food	Itching of fingers, hands, and forearms; stomach distress; lung and kidney damage; increased risk of nasal and lung cancers
Selenium	Se	Erosion of natural deposits; discharge from oil and metal refineries; coal burning; contaminated food and water	Irritation of the mucous membranes; bronchitis, bronchial pneumonia; loss of hair and nails; tooth decay; circulatory problems; reduced mental alertness
Zinc	Zn	Coal and waste burning; sewage sludge; metal smelters; mining, dust from contaminated soils; contaminated food and water; galvanized pipes	Stomach discomfort; skin irritations; anemia; damage to pancreas

Information from U.S. Natural Resource Conservation Service 2000; additional data from multiple sources.

LAKE EUTROPHICATION

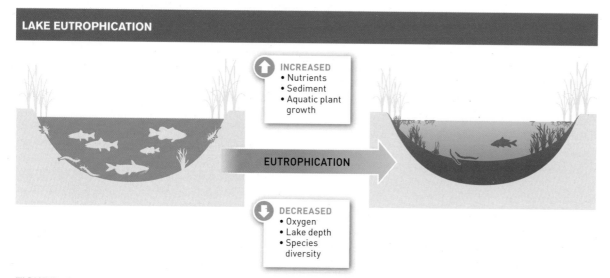

FIGURE 13.7 The natural process of lake eutrophication, which transforms the ecosystems from low to high productivity, takes place, generally, over long periods of time as the result of inputs of nutrients and sediments.

A CONSEQUENCE OF CULTURAL EUTROPHICATION

(Jim W. Grace/Science Source)

FIGURE 13.8 Accelerating eutrophication by excessive nutrient additions to lakes can lead to such high levels of primary production that decomposition by decaying biomass may reduce oxygen concentrations to levels below the lethal limits of many species of fish.

cultural eutrophication An accelerated process of eutrophication resulting from human activities (e.g., sewage disposal, agriculture) that increase the rate of nutrient addition to ecosystems; generally results in excessive algal or plant production, depletion of dissolved oxygen in aquatic ecosystems, and loss of biodiversity.

transboundary pollution The transport of pollutants by wind and water around the biosphere, across geographical and political borders.

aerosol Tiny particles of solid or liquid material suspended in air or other gas.

can accelerate eutrophication through the excessive addition of nutrients, such as sewage and fertilizer, either directly to a body of water or indirectly through runoff from the surrounding landscape. This condition is generally referred to as **cultural eutrophication.** Cultural eutrophication leads to excessive algae and plant growth, the depletion of dissolved oxygen in aquatic ecosystems, and the loss of biodiversity (**Figure 13.8**).

⚠ **Think About It**

1. What are the major differences between the sulfur cycle (see Figure 13.4) and the phosphorus cycle (see Figure 8.6, page 235)?

2. Of the vast number of potential pollutants, what properties would lead the U.S. EPA to choose just six as criteria pollutants?

3. Oxygen is important to the health of both aquatic and soil ecosystems. Why, then, is adding large amounts of organic matter considered a form of pollution in aquatic ecosystems, but not generally in soils?

13.3 Atmospheric and aquatic transport eventually move pollutants around the planet

While the concentrations of pollutants in local areas can build up to very high levels (see introduction, page 386), those pollutants won't necessarily stay in place. Wind and water transport pollutants around the biosphere, crossing geographical and political borders—a phenomenon called **transboundary pollution.** For instance, heavy metals and pesticides released in Europe and North America have been found in both terrestrial and aquatic ecosystems far to the north in the Arctic.

Atmospheric Circulation

The atmosphere and the oceans are circulating constantly and redistributing material, including pollutants, throughout the biosphere (see Figure 8.2, page 232). The 1991 eruption of Mount Pinatubo in the Philippines, for example, provided an unprecedented opportunity to study the dispersal of materials by atmospheric circulation. There had been much larger volcanic eruptions within historic times, but never before had there been satellites in place equipped with sensors that could document the spread of materials injected into the atmosphere by the eruption. Mount Pinatubo injected approximately 20 million metric tons of SO_2 into the stratosphere, which were gradually converted chemically into 30 million metric tons of sulfuric acid and water aerosols.

An **aerosol** consists of tiny particles of solid material or tiny liquid droplets suspended in air or other gas. In the case of the stratospheric aerosols resulting from the Pinatubo eruption, the droplets were made up of a mixture of sulfuric acid and water. These aerosols were initially confined to a tropical band encircling Earth, but they spread through the stratosphere over a period of two years (**Figure 13.9**). Gradually, atmospheric circulation and gravity removed these acidic aerosols from the stratosphere into the troposphere (the lower atmosphere) and to the surface of Earth.

Pollutants emitted into the lower atmosphere are also widely dispersed across Earth. During his voyage on the *H.M.S. Beagle,* Charles Darwin noted the fallout of sand and dust on the ship far out in the North Atlantic Ocean off North Africa. Despite the ship's great distance from Africa, he inferred that the material was being blown by prevailing winds from the Sahara Desert. We know

OBSERVATION OF GLOBAL-SCALE CIRCULATION FOLLOWING THE ERUPTION OF MOUNT PINATUBO, PHILIPPINES

1991 eruption of Mount Pinatubo, Philippines

JUNE 15 TO JULY 29, 1991

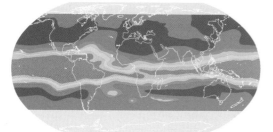

AUGUST 15 TO SEPTEMBER 24, 1993

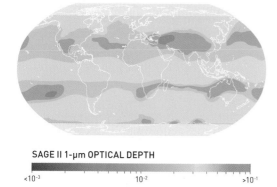

SAGE II 1-μm OPTICAL DEPTH

<10⁻³   10⁻²   >10⁻¹

FIGURE 13.9 Mount Pinatubo erupted on June 15, 1991, and within two years the products of that eruption had spread throughout the globe. (After McCormick et al., 1995)

now that Darwin was correct and that, in fact, dust is transported from the Sahara Desert all the way across the Atlantic to the Amazon River Basin (**Figure 13.10**). The atmosphere can also transport other pollutants long distances. For example, ozone and mercury produced by the burning of coal in China are deposited across North America.

DUST FROM THE SAHARA DESERT

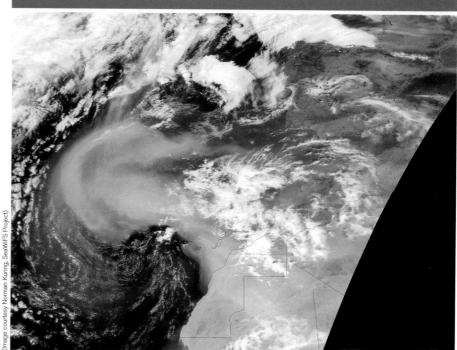

FIGURE 13.10 Some of this dust, which in high enough concentration is a particulate pollutant, originating in the Sahara Desert will be transported by winds all the way across the Atlantic Ocean and deposited in the Amazon River Basin.

Oceanic Transport

The 2011 nuclear accident at Fukushima, Japan, provided dramatic evidence of oceanic transport. The tsunami that disabled the Fukushima nuclear power plant caused immense destruction of property and much of the debris created by the tsunami was dragged out to sea with the receding water. The massive collection of debris, which contained whole ships and largely intact houses, was transported with oceanic currents across the Pacific Ocean to the coasts of Canada and the United States (**Figure 13.11**).

Pollutants can also be transported across the oceans with the help of organisms. For example, radioactive isotopes leaked during the nuclear accident at Fukushima were found in juvenile bluefin tuna caught off the California coast approximately 1 year later (**Figure 13.12**). These young tuna had taken up radioactive cesium-134 while swimming in the waters off Japan and then swam across the Pacific Ocean to their feeding grounds off the West Coast of North America. The levels of cesium-134 in the tuna did not present a health hazard, but it demonstrates how a contaminant can be quickly transported halfway around the planet.

Watersheds

Surface water and groundwater carry pollutants. A **watershed,** also known as a drainage basin or catchment,

How have satellites changed the contributions of imagination versus measurement in studies of large-scale environmental processes?

Given what we now know about atmospheric and oceanic transport, are the concepts of local, regional, and global pollution still useful?

watershed (catchment, drainage basin) The land area from which an aquifer or river system acquires its water; also defined as the dividing line between catchments or drainage basins.

DEBRIS WASHED OUT TO SEA BY FUKUSHIMA TSUNAMI

(U.S. Navy photo by Mass Communication Specialist 3rd Class Alexander Tidd)

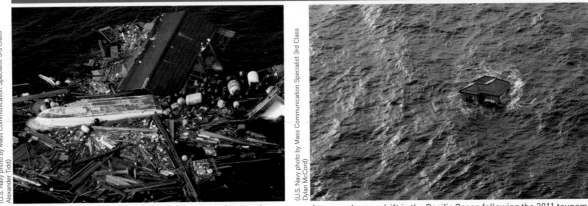

Floating debris from March 13, 2011, earthquake and tsunami off northern Japan

(U.S. Navy photo by Mass Communication Specialist 3rd Class Dylan McCord)

Japanese home adrift in the Pacific Ocean following the 2011 tsunami

FIGURE 13.11 The tsunami that disabled the Fukushima nuclear power plant in March 2011 washed millions of tons of floating debris out into the Pacific Ocean, much of which was transported across the North Pacific by oceanic currents. The transported materials included entire houses.

is the land area drained by a water system, such as the Chesapeake Bay watershed (**Figure 13.13**). Precipitation in a watershed has several potential fates. Some precipitation immediately evaporates and returns to the atmosphere as water vapor. A portion runs off the land as surface flow, eventually entering waterways, ranging from small streams to rivers. Some precipitation infiltrates into the soil, where it may be taken up by terrestrial plants and stored as tissue moisture or transpired as water vapor back into the atmosphere. Yet another fraction of the precipitation, which soaks into the ground, becomes groundwater (see Chapter 6, page 159). Flows of surface water and groundwater determine where the pollutants will move and, as shown by the map of the Chesapeake Bay watershed, these flows often cross political boundaries.

Airsheds

The concept of an airshed is similar to a watershed but related to movements of air rather than water. We

PACIFIC BLUEFIN TUNA, *THUNNUS ORIENTALIS*

(© Mark Conlin/Alamy)

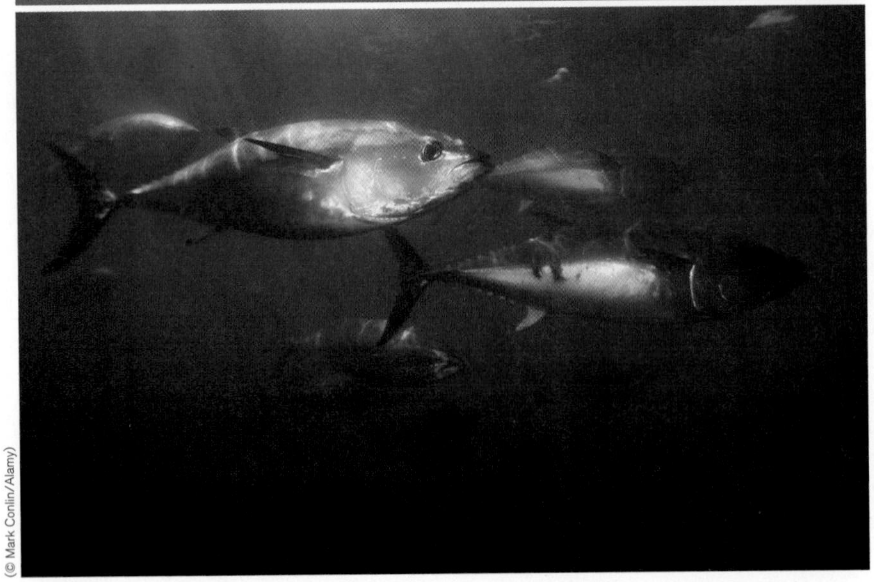

FIGURE 13.12 Pacific bluefin tuna spawn in the western Pacific Ocean off the coast of Asia and then migrate to the eastern Pacific Ocean to feed in the highly productive coastal waters off North America. The juvenile bluefin tuna developing in the waters off Japan, following the Fukushima nuclear accident, absorbed radioactive material leaked during the disaster and brought it with them in their transpacific migration.

THE CHESAPEAKE BAY WATERSHED

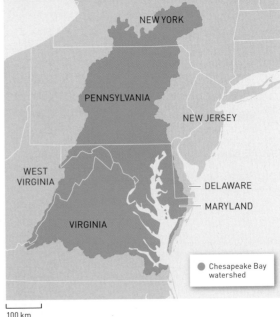

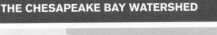

FIGURE 13.13 Water draining into the Chesapeake Bay from its watershed originates in several states.

AIRSHED FOR DEPOSITION OF NITROGEN OXIDES IN CHESAPEAKE BAY

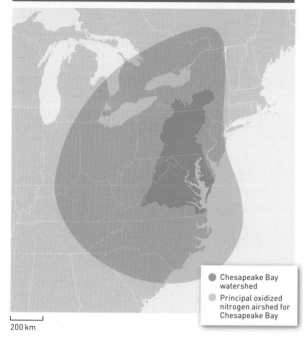

- ● Chesapeake Bay watershed
- ○ Principal oxidized nitrogen airshed for Chesapeake Bay

200 km

FIGURE 13.14 The airshed for Chesapeake Bay is much larger than its watershed. (Paerl et al., 2002)

define an **airshed** as a part of the atmosphere that has a consistent airflow, primarily due to the location of mountains and prevailing wind patterns. Since the air mass within an airshed typically behaves in a united and orderly way, a focus on airsheds can be used to track and manage air pollution. **Figure 13.14** shows the airshed for the deposition of nitrogen oxides in the Chesapeake Bay, along with its watershed. Because it is less confined, the airshed is generally much larger than the watershed.

The large area of airsheds has significant consequences for managing pollution. For example, in December 2013, eight Mid-Atlantic and Northeast states filed a petition with the EPA to require nine Midwest and Appalachian states, which were upwind of them, to reduce air pollution within their borders. The states issuing the petition claimed that prevailing winds were transporting pollutants to them from outside their borders and that

the upwind states are part of an airshed that crosses many state boundaries.

⚠ Think About It

1. How are atmospheric and oceanic transport linked physically?

2. How are watersheds and airsheds similar? In what ways do they differ?

3. In the United States, does it make more sense, environmentally, to regulate pollution at the state level or at the scale of watersheds and airsheds? Explain.

13.1–13.3 Science: Summary

Pollution is contamination of the environment by substances or physical conditions, such as noise or light, at levels harmful to living organisms. Sources of pollution are generally divided into point and nonpoint sources. Some pollutants, called primary pollutants, are harmful in the form in which they are released into the environment, while other substances, called secondary pollutants, are harmful following chemical reactions in the environment. The EPA has singled out a few significant air pollutants, called criteria pollutants, for monitoring and management. Heavy metals can be a significant source of soil and water pollution. Excessive additions of organic matter to aquatic ecosystems will increase biochemical oxygen demand. Additions of excessive nutrients lead to eutrophication. Movements of air, water, and organisms move pollutants around the biosphere. Surface and groundwater flows are avenues for the dispersal of pollutants within watersheds. Airsheds can be used to track and manage regional air pollution.

airshed A concept equivalent to a watershed or drainage basin but related to movements of air rather than water; typically behaves in an orderly way, thus can be used to track and manage air pollution.

13.4–13.7 Issues

The problem of pollution is not a new one for humanity. Traces of heavy metals, for instance, still linger in the soils around Roman smelters dating back more than 2,000 years. But the scope and diversity of pollutants increased with the Industrial Revolution and the growing human population.

One of the most infamous examples of pollution comes from the small town of Sudbury, Ontario. In the early 20th century, the nickel and copper mining industry there was booming. Metals were extracted from their ores through a process called *open bed roasting,* in which ore was heaped on a pile of burning logs that would smolder for months, releasing gaseous fumes containing sulfur dioxide, which in turn caused acid deposition. When a reporter visited in 1908, he called Sudbury "one of the most unattractive places under the sun, for the sulphur fumes . . . have destroyed vegetation in the whole locality, leaving the rocky hills bare of trees and the streets and lawns innocent of a blade of grass." By the 1970s, the land was left so desolate and devoid of life that NASA used it as a place to practice landing on the moon. Today, in the middle of the Anthropocene era, it is almost impossible to find a place on Earth that our pollution has not touched.

13.4 Air pollution exacts major health-related and economic tolls

Pollution damages ecosystems and human infrastructure, both of which result in significant economic costs. It also seriously impacts human health. According to the World Health Organization (WHO), air pollution kills over 7 million people each year—more than malaria and HIV combined. Of those, less than half of the fatalities result from outdoor pollution, primarily in cities, and the remainder result from **indoor air pollution** due to, for example, the fumes of wood and coal burned for heat and cooking. Many people are exposed to both sources of pollution.

The Toll of Indoor Air Pollution

Pollution is not something that only happens outside. Less air exchange indoors allows a variety of pollutants to build up inside buildings, potentially rising to unhealthy concentrations. The indoor pollutants of concern include VOCs, carbon monoxide (CO), radon, asbestos, and tobacco smoke (**Figure 13.15**). **Sick building syndrome** is a circumstance in which many building occupants

indoor air pollution A serious threat to human health resulting from the buildup of pollutants in the indoor environment.

sick building syndrome A circumstance in which many building occupants experience symptoms of illness (e.g., headaches, respiratory and eye irritation, nausea) for which no specific cause has been identified.

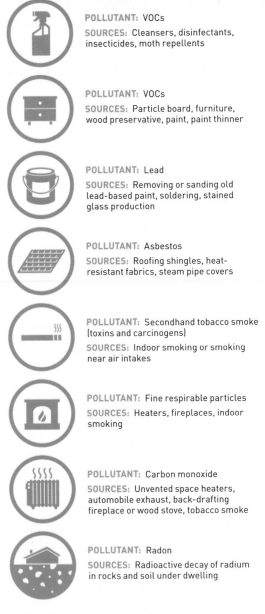

HOUSEHOLD SOURCES OF INDOOR AIR POLLUTION

POLLUTANT: VOCs
SOURCES: Cleansers, disinfectants, insecticides, moth repellents

POLLUTANT: VOCs
SOURCES: Particle board, furniture, wood preservative, paint, paint thinner

POLLUTANT: Lead
SOURCES: Removing or sanding old lead-based paint, soldering, stained glass production

POLLUTANT: Asbestos
SOURCES: Roofing shingles, heat-resistant fabrics, steam pipe covers

POLLUTANT: Secondhand tobacco smoke (toxins and carcinogens)
SOURCES: Indoor smoking or smoking near air intakes

POLLUTANT: Fine respirable particles
SOURCES: Heaters, fireplaces, indoor smoking

POLLUTANT: Carbon monoxide
SOURCES: Unvented space heaters, automobile exhaust, back-drafting fireplace or wood stove, tobacco smoke

POLLUTANT: Radon
SOURCES: Radioactive decay of radium in rocks and soil under dwelling

FIGURE 13.15 The many potential sources of indoor air pollution can easily reduce indoor air quality to unhealthy levels.

experience symptoms of illness, such as headaches, respiratory and eye irritation, and nausea, for which no specific cause has been identified. Sick building syndrome has been frequently observed in new buildings

GLOBAL DEATHS FROM INDOOR POLLUTION

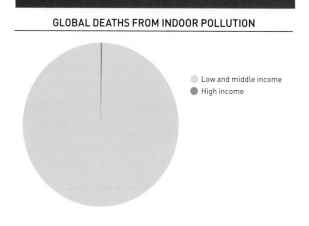

Low and middle income
High income

DEATHS IN LOW- AND MIDDLE-INCOME COUNTRIES

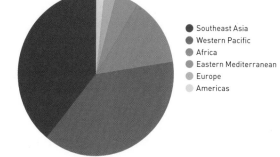

Southeast Asia
Western Pacific
Africa
Eastern Mediterranean
Europe
Americas

FIGURE 13.16 More than 4 million deaths each year from respiratory infections, chronic diseases, and cancers are attributed to exposure to indoor pollution primarily within low- and middle-income countries. (WHO, 2014)

in which many of the recently constructed components are still releasing significant amounts of VOCs from adhesives, plastics, and other materials.

Indoor air pollution is especially severe in developing countries. Nearly 3 billion people in developing countries rely on open fires or inefficient stoves, where they burn biomass, animal dung, or coal indoors. The resulting indoor air pollution takes a tremendous health toll. The WHO estimates that in 2012 alone, over 4 million people exposed to indoor air pollution, primarily in low- and middle-income countries, died prematurely as a result of various ailments, including lung cancer, pneumonia, and stroke (**Figure 13.16**).

Economic Costs of Air Pollution

The damage caused by air pollution, including ill health and shortened lives, comes with real economic costs. These costs can be estimated in terms of deterioration of infrastructure, such as buildings and bridges, reduced agricultural or forest productivity, loss of valuable commercial or recreational fisheries, and so forth. Other costs are related to the impacts of pollution on human health and longevity. For example, the loss of work time due to chronic illness or premature death reduces worker productivity and generates costs related to health care. These and many other types of damages, referred to as externalities in Chapter 2 (see page 49), are estimated to significantly reduce gross domestic product (GDP), the economic bottom line for nations. These costs have not been reliably estimated for all pollutants or for every economy. However, a substantial amount of work has been done to estimate the costs for some of the major sources of pollution in the United States and China, the two largest national economies.

Costs of Air Pollution in the United States

Nicholas Muller and Robert Mendelsohn of the Yale School of Forestry and Environmental Studies estimated the costs of damages created by six common air pollutants: ammonia, fine and coarse particulates, sulfur dioxide, nitrogen dioxide, and VOCs. Making such an estimate for the lower 48 United States was no small task. First, Muller and Mendelsohn had to calculate the concentrations of pollutants across the country and then, using population information, estimate the number and level of exposures of people to the pollutants. Next, they converted exposure levels to likely effects on human mortality, disease rates (morbidity), agriculture, timber production, visibility, structural materials, and recreation. The final step in the analysis was to estimate the economic value of these physical damages. They repeated this analysis for each of the six pollutants at 10,000 locations across the country.

Muller and Mendelsohn's calculations resulted in an estimate of annual damages by the six pollutants, which they called *gross annual damages,* or GAD. Summing their estimates for the six common pollutants, Muller and Mendelsohn estimated the GAD for the United States in 2002—the year on which their analysis was focused—at $75 billion to $280 billion. Their estimate covered a broad range of costs because there are numerous uncertainties for the many calculations that went into their estimate. A major uncertainty was the value to place on human mortality risk and how that value changes with age. Another critical uncertainty concerns how exposure to pollutants affects mortality rates. However, regardless of where the actual costs of pollution fall within this range, they are substantial. Muller and Mendelsohn's estimates placed those costs at 0.7 to 2.8% of U.S. GDP in 2002. The estimated costs of pollution in China are also significant.

Costs of Air Pollution in China

China's economic growth can be described as nothing less than explosive. However, with this growth has come a

In today's increasingly interconnected world, what environmental damages are external to the global economy?

ACID RAIN AND FOREST DESTRUCTION

(Richard Packwood/Getty Images)

FIGURE 13.17 Dieback of forest trees, such as these spruce trees, has been an all too common response to acid rain around the world.

massive increase in pollution (see introduction, page 386). Concerned about rising pollution levels, China approached the World Bank in 2003 about studying the cost of pollution to the country in terms of impacts on health and nonhealth sectors of the economy. In its study, the World Bank team worked closely with China's ministries of agriculture, health and water resources and that nation's own State Environmental Protection Administration.

One of the key findings of the 2007 study was that rising air pollution accounts for increases in lung disease, including cancer, leading to increased rates of absenteeism in schools and the workplace. Increasing levels of water pollution correlate with higher levels of diarrhea, especially in children under 5, and cancers. The report concludes that by degrading water quality, pollution is worsening China's chronically short water supplies. According to the World Bank study, the combined costs of air and water pollution in 2003 amounted to approximately $100 billion, or about 5.8% of China's GDP that year. The estimated costs were 4.3% of GDP for health-related issues and 1.5% for non–health-related aspects of the economy. However you divide these costs, they are major and will likely continue to soar unless China's rising economic fortunes can be decoupled from increasing pollution.

⚠ Think About It

1. How might construction companies increase their business share by promoting the use of low-VOC materials in their buildings?

2. What are the potential negative consequences of quantifying pollution's impact in terms of GDP?

3. What factors may contribute to a higher impact of pollution on China's contemporary economy compared with that of the United States?

13.5 Acid rain is a major source of damage to aquatic and terrestrial ecosystems

From the beginning, almost everyone knew who was to blame for the acid rain in Sudbury, Ontario. In 1918 a group of farmers there sued and later won a lawsuit against the Canadian Copper Company for damage done to their crops by fumes from the company's smelting operations. Nevertheless, broader recognition of the problem of acid rain would take another century as scientists started documenting highly acid rains far from population centers and because of its conspicuous damage to forests.

How Acid Rain Harms Ecosystems

Beginning in the mid-1950s, published observations of acid rain in isolated locations came from Scandinavia. A decade later, scientists speculated on the potential for acid rain to cause damage to ecosystems far from the source of air pollutants. And soon there were reports of reduced tree growth and dying forests in Northern Europe, the United States, and Canada (**Figure 13.17**). In 1960 Eville Gorham and Alan Gordon demonstrated that the farther plant communities were from a smelter near Sudbury, Ontario, the greater the plant diversity, leveling off at a distance of 25 kilometers or more from the smelter (**Figure 13.18**).

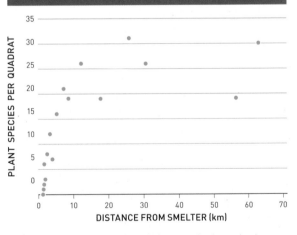

SMELTER IMPACT ON TERRESTRIAL VEGETATION

(Graph: x-axis "DISTANCE FROM SMELTER (km)" from 0 to 70; y-axis "PLANT SPECIES PER QUADRAT" from 0 to 35)

FIGURE 13.18 The number of plant species in randomly located 2-by-20-meter quadrats increased rapidly with distance from the Falconbridge smelter near Sudbury, Ontario, reaching a plateau after a distance of about 25 kilometers from the smelter. (Data from Gorham & Gordon, 1960)

GLOBAL HOTSPOTS FOR SULFATE (SO₄) DEPOSITION

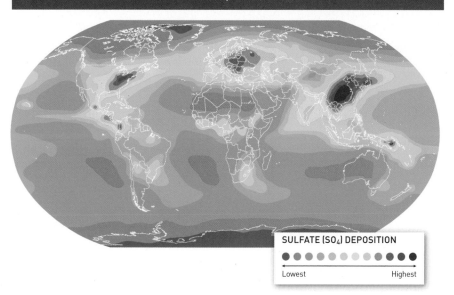

SULFATE (SO₄) DEPOSITION

Lowest   Highest

FIGURE 13.19 Sulfate deposition from industrial sources, which is indicative of acid rain, was highest in northeastern North America, Europe, and Eastern Asia. As shown in the figure, sulfate deposition in China far exceeds that of other regions. (After Dentener et al., 2006)

Today, most acid rain occurs in northeastern North America, Europe, and Eastern Asia (**Figure 13.19**), largely because these are the world's largest consumers of fossil fuels (see Figure 9.19, page 275).

Regions such as the northeastern United States, much of Canada, and Scandinavia are particularly vulnerable because their soils have a low capacity for buffering the effects of acids as a result of their low base content. A **base** is a substance that has the capacity to neutralize acids. In general, **buffering capacity** is a measure of the ability to neutralize acid. Buffering capacity is higher for soils, or waters, with higher base concentrations because they can absorb more acid without changing pH. Soils in other regions, such as the U.S. Midwest and Southwest, generally have higher base content and thus a greater capacity to neutralize acids; they are therefore less threatened by acid rain.

Acid rain harms trees by depleting soil nutrients, which are dissolved and washed away in acidic soil water. Furthermore, acid rain releases aluminum from soils, which is toxic to plants in high enough concentrations. Consequently, with long exposure to acid rain, forest soils are slowly stripped of essential plant nutrients and can become increasingly toxic. Exposure to acidic clouds and fogs can also cause damage to the leaves and needles of trees. Plants stressed by such conditions become vulnerable to a number of other potential sources of mortality, including insects, disease, drought, and cold.

Impacts on Aquatic Ecosystems

Acid rain also impacts organisms that live in water. When acidic water seeps into soils and infiltrates the shallow groundwater, it releases aluminum. When this runoff water flows into streams and lakes, its pH is low and its aluminum content is high—conditions that are toxic to many organisms, including fish and a wide variety of invertebrate species. According to the EPA, freshwater clams and snails are eliminated at a pH of 5.5. At a pH of 5.0, bass, crayfish, and mayflies do not survive, and at a pH of 4.5, trout and salamanders die out. At a pH of 4.0, of the vertebrate animals normally found in the community, only frogs survive. In general, the early life stages of aquatic organisms, such as eggs and larvae, are more vulnerable to acidification of their environments than are adults.

You might be wondering whether acid rain impacted the many lakes around Sudbury, Ontario. Gorham and Gordon also surveyed the aquatic plants living in study lakes near Sudbury. They found that the sulfate concentrations in the lakes decreased with distance from metal smelters and that the number of aquatic plant species increased with distance from smelters. By plotting sulfate concentration against number of aquatic plant species in these lakes, they determined that aquatic plant species numbers decline as sulfate concentration increases (**Figure 13.20**). The concentrations of sulfate in these lakes, however, didn't seem high enough to kill the plants, so Gorham and Gordon wondered whether another pollutant was responsible. Indeed, they found that copper, a metal toxic to plants at higher concentrations, was highest in snow near the smelters and declined with distance (**Figure 13.21**).

⚠ Think About It

1. How would you explain the fact that the earliest discoveries of acid deposition impacts were in terrestrial ecosystems, whereas the phenomenon

Who should compensate for damage to ecosystems by acid rains?

What criteria would you use to choose organisms as indicators of the impacts of acid rain on aquatic or terrestrial ecosystems?

base A substance that has the capacity to neutralize acids; bases release hydroxide ions (OH⁻) and react with acids to form a salt and water.

buffering capacity A measure of the ability of a solution to neutralize acid.

FIGURE 13.20 (a) The sulfate concentrations in lakes near Sudbury, Ontario, decreased with distance from metal smelters. (b) In an opposite pattern, the number of aquatic plant species in the same lakes increased with distance from smelters. (c) Combining the patterns in parts (a) and (b) reveals a negative relationship between sulfate concentrations in these lakes and number of aquatic plant species. (Data from Gorham & Gordon, 1963)

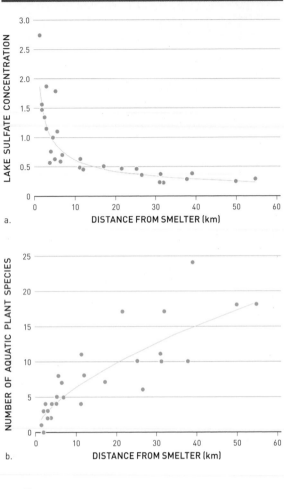

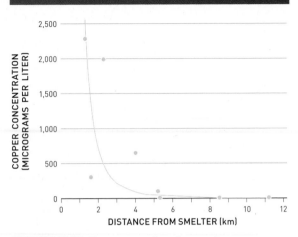

FIGURE 13.21 The concentration of copper in snow collected at increasing distances from the smelter at Falconbridge, Ontario, decreased from more than 2,000 micrograms per liter to less than 5 micrograms per liter. (Gorham & Gordon, 1963)

13.6 Persistent pollutants enter the human food chain

Persistent pollutants can cause environmental problems years or even decades down the line. Even after we have recognized the threat of certain chemicals and begin to control their use, we are still dealing with their fallout.

PCBs and Hudson River Fish

Polychlorinated biphenyls, or PCBs, are a persistent pollutant that has been linked to various cancers, lower sperm counts, and learning disabilities. Before being banned in 1979, approximately 675 million kilograms (1.5 billion pounds) of PCBs were used in the manufacture of a variety of products from microscope oils to refrigerators. They were dispersed from urban settings to the high Arctic, and the most notorious case of PCB contamination comes from New York State.

Over the three decades from 1947 to 1977, a General Electric manufacturing plant released 585,000 kilograms (1.3 million pounds) of PCBs into the Hudson River at Hudson Falls, approximately 320 kilometers (200 miles) upstream of New York City. By the 1970s, fish in the river were so toxic they were deemed unsafe for human consumption. In 1976 New York's Department of Health advised women of child-bearing age and children under age 15 not to eat any fish from the Hudson River below the GE plant at Hudson Falls and that no one eat fish between Hudson Falls and the Federal Dam at Troy, New York, 80 kilometers (50 miles) downriver (**Figure 13.22**).

Following their ban in 1979, General Electric stopped discharging PCBs into the Hudson River and levels of

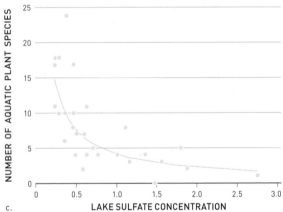

of cultural eutrophication was first documented in aquatic ecosystems?

2. What processes connect the impacts of acid rain in terrestrial ecosystems to impacts on aquatic ecosystems?

3. What are the mechanisms by which acid rain impacts ecosystems?

CONTAMINATION OF A RIVER SYSTEM AND FISH POPULATION BY PCBs

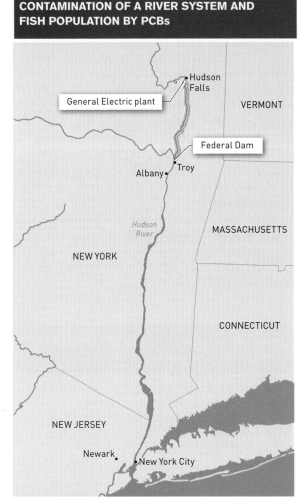

VERMONT

Hudson Falls

General Electric plant

Federal Dam

Albany • Troy

Hudson River

MASSACHUSETTS

NEW YORK

CONNECTICUT

NEW JERSEY

Newark •

• New York City

Striped Bass, *Morone saxatilis*

(Barrett & MacKay/Getty Images)

FIGURE 13.22 Although the most contaminated river sediments were found within 80 kilometers below Hudson Falls, some of the million-plus pounds of PCBs discharged there eventually found their way down the entire lower portion of the river system. Striped bass, *Morone saxatilis,* one of the most prized commercial and sport fishes along the Atlantic Coast, cannot be fished commercially in areas near the Hudson River due to the possibility of PCB contamination.

PCB concentration began to decline in the river's fish. However, concentrations remained above safe levels as PCBs continued to ooze from the riverbed sediments into the food web. Then in 1983 the EPA declared the 320 kilometers of the Hudson River above New York Harbor a Superfund site. In 1985 the coastal areas off western Long Island and New York Harbor were closed to commercial fishing for striped bass, *Morone saxatilis* (see Figure 13.22), a commercially important fish in the region. In the face of legal challenges by GE, it was more than two decades before steps to clean up the sediments in the Hudson River began in earnest in 2009. The full human and economic costs of this single episode of pollution are still being calculated.

Heavy Metals and Agriculture

Mining, metal working, and especially combustion of coal all transfer heavy metals from deep geologic formations to the surface biosphere, where they can cause long-term contamination. Heavy metals can also be introduced into soils at mining sites and with metal smelting. From there, these potent toxins and carcinogens can enter the human food chain. For instance, a survey of lead contamination of vegetable farms in Fujian, China, showed that the concentration of lead in Chinese white cabbage, *Brassica chinensis,* increased with increasing lead concentration in soils (**Figure 13.23**). The problem of heavy metal contamination of food crops is widespread in China, especially near centers of metal mining and processing.

Similar studies done elsewhere in the world, from the United States to India, have shown that vegetables grown

How might the precautionary principle play a role in preventing the release of hazardous POPs into the environment?

UPTAKE OF HEAVY METALS BY A CROP PLANT

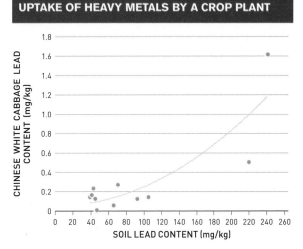

FIGURE 13.23 A survey of vegetable-growing areas in Fujian, China, revealed that the edible parts of Chinese white cabbage (*Brassica chinensis*), growing on soils with higher lead content, contained elevated concentrations of lead. The average concentrations of lead in cabbage in four of the fields exceeded the maximum permissible lead content in food, 0.2 mg/kg, allowed in China. (Data from Huang et al., 2012)

VARIATION IN LEAD CONCENTRATION IN THE EDIBLE TISSUES OF CROP PLANTS

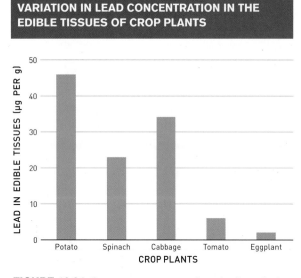

FIGURE 13.24 Crops grown on contaminated soil vary in the extent to which they concentrate heavy metals in their edible tissues. Root crops (e.g., potatoes) and leafy greens (e.g., spinach and cabbage) generally concentrate more heavy metals in their edible tissues, compared with crops producing edible fruits (e.g., tomatoes and eggplant). (Data from Singh et al., 2012)

Why is the problem of heavy metal contamination one that will not soon go away?

on contaminated soils will accumulate heavy metals in their tissues. Leafy green vegetables, such as spinach and cabbage, and some root crops are particularly apt to accumulate higher levels of heavy metals in the edible parts of the plant, compared with the fruits of crops such as tomatoes, peppers, and eggplant (**Figure 13.24**). Over the long term, heavy metals can build up in soils to the point that they are no longer safe for agricultural production.

HEAVY METAL CONCENTRATIONS IN SOILS NEAR SUDBURY, ONTARIO

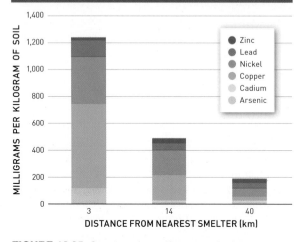

FIGURE 13.25 Concentrations of heavy metals decrease significantly with distance from metal smelters, the major source of heavy metal contamination of soils in the region. (Data from Feisthauer et al., 2006)

In Sudbury, Ontario, researchers have studied how heavy metals harm the native flora and fauna. Heavy metals are highest in soils at 3 kilometers from the nearest smelter in the vicinity of Sudbury, with a progressive decline in concentrations at distances of 14 to 40 kilometers (**Figure 13.25**). The researchers wondered how organisms would fare when raised in these areas. They found that northern wheatgrass, *Elymus lanceolatus,* developed nearly 10 times less mass when grown near the smelter than when grown 40 kilometers away (**Figure 13.26**). In addition, red earthworms, *Eisenia andrei,* which aerate and mix the soil, survived when they grew near the smelter, but they failed to reproduce.

⚠ Think About It

1. What factors make soil contamination by heavy metals a serious concern?

ASSAYING THE IMPACT OF HEAVY METALS ON TERRESTRIAL ORGANISMS

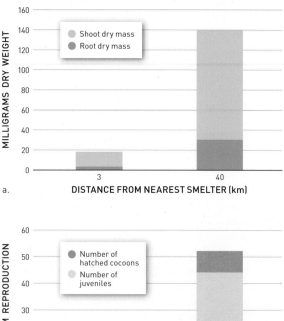

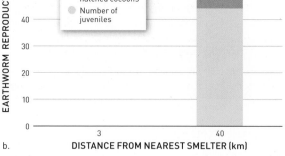

FIGURE 13.26 (a) Growth of northern wheatgrass, *Elymus lanceolatus,* was reduced when grown on soils collected near a smelter in the vicinity Sudbury, Ontario, with high levels of heavy metals. (b) In addition, earthworms, *Eisenia andrei,* did not reproduce when living in those same soils. (Data from Feisthauer et al., 2006)

2. What can be done to pursue economic development without impairing the capacity of soils to produce healthy food?

3. How might the release of heavy metals into the environment be factored into the costs of using coal as an energy source?

13.7 Organic matter and nutrient pollution can disrupt local and distant ecosystems

When too much organic matter and nutrients pollute ecosystems, they can alter their function and harm biodiversity. Around the world, nutrient pollution has been blamed for harmful algal blooms and dead zones in oceans and lakes.

Organic Pollution of Aquatic Ecosystems

Concentrated animal feeding operations, or CAFOs, are pens or buildings where livestock, including chickens, cattle, and pigs, are housed and fed to support industrial-scale meat production. Although CAFOs can increase the efficiency of meat and milk production, they have become a significant point source of air and water pollution. Three causes for concern are the large size of some CAFOs, the huge amount of waste they produce, and the rapid increase in the number of large CAFOs in the United States. A report by the U.S. Government Accountability Office reported that CAFOs in the United States are feeding up to 2 million chickens and as many as 800,000 hogs at one time.

The amount of waste produced by such operations can be staggering. For instance, the 800,000 hogs in the cited example produce more than 1.6 million tons of manure annually, 1.5 times the amount of sanitary waste produced by the 1.5 million people living in Philadelphia, Pennsylvania. Between 1982 and 2002, the number of animals raised in large CAFOs in the United States increased by more than 240%, from 257 million to nearly 900 million. Where the organic wastes from CAFOs are not managed properly, they can severely impact aquatic ecosystems (see Figure 13.5, page 394). Even where organic wastes from CAFOs are well managed, however, they are commonly a significant source of nutrient additions to groundwater and aquatic ecosystems and can be a source of eutrophication.

Nutrient Enrichment and Eutrophication

The industrial production of fertilizers and the burning of fossil fuels have more than doubled the inputs to the global nitrogen cycle since the late 20th century, causing cultural eutrophication. The extra nitrogen is either

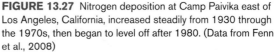

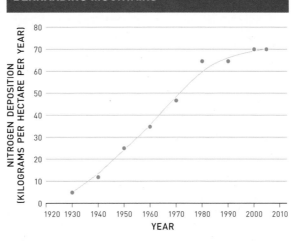

NITROGEN DEPOSITION IN THE SAN BERNARDINO MOUNTAINS

FIGURE 13.27 Nitrogen deposition at Camp Paivika east of Los Angeles, California, increased steadily from 1930 through the 1970s, then began to level off after 1980. (Data from Fenn et al., 2008)

dissolved in water or spewed into the atmosphere as nitrogen oxides and subsequently deposited on soil with rain or snowfall or through dry deposition as a salt or gas. For instance, atmospheric nitrogen deposition in the San Bernardino Mountains east of Los Angeles has increased 14 times over the last 75 years (**Figure 13.27**).

Nitrogen enrichment of terrestrial ecosystems has shifted the composition and function of ecosystems. Some organisms, such as lichens and mycorrhizal fungi, are particularly sensitive and die off with increased nitrogen deposition (**Figure 13.28**). For example, between 1958 and 1987, the number of species of mycorrhizal fungi in soil samples collected in the San Dimas Experimental Forest in the Los Angeles Basin decreased from 29 to 7, a loss of 76% of the species in the community. Changes in species composition can also result in negative effects on ecosystem function. For instance, in some arid ecosystems in California, nitrogen deposition has fostered the increased biomass of nutrient-loving invasive plant species. Because of their high biomass, wildfire risk is now significant where it was once infrequent.

Dead Zones in Coastal Areas

The excess nutrients deposited on terrestrial ecosystems often find their way to coastal waters, with disastrous consequences. Much of the phosphorus, nitrogen, and other nutrients that humans have added to the biosphere eventually end up in the rivers and streams that drain the continents. Consequently, these nutrients contribute to the cultural eutrophication of coastal ecosystems, where they stimulate high levels of primary production. As in eutrophic lakes, high levels of decomposition follow high levels of biomass production in coastal ecosystems, which depletes dissolved oxygen

In what ways could a substantial reduction in the amount of meat in the U.S. diet—for example, a 50% decrease—help the environment?

How are the effects of excess addition of nutrients to an ecosystem analogous to the dose–response relationship discussed in Chapter 11 (see page 346)?

ORGANISMS SENSITIVE TO NITROGEN DEPOSITION

(© Arco Images GmbH/Alamy)

Are coastal "dead zones" actually dead, that is, entirely devoid of life, or is the term misleading?

FIGURE 13.28 Many lichen species, such as this *Alectoria sarmentosa*, are very sensitive to nitrogen levels. Thus, the composition and diversity of lichens can be radically altered by excessive nitrogen deposition. Consequently, they can be used as indicators of nitrogen pollution.

levels—a condition called *hypoxia.* The result is the creation of extensive "dead zones" in coastal areas around the world, including a huge one where the Mississippi River, which drains the extensive agricultural areas of the Midwest and Great Plains, flows into the Gulf of Mexico (**Figure 13.29**). Intensive agriculture has spread cultural eutrophication to coastal areas where the process is killing rich communities of marine organisms, including commercially valuable fish and shellfish populations.

⚠ **Think About It**

1. How is the problem of ecosystem eutrophication an example of a Tragedy of the Commons (see Chapter 2)?

2. Compare cultural eutrophication of lakes with the formation of "dead zones" in coastal regions (see Figure 13.8, page 396).

3. How does the fact that much of the eutrophication of both aquatic and terrestrial ecosystems results from nonpoint sources of pollution complicate developing a solution to the problem?

DEAD ZONE IN THE GULF OF MEXICO

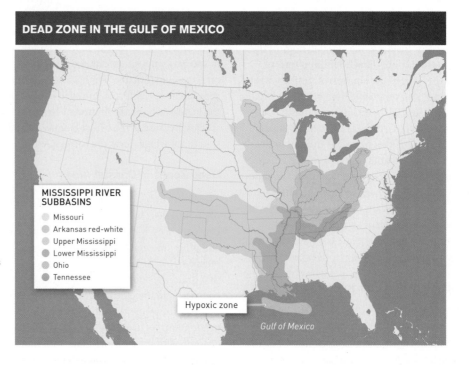

MISSISSIPPI RIVER SUBBASINS
- Missouri
- Arkansas red-white
- Upper Mississippi
- Lower Mississippi
- Ohio
- Tennessee

Hypoxic zone

Gulf of Mexico

FIGURE 13.29 Eutrophication of coastal waters resulting mainly from nutrient additions with runoff from the Mississippi River Basin has produced an extensive area of bottom water that is largely devoid of dissolved oxygen, a condition called hypoxia. The process excludes or kills all fish and shellfish in the area, a situation that is ecologically and economically devastating to the region.

13.4–13.7 Issues: Summary

The World Health Organization estimates that more than 7 million people die of air pollution each year. Other damages caused by pollution come with real economic costs related to deterioration of infrastructure, reduced agricultural or forest productivity, loss of valuable fisheries, and impaired human health and longevity.

The impacts of acid rain have grown with increased burning of fossil fuels, particularly in highly industrialized regions. Damage to terrestrial vegetation, especially to forests in North America and Europe, was one of the first recognized impacts of acid rain, followed by devastated stream and lake ecosystems across vast areas.

The accumulation of persistent organic pollutants, such as DDT and PCBs, and heavy metals in soils and sediments—which, in turn, make it into the human food chain—is of increasing concern. PCBs in the Hudson River are an especially well-studied example of ecosystem contamination by a persistent organic pollutant. Releasing large amounts of organic matter and nutrients from domestic sewage, aquaculture, and agriculture can have negative impacts on the structure and functioning of terrestrial and aquatic ecosystems.

Concentrated animal feeding operations, or CAFOs, have become a particular source of pollution concerns in the United States. Adding excessive nutrients to ecosystems leads to cultural eutrophication, which generally results in excessive algal or plant production, depletion of dissolved oxygen, and loss of biodiversity in aquatic ecosystems. Eutrophication of terrestrial ecosystems can change the species composition and reduce the biodiversity of those ecosystems. Excessive nutrient inputs to coastal ecosystems have produced coastal "dead zones" around the world.

13.8–13.12 Solutions

Sudbury stands as a lesson of what can occur when industry operates without checks and balances. Today, Sudbury is no longer the moonscape it once was. In recent decades, emissions of contaminants have dropped by 90%, soils have been improved, and more than 12 million trees have been planted. In the process, the people of Sudbury have built a much more diversified economy, which includes financial and business services, tourism, health care and research, education, and government. Sudbury has also become a regional center for the arts and culture. The reversal of fortunes by the people of Sudbury shows that it is possible to solve the immense problems posed by pollution.

Think about Earth's air and water as one of the planet's shared community resources. As we saw in Chapter 2, Garrett Hardin proposed that unregulated use of such resources would lead eventually to environmental damage, which he called a Tragedy of the Commons (see page 49). This is exactly what we have seen on the ground in Sudbury and in the air above Beijing. Preventing harm to these community resources requires, first and foremost, environmental regulations, which will ultimately spur new technologies and practices.

13.8 Environmental regulation and international treaties have played important roles in reducing pollution in North America

By the early 18th century, waterways in and around cities in North America became choked with pollution ranging from tannery chemicals to untreated sewage. It was not until 1948 that the Federal Water Pollution Control Act was passed to protect interstate waters, such as the Mississippi River and the Great Lakes, which supported fisheries and drinking water needs. Because there was no central environmental authority, the Surgeon General of the U.S. Public Health Service was put in charge of developing programs to improve the quality of surface waters and groundwaters.

TABLE 13.5
MISSION OF THE U.S. ENVIRONMENTAL PROTECTION AGENCY, ESTABLISHED BY PRESIDENT RICHARD M. NIXON IN 1970

- Establish and enforce standards of environmental protection consistent with national environmental goals.
- Conduct research on pollution and on pollution control, and monitor pollution to strengthen environmental protection programs and to recommend policy changes.
- Assist others, through grants, technical assistance, and other means, to control environmental pollution.
- Assist the Council on Environmental Quality and the president in developing new policies for environmental protection.

The long-term goal of the CWA has been described as striving to make the waters of the nation "fishable and swimmable." How well might these qualities indicate the overall health of aquatic ecosystems?

What would be the likely consequences of environmental legislation without an overseeing agency such as the U.S. EPA?

More than 20 years later, the United States created its first federal agency dedicated to the environment. Richard Nixon was running for president in 1968 and, recognizing growing popular concerns about pollution, made environmental issues part of his presidential campaign. After Nixon won the election, Congress passed the National Environmental Policy Act, generally known as NEPA; but Nixon went further, creating the U.S. Environmental Protection Agency, or EPA, through a 1970 executive order (Table 13.5). New federal laws to strengthen pollution controls soon followed.

The Clean Water Act

In 1972 the Federal Water Pollution Control Act, commonly known as the Clean Water Act (CWA), was thoroughly revised to "restore and maintain the chemical, physical, and biological integrity of the Nation's waters" (EPA, 2012a). The act explicitly protects wetlands and estuaries from pollution and from dredging and filling.

In administering and enforcing the CWA, the EPA works closely with state and tribal governments to develop water quality standards, which play a pivotal role in judging how much of a pollutant may be discharged into a water body and which waters are polluted. States must survey their waters to determine which exceed the limits set by the standards and are categorized as "impaired." States are also responsible for restoring impaired streams by reducing concentrations of pollutants.

The CWA has other important mechanisms for limiting pollution. For instance, if a paint manufacturer wants to discharge waste into U.S. waterways, it must apply for a permit. Permit applicants must demonstrate that the best available technology will be used to reduce the amount of pollutants discharged. The CWA also provides federal money for building or improving sewage treatment plants.

The CWA has continued to evolve. For example, the 1972 amendments to the CWA addressed only point sources of pollution, but its scope was expanded in 1987 to include nonpoint sources as well. In addition, the EPA has shifted from a case-by-case and pollutant-by-pollutant approach to an integrated watershed perspective on pollution regulation.

The Clean Air Act

The Air Pollution Control Act, the first U.S. federal law to address air quality, was passed in 1955. The role of the federal government in controlling air pollution expanded with the Clean Air Act of 1963, and the Air Quality Act of 1967 established air-monitoring studies and point source inspections. The biggest step in pollution control came with the Clean Air Act (CAA) of 1970. This law authorized federal and state governments to regulate both stationary (e.g., electrical power plants) and mobile sources (mainly motor vehicles) of air pollution. According to the U.S. EPA, the 1970 CAA and 1990 amendments gave the EPA broad authority to regulate air pollution in the United States (Table 13.6).

United States–Canadian Cooperation on Acid Rain

Recognizing their common need to address the issue of acid rain, Canada and the United States signed the U.S.–Canada Air Quality Agreement in 1991. The agreement is organized around three parts, called *annexes*. In Annex 1, the United States and Canada committed to limits on emissions and timetables for reducing emissions that cause acid rain. The two countries also agreed to notify and consult with each other regarding sources of acid deposition within 100 kilometers (62 miles) of the border.

Annex 2 focuses on scientific, technical, and economic activities and research to improve understanding of acid rain. Although the agreement was stimulated by the problems associated with acid rain common to the two

TABLE 13.6
AUTHORIZATIONS GRANTED TO THE EPA BY THE CLEAN AIR ACT OF 1970 AS AMENDED IN 1990

- Establish "National Ambient Air Quality Standards" or NAAQS.
- Require "State Implementation Plans" to achieve NAAQS.
- Establish "New Source Performance Standards" for new and modified stationary sources of pollution.
- Establish "National Ambient Air Quality Standards" for hazardous air pollutants.
- Require controls on emissions of pollutants by motor vehicles.
- Develop programs for acid rain control.
- Control 189 toxic pollutants.
- Develop a program for phasing out the use of chemicals that deplete the ozone layer.

Information from EPA, 2013.

countries, the agreement provided a flexible framework that could be used to address other cross-border problems of air pollution. In 2000 it was expanded to address transboundary ozone pollution in Annex 3. The keys to the success of this U.S.–Canada Air Quality Agreement are recognition of common interest and, second, free and frequent communication and cooperation.

Regulatory Mechanisms to Control Pollution

Most of the tools that the EPA uses to protect our air, water, and soil are command-and-control regulations (see Chapter 2) that restrict the type and amount of pollution that can be released, the pollution-control technologies that must be used, and the type of environmental monitoring that must be implemented. For instance, since 1975, the EPA has set federal standards for the emissions of cars and trucks and generally required that they be outfitted with catalytic converters, which reduce the emission of carbon monoxide, unburned fuel, and, sometimes, nitrogen oxides (NO_x). Around the same time, the agency mandated reductions of lead in gasoline, banning the toxic metal, as a fuel additive, completely in 1995.

The EPA has also experimented with market-based approaches (see Chapter 2), pioneering the approach with its Acid Rain Program in the 1990s, following passage of the amendments to the Clean Air Act and U.S.–Canadian Air Quality Agreement. At the time, the agency sought to reduce emissions of SO_2 by 10 million tons and emissions of NO_x by 2 million tons below 1980 levels. The first phase of the Acid Rain Program, which began in 1995, focused on SO_2 reductions at 445 coal-fired power plants in the eastern part of the country. The second phase of the SO_2 reduction program, which was initiated in 2000, included over 2,000 generators fueled mainly by coal but also by oil and gas.

The NO_x reduction program also occurred in two phases. The EPA allowed flexibility on how electric utilities reduced their SO_2 and NO_x emissions. Companies could achieve these reductions through energy conservation, using more renewable energy sources such as wind and solar, installing pollution control technology, or switching to low-sulfur fuels. The program also permitted emission allowance trading, a market-based system for reducing SO_2 emissions. In this system, the EPA grants an electrical utility certain SO_2 emissions allowances for each of its electrical generators. If a particular generating unit emits less SO_2 than it is allowed, the utility may trade the unused allowances with other units in its system of generators. Alternatively, a utility can sell allowances to other utilities or save the allowances to be used to cover emissions in the future. This market-based strategy for pollution control has become a model for solving other environmental problems, such as fisheries (Chapter 8), water rights (Chapter 2), and carbon emissions (Chapter 14).

Economic Benefits of Pollution Control

Regulations to control air pollution have been controversial due to the potential cost for industry. When controls on acid rain were proposed, for instance, the Edison Electric Institute, a lobbying group, predicted that the costs of controlling sulfur dioxide during just Phase I of the Acid Rain program would be $4 to $5 billion annually. Industry members projected that consumer utility bills would increase by 20% to 40%. In contrast, the EPA predicted annual costs would amount to $1 billion. In a review of the program, Don Munton, professor of international studies at the University of Northern British Columbia, found that these costs were overestimates, as he explained in the 1998 article "Dispelling the Myths of the Acid Rain Story." The first years of the program cost only $836 million annually and utility bills actually increased an average of 2% to 4%. Munton concluded his review by stating that the Acid Rain Program was "a bargain."

Other evidence is mounting that pollution controls can provide economic benefits that outweigh their costs. In a 2005 paper in the *Journal of Environmental Management,* Lauraine Chestnut and David Mills estimated that the total cost of Phase I and Phase II of the U.S. Acid Rain Program amounted to $3 billion annually (**Table 13.7**). The benefits included economic

How might estimates of costs and benefits of environmental regulation be influenced by the financial interests of those making the estimates?

TABLE 13.7
ALL COSTS AND BENEFITS OF THE U.S. ACID RAIN PROGRAM IN 2010

Costs	Millions of U.S. (2000) dollars
Controls of SO_2	$2 billion
Controls of NO_x	$1 billion
Total costs	**$3,000,000,000**

Benefits	
Reduced mortality from fine particulate pollutants, ≤2.5 μm (PM2.5), in the U.S. and Canada	$107 billion
Reduced morbidity from fine particulate pollutants, ≤2.5 μm (PM2.5), in the U.S. and Canada	$8 billion
Reduced mortality from ozone in eastern U.S.	$4 billion
Reduced morbidity from ozone in eastern U.S.	$300 million
Improved visibility in parks	$2 billion
Improved recreational fishing in New York State	$65 million
Adirondack ecosystem improvements	$500 million
Total benefits	**$121,865,000,000**

Data from Chestnut and Mills, 2005.

gains from reduced premature mortality and reduced chronic disease (morbidity) in the United States and Canada, as a result of controlling fine particulate pollution and ozone.

To these benefits, the authors added the economic gains resulting from improved visibility in U.S. parks, improved recreational fishing in New York, and ecosystem recovery in the Adirondacks. The sum of these annual economic benefits totaled nearly $122 billion, approximately 40 times the costs.

That's the story for acid rain, but what about the total economic benefits of the Clean Air Act? In 2011 the EPA estimated that by 2020 the costs of pollution control would amount to $65 billion, but that those costs would be dwarfed by the economic benefits, primarily through a reduction of premature mortality, which would have totaled nearly $2 trillion. In economic and environmental terms, it appears that it pays to invest in pollution control.

⚠ Think About It

1. What aspects of the Clean Water Act imply recognition of the ecological services rendered by wetlands and estuaries?

2. What are the advantages and disadvantages of federal and state governments sharing in the process of setting pollution standards?

3. Could U.S.–Canada cooperation on pollution control be expanded into global-scale cooperation to address pollution issues? Explain your position.

4. How might conclusions reached regarding the economic viability of pollution control programs be influenced by whether one takes a long-term (e.g., value to future generations) or short-term (e.g., quarterly profits to a company) perspective?

TEMPORAL TRENDS IN EMISSIONS OF POLLUTANTS THAT CAUSE ACID RAIN

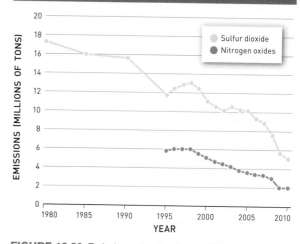

FIGURE 13.30 Emissions of sulfur dioxide (SO_2) and nitrogen oxides (NO_x) from electrical power plants in the United States decreased significantly following regulation by the Clean Air Act. (Data from the EPA, 2015a)

13.9 Control measures have reduced emissions of pollutants and acid rain

When you consider the scale of the United States and Canada, it might seem like an impossible task to rein in damaging pollution across the continent. But the story of acid rain shows that it's sometimes possible to solve such vast environmental challenges in a relatively short timeframe.

Reduced Emissions and Decreased Acid Rain

Over the course of 20 years, the EPA's Acid Rain Program was a success. By 2010 sulfur dioxide emissions were

pH OF PRECIPITATION ACROSS THE UNITED STATES IN 1994 AND 2009

FIGURE 13.31 (a) In 1994 precipitation was highly acidified (pH <5.3) in the eastern half of the country, especially in the upper Midwest and Northeast. (b) By 2009, however, acid rain control programs resulted in substantial reductions in the acidity of rain across the country. (Maps from National Atmospheric Deposition Program, http://nadp.sws. uiuc.edu/)

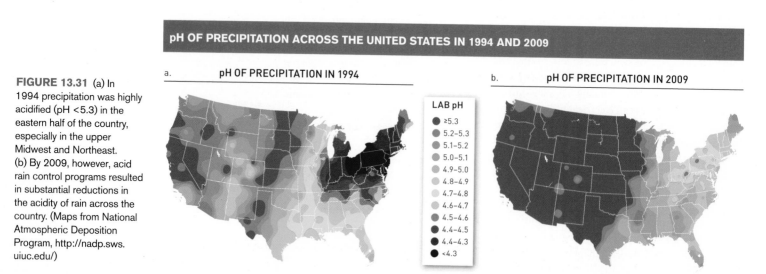

12 million tons lower than in 1980, and emissions of nitrogen oxides were reduced by 4 million tons between 1995 and 2010 (Figure 13.30). One of the major contributors to these decreases was that electrical utilities switched from using high-sulfur coal from Appalachia to low-sulfur coal from the western United States, particularly from Wyoming's Powder River Basin (see Figure 9.31, page 287). As predicted, acid rain also decreased (Figure 13.31). Precipitation in the eastern portion of the country became less acidic between 1994 and 2009 and acid rain (pH <5.3) was essentially eliminated from the central part of the country to the Pacific Coast.

Recovery of Aquatic Ecosystems

With reduced acid rain, lakes and streams that had been acidified during the 1970s and 1980s began to recover. Consider an example from the Adirondack region of New York State. Prior to the start of the Acid Rain Program, 284 lakes out of nearly 2,000 being monitored there were highly acidified. Within 2 to 3 years, that number dropped by 32% to 192 (Figure 13.32). By 2007, 132 of the originally acidified lakes had recovered. Streams in the Adirondack region are also recovering, although at a slightly slower rate.

Up near Sudbury, Clearwater Lake is also showing remarkable recovery. Beginning in 1972, pollution controls decreased the SO_2 emissions of the smelters in the Sudbury areas by more than 90%. In response, the pH of Clearwater Lake, and of other lakes in the area, increased (Figure 13.33). After 1999 the pH in

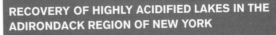

RECOVERY OF HIGHLY ACIDIFIED LAKES IN THE ADIRONDACK REGION OF NEW YORK

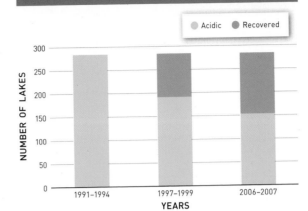

FIGURE 13.32 At the beginning of the EPA's Acid Rain Program in 1995, 284 lakes in the Adirondack region were highly acidified. By 2007, 132 of those lakes, or 46%, had recovered in response to decreased acid rain over the region. (Data from Waller et al., 2012)

What role do you think the flexibility allowed to utility companies by the EPA played in the successful reduction of SO_2 and NO_x emissions? Explain.

Clearwater exceeded 6.0, a threshold allowing for full recovery of the ecosystem. The concentrations of heavy metals, such as nickel and copper, have also decreased following pollution control (Figure 13.34).

However, recovery of the biological community has been uneven. The phytoplankton and zooplankton of many lakes have bounced back, but neither invertebrates

TRENDS IN SO_2 EMISSIONS AND LAKE pH

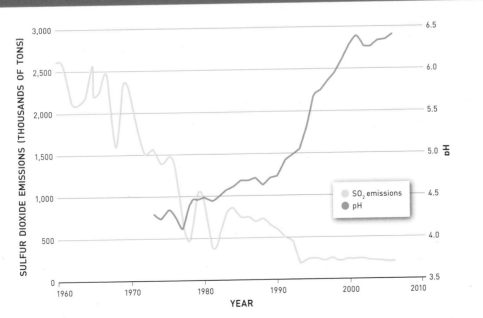

FIGURE 13.33 Total annual sulfur dioxide emissions from the smelters near Sudbury, Ontario, in thousands of metric tons and pH in Clearwater Lake, located 13 kilometers from Sudbury. (Data from Keller, 2009)

SCIENCE   ISSUES   SOLUTIONS

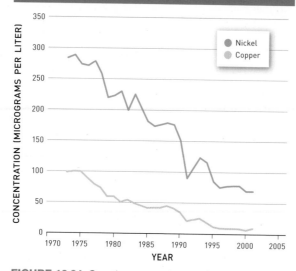

TRENDS IN NICKEL AND COPPER CONCENTRATIONS IN CLEARWATER LAKE NEAR SUDBURY, ONTARIO

FIGURE 13.34 Over the course of nearly three decades, following reduced pollution from the metal smelters in the area, the concentrations of both nickel and copper in Clearwater Lake decreased significantly. (Data from Girard et al., 2006)

Why might non-native species have an advantage over native species in the face of a disturbance, such as acid rain?

that live at the bottom of the lake nor fish are back to natural levels yet. Factors that may retard recovery in certain areas include competition with established acid-tolerant species. Once a community of organisms is established, in this case of acid-tolerant species, they can persist in an ecosystem and compete with individuals of less tolerant species when physical conditions change. Predation by high numbers of invertebrate predators, which have been released from control by plankton-feeding fish eliminated by acid rain, can also reduce the chance of successful colonization by small organisms. In addition, limited dispersal slows colonization by some organisms. Another problem has been high heavy metal concentrations and reduced concentrations of calcium, Ca^{2+}.

Recovery of Terrestrial Ecosystems

Restoring the forests that once covered the Sudbury landscape is a huge challenge—much greater than restoring a forest following logging or fire. The main reason for the difficulty is that the damage done to these ecosystems has seriously impaired the very foundation of terrestrial ecosystems: their soils. Succession in such circumstances takes a very long time unless helped along by active restoration methods. One of the main treatments used at restoration sites in the most disturbed areas has been spreading crushed limestone on the soils. Limestone adds the critical Ca^{2+} ions stripped from the soils by acid rain and can neutralize acidic soils, resulting in higher soil pH. Liming at the restoration sites

has raised average soil pH to 6.3, compared with 4.3 at unrestored sites. Some sites were limed and planted with a single species of pine. Others were limed, fertilized, and planted with three to five tree species.

As a result of active restoration, the number of plant species at the restored sites averaged more than 3 times greater, compared with unrestored sites in the most disturbed zone (Figure 13.35). The number of plant species at the most species-rich restoration site was comparable to that at the low disturbance site, approximately 36 kilometers from the decommissioned smelter. However, approximately 30% of the plant species at restored sites were invasive non-native species, which have high dispersal rates and had colonized the restoration sites on their own. Meanwhile, all plant species at unrestored, control sites were native. Researchers suggest that it may take decades or even centuries for the plant communities at the restored sites to return to a composition close to the original community before disturbance.

⚠ Think About It

1. What do the successful reductions in acid rain and recovery of some ecosystems indicate about the scientific understanding of the causes and effects of acid rain?

2. Why might the ecosystems impacted by acid rain never return to their state prior to acid

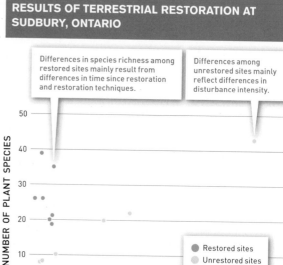

RESULTS OF TERRESTRIAL RESTORATION AT SUDBURY, ONTARIO

Differences in species richness among restored sites mainly result from differences in time since restoration and restoration techniques.

Differences among unrestored sites mainly reflect differences in disturbance intensity.

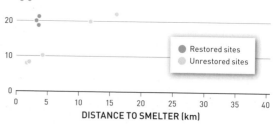

FIGURE 13.35 Decades of natural recovery and active restoration near a decommissioned smelter have shown that active restoration speeds the development of species-rich plant communities in the most disturbed sites. (Data from Rayfield et al., 2005)

rain impacts, especially in terms of their species composition?

3. What can working toward restoration of damaged ecosystems teach us about the ecology of these systems?

13.10 New technologies can reduce indoor air pollution

Indoor air pollution remains as much or more of a threat to human health as outdoor air pollution. According to the World Health Organization (WHO), more people die every year from indoor air pollution than from outdoor air pollution. The good news, though, is that we have the technological know-how to solve the problem, as long as governments, companies, and humanitarian organizations work toward that end.

Healthier Cooking Technologies

Living in a smoke-filled house isn't pleasant, nor is it healthy. One of the most beneficial improvements to environmental health came with the invention of ways to ventilate homes. The first approach was likely a simple hole in the ceiling of a hut or teepee. A further reduction in indoor air pollution was achieved with the invention of the chimney attached to a fireplace or stove.

In developing countries, such as India and Bangladesh, where the problem of indoor air pollution is most severe, the health of local populations could be greatly improved with a few relatively low-tech solutions. One of the simplest ways to reduce exposure to smoke from cooking and heating fires is to thoroughly dry fuels, such as wood or animal dung, before use. Another simple solution is to provide improved stoves that use less fuel for cooking, reduce cooking time, and, in the process, reduce the amount of smoke generated. Combining better stove design with effective chimneys or hoods for collecting and venting smoke can reduce indoor pollution still further.

Other approaches involve switching from dung, wood, and coal to cleaner fuels, such as liquid petroleum gas, biogas generated from dung, or electricity. Solar cookers have also been developed that can be used in areas with abundant sunlight (**Figure 13.36**). Governments and nongovernmental organizations are working very hard to bring some of these solutions to bear in the regions where health is most significantly impacted by indoor air pollution.

Reducing Indoor Air Pollution in Modern Buildings

Modern buildings have a different set of indoor pollutants, for example, volatile organic compounds (VOCs). The most direct way to prevent the buildup of pollutants in the indoor environment is to maintain

(© Joerg Boethling/Alamy)

FIGURE 13.36 Solar cooking technology is being developed to help relieve the widespread problem of indoor air pollution in developing countries. Here, a woman in West Africa prepares food using a solar cooker.

adequate ventilation rates by ensuring that the heating, ventilating, and air conditioning (HVAC) system meets industry standards (**Figure 13.37**). The HVAC system needs to be serviced routinely and air filters must be cleaned or replaced at prescribed intervals. A proactive approach, increasingly used in so-called green architecture, is to reduce the use of building materials or furnishings that are a significant source of hazardous pollutants. Any potential sources of pollutants should be stored away from work areas and vented to the outside. Perhaps most importantly, building occupants and managers need to communicate freely and be informed of building operations and potential air-quality issues.

⚠ Think About It

1. What is the major impediment to fixing the indoor pollution problem in the most affected populations?

2. Why does indoor pollution take a larger toll on human health than outdoor pollution?

MAINTAINING INDOOR AIR QUALITY IN LARGE MODERN BUILDINGS

PRIMARY FUNCTIONS OF A MODERN HVAC SYSTEM

TEMPERATURE CONTROL
Important for maintaining comfort and productivity of occupants; cooling is especially important in large buildings.

FRESH AIR CIRCULATION
Essential for maintaining air quality, including preventing a buildup of carbon dioxide, which has been connected to higher frequency of sick leave in office buildings.

AIR FILTRATION
Reduces particulates in building air, including dust, pollen, and mold spores, which are potential sources of respiratory illness.

PRIMARY COMPONENTS OF A MODERN HVAC SYSTEM

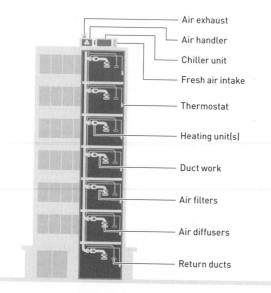

- Air exhaust
- Air handler
- Chiller unit
- Fresh air intake
- Thermostat
- Heating unit(s)
- Duct work
- Air filters
- Air diffusers
- Return ducts

FIGURE 13.37 Large modern buildings maintain the quality of the indoor environment using complex mechanical heating, ventilating, and air conditioning (HVAC) systems.

13.11 Soils and sediments contaminated by hazardous wastes can be cleaned using a variety of techniques

Surprisingly, the cleanup of a severely polluted site, such as Love Canal (see Chapter 12), fell outside any existing legislative authority at the time, including the Resources Conservation and Recovery Act, which is focused on current and future management of hazardous wastes. Authority to address historical contamination required additional legislation. As we saw in Chapter 12, the U.S. Congress passed the Comprehensive Environmental Response, Compensation, and Liability Act (CERCLA), or Superfund Law, in 1980. The first goal of the Superfund program is to identify sites where contamination of air, soil, and water by hazardous substances has been sufficient to threaten human health or harm the environment. Sites sufficiently contaminated are placed on a "National Priorities List," which in 2011 included more than 1,350 sites in the United States. Once these sites are identified, the second objective of the program is to reduce threats to human health and the environment, generally by cleaning the site of hazardous substances. The third goal is to discover those responsible for the contamination and seek payment for site cleanup.

This groundbreaking piece of legislation gave the EPA the authority to assign responsibility for a contaminated site and demand financial assistance in its remediation. This law was the beginning of efforts all over the country

AREA SLATED FOR DREDGING ON THE HUDSON RIVER

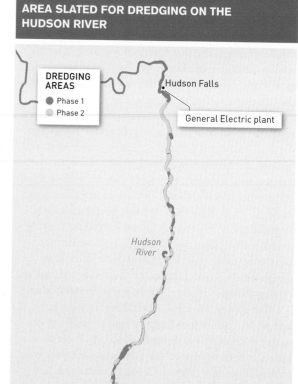

DREDGING AREAS
● Phase 1
● Phase 2

Hudson Falls

General Electric plant

Hudson River

Federal Dam

Troy

5 km

FIGURE 13.38 Most of the PCBs lodged in sediments in the Hudson River occur along a 64-kilometer (40-mile) section below Hudson Falls.

to clean up toxic waste sites. In the case of the Love Canal, Occidental Petroleum, which owned the Hooker Chemical company, agreed to pay $129 million in restitution. The most toxic area was dug up and reburied with a plastic liner and the neighborhood's empty streets were enclosed by a chain link fence to prevent entry.

Persistent Pollutants in the Hudson River

The Superfund Act has also led to the cleanup of many other toxic sites around the country. In 2002 the EPA and General Electric signed an agreement to begin removing PCBs from a 64-kilometer (40-mile) section of the upper Hudson River (Figure 13.38). The first steps in the process involved sampling the length of the river to identify the locations of the most contaminated sediments, building the processing and transportation facilities, and developing detailed plans for the massive project. During the early stages of the work, public concerns were identified and they influenced the way the project was carried out. For example, barges and trains were used to transport dredged sediments out of the Hudson River Valley to address concerns over increased traffic congestion on highways. Also, in response to fears of future contamination, all parties agreed to store toxic sediments outside of the Hudson River Valley in approved hazardous waste repositories.

The plan for dredging was divided into two phases. Phase 1 would involve one season (May to November) of dredging followed by a year spent evaluating the results of the dredging operations. The evaluations, peer-reviewed by independent experts, would be used to determine whether dredging had removed PCBs. Reviewers would also assess whether the dredging had mobilized unacceptable amounts of PCBs into the river ecosystem, which was a serious concern raised by the public and by environmental scientists. Based on the findings, adjustments in techniques could be made to improve the cleanup process. If the results of Phase 1 were acceptable, the project would move into Phase 2, which would involve dredging the remainder of the site.

Phase 1, completed in 2009, resulted in the removal of approximately 215,000 cubic meters (283,000 cubic yards) of PCB-contaminated sediments from a 9.7-kilometer (6-mile) section of river near Fort Edward on the upper Hudson. While the amount of material removed from the river was enormous, the operation was done with precision, with the location and depth of dredging guided by a satellite navigation system. The work employed over 500 people and dredging went on 24 hours a day, 6 days a week, with up to 12 dredging crews working simultaneously (Figure 13.39). The dredged material, which filled 626 hopper barges measuring 59.5 meters by 10.7 meters (195 feet by

Why was it critical that independent experts review the dredging process and results?

DREDGING ON THE HUDSON RIVER: A COMPLEX PROCESS

(USEPA)

FIGURE 13.39 The removal of PCBs from the Hudson River is a massive undertaking involving heavy equipment, long-distance transport, and long-term storage. The process also requires great care and precision to avoid the release of dangerous levels of PCBs from contaminated sediments as they are being removed.

35 feet), was transported by rail to a disposal site in Texas. To reduce the possibility of releasing any residual PCBs, the dredged areas were later covered, or "capped," with 150,000 tons of clean fill material. During dredging operations, the river was continuously monitored to assure that PCB concentrations did not exceed the 500 parts per trillion safe limit specified by the U.S. Safe Drinking Water Act.

Based on independent expert review of the process and results of Phase 1 in 2010, along with comments from the public, the EPA went forward with plans for Phase 2. Planners project that Phase 2 will remove an additional 1.8 million cubic meters (2.4 million cubic yards) of contaminated sediments. Phase 2, scheduled for completion in the fall of 2015 at an estimated cost of approximately $750 million, will be followed by another phase involving habitat restoration along the shoreline and in dredged areas, as well as data collection to evaluate the success of the project.

Bioremediation

Rather than cleaning up a polluted site by moving massive amounts of contaminated soil or sediment to a hazardous waste facility, environmental scientists can use organisms to decontaminate soils, sediments, and groundwater aquifers. This approach to pollution cleanup, called **bioremediation,** can save physical work and money.

When the bioremediation process involves plants, it is called **phytoremediation** (Figure 13.40). Scientists have identified hundreds of **hyperaccumulator** plants that accumulate heavy metals in their tissue. Consider that excavating 30 centimeters (1 foot) of soil contaminated with lead from 4 hectares (10 acres) would require removing 18,200 metric tons (20,000 tons) of soil. By

contrast, using plants to extract the same amount of lead from the soil would require safely disposing of just 455 metric tons of plant biomass, or 1/40th the amount of soil that would otherwise have been removed. In addition, the monetary costs would be a small fraction of what it would cost to excavate, transport, and store more than 18,000 metric tons of contaminated soils.

In places where groundwater has been contaminated with solvents, gasoline, and other organic compounds, bioremediation has been used to decontaminate aquifers. Many organic chemicals that were once believed to be resistant to breakdown by microbes can be metabolized by some component of a microbial community. Sometimes they need to be coaxed to multiply and do their work by adding nutrients, reducing or increasing oxygen availability, or adding particular energy sources (e.g., sugar) to the aquifer (Figure 13.41). Using such techniques, environmental scientists have successfully decontaminated polluted aquifers in situations previously considered physically or economically impossible.

⚠ Think About It

1. How might proximity to New York City have influenced the effort—funds allocated—to clean up the Hudson River?

2. How can the many thousands of square kilometers of soils that have been contaminated by heavy metals as a result of metal smelting and coal burning be effectively treated?

3. How do phytoremediation of soils and bioremediation of aquifers depend on Earth's biodiversity?

What, besides lead, is taken away if excavation and removal are used to treat a contaminated soil?

bioremediation An approach to pollution cleanup that employs organisms, generally microbes or plants, to decontaminate soils, sediments, and groundwater aquifers in place.

phytoremediation Bioremediation using plants to clean up contaminated sediments or soils. *See bioremediation.*

hyperaccumulators Plants that accumulate heavy metals in their tissue.

PHYTOREMEDIATION OF SOILS CONTAMINATED WITH HEAVY METALS

❶ Planting hyperaccumulator plants on site contaminated with heavy metals.

❷ Growth and uptake of heavy metals by plants.

❸ Harvest and hauling of plant biomass heavily laden with heavy metals.

❹ Composting and disposal of heavy metal-contaminated compost at hazardous waste site.

FIGURE 13.40 During the process of phytoremediation, plants that accumulate large amounts of heavy metals are planted on contaminated soils. Eventually, the plant biomass containing the heavy metals that they have taken up is harvested and composted or burned, and the heavy metals are then collected and disposed of or recycled.

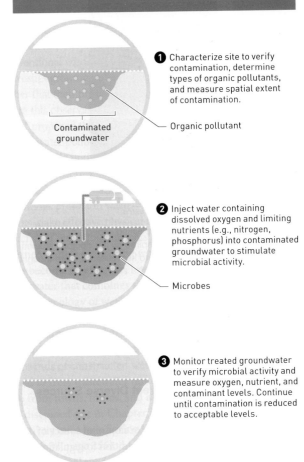

❶ Characterize site to verify contamination, determine types of organic pollutants, and measure spatial extent of contamination.

Contaminated groundwater — Organic pollutant

❷ Inject water containing dissolved oxygen and limiting nutrients (e.g., nitrogen, phosphorus) into contaminated groundwater to stimulate microbial activity.

— Microbes

❸ Monitor treated groundwater to verify microbial activity and measure oxygen, nutrient, and contaminant levels. Continue until contamination is reduced to acceptable levels.

FIGURE 13.41 Groundwater contaminated with various organic chemicals (e.g., gasoline) has been successfully cleaned by creating environmental conditions (e.g., adding an energy source such as sugar) favorable to microbes capable of breaking down the specific pollutant.

13.12 There are many effective ways to reduce organic and nutrient pollution

Treating waste that contains large concentrations of organic matter and nutrients draws on both engineering principles and biological science. The methodology applied depends on whether the environment is urban or rural and whether it is found in a developed or developing country. The largest volumes of waste requiring treatment are produced in urban settings.

Sewage Treatment in Cities

Since city sewers are a major point source of organic and nutrient pollution, governments around the world monitor and regulate sewage treatment facilities. The EPA, for instance, requires all publicly owned sewage treatment facilities to provide a minimum of secondary treatment to wastewater, which removes some organic matter and nutrients from the water (Chapter 6). Members of the European Union and other developed countries have wastewater treatment standards that are similar to or higher than those of the United States. For example, most wastewater in Scandinavia receives tertiary treatment, which further reduces nutrient content and pathogens. In contrast, most wastewater in developing countries still receives little or no treatment before it gets discharged into the environment.

Septic Systems in Rural Areas

Rural residents across the United States and elsewhere rely on septic systems for treatment of domestic wastewater. Fortunately, a properly maintained septic system can effectively treat wastewater and prevent pollution of surface and groundwater. The most commonly used system in the United States directs all wastewater from a residence into a septic tank, which is watertight and generally buried (Figure 13.42). The appropriate volume of a septic tank depends on the size of the home it serves, but typical volumes range from 3,790 to 5,685 liters (1,000 to 1,500 gallons). As wastewater enters the septic tank, solids settle out, forming sludge, while grease and oils float to the surface. Naturally occurring bacteria will decompose most of the organic material in the septic tank, whereas any that resists decomposition will settle to the bottom of the tank as sludge. Sludge will gradually build up over time in septic tanks, so the tanks need to be pumped periodically to keep the septic system functioning properly. The

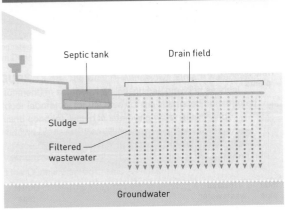

Septic tank   Drain field

Sludge

Filtered wastewater

Groundwater

FIGURE 13.42 In a septic system, the septic tank serves as a site for settling of solids and bacterial breakdown of organic matter. Liquids from the system are dispersed and further processed and purified in the drain field.

Answer the following questions for each chapter section and then answer the Central Question.

Central Question: How can we control and reduce environmental pollution?

13.1–13.3 Science

- What are the main sources of pollutants?
- What types of pollutants do humans produce?
- How do pollutants move around the planet?

13.4–13.7 Issues

- What are the impacts of air pollution on health and economies?
- How does acid rain damage aquatic and terrestrial ecosystems?
- How do persistent pollutants enter the human food chain?
- In what ways do organic matter and nutrient pollution disrupt ecosystems?

Pollution and You

While pollution issues may seem huge in scope and beyond an individual's reach, there are, in fact, many things that each of us can do to help address pollution issues.

☐ Stay informed, act informed.

Keep up to date on local, regional, national, and global pollution issues. Many cities post air-quality indexes online or monitor local water health. Go to http://airnow.gov to find local air-quality scores. When air quality is bad, opt to carpool or take public transportation with others to reduce emissions. Decrease your use of chemicals like fertilizers and harsh household cleaners that can end up in the water supply. As an informed citizen, you can vote on pollution-related issues and urge your community to improve public transportation and water quality.

☐ Use healthy home practices.

Small, dust-sized particles in the home can include a variety of pollutants including lead, pollen, and dust mites. Clean all surfaces that collect dust, including floors, walls, and upholstered furniture, with a vacuum with a HEPA filter. Follow vacuum cleaning with mopping with plain water, which will clean dust missed by the vacuum. Dispose of pet waste properly; in high quantities, it can reduce the oxygen load of water and spread disease.

☐ Keep smoking out.

Because cigarette smoke contains thousands of harmful chemicals, it represents one of the most significant threats to the indoor environment. As a result, one of the critical steps you can take to protect the air quality in your home is to make it a no-smoking area. Some of the well-documented dangers of exposure to secondhand smoke include increased risk of respiratory infections, asthma, and cancer. Exposure to secondhand smoke is especially dangerous to the health of children and infants.

☐ Test for radon.

The U.S. EPA estimates that radon, a radioactive gas produced during the natural decay of uranium, causes more than 20,000 deaths from lung cancer in the United States each year. While it occurs in nearly all geologic formations, radon is more abundant in some settings than others. Radon generally enters the home from the ground and can build up to unhealthy concentration in homes of any age. Testing for radon, which is easy using inexpensive, widely available kits, can give you the information you need to determine whether corrective action is needed.

13.8–13.12 Solutions

- **What impact have environmental regulation and international treaties had on reducing pollution in North America?**

- **What factors have reduced emissions of pollutants and acid rain?**

- **How can technology reduce indoor air pollution?**

- **How can soils and sediments contaminated by hazardous wastes be cleaned?**

- **What are some ways to reduce organic and nutrient pollution?**

Answer the Central Question:

_____

_____

_____

_____

_____

_____

_____

_____

Chapter 13

Review Questions

1. What is pollution?
a. Any environmental change caused by humans
b. Environmental damage caused by industrial activity
c. Release of chemicals into water, air, or soil
d. Alteration of the environment that harms living organisms

2. Which of the following could be considered a point source of pollution?
a. Nutrient-rich runoff from suburban lawns
b. Discharge of a city sewage treatment plant into a river
c. Smoke emitted by several forest fires
d. Increased particulates in air as a result of regional drought

3. A decrease in pH from 7 to 6 results in which of the following?
a. A 50% decrease in hydrogen ion concentration
b. A doubling of hydrogen ion concentration
c. A 10-fold increase in hydrogen ion concentration
d. A 10-fold decrease in hydrogen ion concentration

4. What was the rationale for choosing the so-called criteria pollutants?
a. They were chosen randomly from a long list of pollutants.
b. They are common pollutants that are hazardous to human health.
c. They are among the most toxic substances known.
d. They were chosen as a result of a poll of concerned citizens.

5. What is the main difference between natural and cultural eutrophication of lakes?

a. Natural eutrophication is a more gradual process.
b. Natural eutrophication does not increase the nutrients in lakes.
c. Natural eutrophication does not affect lake depth.
d. Natural eutrophication does not increase lake primary production.

6. What are some of the costs of air pollution?
a. Millions of premature deaths
b. Deteriorating infrastructure
c. Reduced ecosystem production
d. All of the above

7. Which type of crop plant is least likely to accumulate high concentrations of heavy metals?
a. Root crops such as potatoes
b. Leafy vegetable crops such as lettuce
c. Crops producing an edible fruit, such as tomatoes
d. All crops accumulate heavy metals to the same extent

8. What is the outcome of efforts to control acid rain in the United States?
a. There has been no measureable change in acid rain.
b. The acidity of rain has actually increased across the United States.
c. Acid rain has decreased in the eastern United States but not in western states.
d. Acid rain has decreased across the entire United States.

9. Which of the following likely motivates current efforts by the EPA to reduce mercury pollution by coal-fired power plants in the United States?

a. Mercury is a potent neurotoxin.
b. Mercury contamination of soils is difficult to clean up.
c. Fish and food crops accumulate mercury in their tissues.
d. All of the above

10. Where are constructed wetlands likely to be most useful for treating wastewater?
a. For treating wastes from large urban centers
b. For treating farm wastes in sparsely populated rural areas
c. For treating wastes from large CAFOs
d. For treating diffuse sources of nonpoint sources of pollution

Critical Analysis

1. Is massive pollution an inevitable consequence of economic development? Explain.

2. What are the relative advantages and disadvantages of command-and-control versus market-based approaches to controlling pollution?

3. How do cooperative agreements between Canada and the United States to control air pollution relate to the concept of airsheds?

4. Evaluate the evidence that smelting activity near Sudbury, Canada, impacted terrestrial and aquatic ecosystems.

5. Discuss the various ways in which nutrient enrichment has impacted terrestrial and aquatic ecosystems.

Find additional resources and links online at www.macmillanhighered.com/launchpad/molles1e.

Central Question: How can we mitigate and adapt to the environmental and social impacts of climate change?

Explain the factors that control climate and global temperatures.

SCIENCE

CHAPTER 14

Global Climate Change

Analyze the causes and impacts of a warming global climate.

ISSUES

Discuss the local and international tactics that could mitigate global climate change.

SOLUTIONS

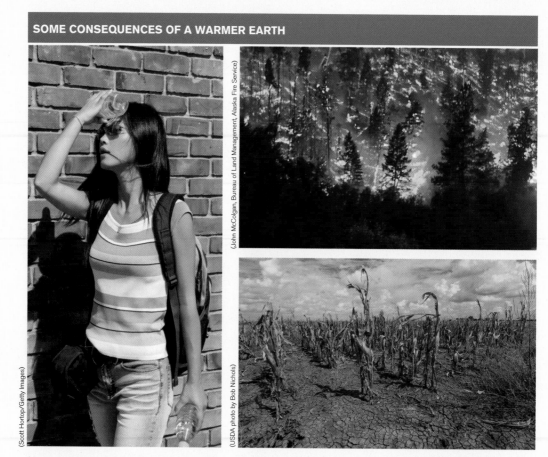

SOME CONSEQUENCES OF A WARMER EARTH

(John McColgan, Bureau of Land Management, Alaska Fire Service)

(Scott Hortop/Getty Images)

(USDA photo by Bob Nichols)

Heat waves are setting temperature records and impacting larger and larger areas around the world. High temperatures combined with drought have been conducive to large wildfires of unprecedented magnitude. Drought has had severe impacts on agricultural production in regions such as the midwestern United States.

Tracking Wildfires in the West

Raging fires and extreme weather events could become more common with a changing global climate

At 7 A.M. on June 23, 2012, a jogger was running along the Waldo Canyon Trail in the mountains above Colorado Springs, Colorado, when he smelled smoke. He veered off the trail to investigate and found a smoldering fire in the woods. After he reported the fire to the local sheriff's department, high winds and drought conditions in the forest caused the fire to spread over 600 acres in several hours' time, leading to evacuations of several nearby communities. By the time firefighters finally contained the Waldo Canyon Fire, two and a half weeks later, it had burned 7,384 hectares (18,247 acres) and 346 homes, killing two people. It ranked as the most destructive fire

in Colorado's history, resulting in insurance claims of more than $450 million. Although the fire may have been started by an arsonist, another suspect has been singled out for its rapid spread and devastating impact: climate change.

That year, the wildfire season in the West came on the heels of a period of unrelenting heat. During the 12 months from August 2011 to July 2012, land temperatures in the 48 contiguous United States were the warmest in 117 years of record-keeping. Across Colorado, wildfires blackened nearly 67,000 hectares (165,000 acres) and destroyed over 600 homes. In Montana and New Mexico, they consumed another 529 homes. In Utah and Wyoming, they forced the shutdown of natural gas fields, interrupting the flow of critical energy supplies. All told, wildfires in the United States in 2012 burned more than 1.7 million hectares (4.1 million acres).

Abnormally high temperatures in the United States had other impacts as well. For instance, cattle had so little healthy pasture that the USDA allowed ranchers to graze their cattle on conservation lands set aside for erosion control and wildlife habitat. Approximately half of the nation's corn crop and one-third of the soybean crop had failed or were near failing—an episode that would play out in the global economy as an increase in food prices. Reduced farm income would hurt a wide range of businesses located in agricultural regions.

> "Preservation of our environment is not a liberal or conservative challenge, it's common sense."
>
> President Ronald Reagan, State of the Union address, January 1984)

Climate scientists modeling future climates believe that the summer of 2012 may provide a preview of some of the environmental and economic consequences of climate change. In fact, they have concluded that by mid-century, if present trends continue, the western United States would be subject to droughts worse than any occurring in the previous 1,000 years. Human action has played a significant role in changing Earth's climate, particularly by increasing the concentrations of gases in the atmosphere that trap the Sun's energy, leading to a temperature increase of almost 1°C since 1880. Climate scientists predict that climate change will include a higher frequency of heat waves, droughts, and other weather extremes along with the loss of the polar ice caps and a rise in sea level.

By the end of the 21st century, climate models suggest that the temperature of Earth's surface will rise another 2 to 3°C. "Warming of the climate system is unequivocal, and since the 1950s, many of the observed changes are unprecedented over decades to millennia," wrote the authors of the fifth assessment of the Intergovernmental Panel for Climate Change (IPCC), published in 2014. "It is extremely likely that human influence has been the dominant cause of the observed warming."

The good news is that once we recognize that we are significant contributors to climate change, there are steps we can take to reduce the problem. However, as we address this issue, we will need to avoid causing other forms of disruption, both environmental and economic.

Central Question

How can we mitigate and adapt to the environmental and social impacts of climate change?

(Jean-Louis Klein & Marie-Luce Hubert/Science Source)

14.1–14.4 Science

climate The average weather occurring across a region over a long period, including average temperatures, precipitation, and so forth.

weather Atmospheric conditions, temperature, humidity, cloud cover, rainfall, etc. at a particular place and time (e.g., conditions during a particular day or month).

The seasons occur so predictably that it's natural to assume that the **climate**—the average **weather** occurring across a region over a long period, including average temperatures, precipitation, and so forth—has always been what it is today and will continue to be that way. If, as child, you skied down a snow-covered slope in December or took a dip in a cool stream on a July day, you expect that the next generation will be able to do the same. Indeed, most of human history has occurred during a period of relative stability, but Earth's climate has undergone spectacular changes over the course of geologic history. Some 20,000 years ago, Wisconsin and New York state were covered by a massive glacier stretching down from the north that melted during a rapid period of warming, leaving behind the Great Lakes and other geographic features. Similarly, about 10,000 years ago, parts of the Sahara desert were covered in grass and trees.

Today, we recognize that the climate is changing in new and sometimes unpredictable ways due to the release of greenhouse gases and other human activities. Many ski slopes are no longer receiving the snowfall they once did, and some freshwater streams are drying up in

?

What is the difference between climate and weather?

PORTRAITS OF EARTH AND ITS PLANETARY NEIGHBORS

(NASA/JPL)

Mercury · Venus · Earth · Mars · Jupiter · Saturn · Uranus · Neptune

FIGURE 14.1 Mars, approximately 228 million kilometers (km) from the Sun, is the smallest of the three planets discussed here. Earth is approximately 78 million km closer to the Sun and twice the diameter of Mars. Venus is approximately the same size as Earth, around 40 million km, or 30%, closer to the Sun. However, the average temperature of Venus is more than 30 times higher than Earth's.

the summertime. But to really understand how Earth is changing and what makes Earth's climate so unique and fragile, we have to take a journey through the upper atmosphere and to our neighbors in the solar system.

14.1 The atmosphere exerts key controls on planetary temperatures

Our solar system contains eight planets, but life is only known from our own, Earth, which is the third planet from the Sun. You might wonder why and how so much life ended up here instead of, say, on our two planetary neighbors, Venus and Mars (**Figure 14.1**).

In August 2012, NASA's wheeled rover *Curiosity* landed on Mars and sent back bleak images of a red, stone- and sand-strewn landscape stretching to the horizon (**Figure 14.2**). Those stark, lifeless views were hardly surprising because the average surface temperature is about –65°C (–85°F), a place where even a polar bear would have trouble staying warm. It still might be possible that there are microbes hidden in some icy crevice, but they would have to be highly resistant to freezing. As for Venus, life as we know it would be impossible, because it's about as hot as a traditional wood-fired pizza oven: The surface temperature averages 464°C (867°F).

Sitting comfortably between the temperature extremes of Mars and Venus, we find Earth, with an average global temperature of about 15°C (59°F), a climate to which its inhabitants are well attuned. In their 2012 book *The Goldilocks Planet: The 4 Billion Year Story of Earth's Climate*, authors Jan Zalasiewicz and Mark Williams

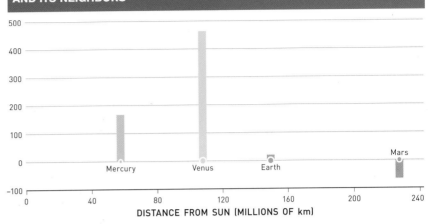

DISTANCE FROM THE SUN AND AVERAGE TEMPERATURES OF EARTH AND ITS NEIGHBORS

FIGURE 14.3 The average temperatures of Mercury, Earth, and Mars show a fairly regular decrease with increased distance from the Sun. The very high temperature of Venus, appearing as a dramatic exception to this pattern, suggests that distance from the nearest source of radiant energy is not the only factor influencing planetary temperatures.

compare Earth to the bowl of porridge in the classic fairy tale of Goldilocks and the three bears. They write that our planet was neither too cold nor too hot but "just right" for humanity. What do you think produces the differences in temperature among these planets?

Distance from the Sun explains some of the differences. However, whereas Venus is closer than Earth to the Sun, Venus is 2.8 times hotter than the planet Mercury, which is the closest planet to the Sun (**Figure 14.3**). Clearly, distance to the Sun is not all that accounts for how climates differ.

Atmosphere and Planetary Temperature

Earth's atmosphere is a layer of gases that stretches from the surface of Earth to the edge of space, some 500 kilometers above the surface. Our atmosphere is made up mainly of nitrogen and oxygen, which are basically transparent to visible sunlight. On a clear day, most of the Sun's beams penetrate the atmosphere like a glass window and hit Earth's surface, where two things happen. Some of that light is immediately reflected back toward the sky, particularly when it has hit a bright surface like fresh, white snow. Most of it warms Earth's surface like a parking lot on a summer day, and that energy is slowly re-emitted not as visible light, but as infrared radiation, one avenue for transmission of heat energy. Consider that even after the Sun has set on a particularly hot day and there is no more visible light, you can still feel the heat radiating from the ground.

Infrared radiation has longer wavelengths than visible light, which means it has different properties. Most of it does not pass back through the atmosphere into outer space, but is rather absorbed by clouds

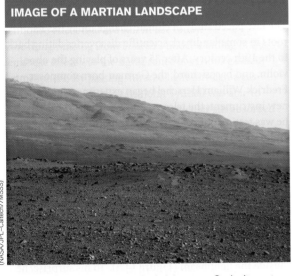

FIGURE 14.2 The mobile robotic laboratory *Curiosity* sent this panoramic photo of a landscape from its point of landing on Mars within the Gale Crater. The rim of the crater is seen in the distance.

pointed out, carbon dioxide emissions from the burning of coal rivaled emissions from volcanoes, and increased burning of coal could lead to future global warming.

He predicted that a further doubling of atmospheric carbon dioxide would raise global temperatures 5° to 6°C, while a halving of those levels would decrease global temperatures by 4° to 5°C. Arrhenius was generally correct about the impact of carbon dioxide and other atmospheric gases on global temperatures. However, he had no satisfactory explanation for how cycling of the climate between ice ages and warmer periods is produced. We now know that these climate cycles are triggered by variation in the tilt of Earth's axis and the shape of its orbit around the Sun.

⚠ **Think About It**

1. Imagine looking at the world with an infrared camera. What objects would glow most brightly?

2. Can you find examples of volcanoes changing the climate in the past?

14.3 Global temperatures and atmospheric CO$_2$ concentrations have varied cyclically

Over the course of Earth's history, there have been many ice ages, which have occurred with striking regularity. Although carbon dioxide concentrations in the atmosphere play a key role in the changing climate, scientists now understand that a number of factors interact to set the clock for these cycles.

The Climate Record in Ice

As snow accumulates in Earth's cold places on high mountains and at high latitudes, it compresses lower layers and transforms snow into ice, trapping a sample of air from the distant past. It turns out that when temperatures are higher, water vapor in the atmosphere contains more deuterium, a heavy isotope of hydrogen. That means climate scientists can use these ice samples to measure both the past concentrations of gases in the atmosphere and to estimate historical temperatures.

European teams recently drilled more than 3.2 kilometers (more than 2 miles) deep into the Antarctic ice cap (**Figure 14.7**) using the ice core they retrieved to reconstruct an 800,000-year record of the climate. Because atmospheric gases are present at extremely low concentrations in the atmosphere, scientists report their measurements in units of parts per million (ppm). In this study, the scientists found that there were between 170 and 300 molecules of CO$_2$ for every million gas

molecules in the ancient atmosphere. In other words, there were between 170 and 300 ppm CO$_2$. Today, Earth's atmosphere contains more than 400 ppm CO$_2$.

Carbon dioxide has fluctuated over time. When temperatures were low, carbon dioxide levels were also low, and when temperatures were higher, carbon dioxide levels were elevated. One of the most notable features of the record of glacial and interglacial periods is their regular occurrence at approximately 100,000-year intervals. Over the last 800,000 years, there have been exactly nine ice ages. What accounts for these climate cycles? It turns out they coincide with changes in various features of Earth's orbit and rotation on its axis.

Role of Earth Orbital Cycles

In the 1920s and 1930s, Milutin Milankovitch, a Serbian astronomer, began studying the theory that ice ages were caused by periodic changes in Earth's rotation on its axis and orbit around the Sun. He hypothesized that the amount of Sun hitting Earth during the summer is critical to the beginning or ending of an ice age because it dictates whether snow accumulates year after year. Milankovitch reasoned that when solar radiation was weak and temperatures cooler during Northern Hemisphere summers, snow would begin to accumulate, initiating an ice age.

Milankovitch identified three major aspects of Earth's orbital cycles that could influence solar inputs. The first aspect is variation in the shape of Earth's elliptical orbit around the Sun, which lengthens and shortens on a 100,000-year cycle. This variation in **eccentricity** affects solar inputs because it changes Earth's distance from the Sun. Presently, Earth's low eccentricity produces an input of solar radiation that is about 6% lower on July 4, when Earth is farthest from the Sun, than on January 3, when the planet is closest to the Sun. At its highest eccentricity, this difference in solar inputs is 20% to 30%.

The second aspect is the cycle in the tilt of Earth's axis of rotation, from a minimum of 21.5° to 24.5° every 41,000 years. The tilt cycle produces variation in heating of the planet's two hemispheres at high latitudes. A low axial tilt decreases the amount of insolation at high latitudes, producing cooler summers, allowing snow and ice to build up, as well as warmer winters. During periods with a higher axial tilt, winter snow melts during the hotter summers at high latitudes. Currently, with an axial tilt of 23.5°, Earth is midway between these two extremes.

The third aspect identified by Milankovitch is that Earth wobbles on its axis on a 26,000-year cycle. Earth's axial wobble produces the **precession of the equinoxes,** which causes the equinoxes to occur at different points in Earth's orbit over time, repeating itself every 26,000 years. Changing the positions of the equinoxes and solstices in Earth's orbit changes the amount of sunlight received in the Northern and Southern Hemispheres. Precession of

eccentricity Variation in the shape of Earth's orbit around the Sun.

precession of the equinoxes Slow drift in the position in Earth's orbit at which the quinoxes occur, a cycle repeating itself approximately every 26,000 years.

800,000-YEAR CARBON DIOXIDE AND TEMPERATURE RECORD

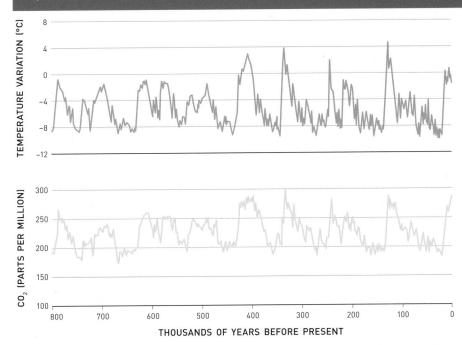

(CarstenPeter/National Geographic/Getty Images)

FIGURE 14.7 Cores of ice drilled from the Antarctic ice cap by several research teams (photo) provide insights into past climates and concentrations of atmospheric carbon dioxide. The two graphs show that increases and decreases in air temperature (upper panel) correspond to increases and decreases in levels of atmospheric carbon dioxide (lower panel) recorded in Antarctic ice over the past 800,000 years. (Data from Lüthi et al., 2008)

the equinoxes can create more dramatic seasons (warmer summers, colder winters) or less dramatic seasons (cooler summers, warmer winters). At present, Earth is closest to the Sun during the winter solstice in the Northern Hemisphere and farthest from the Sun during the Northern Hemisphere's summer solstice, which reduces the seasonal contrast in the Northern Hemisphere.

These cycles are today referred to collectively as **Milankovitch Cycles** and seem to explain the occurrence of ice ages over the last 800,000 years (**Figure 14.8**). It's important to understand how the Milankovitch Cycles affect Earth's climate, but planetary motions do not explain the unprecedented rise in global temperatures over the last century.

Milankovitch Cycles
Cyclic changes in the shape of Earth's orbit, tilt in its axis, and precession of the equinoxes that produce variation in Earth's climate.

MILANKOVITCH CYCLES

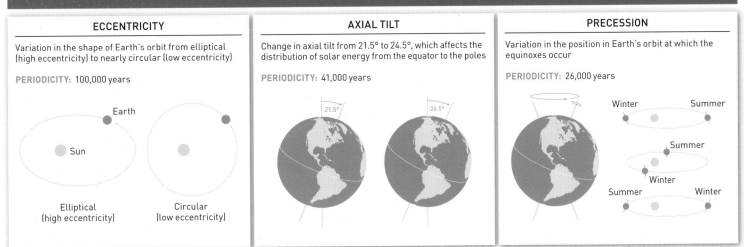

FIGURE 14.8 Aspects of the Milankovitch Cycles affect the input of solar energy to Earth.

FIGURE 14.9 At the end of the last glacial maximum, increased global temperatures (orange line) followed rising atmospheric CO_2 concentrations (yellow line). Meanwhile, Antarctica temperatures (blue line) either rose before CO_2 concentrations rose or coincided with rising atmospheric CO_2. (Data from Shakun et al., 2012)

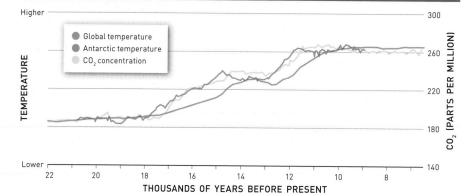

Hemispherical Differences in Timing of Warming

Although CO_2 and temperature are tightly coupled in the Antarctic ice record, scientists discovered that temperature increases there happen before carbon dioxide levels increase. That fact may seem puzzling, but as we shall now see, carbon dioxide is part of a global feedback loop, meaning it can be both a contributing cause and a consequence of a warming climate. In order to understand this concept, we need to zoom out of Antarctica and take a look at other forces that shape Earth's climate as a whole.

Carbon dioxide clearly plays a key role in shaping Earth's climate over geologic time. Scientists have begun to understand how it interacts with the Milankovitch Cycles and why we should be concerned about future carbon dioxide emissions. In 2012 Jeremy Shakun of Harvard and Columbia's Lamont–Doherty Earth Observatory decided to focus on the last ice age in greater detail to understand how the greenhouse effect interacted with Milankovitch Cycles.

Rather than studying just an Antarctic ice core, he and his research team reconstructed temperature records from both the Northern and Southern Hemisphere using a variety of approaches. For instance, the tiny shells of some marine plankton contain both magnesium and calcium, but the ratios of these two elements are known to correlate closely with temperature. Rather than taking a core from the Antarctic ice, the researchers took cores from the ocean bottom, which contains these shells going back in time. On land, they could take cores from lake bottoms, which are packed with pollen from plants, which reflect the type of ecosystem present at a specific time. Examining these various cores, Shakun confirmed the Antarctic results that in the Southern Hemisphere warming occurred before CO_2 increases. However, in the Northern Hemisphere, and on Earth as a whole, temperature began to rise five centuries *after* increases in CO_2 (**Figure 14.9**).

Shakun proposed that Milankovitch Cycles had initiated the end of the last glacial period and set off a complicated reaction. First, greater sunlight in the Northern Hemisphere led to the melting of glaciers about 19,000 years ago. The influx of freshwater into the Atlantic Ocean weakened the Atlantic Meridional Overturning Current (AMOC), an ocean current that normally transports heat from the Southern Hemisphere to the north (**Figure 14.10**). This weakening trapped heat in the Southern Hemisphere, including Antarctica. As the southern oceans warmed, the solubility of the CO_2 they held decreased, resulting in a release of CO_2 to the atmosphere, much like CO_2 is released from a carbonated beverage as it warms. This massive release of CO_2 by the southern oceans amplified the warming of Earth via the greenhouse effect, speeding the end of the last ice age.

Thus, while the timing of ice ages and interglacial periods is set by Milankovitch Cycles, the speed of the transition may be determined by carbon dioxide in the atmosphere. As we shall see in the next section, when we think about Earth's climate on a human timeline—hundreds or, perhaps, thousands of years—carbon dioxide in the atmosphere remains our fundamental concern for the future.

⚠ Think About It

1. Because higher carbon dioxide concentrations are correlated with higher temperatures in the 800,000-year ice record, would it be accurate to conclude that carbon dioxide caused increases in Earth's temperature?

2. Would a planet without Milankovitch Cycles have ice ages?

3. What other biological phenomena capture information on climates from the past?

ATLANTIC MERIDIONAL OVERTURNING CURRENT (AMOC) AND WARMING OF THE SOUTHERN HEMISPHERE

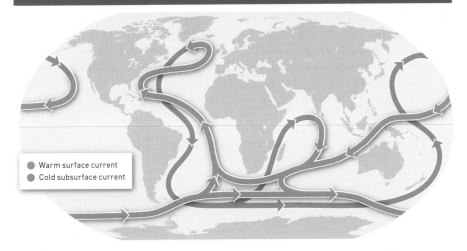

- Warm surface current
- Cold subsurface current

FIGURE 14.10 The AMOC (orange arrows) transports heat from the Southern Hemisphere to the Northern Hemisphere. Climate scientists propose that large amounts of freshwater runoff in the Northern Hemisphere from melting glaciers about 19,000 years ago weakened this flow of heat-conveying water to the Northern Hemisphere, resulting in early warming of the Southern Hemisphere.

14.4 Atmospheric CO_2 appears to be the thermostat controlling global temperatures

Because of the global scale of the questions posed by climate scientists today, the vast amounts of data involved, and the inability to conduct conventional experiments, computers and computer models are essential to climate research. Climate models are built on systems of mathematical equations, and scientists use them to refine studies of a particular problem and test their hypotheses. In the early 2000s, scientists at the NASA Goddard Institute for Space Studies in New York City developed one such model, called ModelE (GISS ModelE), to simulate the interaction between Earth's atmosphere and its oceans, land, and ice based on fundamental physics.

Using the program, they could take a snapshot of the impact of different gases in the atmosphere on global temperatures. They found that the most important gas in the atmosphere for absorbing infrared radiation and directing heat back to Earth's surface is water vapor, which accounted for half of the greenhouse effect. Next, clouds absorb one-quarter of this energy. Finally, CO_2 represents just one-fifth of the greenhouse effect, followed by other greenhouse gases such as methane and nitrous oxide (**Figure 14.11**).

You might conclude that carbon dioxide is a minor player in global temperatures, but the story is a little more complicated. As Arrhenius noted more than a century earlier, water vapor is much more variable than atmospheric CO_2 in time and space. That's because warm air absorbs more water vapor than cold air, which is why it can be so much more humid during a Florida

summer than during a Colorado winter. Consequently, as CO_2 emissions notch up global temperatures, the amount of water vapor in the atmosphere also increases, resulting in a positive feedback loop that amplifies the greenhouse effect (see Chapter 2, page 48). For instance, if CO_2 concentrations in the atmosphere increase and global temperatures increase, the amount of water vapor in the atmosphere will also increase, resulting in higher

RELATIVE INFLUENCES ON THE GREENHOUSE EFFECT

PERCENT OF GREENHOUSE EFFECT

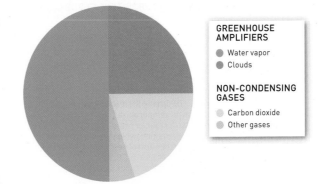

GREENHOUSE AMPLIFIERS
- Water vapor
- Clouds

NON-CONDENSING GASES
- Carbon dioxide
- Other gases

FIGURE 14.11 Water vapor and clouds jointly produce approximately 75% of the greenhouse effect, while carbon dioxide and other gases, such as methane and nitrous oxide, are responsible for 20% and 5%, respectively. Unlike water vapor, carbon dioxide does not condense into liquid and precipitate at temperatures found in Earth's atmosphere. Therefore, like a thermostat, carbon dioxide (along with some other trace gases) sets the level of the greenhouse effect, while water vapor and clouds amplify their effects. (Data from Lacis et al., 2010)

rates of decomposition, which releases CO_2 into the atmosphere. This is only one of several positive feedbacks that can increase the greenhouse effect (**Figure 14.12**). The reverse, a **negative feedback,** can also occur. A form of negative feedback is one in which an increase in some factor in a system, such as a climate system, produces a decrease in that factor within the system. For example, increased atmospheric CO_2, higher temperatures, and increased moisture can stimulate plant growth, which will remove CO_2 from the atmosphere.

So how big of an impact does carbon dioxide have on Earth's climate? It's not possible to run an experiment where we take CO_2 out of the atmosphere in real life, but we can model it. In one simulation, the Goddard team eliminated CO_2 and other greenhouse gases and watched the response over time. During the first year, global temperatures in the model dropped by nearly 5°C, and after 50 years temperatures had dropped nearly 35°C to −21°C, leaving only one-third of the world's oceans free of ice (**Figure 14.13**).

Furthermore, as Arrhenius had predicted so long ago, atmospheric water vapor dropped precipitously by 90%. So, even though CO_2 is only the third largest contributor to Earth's greenhouse effect, the Goddard research team

negative feedback An increase in some factor in a system, such as a climate system, produces a decrease in that factor within the system.

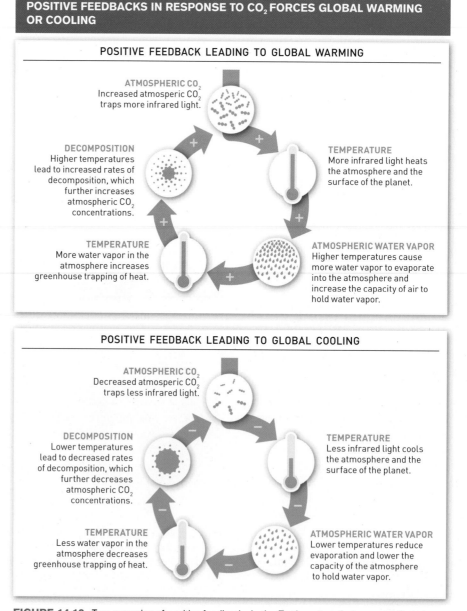

FIGURE 14.12 Two examples of positive feedbacks in the Earth system that can contribute to global climate change.

CONSEQUENCES OF POSITIVE FEEDBACK FOLLOWING REMOVAL OF NON-CONDENSING GASES FROM EARTH'S ATMOSPHERE

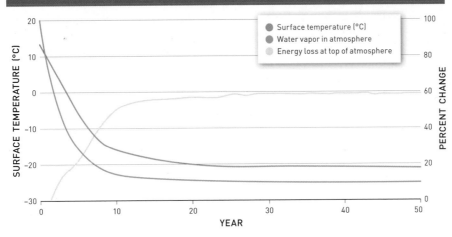

FIGURE 14.13 The GISS ModelE shows that removing key greenhouse gases, including CO_2, methane (CH_4), and nitrous oxide (N_2O), from the atmosphere would lead to a rapid decline in water vapor, increased loss of infrared heat radiation at the top of the atmosphere, and precipitous decline in planetary temperature. (After Lacis et al., 2010)

concluded that it is the primary determinant of global temperatures, with atmospheric water vapor levels following along as temperatures increase or decrease. Because we live on a Goldilocks planet, a temperature change of just a few degrees can have startling consequences for life on Earth, as we shall see in the next section.

⚠ Think About It

1. You might consider the increased burning of fossil fuels a massive experiment with Earth's climate. What is missing in order to test the effect of this treatment? (Hint: What are the key parts of a well-designed experiment?)

2. Besides water vapor and carbon dioxide, what might be some other potential sources of positive or negative feedback in the climate system?

14.1–14.4 Science: Summary

The composition of planetary atmospheres accounts for much of the differences in temperatures among the planets in the solar system.

Based on two centuries of scientific research, we know that trace gases in the atmosphere, including water vapor, carbon dioxide, and nitrous oxide, absorb infrared light and heat the planet's surface—the greenhouse effect.

Ice cores and other lines of evidence show periods of global low temperatures coinciding with low CO_2 concentrations, and high temperatures with high CO_2 concentrations. Such cycling also coincides with cycles in Earth's orbit around the Sun, wobble on its axis, and tilt in its axis of rotation.

CO_2 acts as a thermostat for global temperatures: As CO_2 levels rise, Earth's temperature slowly increases, which adds water vapor to the atmosphere, causing temperatures to increase even more in a positive feedback loop. Even small changes in CO_2 levels can result in climate changes with serious impacts on Earth's biodiversity and on human, social, and economic systems.

14.5–14.8 Issues

In the early days of our species, it was inconceivable that we could alter the climate system. Our numbers were too few and our impacts immaterial. However, a growing body of scientific evidence demonstrates that human activity is responsible for increasing the levels of greenhouse gases in Earth's atmosphere. In response, global temperatures are increasing, accompanied by other aspects of climate change. The changing climate has already caused shifts in species' geographic distributions and damaged important natural ecosystems. Unchecked, climate change threatens further impacts, ranging from species' extinctions and agricultural failures to the displacement of human populations.

14.5 Precise measurements reveal that fossil fuel burning is the main cause of increased atmospheric CO_2 levels

On May 18, 1955, Charles David Keeling, a young postdoctoral researcher, set up camp near a coastal river in California's Big Sur State Park.

Keeling was a young chemist and outdoor enthusiast interested in learning how carbon dioxide moved between water and air, as part of a geological study of limestone. There in the sheltering redwoods, Keeling was having such a good time, he took water and air samples every few hours throughout the day and night, far more than he needed. He went back to his university, the California Institute of Technology, with some surprising findings—not just about limestone, but about Earth's atmosphere itself.

First, he found that the CO_2 concentration in the forest air changed on a daily cycle. It decreased during the day, as plants took in CO_2 to make sugar during photosynthesis, and increased during the night due to respiration by plants and other organisms. Second, he determined that CO_2 concentrations in air were about the same everywhere he made measurements in the afternoon: 310 parts per million (ppm). No one had ever measured CO_2 in the atmosphere so consistently and so precisely before.

Word of Keeling's groundbreaking work soon found its way to Harry Wexler at the U.S. Weather Bureau. Wexler offered him a chance to measure atmospheric CO_2 around the world, including at a new observatory located at 3,397 meters (11,138 feet) on Mauna Loa, a volcano on the island of Hawaii (**Figure 14.14**).

Mauna Loa and the Keeling Curve

At the time Keeling began his measurements, there was widespread disagreement in the scientific community about the atmospheric effects of fossil fuel burning. The prevailing idea was that the vast oceans, which actively exchange CO_2 with the overlying air, would be capable of rapidly absorbing excess atmospheric CO_2. Scientists saw Keeling's precise measurements as a way to settle the debate.

Keeling's first two years at Mauna Loa produced a sawtooth plot showing CO_2 rising and falling with each season. The reason for this sawtooth pattern is that the concentration of CO_2 reaches a maximum during each

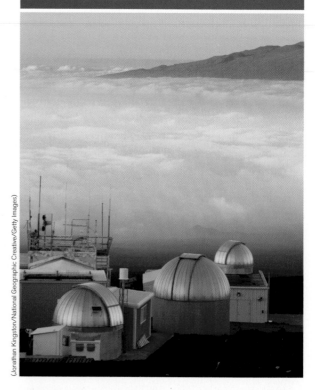

MAUNA LOA CO_2 OBSERVATORY

(Jonathan Kingston/National Geographic Creative/Getty Images)

FIGURE 14.14 The CO_2 observatory was located on Mauna Loa because its location far from sources of air pollution would produce better-quality data. The concentrations of atmospheric carbon dioxide have been monitored at the Mauna Loa Observatory for over half a century.

MEASUREMENTS OF ATMOSPHERIC CARBON DIOXIDE FROM MAUNA LOA, HAWAII

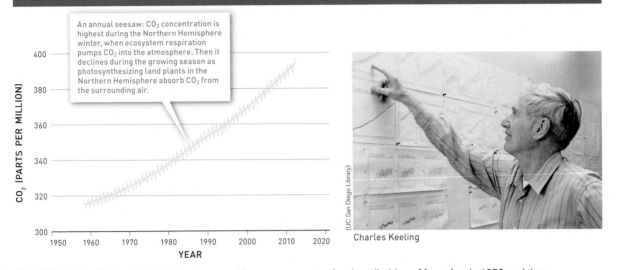

An annual seesaw: CO_2 concentration is highest during the Northern Hemisphere winter, when ecosystem respiration pumps CO_2 into the atmosphere. Then it declines during the growing season as photosynthesizing land plants in the Northern Hemisphere absorb CO_2 from the surrounding air.

Charles Keeling

(UC San Diego Library)

FIGURE 14.15 Charles David Keeling began making measurements of carbon dioxide on Mauna Loa in 1958, and those measurements have continued to the present. The result was the Keeling Curve, showing for the first time that atmospheric carbon dioxide concentrations were indeed increasing, as Svante Arrhenius in 1896 predicted they would. This longest continuous record of atmospheric CO_2 has provided innumerable insights into the global carbon cycle and how humans may be affecting Earth's atmosphere. (Data from the Earth System Research Laboratory of NOAA)

Northern Hemisphere winter as respiration by all the organisms that make up ecosystems pump CO_2 into the atmosphere. Then it declines during the Northern Hemisphere growing season as plants come out of dormancy and rates of photosynthesis increase across the northern continental landmasses. As Keeling continued taking these measurements over the next 50 years, he saw the sawtooth curve gradually rising at Mauna Loa. By 2012 atmospheric CO_2 had increased by approximately 25%. This striking graph is what we now call the Keeling Curve (**Figure 14.15**).

The dramatic modern increase in atmospheric CO_2 becomes especially apparent when we look back in time, again using the ice record. As shown in **Figure 14.16**, atmospheric CO_2 oscillated around an average of about 280 ppm for approximately 1,800 years and then began rising with increased fossil fuel use during the Industrial Revolution, which began in the late 18th century, just as

Why was Keeling's carbon dioxide detector set in Hawaii of all places?

RECORD OF ATMOSPHERIC CARBON DIOXIDE FROM THE YEAR 1 TO 2011 CE

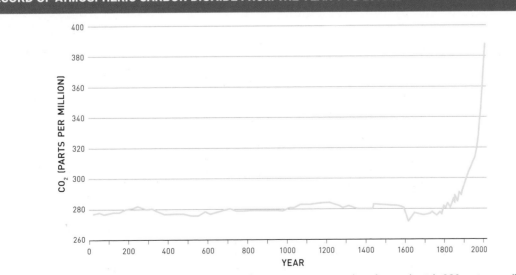

FIGURE 14.16 Following 18 centuries of oscillating around an average concentration of approximately 280 parts per million (ppm), atmospheric carbon dioxide concentration increased rapidly after 1800, increasing over 100 ppm by the year 2011. (Data from Etheridge et al., 1996; MacFarling Meure et al., 2006; Scripps CO_2 Program, http://scrippsco2.ucsd.edu)

DECLINE IN CONCENTRATION OF ¹³C RELATIVE TO ¹²C IN ATMOSPHERIC CO_2

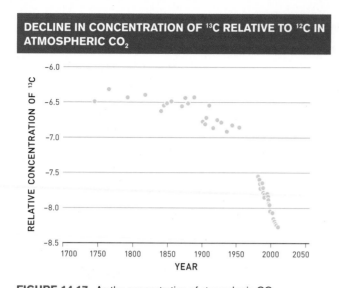

FIGURE 14.17 As the concentration of atmospheric CO_2 increased after 1800, the relative concentration of ^{13}C in atmospheric CO_2 has declined, indicating that the buildup of atmospheric CO_2 is largely the result of the burning of fossil fuels and biomass, which contain less ^{13}C relative to ^{12}C. (Data from Friedli et al., 1986, and Scripps CO_2 Program)

Svante Arrhenius predicted it would—except it has risen even faster than Arrhenius anticipated.

Carbon Isotopes and Fossil Fuels

Because fossil fuel emissions are not the only source of carbon dioxide in Earth's atmosphere, alternative hypotheses are that the rise in CO_2 concentrations might be due to volcanic eruptions or might result from emissions from the oceans. Volcanic eruptions can be discounted as the source of increased atmospheric carbon dioxide since fossil fuel burning releases over 100 times more CO_2 to the atmosphere. How about the oceans as the source? Scientists can answer this question by looking at the ratio of two carbon isotopes in the atmosphere, ^{13}C to ^{12}C, which Keeling had also measured in Big Sur.

Because plants preferentially take in the lighter ^{12}C isotope from the air, plant biomass, and therefore the CO_2 emitted by plants during respiration, has a higher relative concentration of ^{12}C than occurs in the atmosphere. At Big Sur, Keeling could see that the ^{13}C isotope decreased in relative concentration at night, when the plants respired and weren't absorbing carbon dioxide during photosynthesis, and increased in relative concentration during the day when the plants photosynthesized, removing ^{12}C-rich carbon dioxide from the atmosphere. These carbon isotope measurements proved to Keeling that respiration by plants was primarily responsible for the buildup of CO_2 in Big Sur during the forest night.

The relative concentrations of ^{13}C and ^{12}C can also be used to determine the source of increased atmospheric CO_2 over the past two centuries. Because fossil fuels formed over millions of years out of compressed plant biomass, the carbon dioxide released when fossil fuels combust also contains relatively low concentrations of the ^{13}C isotope compared with that in the atmosphere. In contrast, carbon dioxide in the oceans has the same ratio of ^{13}C to ^{12}C as the atmosphere. If atmospheric CO_2 increases are the result of fossil fuel burning, which is relatively rich in ^{12}C, the relative concentration of ^{13}C should begin to decline after the Industrial Revolution in 1800—and that is exactly what has happened (**Figure 14.17**).

Meanwhile, other greenhouse gases—specifically, methane, CH_4, and nitrous oxide, N_2O—have also increased since the Industrial Revolution (**Figure 14.18**). Some of these gases come from agricultural sources in

INCREASES IN GREENHOUSE GAS CONCENTRATIONS

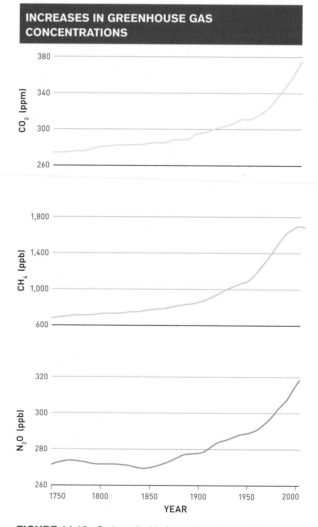

FIGURE 14.18 Carbon dioxide is not the only greenhouse gas that has increased during the past two centuries. The atmospheric concentrations of both methane (CH_4) and nitrous oxide (N_2O) have also increased. (Data from MacFarling & Meure, 2006)

addition to fossil fuel burning. For instance, flooded rice paddies and livestock such as cattle and buffalo produce significant quantities of climate-warming methane.

⚠ Think About It

1. Looking at the Keeling Curve, how many years did it take to be sure that carbon dioxide was rising?

2. How do we know that the increase in carbon dioxide over the last 50 years has resulted from fossil fuels?

14.6 As CO₂ levels have risen in modern times, global temperatures have increased significantly

Because the burning of fossil fuels over the last 200 years has increased carbon dioxide concentrations in the atmosphere, Earth should be warming via the greenhouse effect. Indeed, four climate research groups—one in Japan, one in the United Kingdom, and two in the United States—have shown that global temperatures have increased significantly since 1880 by about 0.85°C (1.5°F) (**Figure 14.19**).

The relatively smooth increase in atmospheric concentration of CO_2 (see Figure 14.17) contrasts sharply with the considerable variability in global temperatures shown in Figure 14.19. This contrast suggests that factors other than greenhouse gases significantly affect global temperatures, at least in the short term. One of the main sources of this discrepancy is the El Niño Southern Oscillation (ENSO), which we reviewed in Chapter 6 (see page 161). For instance, the peak in global temperature around 1942 was associated with a strong El Niño, and the subsequent declines that followed were associated with several La Niña episodes in the 1950s. Similarly, the exceptionally strong El Niño of 1998 brought about a spike in global temperatures and was followed by a La Niña episode, which reduced global temperature. Such influences are not surprising, since ENSO events involve the massive transport of heat stored in the Pacific Ocean. However, these short-term variations do not affect the long-term trends of increasing global temperature that have been documented.

The increase in global temperature becomes even more apparent in a plot of temperatures averaged by decades (**Figure 14.20**). By removing the higher-frequency, year-to-year variation in temperatures, the plot becomes smooth. Figure 14.20 shows that record high temperatures were recorded from the 1980s through the 2000s, with each succeeding decade setting a new record for high temperatures. Global temperatures for 2010 to 2014 continued to be well above average, despite a prolonged La Niña at the beginning of this new decade. In fact, the average global temperature in 2014 was the warmest recorded since 1880.

Consequently, the warming of the globe shows no signs of flagging. By the end of the 21st century, climate models suggest that the temperature of Earth's surface is expected to rise between 2° and 3°C, according to the IPCC.

INDEPENDENT ESTIMATES OF GLOBAL TEMPERATURES BY FOUR CLIMATE RESEARCH GROUPS

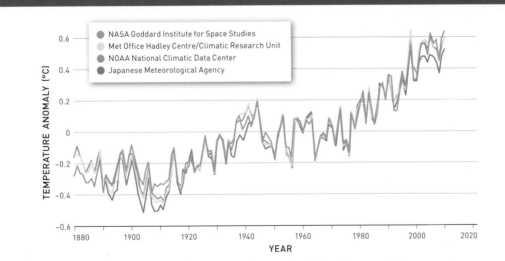

FIGURE 14.19 Estimates of global temperatures made independently all show the same basic pattern of temperature increases since 1880, with the first decade of the 21st century being the warmest on record. (Data from NASA Earth Observatory, 2011)

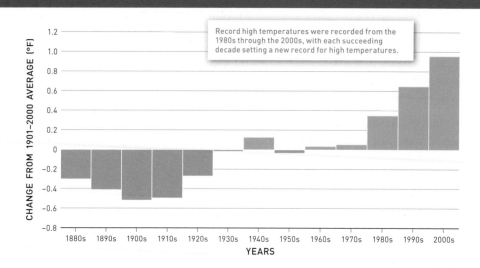

GLOBAL TEMPERATURES AVERAGED BY DECADE

Record high temperatures were recorded from the 1980s through the 2000s, with each succeeding decade setting a new record for high temperatures.

FIGURE 14.20 Plotting global temperature changes since 1880 as decadal averages removes annual fluctuations in the record, revealing clearly the record-breaking temperatures of the 1980s, 1990s, and 2000s. (After 2009 State of the Climate Highlights, www.ncdc.noaa.gov/bams-state-of-the-climate.)

Reanalysis and Confirmation of the Temperature Record

Although it may seem like a trivial matter to measure air temperature in a single spot over time, several criticisms have been leveled at attempts to test whether Earth as a whole is really warming.

Some critics have pointed out that weather stations are mostly located in urban areas, which have become "urban heat islands" over time because asphalt and cement absorb a lot of heat. In addition, the quality of the monitoring process varies from station to station, which raises the possibility that these records are inaccurate. Some critics argue that only data from the best sites should be included in global estimates. Others say the opposite: Climate researchers have been too selective in their choice of data sources. If the records from more weather stations were included, they argue, the warming pattern would disappear.

To settle the question, Richard A. Muller at the University of California at Berkeley organized the Berkeley Earth Surface Temperature (BEST) study. Muller, who was himself a skeptic of global temperature estimates, compiled weather records from 36,000 land-based weather stations, approximately 5 times the number of stations included in other temperature analyses. These records included 1.6 billion temperature measurements from around Earth, going back to 1753, which extended the analysis over 100 years farther back in time than the plot in Figure 14.20. In 2011 Muller reported that temperatures on land have risen 1.5°C in

the past 250 years and 0.9°C in the most recent 50 years. "I was not expecting this, but as a scientist," Muller later told the newspaper *The Telegraph* when he published his findings in 2012, "I feel it is my duty to let the evidence change my mind" (**Figure 14.21**). He

INDEPENDENT CONFIRMATION OF LAND TEMPERATURE RECORD

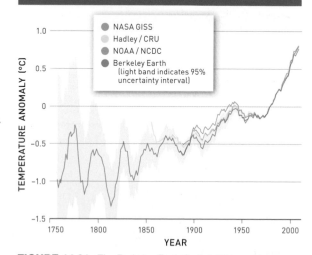

- NASA GISS
- Hadley / CRU
- NOAA / NCDC
- Berkeley Earth (light band indicates 95% uncertainty interval)

FIGURE 14.21 The Berkeley Earth Surface Temperature (BEST) research group independently confirmed global warming of temperatures over land using a much larger sample of meteorological stations and controlling for urban heat island effects. (Data from BEST, http://berkeleyearth.org/)

went on to say that the best statistical fit to the rise in Earth temperatures was the human-caused increase in atmospheric CO_2.

⚠ Think About It

1. Why is it important that four scientific groups analyzed the temperature data?

2. What are some reasons why scientists might be skeptical of global warming?

3. What are some reasons why politicians might claim to be skeptical of global warming?

14.7 Rising temperatures have been accompanied by diverse changes in the Earth system

The increase in greenhouse gases and global temperatures over the last century has set into motion a variety of changes to Earth's systems. Some of the most striking consequences of climate change include the melting of the ice caps, the warming of the oceans, the shifting of species' geographic ranges and seasonality, and the die-off of coral reefs and forests.

Melting Ice

In mid-September 2012 sea ice around the North Pole reached its lowest extent since satellite monitoring began in 1979 (**Figure 14.22**). For the first time in history, cargo ships could pass unimpeded through the long-sought Northwest Passage through the northern islands of Canada and the Northeast Passage along Scandinavia and Russia.

As reflective ice is replaced with open water, more heat will be stored in the world's oceans, which will pump more water vapor into the atmosphere. Both changes will amplify global warming. Meanwhile, polar bears and ring seals, for which sea ice is essential habitat, are two potential victims of climate change. The opening of Arctic waters has also set in motion an economic race, between Russia, Canada, the United States, and other nations, to claim rights to mineral wealth under the Arctic seafloor, particularly oil and natural gas deposits.

On land, the ice sheets, glaciers, and ice caps are also melting at a rapid rate, causing sea level to rise. In July 2012 a ridge of warm air stalled over Greenland, melting the surface of its ice sheet over large areas. Four days later, the Petermann Glacier on Greenland split and shed a gigantic ice island with a surface area of 120 km²

EXTENT OF ICE COVER IN THE ARCTIC OCEAN

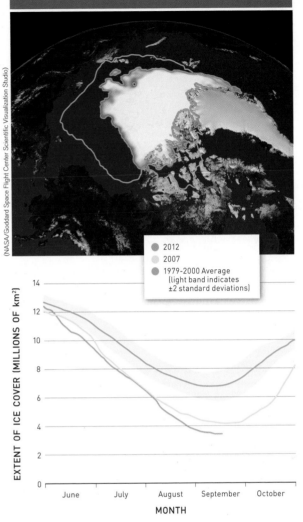

(NASA/Goddard Space Flight Center Scientific Visualization Studio)

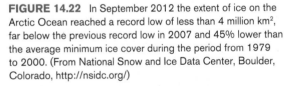

- 2012
- 2007
- 1979–2000 Average (light band indicates ±2 standard deviations)

FIGURE 14.22 In September 2012 the extent of ice on the Arctic Ocean reached a record low of less than 4 million km², far below the previous record low in 2007 and 45% lower than the average minimum ice cover during the period from 1979 to 2000. (From National Snow and Ice Data Center, Boulder, Colorado, http://nsidc.org/)

(46 mi²)—about twice the size of Manhattan Island (**Figure 14.23**). Warming of the oceans also contributes to sea level rise, since water expands as it heats up.

The human and economic impacts of sea level rise will be substantial, since the human population is concentrated in coastal areas with their ports and beaches. Over the last century, the oceans have risen about 20 centimeters (8 inches), but scientists predict the rate of rise will speed up significantly over the next century, along with temperatures. Rising sea levels have already forced the movement of communities from low-lying areas on the tiny Pacific island nation of Tuvalu, and it threatens

ICE LOSS FROM GREENLAND'S ICE SHEET

(NASA Earth Observatory image by Jesse Allen and Robert Simmon, using data from NASA/GSFC/ METI/ERSDAC/JAROS, and U.S./Japan ASTER Science Team)

FIGURE 14.23 A large ice island of approximately 120 km² broke off from the Petermann Glacier on Greenland on July 16, 2012.

HEAT ABSORPTION BY COMPONENTS OF THE EARTH SYSTEM

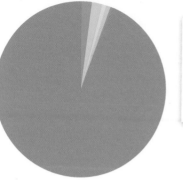

EARTH SYSTEM COMPONENT
- Oceans
- Atmosphere
- Continents
- Glaciers & ice caps
- Arctic sea ice
- Ice sheets

FIGURE 14.25 When climate scientists inventory the absorption of heat as Earth warms, they find that the oceans are overwhelmingly the major absorber of heat energy. This is because the oceans cover a large area and encompass a very large volume of water, which has a high capacity for heat storage. (Data from State of the Climate Highlights, www.ncdc. noaa.gov/bamsstate-of-the-climate)

that nation's very existence (**Figure 14.24**). If present trends continue, sea level rise may force tens of millions of people living in coastal areas to relocate.

The impacts of the melting ice will not be felt equally around the globe. Recent research shows that, compared with many other regions, sea level has been rising more

rapidly along the northeastern coast of the United States. Consequently, Boston, New York, Philadelphia, and Baltimore may be among the first major urban and economic centers to suffer damages from coastal flooding.

Warming Oceans

The global ocean is the largest heat sink for the warming Earth, dwarfing all other parts of the Earth system (**Figure 14.25**). However, it's not just that the global ocean is the planet's largest absorber of heat. One would expect that of such a large mass of water. The significant thing, from a climate change perspective, is that the ocean's heat content is increasing (**Figure 14.26**). One result of a warmer ocean is the loss of photosynthesizing plankton, which account for half of primary production on Earth. This decline in production would lead to declines in the fish populations that humans depend on.

How could ocean warming result in lower marine primary production? As we saw in Chapter 8, primary production in the oceans is controlled mainly by the availability of inorganic nutrients, such as iron, which are generally found in greatest supply where they can be renewed by runoff from land; by upwelling; and by mixing of deep, nutrient-rich water with surface water (see page 232). But when the oceans heat up, distinct thermal layers in the water column form, which prevent mixing of surface water with nutrient-rich deeper and cooler waters. Thus, critical nutrients do not reach marine life closer to the surface.

Scientists are already observing losses in the ocean's phytoplankton biomass at a rate of about 1% per year. A

SEA LEVEL RISE THREATENS ISLAND NATIONS

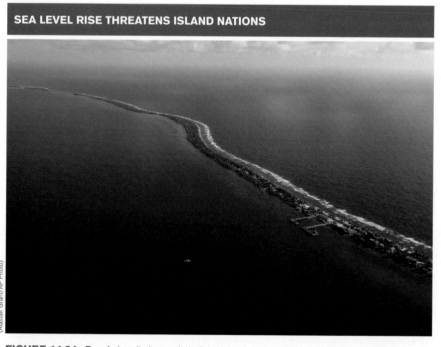

(Alastair Grant/AP Photo)

FIGURE 14.24 Funafuti atoll, the capital of the island nation of Tuvalu, which is situated approximately halfway between Australia and Hawaii, would be made uninhabitable by even moderate sea level rise.

A WARMING OCEAN

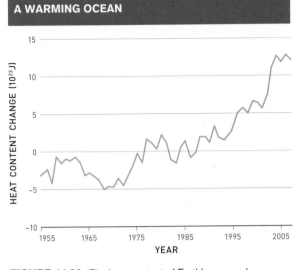

FIGURE 14.26 The heat content of Earth's oceans has increased approximately 10-fold in the past half-century. (After 2009 State of the Climate Highlights, www.ncdc.noaa.gov/bams-state-of-the-climate)

2010 article in *Nature* reported declines in phytoplankton biomass in 8 of 10 ocean regions since 1899. Another study found that the ocean's least productive waters grew by 6.6 million km² between 1998 and 2006 (**Figure 14.27**).

Shifts in Species' Ranges and Seasonality

Owing to climate change, spring comes earlier every year and winter later, a phenomenon that has major biological impacts. Many plants that once bloomed in May are now blooming in April, and hibernating mammals that once emerged from their dens in March are emerging in February. In the eastern North Pacific, gray whales (*Eschrichtius robustus*) travel farther north in the Arctic Ocean during the summer, stay at their northern feeding grounds later in the year, and sometimes overwinter in northern waters.

This is an example of how geographic ranges of species are shifting—some moving up mountain ranges and into higher, cooler latitudes. For instance, butterfly communities in the Sierra Guardarrama in central Spain shifted uphill 293 meters (961 feet) in the three decades between 1973 and 2004. In Massachusetts, scientists have documented shifting ranges and population changes in 100 species of butterflies. Between 1992 and 2010, populations of many butterfly species near their northern limits increased in abundance with warmer temperatures, while northern species near the southern limits of their ranges decreased in abundance as temperatures grew too hot (**Figure 14.28**).

Movements northward and upward in elevation have also been documented in plants, birds, mammals, fish, and a wide range of insects, spiders, and other invertebrates. As some have debated the reality of global warming, Earth's inhabitants have been voting with their feet, wings, and fins.

Dying Forests and Corals

While some groups of organisms are shifting distributions in response to climate change, other, less mobile organisms are dying off. This mortality is especially

CONTRAST IN OCEAN COLOR AND PRODUCTIVITY

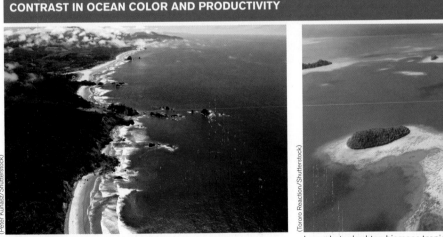

Phytoplankton-rich temperate ocean

Low-phytoplankton biomass tropical ocean

FIGURE 14.27 The chlorophyll of abundant phytoplankton is reflected in the greenish cast of this highly productive area of temperate ocean. Meanwhile, the crystalline blue of these tropical waters signals the low biomass of phytoplankton and low primary production. The ocean's least productive waters, such as this example, are increasing in extent, while the area of the more productive waters declines.

BUTTERFLY RESPONSES TO CLIMATE WARMING IN MASSACHUSETTS

(Barrett & MacKay/All Canada Photos/Getty Images)

Atlantis fritillary, *Speyeria atlantis*

(Jeffrey S. Pippen, www.jeffpippen.com)

Frosted elfin, *Callophrys irus*

FIGURE 14.28 The atlantis fritillary, *Speyeria atlantis,* which is near the southern boundary of its geographic distribution has been declining in abundance in Massachusetts. Meanwhile, the population of the frosted elfin, *Callophrys irus,* which is near the northern edge of its distribution, has increased 10-fold since 1992. (Data from Breed et al., 2013)

noticeable where it involves foundation species (see Chapter 4, page 104)—for example, reef-building corals and forest trees (**Figure 14.29**).

In June 2010 abnormally warm waters swept through the shallow water reefs of the Caribbean, causing a devastating coral bleaching event. Coral bleaching occurs when reef-building corals that are under stress, for example, from high temperatures, expel symbiotic algae living in their tissues, leaving behind a pale structure. Under nonstressful conditions, these symbiotic algae provide energy in the form of sugars to the corals in exchange for nutrients and physical protection. Although newly bleached corals are not dead, their chance of dying is significant, especially if the source of stress continues. Because these reefs act as nurseries and habitats for fish, octopus, and other marine species, their disappearance has a rippling impact on marine ecosystems. If global temperatures increase 2°C, as anticipated, scientists estimate that 70% of coral reefs will be seriously damaged. In addition to heat, more carbon dioxide in the ocean causes the water to become more acidic, making it more difficult for coral-building organisms to form their calcium carbonate structures.

As we learned earlier, wildfires in the western United States during 2011 and 2012 were among the largest and most costly ever recorded. The stress induced by high temperatures and drought can kill standing trees, increasing dry tinder and, as a result, the intensity of wildfires. In times past, forests would reestablish on burned areas. However, forest ecologists predict that the high intensity of recent fires may cause these and other forests around the world to be replaced with other, more drought-tolerant woodlands or shrublands.

Why have there been so many major blizzards and snowfalls in recent U.S. winters if the overall climate is warming? Shouldn't there be fewer?

⚠ Think About It

1. We tend to think only of the negative consequences of climate change, but are there any economic benefits?

2. How might coastal communities adapt to rising sea levels?

14.8 Climate change can lead to a wide range of societal costs

Climate scientists have been predicting that a warmer Earth will be one in which there are many more extreme weather events, including stronger winds, more frequent torrential rains and floods, and higher temperatures during heat waves.

Take, for instance, those scorching hot days of summer, when it's impossible to get anything done without the air conditioning on full blast. Not only are these scorchers getting more common, they are affecting a wider geographic area. James Hansen of NASA's Goddard Institute for Space Studies compared the 30 years from 1951 to 1980, which were used as a reference period, to the most recent decade of 2001 to 2011. In the past, only about one-third of summers had average temperatures that would be considered significantly hotter than typical. However, during the decade from 2001 to 2011, that number had increased to 75% of summers. Meanwhile, the area of the Earth experiencing super-hot summers, with temperatures frequently reaching levels high enough for weather forecasters to warn of a dangerous heat wave,

IMPACT OF WARMING ON CORAL REEFS AND FOREST TREES

(Rainer von Brandis/E+/Getty Images)

Bleaching coral reef

(Ethan Miller/Getty Images)

Dying pine forest

FIGURE 14.29 As the oceans warm, reef-building corals are suffering elevated mortality, shown in this photograph as bleached, whitish corals (top). Meanwhile, on land, elevated temperatures and drought have combined to induce physiological stress in these lodgepole pines, *Pinus contorta,* in northern Colorado. Warmer winters have also reduced the mortality of beetles that attack the trees. In a state of stress, the trees did not mount sufficient defense against a massive beetle attack, which was the direct cause of death across this swath of mountain landscape.

has increased from much less than 1% of Earth's surface to over 10%, a greater than 10-fold increase.

One of the most serious societal consequences of a warming Earth comes not from the heat itself, but from changes in the way water moves through our landscapes, farmland, and ecosystems. Scientists expect an increase in the depth and frequency of droughts, interrupted by short bursts of torrential rains and floods. The first decade of the 21st century was marked by a number of such events, and their economic costs are considerable. A 2010 study from the Pew Trust estimated that the global cost of climate change will range from $5 trillion to $90 trillion by 2100 (**Figure 14.30**).

Disruptions to Agriculture

Rising temperatures will harm our food system, which will already be stretched thin by a growing human population. According to the Food and Agriculture Organization of the United Nations, global food demand will increase by 70% by the year 2050. Meanwhile, the International Food Policy Research Institute has predicted that climate change will result in less food available per capita across the globe than there was in the year 2000.

We are already seeing the impact of extreme heat events on agriculture today. In 2012 the drought in the United States resulted in low levels of soil moisture during the growing season (**Figure 14.31** on page 451). Low soil moisture, in turn, reduced areas of healthy corn and soybean production by one-half and one-third, respectively. Similarly, a 2010 heat wave in Russia decreased grain production by 30%, and a 2011 heat wave in France decreased that nation's grain harvest by 12%.

Droughts also affect livestock farmers. During the 2012 drought in the United States, over half of rangelands and pastures across the country were in poor to very poor condition, forcing many farmers to sell off cattle. High feed prices led to other farmers selling off hogs. Unless agriculture can adapt to a rapidly changing climate, such shortfalls in production will only worsen.

Impacts on Human Health

Extreme heat can be deadly. The summer of 2003 was the hottest one in Europe in the last 500 years. During that heat wave, an estimated 70,000 of people died in 16 countries. The elderly, the infirm, and the poor without adequate air-conditioning systems are typically the victims of such heat events. In 2012 the U.S. Centers for Disease Control and Prevention (CDC) reported that the number of heat-related deaths in the country averaged approximately 700 annually, more than the average total number of deaths in the United States from hurricanes, tornadoes, floods, and earthquakes. With global warming, scientists at the CDC predict that by 2050 annual heat-related deaths will rise to between 3,000 and 5,000.

Climate change is also predicted to increase the prevalence of certain infectious diseases. For instance, insect-borne tropical diseases, such as malaria and dengue fever, will become more prevalent in temperate environments. Dengue fever has recently been reported

MAPPING A SAMPLE OF EXTREME WEATHER

YEAR		METEOROLOGICAL RECORD-BREAKING EVENT	IMPACT / COSTS
2000	1	Wettest autumn on record since 1766	£1.3 billion
2002	2	Highest daily rainfall record in Germany since at least 1901	Flooding of Prague and Dresden, U.S. $15 billion
2003	3	Hottest summer in at least 500 years	Death toll exceeding 70,000
2004*	4	First hurricane in the South Atlantic since 1970	Three deaths, U.S. $425 million
2005	5	Record number of tropical storms, hurricanes, and category 5 hurricanes since 1970	1,836 deaths, U.S. $110 billion (Hurricane Katrina was the costliest U.S. natural disaster)
2007	6	Strongest tropical cyclone in the Arabian Sea since 1970	Biggest natural disaster in the history of Oman
	7	May–July wettest since records began in 1766	Major flooding, £3 billion
	8	Hottest summer on record in Greece since 1891	Devastating wildfires
2009	9	Heatwave breaking many station temperature records (32–154 years of data)	Worst brushfires on record, 173 deaths, 3,500 houses destroyed
2010	10	Hottest summer since 1500	500 wildfires around Moscow, grain-harvest losses of 30%
	11	Rainfall records	Worst flooding in Pakistan's history, nearly 3,000 deaths, 20 million people affected
	12	Highest December rainfall recorded since 1900	Flooding of Brisbane, 23 deaths, estimated U.S. $2.55 billion
2011	13	Most active tornado month on record (April) since 1950	Tornado hit Joplin causing 116 deaths
	14	January–October wettest on record since 1880	Severe floods when Hurricane Irene hit
	15	Most extreme July heat and drought since 1880	Wildfires burning 3 million acres (preliminary impact of U.S. $6–8 billion)
	16	Hottest and driest spring on record in France since 1880	French grain harvest down by 12%
	17	Wettest summer on record (The Netherlands, Norway) since 1901	(not yet documented)
	18	72-hour rainfall record (Nara Prefecture)	73 deaths, 20 missing, severe damage
	19	Wettest summer on record since 1908	Flooding of Seoul, 49 deaths, 77 missing, 125,000 affected

FIGURE 14.30 The first decade of the 21st century brought a host of record-breaking weather events that climate scientists have attributed to global warming. (Data from Coumou & Rahmstorf, 2012)

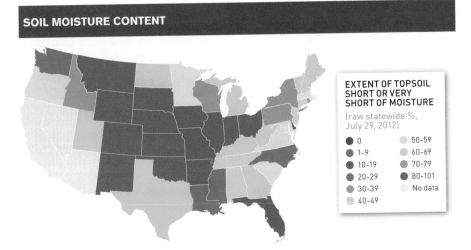

SOIL MOISTURE CONTENT

EXTENT OF TOPSOIL SHORT OR VERY SHORT OF MOISTURE
(raw statewide %, July 29, 2012)

- 0
- 1–9
- 10–19
- 20–29
- 30–39
- 40–49
- 50–59
- 60–69
- 70–79
- 80–101
- No data

FIGURE 14.31 The U.S. Department of Agriculture reported on July 29, 2012, that soil moisture was in short supply or very short supply in topsoil across a very large area of the contiguous United States. (Data from NASS, July 30, 2012, www.nass.usda.gov)

in the United States in Florida and south Texas, and in European countries, including France and Croatia. Higher temperatures can shorten the length of insect life cycles, resulting in higher densities of disease vectors, such as mosquitoes. High summer temperatures have also been associated with outbreaks of West Nile fever, which were reported from Romania and Greece in 2010 and across the United States and Canada during 2012.

The torrential rains expected from global warming may also raise the risk of water contamination by a variety of infectious diseases, such as Cryptosporidium and toxin-producing *Escherichia coli*.

⚠ Think About It

1. How might people respond to warmer temperatures to make themselves more comfortable in their apartments and offices?

2. Can you think of how humans might change their diets to cope with changes in agriculture?

14.5–14.8 Issues: Summary

Humans are changing the climate. Led by the findings of Charles Keeling, climate researchers have recorded global increases of CO_2 and temperatures.

Rising global temperatures have set in motion many changes to the Earth system, including the loss of Arctic sea ice; melting ice sheets, glaciers, and ice caps; sea level rise; shifting species' ranges; and the degradation of vital foundation species, such as forest trees and reef-building corals.

Earth is experiencing more extreme weather events, such as heat waves, droughts, and intense storms. Such changes have disrupted agriculture, producing lower harvests. Meanwhile, heat waves and extreme weather events have resulted in tens of thousands of premature deaths and the movement of tropical diseases into higher latitudes.

14.9–14.11 Solutions

The central question of this chapter asks how we might mitigate and adapt to the environmental and social impacts of climate change. The consensus among climate scientists is that the best way to stem climate change is by reducing fossil fuel use and increasing protection of forests and other ecosystems that serve as carbon sinks. The Intergovernmental Panel on Climate Change (IPCC) was established in 1988 to bring together experts on climate change from around the world to review the evidence and make policy recommendations. Two years later, the IPCC released its first report, calling for an international treaty to address the prospect of climate change. In 1997, the heads of state of 192 nations began to sign on to the Kyoto Protocol, which required developed countries to reduce their greenhouse gas emissions below 1990 levels. The reason the Protocol focused on developed countries is that those countries have the highest levels of greenhouse gas emissions per capita. The United States, which has less than 5% of the world's population, accounts for one-fifth of greenhouse gas emissions. In 1997, the United States produced 19.7 metric tons of carbon dioxide per person, compared with Nicaragua, which produced less than a single metric ton per person.

Despite an initial wave of support for the Kyoto Protocol, it never entered into force. The United States initially signed it, but the U.S. Senate did not ratify it due to the treaty's focus on greenhouse gas reductions in developed countries and its potential to handicap the U.S. economy. Indeed, China and India, which represent one-third of the world's population, have some of the fastest growing economies, whose rising fossil fuel consumption remains unchecked. The most significant efforts to mitigate and adapt to the environmental and social impacts of climate change are occurring at national, state, local, and individual levels.

14.9 Developing a road map to reduce carbon emissions

In the two decades since the Kyoto Protocol mapped a path to reducing greenhouse gas emissions, emissions have, in fact, increased by approximately 50%. As anticipated, most of this increase has been the result of economic development in developing countries, such as China and India, where emissions jumped nearly 180% (**Figure 14.32**). These countries are following in

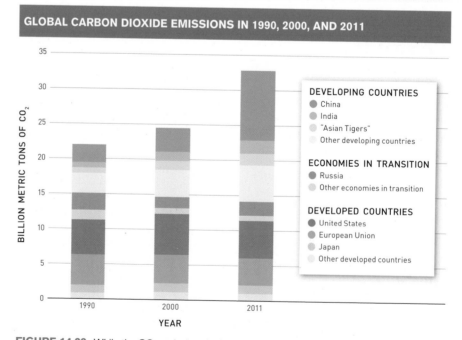

GLOBAL CARBON DIOXIDE EMISSIONS IN 1990, 2000, AND 2011

DEVELOPING COUNTRIES
- China
- India
- "Asian Tigers"
- Other developing countries

ECONOMIES IN TRANSITION
- Russia
- Other economies in transition

DEVELOPED COUNTRIES
- United States
- European Union
- Japan
- Other developed countries

FIGURE 14.32 While the CO_2 emissions in developed countries increased slightly from 1990 to 2011, and those from the economies in transition decreased, CO_2 emissions from developing countries increased by nearly 180%. (Data from Olivier, Janssens-Maenhout, & Peters, 2012)

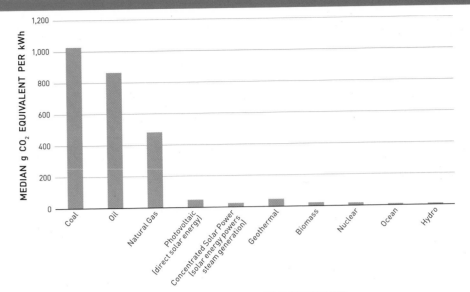

GREENHOUSE GAS EMISSIONS BY ELECTRICITY GENERATION USING VARIOUS ENERGY SOURCES

FIGURE 14.33 Life cycle assessment reveals a wide range of greenhouse gas emissions, with the highest by far resulting from fossil fuels. (Data from IPCC, 2011)

Is it fair to expect developing countries to agree to the same carbon dioxide emissions guidelines, since they did not have the benefit of several decades of unregulated fossil fuel use?

the footsteps of the United States, where about 40% of our greenhouse gas emissions come from electricity generation that powers corporate, industrial, and residential buildings, and another 30% from the transportation sector.

However, developing countries do not necessarily have to look to the United States as a model. Germany, for instance, is a highly developed country with a strong economy that produces about half the greenhouse gas emissions per capita as the United States. How did the Germans do it? They have achieved this, in part, by conserving fossil fuels and replacing oil, natural gas, and coal with renewable energy resources such as wind and solar power. Such a transition requires not only political incentives and advances in engineering, but also a careful accounting of the environmental and economic costs and benefits of various energy sources in different applications.

Evaluating Carbon Footprints of Electrical Generation

The first way we can reduce our emissions is by switching our energy sources to sources with lower CO_2 emissions. In order to do this, we need accurate measurements of each source of energy: How much CO_2 does each release from start to finish?

Consider the fact that it takes about 1 pound of coal to keep a 100-watt light bulb running for about 10 hours. As the carbon in this 1 pound of coal combusts, it combines with oxygen in the air to add about 2.86 pounds of

carbon dioxide to the atmosphere. By contrast, burning enough natural gas to power that same light bulb releases half as much carbon dioxide, making it a relatively cleaner source of energy.

Nevertheless, these simple calculations only tell us how much carbon dioxide is produced during combustion at a power plant. To properly compare greenhouse gas emissions of different energy sources, including both fossil fuels and renewable energy supplies, we need to include other factors in our calculations, such as mining the materials used for solar cells, laying natural gas lines, or dismantling a nuclear power plant at the end of its life. Incorporating all these factors into calculations is called a **life cycle assessment (LCA).** The amount of greenhouse gases emitted in the process of generating electricity varies widely among energy sources, including fossil fuels (**Figure 14.33**).

The result of a life cycle assessment of a technology used for electricity generation is the **carbon footprint,** which is expressed as grams of CO_2 equivalent per kilowatt hour. We need to think in terms of CO_2 equivalents, since other powerful greenhouse gases (like methane or nitrous oxide) can be emitted as by-products when some technologies are deployed and operated. For example, methane can be leaked into the atmosphere during the extraction, distribution, and use of natural gas.

Using LCA for natural gas and coal, we see that natural gas is still much better than coal in terms of emissions of CO_2 equivalents. There is, however, considerable controversy over the benefits of natural gas. Some new estimates claim that gas leaks are sufficient to make the

life cycle assessment (LCA) An estimate of the total environmental impact of a product or technology as a result of activities such as extraction of an energy source (e.g., coal), transport, processing of raw materials, construction, maintenance, dismantling, removal, and recycling or disposal of structures.

carbon footprint The total amount of CO_2 and other greenhouse gases produced over the course of the life cycle of a particular technology, individual, or population—for example, the carbon footprint of the United States.

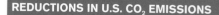

REDUCTIONS IN U.S. CO₂ EMISSIONS

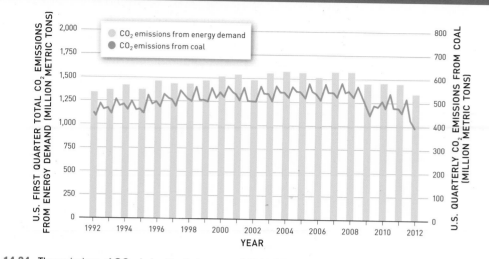

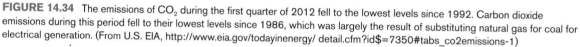

FIGURE 14.34 The emissions of CO_2 during the first quarter of 2012 fell to the lowest levels since 1992. Carbon dioxide emissions during this period fell to their lowest levels since 1986, which was largely the result of substituting natural gas for coal for electrical generation. (From U.S. EIA, http://www.eia.gov/todayinenergy/ detail.cfm?id$=7350#tabs_co2emissions-1)

carbon footprint of natural gas as large as or larger than that of coal. Regardless, as supplies of natural gas have increased and prices have decreased, many generating stations in the United States have switched from using coal to natural gas. Consequently, current estimates of greenhouse gas emissions by the United States fell significantly in 2012. In the first quarter of that year, energy-related CO_2 emissions were the lowest since 1992 (**Figure 14.34**).

Referring back to Figure 14.33, you will also notice that nuclear power has a small carbon footprint. France, for example, which generates 75% of its electricity from nuclear energy, along with about 15% from hydropower, has a very small carbon footprint. However, nuclear power is accompanied by other issues, from potential security risks to environmental contamination to nuclear accidents to issues around the safe disposal of nuclear waste (see Chapter 9).

Building dams for hydroelectric power may seem like a straightforward clean energy solution, but even they produce greenhouse gases. In Brazil, which has experienced a boom in dam-building projects, reservoirs fill with leaves and other organic matter, which release the potent greenhouse gas methane when they decompose—not to mention other major environmental impacts associated with dams.

Solar power, geothermal power, wind energy, and tidal energy are all renewable sources of energy with small carbon footprints, but they have suffered from limited availability and high costs. That seems to be changing, though. In the last three decades, the cost of solar power in the United States has plummeted from

$75 per watt to $0.75 per watt, thanks to advances in technology, government subsidies, and competition from international manufacturers in China. Nevertheless, the cost of electricity from renewable energy is still greater than that produced by coal or natural gas. This means that a transition to renewable energy will likely require assistance from policy makers, which we will discuss later.

Increasing Transportation Efficiency

Developing more efficient transportation systems offers the second major opportunity for reducing greenhouse gas emissions. In the United States, the National Highway Traffic Safety Administration (NHTSA) sets targets for fuel efficiency through the Corporate Average Fuel Economy (CAFE) standards. Those standards are set in miles per gallon of fuel. In response to efforts by the NHTSA, engineering innovation, and increasing fuel prices, the fuel efficiency of the U.S. vehicle fleet has improved considerably over the years (**Figure 14.35a**). And as the amount of gasoline or diesel required for traveling a particular distance decreases, the amount of CO_2 emitted into the environment also declines (**Figure 14.35b**).

By NHTSA estimates, if the 2025 fuel efficiency targets are met, the amount of CO_2 emitted by passenger cars and light trucks new in 2025 will be 40% lower than those meeting 2012 fuel standards. Consumers will realize savings from this greater fuel economy more than 3 times the added vehicle costs for the technology necessary to achieve the fuel economy. As a result of improved fuel efficiency standards, 4 billion fewer barrels of oil will be

AVERAGE FUEL ECONOMY OF PASSENGER CARS IN THE UNITED STATES

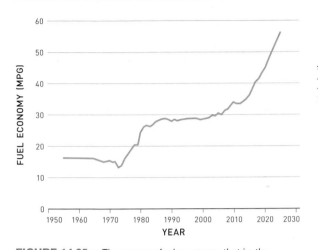

FIGURE 14.35a The average fuel economy, that is, the number of miles of travel per gallon of fuel consumed, of passenger cars in the United States doubled between 1955 and 2012 and is expected to increase significantly over the next decade. (Data from NHTSA, http://www.nhtsa.gov/)

RELATIONSHIP BETWEEN FUEL ECONOMY AND CO_2 EMISSIONS

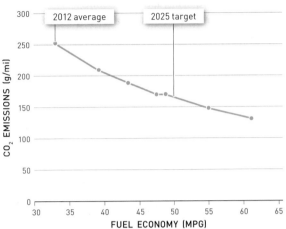

FIGURE 14.35b The fuel economy of motor vehicles is clearly related to their emissions of CO_2. If achieved, the fuel efficiency targets set by the U.S. National Highway Transportation Safety Administration for 2025 would lower CO_2 emissions by passenger cars and light trucks in the country by approximately 40%, compared with average emissions in 2012. (Data from NHTSA, http://www.nhtsa.gov/)

consumed over the lifetimes of vehicles manufactured between 2017 and 2025, which will in turn result in a reduction of CO_2 emissions by approximately 1.8 billion metric tons.

Urban Design Can Reduce Fossil Fuel Use

Beyond improving fuel efficiency, urban planning is another way to decrease the impact of transportation systems in both the developed and developing world. Reducing urban sprawl, living in more compact cities, and telecommuting can decrease our reliance on fossil fuels for transportation. It would also reduce the need for new roads, which require cement production, a significant contributor to carbon dioxide emissions.

In addition, public transportation, such as buses and trains, along with bicycles, produce lower per capita greenhouse gas emissions than private cars. However, people will only take the bus or bike to work if they know it is efficient, clean, and safe. In terms of long-distance travel, life cycle assessments suggest that planes and trains are running neck-and-neck in terms of greenhouse gas emissions, but this depends on the source of airplane fuel and electric power.

Legislating Carbon Emissions

According to the IPCC, the cost for stabilizing the climate represents less than 2% of the global economy, whereas the cost of doing nothing could be as high as 20%. Given those numbers, reducing greenhouse gas emissions may seem like a no-brainer. But one of the reasons why inefficient vehicles remain on the road and companies continue to use coal energy is that they have not been made to pay for the consequences of climate change. Recently, economists and policy makers have designed schemes for forcing emitters to pay the true cost of their energy choices. None of these schemes has a chance to reach its full potential unless all countries are operating on the kind of level playing field that the Kyoto Protocol or another international treaty could provide.

Like alcohol and cigarettes, carbon emissions impose a cost on society, and a tax is one way to recover those costs. Several countries, including Finland and Sweden, have already enacted modest **carbon taxes,** and the city of Boulder, Colorado, famously passed the first tax on carbon emissions from electricity in 2007, charging approximately $7 per ton of carbon. The $1 million the city earns every year is used to fund Boulder's climate action plan, which helps encourage solar and wind power. In 2012 the Congressional Research Service pointed out that a tax rate of $20 per metric ton of carbon dioxide could cut the nation's $1 trillion annual deficit in half. However, many businesses and economists strongly oppose a carbon tax, believing it would harm the country's economic competitiveness and send business

carbon taxes Tax on carbon emissions.

Appendix A
Basic Chemistry

Understanding the material basis of ecosystems and economic systems is improved by delving beneath the surface layers of matter to consider some details of chemistry, the science concerned with the composition, structure, properties, and reactions of **matter.** A logical entry point into a discussion of basic chemistry is the atom.

Atoms

The basic constituent of all matter is the **atom,** which is the smallest particle of a pure substance that still retains the chemical and physical properties of that substance. The simplest and most abundant atom in the universe is hydrogen. Hydrogen's **nucleus,** the central massive region of an atom, contains a single **proton** orbited by a single **electron** (Figure A.1). Because they have equal and opposite electrical charges, protons (+1) and electrons (−1) are attracted to each other and cancel each other out electrically. As a consequence, the hydrogen atom with its single proton and single electron is electrically neutral [(+1) + (−1) = 0].

Because an electron moves at nearly the speed of light, it forms an *electron cloud* around the nucleus, which is one way to represent the orbit of an electron around the nucleus (Figure A.1a). For convenience, however, we can mentally freeze the position of an electron and picture it as moving in a fixed orbit (Figure A.1b). Notice that the helium atom (Figure A.1c) has two protons (+2 charge) and two electrons (−2 charge), which again exactly balance each other electrically. However, the helium nucleus also includes two **neutrons,** which are equal in weight to protons but have no electrical charge. Therefore, neutrons contribute to **atomic weight** but not to the electrical properties of atoms. Meanwhile, because it takes approximately 2,000 electrons to equal the mass of a single neutron or proton, electrons contribute to the electrical properties of atoms but insignificantly to their mass.

Substances composed of a single type of atom, such as hydrogen or helium, are called **elements.** Elements that we commonly encounter in everyday life include the oxygen in the air we breathe, the aluminum in a soft drink can, and the copper in a penny. The elements are traditionally arranged into a table called the *periodic table,* which orders elements by number of protons, or **atomic number,** and groups elements into families of substances with similar chemical properties (Figure A.2).

HYDROGEN

a. Hydrogen consists of a nucleus of a single proton surrounded by an electron cloud, the space in which the single electron of hydrogen moves.

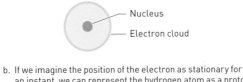

Nucleus
Electron cloud

b. If we imagine the position of the electron as stationary for an instant, we can represent the hydrogen atom as a proton orbited by a tiny electron.

NUCLEUS
⊕ Protons: 1 (positive charge)
◯ Neutrons: 0 (no charge)

⊖ Electrons: 1 (negative charge)

Atomic number: 1

HELIUM

c. Helium consists of a nucleus of two protons and two neutrons orbited by two electrons.

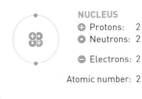

NUCLEUS
⊕ Protons: 2
◯ Neutrons: 2

⊖ Electrons: 2

Atomic number: 2

FIGURE A.1

The atomic structure of elements provides clues regarding their chemical properties. Table A.1 shows the atomic structures of 11 biologically important elements. Just six of these—carbon, hydrogen, nitrogen, oxygen, phosphorus, and sulfur—make up approximately 99% of the mass of organisms from humans and plants to bacteria. The electrons of each element in the table are illustrated in their orbits, or electron shells. Notice that the innermost electron shell of each element is occupied by a maximum of two electrons, whereas the other electron shells are occupied by a maximum of eight electrons.

The basic chemistry of the elements is strongly influenced by the number of electrons in the outermost

matter Anything that occupies space and has mass; exists in three main physical states: as a solid, liquid, or gas.

atom The smallest particle of a pure substance that still retains the properties of the substance.

nucleus The massive central core of an atom, which is made up of protons and neutrons, around which the atom's electrons move.

proton A subatomic particle, found in the nucleus of atoms, that has a positive charge (+1) and atomic mass of 1.

electron A subatomic particle with a negative charge (−1) and a mass approximately 1/2,000 that of a proton or neutron.

neutron A subatomic particle, found in the nucleus of atoms, that has a mass approximately equal to that of a proton but no electrical charge.

atomic weight The average mass of the atoms in a naturally occurring pure substance (e.g., gold, carbon, oxygen).

element A substance composed of a single type of atom (e.g., hydrogen, helium, iron, or lead).

atomic number The number of protons in the nucleus of an atom of an element; equal to the number of electrons in a neutral atom.

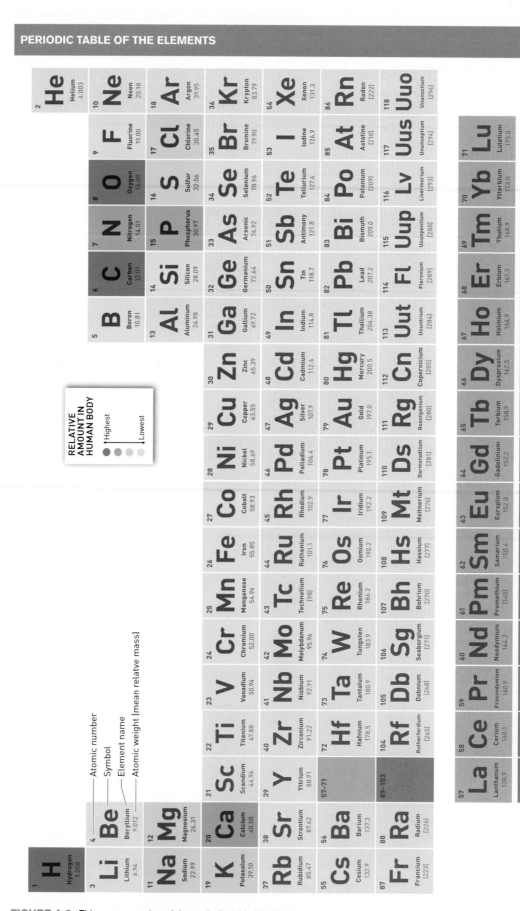

PERIODIC TABLE OF THE ELEMENTS

FIGURE A.2 This representation of the periodic table highlights elements known to be essential for human health. Research continues on other elements that may be essential in trace amounts.

TABLE A.1 THE ATOMIC STRUCTURES AND IMPORTANCE OF 11 BIOLOGICALLY ESSENTIAL ELEMENTS

Symbol	Name	Structure	Biological Importance
H	Hydrogen		Hydrogen, present in all organic compounds (e.g., carbohydrates, fats, and proteins and part of water) is a major component of biological structure.
C	Carbon		Carbon is central to all organic compounds and forms the core of biological structure. Life on Earth is based on the chemistry of carbon.
N	Nitrogen		Nitrogen is an essential part of amino acids and therefore of proteins and amino acids.
O	Oxygen		Oxygen is part of many organic molecules, is a component of water, and is critical for respiration in many organisms.
Na	Sodium		Sodium is important for proper nerve and muscle function in animals. Humans need sodium in the diet, making salt a valuable commodity throughout history.
Mg	Magnesium		Magnesium is central to the structure of chlorophyll in algae and plants and is necessary for the proper functioning of enzymes in animals.
P	Phosphorus		Phosphorus is essential to the structure of RNA and DNA, is part of the energy-bearing molecule ATP, and is important to bone and tooth structure.
S	Sulfur		Sulfur is a key constituent of some amino acids, the building blocks of proteins, and helps determine the structure of enzymes and other proteinss.
Cl	Chlorine		Chlorine as the negatively charged ion chloride, Cl^-, is essential for nerve and muscle and other cell function and is part of the stomach acid HCl in humans.
K	Potassium		Potassium as the positively charged K^+ ion is an important substance in cellular fluids of both plants and animals; it is essential for nerve and muscle function.
Ca	Calcium		Calcium is essential for cell wall structure, in the structure of bones and teeth, and is important to blood clotting.

electron shell. Those elements with few electrons in the outer electron shell tend to give up, or *donate*, electrons to other elements, while elements with nearly a full complement of electrons in their outer electron shell tend to *accept* electrons from other elements. For instance, hydrogen, sodium, and potassium tend to give up the single electron in their outermost shell, and magnesium and calcium give up the two electrons in their outer

FOUR WAYS TO REPRESENT MOLECULES

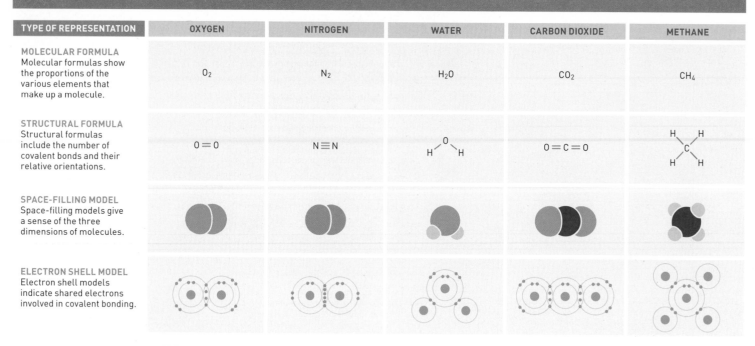

TYPE OF REPRESENTATION	OXYGEN	NITROGEN	WATER	CARBON DIOXIDE	METHANE
MOLECULAR FORMULA Molecular formulas show the proportions of the various elements that make up a molecule.	O_2	N_2	H_2O	CO_2	CH_4
STRUCTURAL FORMULA Structural formulas include the number of covalent bonds and their relative orientations.	O=O	N≡N	H–O–H	O=C=O	(methane structure)
SPACE-FILLING MODEL Space-filling models give a sense of the three dimensions of molecules.					
ELECTRON SHELL MODEL Electron shell models indicate shared electrons involved in covalent bonding.					

FIGURE A.3

molecule A particle consisting of two or more atoms held together by chemical bonds; the constituent atoms may be of the same or different elements.

compound A substance composed of a fixed ratio of two or more elements (e.g., water consists of two hydrogen atoms and one oxygen atom: H_2O); compounds can be broken down into the elements of which they are made through chemical or physical processes.

chemical reaction A process by which new substances are produced via the rearrangement of the atoms undergoing the reaction, generally through the exchange or sharing of electrons.

covalent bond A bond between atoms that share one or more pairs of electrons.

electron shells. Meanwhile, oxygen and sulfur *accept* two electrons to fill their outermost electron shells, and chlorine accepts one. What about an element such as carbon, with an outer shell occupied by four electrons, exactly half the full complement of eight? Such elements tend to share electrons with other elements.

Molecular Structure and Chemical Reactions

As atoms interact, donating, accepting, or sharing electrons, they form substances called **molecules.** Molecules made up of atoms of two or more different elements are **compounds. Figure A.3** shows four common ways to represent molecules. The representation you use to illustrate a molecule will depend on the information you want to convey and on convenience. The *molecular formula,* showing the proportions of each of the elements making up a molecule, is easy to write. The *structural formula* includes the number of bonds and their relative positions. A *space-filling model* gives a sense of three-dimensionality. Finally, the *electron shell model* includes information on the shared electrons, forming the bonds that hold the atoms of a molecule together. As shown in Figure A.3, bonds may involve sharing of one, two, or three pairs of electrons; that is, bonds may be single (e.g., carbon to hydrogen), double (e.g., carbon to oxygen), or triple (e.g., nitrogen to nitrogen).

The discussion so far has included a wide variety of chemical compounds that occur in our own bodies and in

the world around us. How do such chemical compounds form? Compounds form as the products of **chemical reactions** between elements, between compounds, or between compounds and elements. **Figure A.4** outlines a chemical reaction in which methane gas, the main ingredient of natural gas, reacts with oxygen to produce carbon dioxide and water. During the burning of natural gas, the electrons forming the **covalent bonds** in both methane molecules and oxygen molecules are broken and reformed. The bond electrons are rearranged to form covalent bonds between carbon and oxygen and between oxygen and hydrogen, yielding carbon dioxide and water as products of the reaction. This reaction releases considerable heat and light as the electrons involved in bond formation go from higher energy states in the reactants (methane and oxygen) to lower energy states of the products (water and carbon dioxide). The heat from this reaction is commonly used to warm homes and cook food.

The compounds produced by some reactions do not involve covalent bonds. **Figure A.5** shows the reaction between sodium (a highly reactive, soft metal) and chlorine (an irritating, greenish gas). The product of this reaction, sodium chloride (better known as table salt), has properties that are quite different from the elements that combined to form it. The upper panel of Figure A.5 shows the reaction using the molecular formulas for sodium and chlorine, while the lower panel uses electron-shell models to indicate the rearrangements of electrons involved in this chemical reaction.

TWO WAYS TO PICTURE A CHEMICAL REACTION

a. MOLECULAR FORMULA

Methane and oxygen react to form carbon dioxide and water, giving off energy in the process.

Energy

$$CH_4 + 2O_2 \rightarrow CO_2 + 2H_2O$$

Methane   Oxygen   Carbon dioxide   Water

b. ELECTRON SHELL MODEL

During this reaction, the electrons that form covalent bonds go from a higher energy state in methane and oxygen (the reactants) to a lower energy state in carbon dioxide and water (the products).

Electron sharing = covalent bonds

Energy

Covalent bonds

Methane   Oxygen   Carbon dioxide   Water

The change in energy state of the electrons that form covalent bonds is the source of energy given off during the reaction, which we perceive as a flame.

FIGURE A.4

During the formation of sodium chloride, the single electron in the outermost electron shell of sodium moves from the sodium atom to the chlorine atom, releasing energy. In the process, the sodium atom is converted into a sodium **ion** with a charge of +1. During this reaction, the chlorine atom is converted to a chloride ion with a charge of –1.

Where do these charges come from? Sodium atoms have an atomic number of 11, which means that they have exactly 11 protons in their nucleus. When one of the sodium electrons moves to chlorine, the remaining 10 electrons do not quite cancel out the total positive charge of the 11 nuclear protons. Consequently, the sodium ion has a net charge of +1. Similarly, with the addition of an electron, 18 electrons orbit the chloride ion with its 17 protons, producing a net charge of –1. Because they have opposite charges, sodium and chloride ions are attracted to each other. The result of this attraction is an **ionic bond** between sodium and chloride ions.

Cells and Vital Molecules

All the life forms on Earth, from bacteria and algae to whales and redwood trees, are built of individual units called *cells* (**Figure A.6**). The structure and functioning of cells is the product of the molecules of which they are made. Though atoms can combine in innumerable ways to form endless varieties of molecules, only a few types of molecules are required to form the great diversity of life on Earth. Those key molecules are the basis for the structure, energetics, and reproduction of all organisms.

Examples of the main molecules making up living systems are arrayed around the animal cell illustrated in Figure A.6. Cells are separated from the surrounding environment by a membrane that is selectively permeable. Within a cell are a variety of organelles, also bound by a membrane, that perform essential functions in the life of the cell. The cell membrane is constructed of **lipids** with embedded **proteins,** consisting of chains of amino acids. The cell nucleus contains **DNA (deoxyribonucleic acid),** the repository of genetic information in the cell. That information is carried to the ribosomes, where proteins are assembled. The mitochondria are sites where the energy in sugars and fats are converted to **ATP (adenosine triphosphate),** a common energy source for cell processes.

Summary

Everything in the universe is composed of matter, which exists as a solid, liquid, or gas. The basic constituent of

THE FORMATION OF AN IONIC COMPOUND

a. MOLECULAR FORMULA

Sodium and chlorine react to form sodium chloride, an ionic compound.

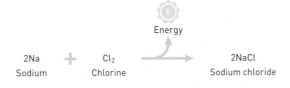

Energy

$$2Na + Cl_2 \rightarrow 2NaCl$$

Sodium   Chlorine   Sodium chloride

b. ELECTRON SHELL MODEL

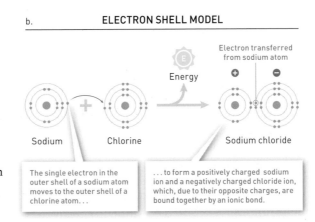

Electron transferred from sodium atom

Energy

Sodium   Chlorine   Sodium chloride

The single electron in the outer shell of a sodium atom moves to the outer shell of a chlorine atom. . .

. . . to form a positively charged sodium ion and a negatively charged chloride ion, which, due to their opposite charges, are bound together by an ionic bond.

FIGURE A.5 The chemical reaction between sodium (a soft metal) and chlorine (a greenish gas), to form sodium chloride, common table salt.

ion An atom, or group of atoms, with a net positive or negative charge (e.g., chloride, Cl^-, and sodium, Na^+, ions).

ionic bond A chemical bond involving the attraction between two oppositely charged ions.

lipids Organic molecules composed of long chains of carbon atoms bonded mainly to hydrogen (e.g., fats, oils, or waxes); important components of cell membranes; function as energy storage molecules in animals and plants.

proteins Long chains of amino acids (i.e., molecules consisting mainly of carbon, nitrogen, hydrogen, and oxygen) that control rates of chemical reactions, provide structural support, and perform many other functions.

DNA (deoxyribonucleic acid) The carrier of genetic information in cells, consisting of two complementary chains of molecules, called nucleotides, wound in a double helix; this source of biological inheritance, passed from parents to offspring, directs the development and functioning of an organism.

ATP (adenosine triphosphate) A molecule made up of adenine bonded to three phosphate molecules that releases energy when one of the bonds between phosphate molecules is broken; ATP is the main energy source for cells.

THE CELL AND KEY MOLECULES RESPONSIBLE FOR THE STRUCTURE AND FUNCTIONING OF LIVING SYSTEMS

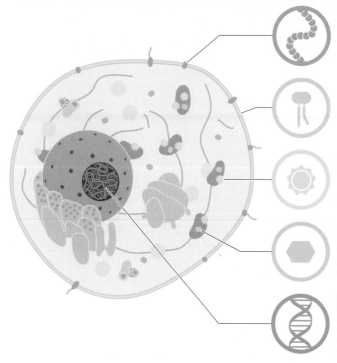

PROTEIN
Proteins consist of long chains of molecules called amino acids and perform a wide range of tasks (e.g., facilitate biochemical reactions, aid the immune response by targeting foreign material for destruction, and provide key structural elements, such as connective tissue).

PHOSPHOLIPID
Phospholipids are made up of a fatty acid tail, which is hydrophobic, and a phosphate head, which is hydrophilic. Cell membranes consist of two layers of phospholipids, oriented with the hydrophobic parts facing inward and the hydrophilic parts forming the inner and outer surfaces of the cell membrane.

ATP
ATP (adenosine triphosphate) stores chemical energy in a readily available form that is used as a source of energy to power cell processes, including synthesis of proteins and other molecules, cell division, and cell movement.

CARBOHYDRATE
Carbohydrates are molecules made up of carbon atoms to which hydrogen and oxygen molecules are bound in a ratio of 2:1, as in a water molecule. Cells use some familiar carbohydrates (e.g., sugars and starch) to store chemical energy.

NUCLEIC ACID (DNA)
Nucleic acid (DNA) is a long-chain molecule that encodes instructions necessary for carrying out cellular functions (e.g., synthesis of proteins) and for directing the development of complex organisms. What we call "genes" are those segments of DNA that code for a particular structure or process.

FIGURE A.6 A few types of molecules are responsible for the majority of the structure and functioning of living systems.

all matter, the atom, is the smallest particle of a substance that retains its properties. Atoms react chemically—by donating, accepting or sharing electrons—to form molecules; molecules made up of two or more elements are called compounds. All the life forms on Earth, from bacteria and algae to whales and redwood trees, are built of individual units called cells, the structure and functioning of which are determined by a few classes of molecules, including DNA, lipids, proteins, sugars, and ATP.

Appendix B
The Rock Cycle: Product of a Dynamic Planet

All ecosystems, from oceans to deserts, develop and function on a geologic foundation of rock. However, as we shall see, this geologic foundation is not static but is renewed over very long timescales in a process called the *rock cycle*. We begin our exploration of the rock cycle with a review of Earth structure.

Earth Structure

Earth can be divided into three major layers: core, mantle, and crust (**Figure B.1**). The **crust,** which is made up of relatively low-density rocks and is the most superficial of Earth's layers, forms the continents and ocean floor. Continental crust is 20 to 70 kilometers (12 to 43 miles) thick, whereas the crust making up the ocean floor is only about 8 kilometers (4 miles) thick. Immediately below the crust is the **mantle,** consisting of rocks of higher density than those that make up the crust and extending to about 2,900 kilometers (1,800 miles) below Earth's surface. The mantle, which constitutes the largest portion of Earth's volume, can be divided into upper and lower layers, again differing in density.

The uppermost layer of the upper mantle consists of rigid, relatively brittle rocks, which, along with Earth's crust, make up the **lithosphere.** In contrast, most of the upper mantle has a consistency similar to plastic, and a small portion of it is liquid. This softer portion of the upper mantle is called the **asthenosphere**. The lower mantle is composed of solid rock of higher density than the upper mantle. The **core** of the planet, which extends from about 2,900 kilometers to 6,370 kilometers (1,800 to 3,955 miles) in depth at the center of the planet, is also two-layered. The core appears to consist of an outer liquid layer of molten iron and nickel and a solid, inner iron–nickel center.

Earth is stratified not only by density but also by temperature and pressure. Temperature and pressure are lowest at Earth's surface and increase gradually with depth below the surface, until the highest temperatures and pressures are reached at the core. Geologists

crust The relatively low-density rocks that form the most superficial of Earth's layers, including the continents, where the crust is 20 to 70 kilometers thick, and the ocean floor, which is about 8 kilometers thick.

mantle The layer of Earth between the crust and the core; represents the largest portion of Earth's volume and consists of higher-density rocks than those that make up the crust.

lithosphere The uppermost layer of the upper mantle above the asthenosphere, consisting of rigid, relatively brittle rocks.

asthenosphere The layer of the upper mantle immediately below the lithosphere; has an overall consistency similar to plastic.

core The core of Earth consists of an outer liquid layer of molten iron and nickel and a solid, inner iron–nickel center.

EARTH STRUCTURE

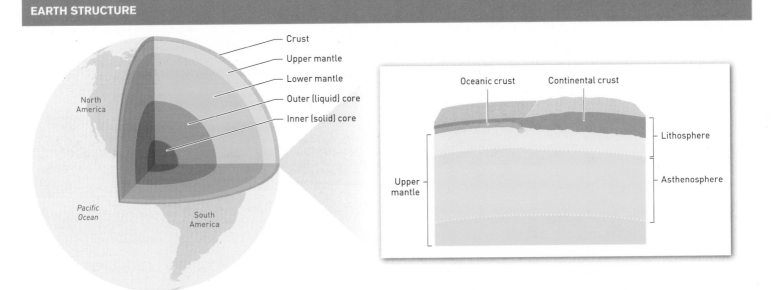

FIGURE B.1 Planetary differentiation resulted in Earth's current layered structure, with highest temperatures and densities prevailing at Earth's core and lowest temperatures and densities at Earth's crust.

estimate that the temperature at the solid inner core may be nearly 8,000°C, whereas the temperature of the liquid outer core may be slightly less than 5,000°C. If temperatures increase with depth below the surface of Earth, how can the upper mantle and upper core be liquid while the lower mantle and inner core are solid? Earth scientists explain this apparent contradiction by pointing out that the higher pressures at greater depth keep materials solid at higher temperatures. The physical structure of Earth forms the basis for important Earth processes. One of the most consequential of those processes is plate tectonics.

Plate Tectonics

The theory of **plate tectonics** explains the mechanisms responsible for some of Earth's most dramatic features and processes, including the formation of ocean basins, continents, and mountains, as well as the geographic distributions of earthquakes and volcanic activity. Because it can account for so many geologic phenomena, the theory of plate tectonics unifies geology in a way similar to how the theory of evolution unifies biology.

There is abundant evidence that the surface of Earth is not one continuous layer of crust overlying the mantle.

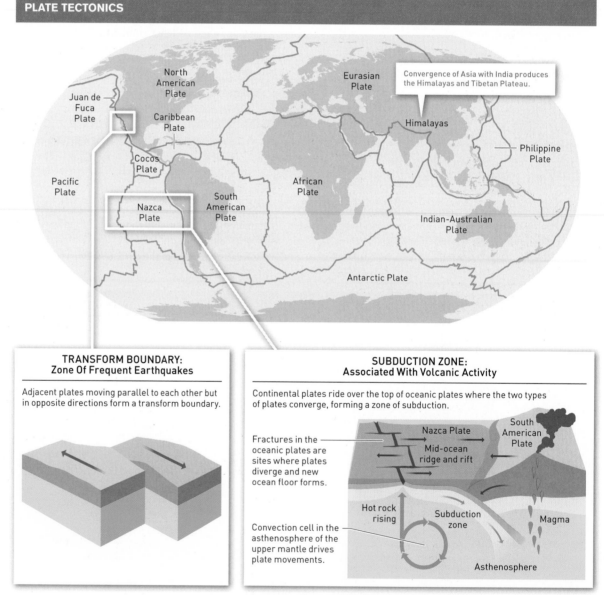

PLATE TECTONICS

plate tectonics A theory proposing that Earth's surface is divided into plates that move on the upper layer of the mantle; explains Earth structure and processes, including the formation of ocean basins, continents, and the geographic distribution of earthquakes and volcanic activity.

FIGURE B.2 Once thought to be fixed in place, the surface features of Earth are now known to be the result of dynamic interactions among the tectonic plates. The theory of plate tectonics accounts for the distributions of phenomena like earthquakes and volcanoes.

Rather, Earth's surface is divided into several plates of lower-density lithosphere that "float" on the more dense and plastic asthenosphere (**Figure B.2**). The plates, which vary in thickness from about 80 to 120 kilometers (49 to 74 miles), consist of a surface layer of crust and a deeper layer of the upper mantle. In general, continental plates are thicker than oceanic plates.

Geologists propose that convective currents in the asthenosphere drive the movements of the plates. These convection currents form as hotter material deeper within the mantle rises toward the surface of Earth. This material rises because it has a lower density than the cooler material above it. However, as the lower-density material rises and moves across the upper layers of the asthenosphere, it cools and, as it does so, its density increases. Eventually, the density of this moving mass of material within the upper asthenosphere increases enough that it sinks, creating a **convection cell,** a pattern

THE ROCK CYCLE

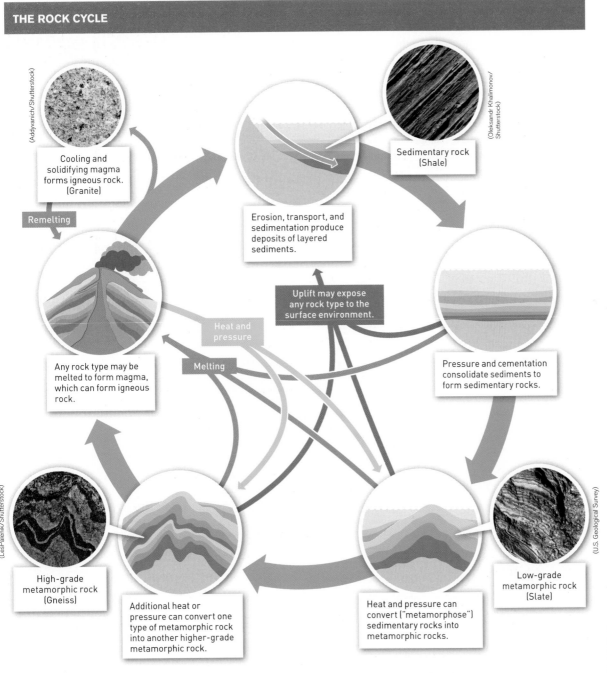

(Addyvanich/Shutterstock)

Cooling and solidifying magma forms igneous rock. (Granite)

Remelting

(Oleksandr Khalimonov/ Shutterstock)

Sedimentary rock (Shale)

Erosion, transport, and sedimentation produce deposits of layered sediments.

Uplift may expose any rock type to the surface environment.

Heat and pressure

Melting

Any rock type may be melted to form magma, which can form igneous rock.

Pressure and cementation consolidate sediments to form sedimentary rocks.

(LesPalenik/Shutterstock)

High-grade metamorphic rock (Gneiss)

Additional heat or pressure can convert one type of metamorphic rock into another higher-grade metamorphic rock.

Heat and pressure can convert ("metamorphose") sedimentary rocks into metamorphic rocks.

(U.S. Geological Survey)

Low-grade metamorphic rock (Slate)

convection cell Pattern of circulation caused by differing temperatures in a liquid or gas due to, for example, varying temperatures in Earth's semi-liquid mantle.

FIGURE B.3 The large-scale geologic phenomenon known as the rock cycle involves several geologic processes that rework the materials of which rocks are composed to form and reform igneous, sedimentary, and metamorphic rocks.

subduction Process in which one tectonic plate moves under another, generally occurring where oceanic plates, which are of higher density, collide with continental plates.

subduction zone Zone where oceanic plates and continental plates collide, forming deep sea trenches and active volcanoes along the continental margin.

rock A natural, solid, inorganic substance formed from one or more minerals.

igneous rock Rock formed as molten rock cools and solidifies.

sedimentary rock Rock formed either as rock fragments deposited by water, wind, or ice are cemented together and solidify or as rock forms through chemical precipitation.

metamorphic rock Rock formed when any type of rock changes as it is subjected to heat and pressure.

rock cycle Geologic processes that convert each of the three major rock types (igneous, sedimentary, and metamorphic) into one of the other types.

of circulation caused by differing temperatures in a semi-liquid, liquid, or gas.

The convection cells within the asthenosphere pull the plates along, putting them in motion. These moving plates are thus literally on a collision course with each other. The fastest ones move only a few centimeters per year, but tremendous forces are involved—forces sufficient to move entire continents and to form and move the floor of entire ocean basins. These same forces raise great mountain ranges, such as the Himalayas and the Rockies, create deep ocean trenches, and produce earthquakes.

Because oceanic plates are of higher density than continental plates when the two plate types collide, the oceanic plate generally moves under the continental plate in a process known as **subduction.** Collisions between oceanic plates and continental plates form deep trenches in the **subduction zone** and active volcanoes along the continental margin.

During Earth's estimated 4.5-billion-year history, the forces of plate tectonics have changed the face of the planet many times. Mountains have been built and worn down by water, wind, and ice. Gigantic continents have formed and broken apart. Submarine volcanic activity has formed oceanic plates that were later driven back into the mantle by subduction and their materials recycled. These various phenomena and processes can be viewed as part of a large-scale process called the rock cycle.

The Rock Cycle

At the center of most Earth processes are **rocks,** naturally occurring solid materials formed as a mixture of one or more minerals. **Igneous rock** forms as molten rock cools and solidifies. **Sedimentary rock** forms either as rock fragments deposited by water, wind, or ice are cemented together and solidify or by **chemical precipitation,** a process by which a dissolved substance goes from dissolved to solid state in a body of water, such as a pond or sea, and settles to the bottom. **Metamorphic rock** forms when any type of rock is transformed by exposure to high heat and pressure. During Earth's history, rocks have been made and remade many times by the **rock cycle** (Figure B.3).

Glossary

abiotic Physical and chemical components of the environment.

abundance (population size) The number of individuals in a population.

acid A substance that releases hydrogen ions upon dissociation when dissolved in water, resulting in reduced pH of the solution; acids neutralize bases.

acid deposition An inclusive term that includes both wet and dry deposition of acids.

acid mine drainage A problematic result of strip mining, in which surface flow of groundwater turns acidic as it percolates through mine wastes (tailings).

acid rain Acidified rainfall. See *acid deposition*.

adaptations Traits favored by natural selection for surviving and reproducing in a particular environment.

additive effect An interaction of two toxic substances wherein their combined toxicity is simply the sum of their individual effects.

aerosol Tiny particles of solid or liquid material suspended in air or other gas.

age structure The proportions of individuals of various ages in a population; the relative proportions of individuals of reproductive and pre-reproductive age indicate whether a population is growing, stable, or declining.

A horizon (topsoil) Soil layer immediately below the O horizon that includes significant amounts of organic matter, generally expressed by dark color.

airshed A concept equivalent to a watershed or drainage basin but related to movements of air rather than water; typically behaves in an orderly way, thus can be used to track and manage air pollution.

allergen A substance that activates the immune system, inducing an allergic reaction.

allopatric (geographic) speciation A process by which new species are formed that occurs as the result of the division of a population into two geographically separate populations; over time, genetic differences arise and accumulate in the two separate populations, eventually leading to reproductive isolation.

ammonification The process by which decomposers break down proteins and amino acids, releasing nitrogen in the form of ammonia (NH_3) or ammonium ion (NH_4^+).

anaerobic An environment without molecular oxygen (O_2).

antagonistic effect An interaction of two toxic substances wherein the toxicity of one chemical is reduced in the presence of the second chemical, which may be used as an antidote.

Anthropocene era A new geologic era dominated by the effects of humans.

anthropocentric Human-centered; for example, human-centered environmental ethics emphasizes impacts on humans.

antibiotics Substances that suppress bacterial growth or attack and that are used in modern medicine in the treatment of bacterial diseases.

aquaculture The controlled growing of aquatic organisms (e.g., fish, shellfish, algae, or plants) as a crop, mainly for food; carried out in marine, brackish water, or freshwater environments.

aquifer A geologic formation containing groundwater; gains water through the process of infiltration and loses water through groundwater flow.

artificial selection A process in which humans "select" which individuals in a population mate to produce descendants with desired characteristics.

asthenosphere The layer of the upper mantle immediately below the lithosphere; has an overall consistency similar to plastic.

atom The smallest particle of a substance that still retains the properties of the substance.

atomic number The number of protons in the nucleus of an atom of an element; equal to the number of electrons in a neutral atom.

atomic weight The average mass of the atoms in a naturally occurring pure substance (e.g., gold, carbon, oxygen).

ATP (adenosine triphosphate) An energy-bearing molecule containing phosphorus used to transport energy within cells.

autotroph See *primary producer*.

background extinction Average rate of extinction occurring over long periods of time between periods of mass extinction.

bacteria Single-celled organisms (singular *bacterium*) lacking a nucleus or other membrane-bound organelles; the vast majority of bacteria are not pathogens.

base A substance that has the capacity to neutralize acids; bases release hydroxide ions (OH^-) and react with acids to form a salt and water.

B horizon A depositional soil layer in which materials transported from the A and E horizons accumulate.

bioaccumulation The absorption and increase in concentration of chemicals in organisms, over time, including potentially toxic chemicals. See *biomagnification*.

biocentric Centered on life in all its forms; for example, biocentric environmental ethics extends moral obligation to all life.

biochemical oxygen demand (BOD) An indicator of the amount of organic matter in water, measured as the quantity of oxygen consumed by microorganisms as they break down the organic matter in a sample of water.

biodegradable A substance that can be decomposed to its chemical constituents by biological processes.

biodiesel A liquid fuel made from vegetable oils and animal fats.

biodiversity Biological variety from genes and species to diversity at the scale of ecosystems and the globe.

biodiversity hotspot A region that supports at least 1,500 endemic plant species, approximately 0.5% of the world total, and that has been reduced in area by at least 70%.

biofuel A liquid or gaseous fuel derived from biomass (e.g., ethanol, biodiesel).

biogeochemical cycle The cyclic path of an inorganic substance, such as phosphorus, nitrogen, or carbon, through the Earth system, including the atmosphere, Earth's crust, oceans, lakes, and rivers; key biological components are producers, consumers, detritivores, and decomposer bacteria and fungi.

biological environment The kinds and diversity of pathogens, predators, parasites, and competitors with which an organism interacts.

biomagnification An increase in concentration of a substance (e.g., heavy metals or fat-soluble chemicals) at sequentially higher trophic levels in a food web. See *bioaccumulation*.

biomass fuel A combustible fuel derived from biological materials (e.g., wood, charcoal, dung).

biome Associations of plants, animals, and other organisms that occur over large areas and that are characterized by distinctive biological structure, especially by characteristic growth forms (e.g., trees, shrubs, or grasses on land; corals, kelp, or mangrove trees in aquatic environments).

bioreactor A system designed to cultivate algae; helps filter wastewater before it enters the environment.

bioremediation An approach to pollution cleanup that employs organisms, generally microbes or plants, to decontaminate soils, sediments, and groundwater aquifers in place.

biotechnology The application of engineering techniques to modify organisms genetically for a particular purpose.

biotic Living components of the environment.

bitumen A flammable, highly viscous or semisolid mixture of hydrocarbons.

black market The exchange of illegal goods and services.

boom A barrier used to contain oil slicks and prevent them from entering sensitive coastal areas.

bottom ash The ash that accumulates at the bottom of an incinerator during the combustion of solid waste.

bottom trawlers Weighted nets dragged along the ocean to catch groundfish (e.g., cod, flounder, scallops, shrimp, crab).

brackish water Natural waters with a salt content intermediate between freshwater and seawater, commonly occurring near the mouths of rivers where freshwater and seawater mix.

brownfield An abandoned industrial site generally contaminated with hazardous waste and unusable without remediation.

Bt Insect-killing crystalline substance produced by the bacteria species *Bacillus thuringiensis*.

buffer zone A zone around a nature reserve or protected area in which limited economic activity is allowed.

buffering capacity A measure of the ability of a solution to neutralize acid.

bushmeat The butchered meat of wild animals, most commonly from African forests.

bycatch Discarded catch and mortality of any organism (e.g., fish, invertebrate animals, birds, dolphins, sea turtles) as a result of contact with fishing gear.

cap and trade Systems for regulating the emission of carbon dioxide and other pollutants.

carbon capture and sequestration A technological strategy to reduce additions of carbon dioxide and other greenhouse gases to the atmosphere by collecting them at the point of emission and storing them in a place or chemical form that removes them from active circulation in the carbon cycle.

carbon credit A permit to produce a certain amount of carbon emissions, which can be traded or sold, if the full emissions allowance is not used.

carbon cycle The cycling of carbon through the Earth system; key biological processes in the carbon cycle include photosynthesis, respiration, and decomposition.

carbon footprint The total amount of CO_2 and other greenhouse gases produced over the course of the life cycle of a particular technology, individual, or population—for example, the carbon footprint of the United States.

carbon sink A part of the Earth system that takes carbon compounds from the atmosphere, through biological or physical processes, and removes them from active circulation.

carbon taxes Tax on carbon emissions.

carcinogen A substance that causes cancer by directly damaging the DNA of cells.

carnivore (predator) An animal that feeds on other living animals (e.g., a lion or a spider).

carrying capacity (K) The number of individuals in a population that an environment can support over the long term.

catch-and-release fishing The practice of releasing fish back into the water after catching them.

catch-per-unit effort A measure of how many fish are caught using a specific piece of gear—a net or a line—for a certain period of time.

cellular respiration A process taking place in cells that requires oxygen. During cellular respiration, molecules, such as glucose, are broken down and energy, water, and carbon dioxide are released.

cellulosic ethanol Ethanol produced from wood and other cellulose-rich materials.

centrally planned economy An economy in which decisions about the production and consumption of goods and services are made by a central authority. See *market economy*.

channelize To engineer a change to the natural form of a stream or river, including straightening, deepening, or widening the channel.

chemical energy A form of potential energy; energy stored in the bonds of molecules, such as sugars, fats, or methane.

chemical reaction A process by which new substances are produced via the rearrangement of the atoms undergoing the reaction, generally through the exchange or sharing of electrons.

child mortality rate The number of infants per 1,000 live births who die before reaching 5 years of age.

C horizon The deepest soil layer, consisting mainly of lightly weathered parent material.

clear-cutting An economically efficient technique whereby an entire area is cleared of its trees.

climate The average weather occurring across a region over a long period, including average temperatures, precipitation, and so forth

climate adaptation A strategy to reduce the impacts and hazards associated with climate change.

climax community The community at the end of a successional sequence that persists until a disturbance disrupts it sufficiently to restart succession.

coal Sedimentary or metamorphic rock high in carbon and energy content formed over millions of years under conditions of high pressure and temperature (lignite, sub-bituminous coal, bituminous coal, anthracite).

cogeneration Generally refers to the use of a single source of energy for multiple purposes.

combined cycle power plants Power plants that combine a gas turbine engine with a steam power plant.

command-and-control regulations Laws and regulations that control activities and industries through the use of subsidies and penalties prescribed by the government.

commercial fishing Catching fish for profit; represents the vast majority of the fish captured around the world.

common-pool resource A resource owned and utilized in common by a community (e.g., a community forest or grazing land).

common property Property owned or controlled by a community, such as an indigenous tribe.

competition Interactions among individuals that depend on the same resources; generally results in reduced growth, reproduction, or survival of one or both competitors.

competitive exclusion principle If two species with identical niches compete for a limited resource (e.g., nectar), one or the other will be a better competitor and will eventually eliminate the other species.

composting A process involving aerobic decomposition of organic material used to recycle garden waste and organic components of municipal solid waste.

compound A substance composed of a fixed ratio of two or more elements (e.g., water, which consists of two hydrogen atoms and one oxygen atom: H_2O). Compounds can be broken down into the elements of which they are made via chemical or physical processes.

Comprehensive Environmental Response, Compensation, and Liability Act (CERCLA) Superfund Law enacted in 1980 to regulate hazardous waste and require companies to dispose of it safely.

conflict of interest Competing interests, including personal, philosophical, and financial interests, that may interfere with an objective judgment.

conservation The preservation, wise use, or restoration of species, ecosystems, or natural resources.

conservation ethic A philosophy of resource management that promotes the efficient use of natural resources to provide the greatest good to the greatest number of people.

conservation of matter A physical law describing how during chemical reactions matter is neither created nor destroyed but conserved.

constructed wetlands Artificial wetland ecosystems, used in the treatment of wastewater, that are constructed in areas where wetlands may not occur naturally.

consumer An organism that meets its dietary needs by feeding on other organisms or on organic matter produced by other organisms. See *heterotroph*.

containment structure A steel and concrete enclosure designed to prevent the release of radioactive material in the case of a serious nuclear reactor accident.

control group A baseline for comparisons.

control rods Long rods made of neutron-absorbing substances, used to control the rate of fission in a nuclear reactor.

convection cell Pattern of circulation caused by differing temperatures in a liquid or gas due to, for example, varying temperatures in Earth's semi-liquid mantle.

Convention on Biological Diversity An international agreement negotiated under the sponsorship of the United Nations Environmental Programme to promote the conservation of biological diversity, the sustainable use of its components, and the fair and equitable sharing of the benefits arising out of the utilization of genetic resources.

conventional-tillage agriculture Tilling a field to break up soil clumps and smooth the soil surface before planting, as well as weeding using specialized machinery.

core The core of Earth consists of an outer liquid layer of molten iron and nickel and a solid, inner iron–nickel center.

Coriolis effect A deflection in the winds from a straight north–south path as a consequence of Earth's rotation on its axis from west to east; deflects winds to the right of their direction of travel in the Northern Hemisphere and to the left in the Southern Hemisphere.

corrosive Capable of causing permanent damage to a variety of surfaces, including living tissue; corrosive substances include strong acids (pH of 2 or less) or strong bases (pH of 12 or greater).

covalent bond A bond between atoms that share one or more pairs of electrons.

criteria pollutants Very common sources of air pollution (e.g., sulfur dioxide) chosen by the EPA to be regulated because they are hazardous to human health and the environment.

critical habitat Areas that are essential for the survival of a listed endangered or threatened species.

crop rotation A method farmers use to maintain soil fertility and reduce the buildup of pests by rotating crops on two-, three-, or four-year cycles.

crust The relatively low-density rocks that form the most superficial of Earth's layers, including the continents, where the crust is 20 to 70 kilometers thick, and the ocean floor, which is about 8 kilometers thick.

cultural eutrophication An accelerated process of eutrophication resulting from human activities (e.g., sewage disposal, agriculture) that increase the rate of nutrient addition to ecosystems; generally results in excessive algal or plant production, depletion of dissolved oxygen in aquatic ecosystems, and loss of biodiversity.

dam A structure that blocks the flow of a stream or river; may be used to reduce downstream flooding or to store water in a reservoir.

data The measurements made during a scientific study.

debt-for-nature swap A transaction wherein a developed nation forgives the debt of a developing nation in exchange for conservation pledges.

decomposer An organism, mainly fungi and bacteria, that breaks down dead plant and animal tissues, promoting the process of decomposition. See *detritivore*.

demanufacturing The dismantling of equipment, especially electronics, into constituent components and scrap metals that can be reused or recycled.

demographic transition A theory proposing that, with improved living conditions, human populations will undergo a gradual change from an earlier state of high death rates and birthrates to a state of low death rates and birthrates, with improved living conditions. The demographic transition model fits the history of today's developed countries well.

demography The statistical study of populations, generally human populations, including their density, growth, age structure, birthrates, and death rates.

denitrification The process by which specialized bacteria in soil and water convert nitrate ions back into nitrogen gas (N_2), which returns to the atmosphere.

density-dependent factors Mechanisms of population control that change with the density of a population (e.g., infectious disease, predation).

density-independent factors Controls on populations that are not affected by population density (i.e., physical aspects of the environment such as drought, floods, and extreme temperatures).

desalination The process of removing salts from seawater or brackish water to form freshwater.

desertification A process of degradation of once fertile lands to a desertlike condition of reduced plant cover and primary production.

detritivore An organism that feeds on dead organic matter (e.g., fallen leaves on the floor of a forest). Detritivores help in the process of decomposition; examples include many insects and earthworms. See *decomposer.*

discharge In an aquifer, the movement of water from the groundwater to a body of surface water (e.g., a river or lake).

disease A condition in which normal biological function is impaired by bacteria, viruses, parasites, improper diet, or pollutants.

dispersant A chemical used in oil spill cleanup that thins and dissolves the thick crude.

distillation A desalination process that uses heat to evaporate water from seawater or brackish water and condenses the resulting salt-free water vapor to produce freshwater.

distribution The geographic range of a species.

disturbance A discrete event (e.g., a fire, earthquake, or flood) that disrupts a population, ecosystem, or other natural system by changing the resources available or by altering the physical environment.

DNA (deoxyribonucleic acid) The carrier of genetic information in cells, consisting of two complementary chains of molecules, called nucleotides, wound in a double helix; this source of biological inheritance, passed from parents to offspring, directs the development and functioning of an organism.

domestication The deliberate change of a wild animal or plant population through selective breeding to better meet the needs of humans.

dose–response assessment A test of the response of an organism to a range in the dose, or concentration, of a potentially toxic substance.

drought An extended period of dry weather during which precipitation is reduced sufficiently to damage crops, impair the functioning of natural ecosystems, or cause water shortages for human populations.

dry casks Steel and concrete structures used for temporary storage of nuclear waste.

eccentricity Variation in the shape of Earth's orbit around the Sun.

ecocentric Centered on entire ecosystems; for example, ecocentric environmental ethics extends moral obligation to the nonliving components of the environment, emphasizing the integrity of whole natural systems.

ecological community All the species—plants, animals, fungi, and microbes—that exist and interact in a given location.

ecological economics A branch of economics that draws on many disciplines in studies of the influence of economic activity on the environment in an attempt to build a conceptual bridge between humans and human institutions and the rest of nature.

ecological footprint The environmental impact of a human population as the area of land and sea needed to produce the resources it consumes and to absorb the wastes it produces.

economic externality A cost or benefit to the environment or to society resulting from the production and use of a product that is not included in the market price of the product.

economic system A network of people, institutions, and commercial interests involved in the production, distribution, and consumption of goods and services.

economics A social science that deals with the production, distribution, and consumption of goods and services, as well as the theory and management of economic systems.

ecosystem The organisms living in a place and the biological, physical, and chemical aspects of the environment with which they interact. Ecosystem ecologists focus much of their research on the flux and transformation of matter and energy.

ecosystem-based management An approach to management of natural resources that considers the entire ecosystem; a departure from earlier, single-species approaches to natural resource management.

ecosystem diversity A measure of the variety and extent of ecosystems in an area.

ecosystem engineer A species, like the beaver, that influences ecosystem structure and processes by altering the physical environment.

ecosystem services The benefits that humans receive from natural ecosystems such as food, water purification, pollination of crops, carbon storage, and medicines.

ecotourism Recreational travel that helps conserve the environment and improves the well-being of local people.

edge effects Environmental conditions occurring near the edge of an ecosystem (e.g., near the edge of an isolated forest fragment); conditions at edges differ from those deep in the ecosystem interior.

E horizon Soil layer between the A and B horizons, from which clays and dissolved materials are transported down the soil profile to the underlying B horizon.

electron A subatomic particle with a negative charge (–1) and atomic mass of 1.

electronic waste (e-waste) A portion of the waste stream consisting of discarded electronic products that typically contain hazardous components (e.g., heavy metals like lead, and other toxins).

element A substance composed of a single type of atom, such as hydrogen, helium, iron, or lead, that cannot be broken down into simpler substances via chemical or physical means.

El Niño A period of warmer than average sea surface temperatures and lower barometric pressure in the eastern Pacific Ocean, favoring the production of storms in the eastern Pacific Ocean.

El Niño Southern Oscillation An oscillating climatic system involving variation in ocean surface temperatures and barometric pressures across the Pacific Ocean.

emigration The movement of individuals out of one area, or country, to another.

endangered species A species whose populations have become so small that they may become extinct in the near future.

Endangered Species Act of 1973 (ESA) Legal protection in the United States for both domestic and foreign endangered species that declared plants and *all* invertebrate animals eligible for protection.

endemic species Local or regional species of organisms found nowhere else on Earth.

endocrine disruptor A chemical that mimics hormones, including female hormones (estrogen and progesterone), male hormones (testosterone), or thyroid hormones.

energy The capacity to do work. See *work.*

energy pyramid A graphic representation of the distribution of energy among trophic levels in an ecosystem. Because large amounts of energy are dissipated at every trophic level, these diagrams take the form of a pyramid.

energy return on energy investment (EROEI) The ratio of the energy content of an energy source (e.g., gasoline) to the amount of energy that must be used in, for example, drilling, transporting, and refining to produce the energy source.

energy subsidies Government policies aimed at lowering the costs of energy production or lowering the cost to energy consumers.

enrichment A nuclear process in which uranium-235 is separated from less valuable uranium-238.

entropy A measure of the amount of disorder in a system.

environment The physical, chemical, and biological conditions that affect an organism.

environmental economics A branch of economics that draws mainly from the field of economics as it assesses and manages the costs and benefits of economic impacts on the environment.

environmental ethics The branch of philosophy that concerns the moral responsibilities of humans with regard to the environment.

environmental hazards Phenomena dangerous to humans, including infectious disease, toxic substances, and pollutants.

environmental health An area of research and action that assesses and attempts to mitigate the physical, chemical, and biological factors in the environment that impact human health.

environmental justice The fair treatment and meaningful involvement of all people in the development, implementation, and enforcement of environmental laws, regulations, and policies.

environmental science Study of the influence of humans on the environment and the effects of the environment on humans; also attempts to find ways of reducing human harm to the environment.

environmentalism An ideological and social movement that advocates the protection of the environment from human harm through political action and education.

equilibrium model of island biogeography The hypothesis that the number of species on an island is determined by a balance between rates of immigration of new species and rates of species extinction on the island, where rates of species immigration and extinction are determined by island size and isolation from sources of immigrants.

erosion A process that removes geologic materials, ranging from clay-sized particles to boulders, from one part of a landscape to be deposited elsewhere; increased rates of soil erosion due to human activity can reduce soil fertility.

euphotic zone A surface layer of the oceans and deep lakes where there is sufficient light to support photosynthetic aquatic organisms.

eutrophication A natural process by which nutrients, especially those that limit primary production, build up in an ecosystem.

evolution A change in the genetic makeup of a population as a consequence of one of several different processes, including natural selection and selective breeding.

exposure assessment An assessment of the population that might be exposed to an agent of concern and of potential routes of exposure.

extinction The loss of all members of a species.

field experiments Experiments in which the experimenter generally controls or manipulates a single factor, the factor of interest, while allowing all other factors to vary normally.

fire regime The frequency and intensity of fires that typically occur in a particular ecosystem.

first law of thermodynamics A physical law concerned with the conservation of energy: Though one form of energy may be transformed to other forms, the total amount of energy in a system plus its surroundings is conserved; that is, the total amount of energy remains the same. See *second law of thermodynamics.*

fisheries collapse The decline in a certain species' annual catch below 10% of its historic catch.

fishery A population of fish or shellfish, and the economic system involved in harvesting the population, often identified by the geographic area where the fish or shellfish are harvested.

flagship species A species that attracts and sustains human interest in protecting ecosystems.

flammable Easily ignited; a flammable substance can ignite and burn easily (e.g., from friction, absorption of moisture, or contact with other waste materials).

flood A river or stream overflowing its banks and inundating the surrounding landscape known as the floodplain.

floodplain The area of land that stretches from a water channel to the valley walls.

flux The rate of flow of materials or energy across a given area (e.g., the flow of water vapor from the ocean's surface to the atmosphere or the flow of radiant energy between an organism and its surroundings).

fly ash Particles formed during combustion that are light enough to become airborne and exit a combustion chamber with exhaust gases, including soot and dust.

food web A set of feeding relationships among organisms indicating the flow of energy and materials in an ecosystem.

forestry The management of forests and woodlands for the harvest of timber or fuelwood.

fossil fuels Fossilized organic material, mainly the remains of ancient photosynthetic organisms that converted the Sun's radiant energy into chemical energy (e.g., coal, oil, natural gas).

foundation species A species that strongly influences community structure by creating environments suitable for other species by virtue of its large size or biomass.

fracking (hydraulic fracturing) A controversial extraction technique that involves drilling horizontally into a rock formation and pumping in a mixture of fluids and sands to fracture it, thus creating a path through which natural gas or oil can flow out.

freshwater Water with a salt content, or salinity, below that of brackish water (i.e., salinity less than 500 mg/l).

fuel rods Tubes containing small pellets of uranium-235 used as an energy source in nuclear reactors.

gas turbine engine Engine that burns natural gas, sending a hot, high-pressure stream of gas through a turbine connected to an electrical generator.

gasification A technique to make carbon capture at coal-fired power plants more effective by creating synthetic natural gas from coal before combustion, allows for the more efficient sequestration of pure carbon dioxide.

genes Stretches of DNA that direct the growth, development, and functioning of organisms.

genetic diversity The sum of the different genes and gene combinations found within a single population of a species and across populations of the same species.

genetically modified (GM) organism (GMO) An organism into which one or more genes have been incorporated using the techniques of biotechnology.

geoengineering Proposed technological approaches to manipulating processes that influence climate, such as carbon dioxide uptake by marine algae, in an effort to mitigate global climate change.

geo-sequestration An approach to carbon capture and sequestration that involves compressing carbon dioxide and pumping it into underground structures such as salt-water aquifers or empty oil and natural gas fields.

gillnetting The practice of placing panels of large mesh net in the water column to catch fish, the size of which depend on the mesh size; fish that cannot pass all the way through the gillnet are ensnared by their gill covers when trying to retreat.

gravitational potential energy The amount of potential energy an object contains due to its mass and height above a reference point, such as Earth's surface.

greenhouse effect The absorbing and reradiating of infrared light by various components of Earth's atmosphere, resulting in higher surface and atmospheric temperatures.

gross domestic product (GDP) The total market value of all the goods (e.g., manufactured articles or agricultural crops) and services (e.g., transportation and banking services) produced within the borders of a nation during some period of time. See *per capita GDP*.

gross primary production The total amount of organic matter produced by the primary producers in an ecosystem over some period of time, for example, per year. See *net primary production*.

groundwater Water found in the pore spaces in rock and sediments beneath Earth's surface; feeds wells, springs, and desert oases, and is Earth's second largest reservoir of freshwater.

groundwater depletion The amount of groundwater pumped from an aquifer in excess of recharge. Groundwater depletion can result in land subsidence, which reduces the capacity of an aquifer to store water and can damage buildings and other infrastructure.

habitat Where an organism usually lives (e.g., a forest, coral reef, or marsh).

habitat corridor A strip of suitable habitat linking protected areas intended to increase the movement of wildlife between protected areas to sustain genetic variation and reduce the likelihood of extinction of protected populations.

habitat fragmentation A subdivision of a formerly continuous habitat into isolated habitat patches as a result of activities such as deforestation, road building, and dam construction on rivers.

half-life The time needed for half of a given amount of a radioactive isotope to decay.

hazardous waste A flammable, reactive, corrosive, or toxic waste capable of causing illness, death, or other harm to humans and other organisms.

herbivore (primary consumer) A consumer whose diet consists entirely of plants or other primary producers, for instance, an elephant or a grasshopper.

heterotroph An organism, incapable of producing its own food, that meets its energetic and nutritional needs by feeding on organic matter produced by plants and other primary producers or on other heterotrophs. See *consumer*.

high-level nuclear waste Radioactive waste, primarily nuclear fuel rods that have been depleted to the point that they can no longer contribute to the efficient production of electricity.

Human Development Index (HDI) An index of national development that includes life expectancy at birth, educational opportunities, and economic productivity.

hydrocarbon An organic molecule made up of carbon and hydrogen only; the simplest hydrocarbon is methane (CH_4), the main component of natural gas.

hydroelectric power Electricity generated by tapping the energy of water being driven downhill under the force of gravity.

hydrokinetic power A form of hydropower in which power stations use the kinetic energy of waves, tidal currents, or river flow to generate electricity.

hydrologic cycle The movement of Earth's water between the oceans, atmosphere, and terrestrial and freshwater environments.

hyperaccumulators Plants that accumulate heavy metals in their tissue.

hypothesis An explanation of an observation, or a set of relationships, based on a limited amount of information; hypotheses are used to guide scientific experiments, observation, and modeling.

igneous rock Rock formed as molten rock cools and solidifies.

immigration The movement of individuals into an area, or country, to which they are not native.

indicator species A species that provides information about the state of the ecosystem in which it lives.

individual transferable quota (ITQ) (catch shares) A guaranteed right to a certain portion (quota) of the catch in a fishery or exclusive rights to certain fishing grounds; may also be granted to a fisheries cooperative or community.

indoor air pollution A serious threat to human health resulting from the buildup of pollutants in the indoor environment.

industrial fishermen Commercial fishers who may travel for weeks at a time and use expensive, technologically advanced gear to process and refrigerate or freeze their catch on board.

instream uses Benefits, such as dilution of sewage and recreational fishing and boating, that result from water flowing in river or stream channels.

integrated multi-trophic aquaculture (IMTA) An approach to aquaculture that involves raising several species of aquatic organisms with complementary feeding habits in close proximity.

Integrated Pest Management (IPM) An approach to managing pests (e.g., insects, pathogens, weeds) that incorporates multiple sources of information to contain pest damage within acceptable limits while trying to minimize harm to people, property, and the environment.

integrated waste management A management strategy that minimizes waste disposal by stressing the importance of reducing waste, reusing materials, recycling, composting, and recovering energy from waste materials.

intercropping Growing two or more crops in the same field.

internal combustion engine Engine (most commonly used in cars, boats, and jet airplanes) in which combustion directly drives a set of pistons or turbines hooked up to a crank arm.

interspecific competition Competition among individuals of different species.

intraspecific competition Competition among individuals of the same species.

invasive species An introduced species that poses a serious threat to native populations.

ion An atom, or group of atoms, with a net positive or negative charge (e.g., chloride, Cl⁻, and sodium, Na⁺, ions).

ionic bond A chemical bond involving the attraction between two oppositely charged ions.

irrigation A system for artificially delivering water to crops so that they can grow in areas with too little precipitation to support them otherwise.

J-shaped (exponential) population growth Population growth that occurs at a constant, or fixed, rate per capita and that produces a characteristic J-shaped pattern of increase in population size over time.

kerogen A waxy substance found in shale and other sedimentary rocks that yields oil when heated; occurrs during an intermediate stage of petroleum formation.

keystone species A species with substantial influence on community structure, despite its low biomass or numbers relative to other species; the influence of keystone species is often exerted through feeding activities.

kinetic energy The energy of a moving object, which is equal to one-half the mass of the object times the square of its velocity.

K-**selected species** Organisms with populations that generally stabilize close to their carrying capacity and are often regulated by density-dependent factors.

laboratory experiments Experiments in which scientists attempt to control, or keep constant, all factors that may influence their study system, while they vary the factor of interest and observe the effect of the variation on the study system.

Lacey Act First passed in 1900 and amended in 2008, this law forbids trade in illegally harvested plants and animals.

land ethic An ecocentric system of environmental ethics proposed by Aldo Leopold to promote the integrity, stability, and beauty of the biological community.

La Niña A period of lower than average sea surface temperatures and higher barometric pressures in the eastern Pacific Ocean, resulting in reduced storm activity in the eastern Pacific Ocean.

leachate Water that has seeped down through the waste in a landfill; flows to a sump in a modern landfill, where it can be pumped out and treated.

life cycle assessment (LCA) An estimate of the total environmental impact of a product or technology as a result of activities such as extraction of an energy source (e.g., coal), transport, processing of raw materials, construction, maintenance, dismantling, removal, and recycling or disposal of structures.

life expectancy at birth The predicted average life span of individuals born during a particular year.

life history Characteristics of a species, such as the age at which individuals begin reproducing, the number of offspring they produce, and the rate at which the young survive.

lipids Organic molecules composed of long chains of carbon atoms bonded mainly to hydrogen (e.g., fats, oils, or waxes); important component of cell membranes; function as energy storage molecules in animals and plants.

lithosphere The uppermost layer of the upper mantle above the asthenosphere, consisting of rigid, relatively brittle rocks.

loam A soil consisting of approximately equal proportions of sand, silt, and clay.

longline fishing The practice of laying out a very long line with hundreds or thousands of baited hooks; used to catch tuna (near the surface) or groundfish (e.g., halibut, cod).

low-level nuclear waste Radioactive waste, including any item that has become contaminated with small amounts of radioactive particles, including instruments, protective suits, or clothing from nuclear facilities.

malaria A disease transmitted by mosquitoes that results from infection by a protozoan parasite of the genus *Plasmodium;* its life cycle uses two hosts: mosquitoes and humans.

mantle The layer of Earth between the crust and the core; represents the largest portion of Earth's volume and consists of higher-density rocks than those that make up the crust.

marine protected areas (MPAs) Protected areas in coastal regions and the oceans that help conserve ecosystems critical for biodiversity (e.g., coral reefs and salt marshes) and sustain populations that supply fish and other marine resources.

market-based approach An alternative to command-and-control regulation that seeks to encourage adherence to social or environmental goals using the principles of supply and demand.

market economy An economy in which decisions about the production and consumption of goods and services are not centralized but made by businesses and individuals, generally acting in their own self-interest. See *centrally planned economy.*

market failure A situation in which free markets do not allocate goods and services efficiently, such as when the price of a product does not include its environmental impact.

mass extinction A period wherein a large proportion of species becomes extinct within a few million years or less.

matter Anything that occupies space and has mass; matter exists in three main physical states: as a solid, liquid, or gas.

maximum sustainable yield (MSY) The maximum harvest of a renewable natural resource that does not reduce future yields (e.g., the sustainable annual catch from a fish population).

metamorphic rock Rock formed when any type of rock changes as it is subjected to heat and pressure.

Milankovitch Cycles Cyclic changes in the shape of Earth's orbit, tilt in its axis, and precession of the equinoxes that produce variation in Earth's climate.

models In science, simplified representations of a system, constructed on a scale more convenient for study than the actual system of interest.

moderator A substance (most commonly pressurized water) used in a nuclear reactor to reduce the speed at which neutrons travel.

molecule Two or more atoms held together by chemical bonds; the constituent atoms may be of the same or different elements.

money A medium of exchange using coins or paper bills.

monoculture A planting of a single variety of crop, generally over a large area, that creates an attractive target for pests and pathogens of the crop.

mountaintop removal mining An extremely destructive coal mining practice that involves clear-cutting of the forests on a mountain and adjacent stream valleys; miners then use explosives to break up the rock overlying the coal deposit, depositing it in the adjacent valleys, which are buried as the coal is exposed.

MRSA (methicillin-resistant *Staphylococcus aureus*) A pathogenic bacterium resistant to the antibiotic methicillin; MRSA originated in hospitals and then spread to the broader community.

mulch A natural or synthetic covering to the soil surface that conserves moisture, reduces soil temperature variation, and decreases growth of weeds.

municipal solid waste (MSW) Solid waste from institutions, households, and businesses, including paper, packaging, food scraps, glass, metal, textiles and other solid discards.

mutation A change in the structure of an organism's DNA, i.e., in its genes.

mutualism Mutually beneficial relationship between organisms.

natural capital The value of the world's natural assets (e.g., minerals, air, water, and living organisms).

natural enemies Predators and pathogens that attack herbivorous insects and other pest organisms.

natural selection A process of interaction between organisms and their environment that results in different rates of reproduction by individuals in the population with different physical, behavioral, or physiological characteristics; can change the relative frequencies of particular genes in the population—that is, in evolution.

negative feedback An increase in some factor in a system, such as a climate system, produces a decrease in that factor within the system.

net primary production The net production of organic matter by the primary producers in an ecosystem, that is, gross primary production less the organic matter used by primary producers to meet their own energy needs. See *gross primary production.*

neurotoxins Toxic substances that attack nerve cells.

neutron A subatomic particle, found in the nucleus of atoms, that has a mass approximately equal to that of a proton but no electrical charge.

niche A description of the physical and biological requirements of a species.

nitrification The conversion of ammonia or ammonium to nitrites (NO_2^-) and nitrates (NO_3^-) by nitrifying bacteria.

nitrogen assimilation The incorporation by plants of nitrate and ammonium into essential nitrogen-containing organic compounds.

nitrogen cycle The process whereby nitrogen passes through and between ecosystems, involving several key actions by microorganisms, including nitrogen fixation, decomposition, ammonification, nitrification, and denitrification.

nitrogen fixation Incorporation of atmospheric nitrogen, N_2, into nitrogen-containing compounds by bacteria, living in association with plants or free living.

non-biodegradable A substance that cannot be decomposed to its chemical constituents by biological processes.

nonpoint sources of pollution Diffuse, and sometimes mobile, sources of pollution (e.g., runoff from an industrial, municipal, or agricultural landscape or the exhaust from automobiles).

nonrenewable energy Sources of energy, including coal, petroleum, natural gas, and nuclear fuels, that are not renewable on timescales

meaningful to human lifetimes and that can be depleted with continued use.

nonrenewable resources Natural resources, such as fossil fuels, that exist in a limited supply and are not renewed on timescales meaningful to humans.

no-till (low-till) agriculture An approach to growing crops involving reduced or no cultivation; creates less soil disturbance and leaves crop residues on the field.

nuclear energy A form of energy released when the nucleus of an atom breaks apart (nuclear fission), or when the nuclei of two atoms fuse (nuclear fusion).

nuclear fission A process in which the bonds holding the protons and neutrons that make up the nucleus of an atom are broken, resulting in the release of a large quantity of energy.

nuclear fusion A process in which the nuclei of two atoms fuse to form a new type of atom, releasing large amounts of energy.

nucleus The massive central core of an atom, which is made up of protons and neutrons, around which the atom's electrons move.

observation Qualitative or quantitative information gathered systematically from the natural world.

O horizon The surface layer of many soils, which is rich in organic matter and a site of active decomposition.

omnivore A consumer that eats both plant and animal material.

open access A property for which there are no restrictions about who may enter and exploit its resources.

overburden The layer of rock overlying a mineral deposit (e.g., coal).

oxbow lake A crescent-shaped lake formed on a river's floodplain by rerouting the main river channel, generally during a flood.

ozone A molecule made up of three oxygen atoms; considered a pollutant in the lower atmosphere, but in the upper atmosphere it shields against potentially harmful rays from the Sun.

pandemic Expansion of a disease affecting a large proportion of a population in a very large geographic area (e.g., across an entire continent).

parasite An organism that lives in or on another organism, called the host; hosts are harmed by the parasite, while the parasite receives various benefits from the host (e.g., food, protection, dispersal of offspring).

parent material The bedrock or unconsolidated deposits, such as windblown sand or silt, from which soil develops.

pathogen An organism that produces illness.

peer review As part of the process of publishing scientific papers, experts in the field of research covered by a prospective scientific paper review the research prior to publication; they check for soundness of the methods, analyses, results, and coverage of the relevant prior publications on the subject.

per capita GDP The market value of the goods (e.g., manufactured articles or agricultural crops) and services (e.g., transportation and banking services) produced within the borders of a nation per individual in its population.

persistent organic pollutants (POPs) Organic chemicals (e.g., PCBs) that remain in the environment indefinitely; can biomagnify through the food web and pose a threat to human health and the environment.

pesticide Generally a chemical substance used to kill destructive organisms, including insects (insecticide), fungi (fungicide), weeds (herbicide), and rodents (rodenticide).

pesticide resistance An evolved tolerance to a pesticide by a pest population as a result of repeated exposure to a pesticide, ultimately rendering the chemical ineffective.

petroleum (crude oil) A mixture of hydrocarbons contained in sedimentary rocks of marine origin; developed from the accumulated remains of algae and zooplankton deposited on the sea floor over millions of years.

pH An indicator of the relative hydrogen ion concentration of a solution. A pH of 7 indicates a neutral solution; a pH of less than 7 is acidic (elevated hydrogen ion concentration); a pH greater than 7 is basic (reduced hydrogen ion concentration).

photoelectric effect The ejection of electrons from a substance (e.g., a metal or semiconductor) in response to stimulation by light energy.

photosynthesis A biochemical process employed by green plants, algae, and some bacteria that uses solar energy to convert water and carbon dioxide into the chemical energy in a simple sugar called glucose.

phytoremediation Bioremediation using plants to clean up contaminated sediments or soils. See *bioremediation*.

pioneer community The earliest community to develop during succession.

plate tectonics A theory proposing that Earth's surface is divided into plates that move on the upper layer of the mantle; explains Earth structure and processes, including the formation of ocean basins, continents, and the geographic distribution of earthquakes and volcanic activity.

point sources of pollution Clear-cut, generally stationary, sources of pollution (e.g., power plants, factories, or sewage outfalls of cities) that are easier to identify, monitor, and regulate.

pollutant A substance (e.g., oil, pesticides) or physical condition (e.g., excessive noise) harmful to living organisms that contaminates air, water, or soil.

pollution Contamination of the environment, generally of air, water, or soil, by substances or conditions (e.g., noise, light) at levels harmful to living organisms; generally a result of human activity but may result from natural processes (e.g., wildfires, volcanic eruptions).

polyculture The growing of multiple domesticated crops that may be intermixed with useful wild species.

population density The number of individuals in a population per unit area.

population doubling time The amount of time required for a population, growing at a particular rate, to double its size.

population ecology Branch of ecology that is concerned with the factors influencing the structure and dynamics of populations, including population size, distribution, and growth.

population momentum Population growth as a consequence of a large number of women reaching childbearing age.

populations All the individuals of a species that inhabit a particular place at the same time.

positive feedback A stimulus in which an increase in some factor in a system, such as an economic system or ecosystem, produces additional increases in that factor within the system or in which a decrease in a factor causes additional decreases.

potential energy The amount of energy an object has due to the configuration of its parts (e.g., a loaded spring), its chemical makeup, or its position in a force field (e.g., Earth's gravitational field).

pot-traps Baited traps used to catch lobster or crab.

precautionary principle A principle advising that precautionary measures should be taken to protect human or environmental health, even if some cause-and-effect relationships related to potential threats are not fully understood scientifically.

precession of the equinoxes Slow drift in the position in Earth's orbit at which the equinoxes occur, a cycle repeating itself approximately every 26,000 years.

preservation ethic An environmental ethic emphasizing the protection of natural ecosystems in their original unspoiled states.

prevailing winds Winds that blow consistently from one direction (e.g., the northeast trade winds blow from the northeast).

primary energy A form of energy that requires only extraction or capture for use (e.g., coal, crude oil, wind).

primary pollutant A substance that is harmful when released into the environment (e.g., carbon monoxide, crude oil).

primary producer (autotroph) An organism, generally a plant or alga, that converts the radiant energy in sunlight to the chemical energy in sugars through the process of photosynthesis.

primary succession Succession on a bare geologic surface, such as a recent lava flow.

private property Property owned by individuals.

protected area A geographically defined area designated or regulated and managed to achieve particular conservation objectives (e.g., national parks, national forests, wildlife refuges).

proteins Long chains of amino acids (i.e., molecules consisting mainly of carbon, nitrogen, hydrogen, and oxygen) that control rates of chemical reactions, provide structural support, and perform many other functions.

proton A subatomic particle, found in the nucleus of atoms, that has a positive charge (+1) and atomic mass of 1.

radiant energy The energy of electromagnetic radiation, including visible light, infrared light, ultraviolet light, microwaves, radio waves, or X-rays.

ranching The practice of raising domesticated livestock for meat, leather, wool, and other products.

reactive Chemically responsive; a reactive substance will readily undergo a violent chemical change when in contact with other substances.

reclamation A process that restores an ecosystem to its natural structure and functioning prior to mining or to an economically usable state.

recycling The process of returning raw materials in waste (e.g., glass, plastics, metal, paper) to the manufacturer for reuse.

renewable energy Sources of energy, including solar, wind, hydrologic, geothermal, and biomass, that can be replenished in a relatively short period of time. Use does not deplete renewable energy sources.

renewable resources Natural resources, such as wood, forage, or fish, that are replaced through natural processes on relatively short timescales and thus can last indefinitely under careful management.

replacement-level fertility The total fertility rate required to sustain a population at its current size, which varies from approximately 2.1 births per woman in the more developed countries to 2.5 or higher in the least developed countries, where mortality rates are higher.

reservoir A body of water, ranging in size from a pond to an ocean, including below-ground deposits of water; constructed dams retain water in artificial reservoirs, which are commonly used to store and divert water for human use.

Resource Conservation and Recovery Act (RCRA) A law passed by the U.S. Congress that banned open dumping of wastes and set standards for solid waste landfills.

resource partitioning Coexisting species use different resources, such as food, nesting sites, and feeding areas.

reverse osmosis A desalination process that uses selectively permeable membranes and pressure to separate salts and water.

R horizon The base of a soil profile composed of consolidated bedrock, immediately below the C horizon.

riparian The transition zone between a river or stream and the terrestrial environment, generally inhabited by a biological community distinctive from adjacent aquatic and upland communities. Riparian zones naturally flood periodically and usually have shallow water tables.

risk The chance of harmful effects to human health or to ecological systems resulting from exposure to any physical, chemical, or biological agent.

risk characterization A qualitative or quantitative estimate of the likelihood that hazards associated with an agent of concern will negatively impact an exposed population.

rock A natural, solid, inorganic substance formed from one or more minerals.

rock cycle Geologic processes that convert each of the three major rock types (igneous, sedimentary, and metamorphic) into one of the other types.

r-selected species Organisms with populations that generally fluctuate widely in size; subject to catastrophic mortality from harsh weather, fires, and other density-independent factors.

runoff The amount of water falling as precipitation that flows off the land as surface and subsurface flow.

run-of-the-river power plants Hydroelectric systems that provide little or no water storage in a reservoir and divert a portion of river flow through pipes that pass directly through a turbine.

salinization The process of salt buildup in a soil.

sanitary landfill A solid waste disposal site consisting of a lined pit constructed and managed in ways to minimize environmental impacts.

saturated zone The layers of rock below the water table, in which the pore spaces in the geologic formation are saturated with water.

science A formal process used to study nature, and the body of knowledge resulting from that process.

seawater Ocean water and the water of seas, such as the Caribbean and Mediterranean seas. The salinity of seawater averages about 34,000 mg/l (34 g/l), but ranges from 30,000 to about 40,000 mg/l (30 to 40 g/l).

second law of thermodynamics With each energy transformation, or transfer, the amount of energy in a system available to do work decreases. In other words, the quality of the energy declines with each energy transfer or transformation. See *first law of thermodynamics*.

secondary pollutant A pollutant formed from the chemical reactions between other pollutants (e.g., ozone in the lower atmosphere).

secondary (consumer) production The amount of consumer biomass, or energy, that goes into growth and reproduction, analogous to net primary production by photosynthetic organisms.

secondary succession Succession following disturbance of an established community that doesn't destroy all living creatures or the soil.

sedimentary rock Rock formed either as rock fragments deposited by water, wind, or ice are cemented together and solidify or as rock forms through chemical precipitation.

selective logging The clearing of land for lumber that focuses on the most mature, high-value trees, leaving the forest ecosystem largely intact.

semiconductor A material that conducts current, but only somewhat, because its properties lie somewhere between those of an insulator and a conductor.

sex ratio at birth The ratio of male to female newborns.

shelterwood harvesting Removes the tallest trees in a series of partial cuts, leaving behind enough of a forest canopy to provide shelter for speedy regrowth of shade-tolerant trees (e.g., red oak, American beech).

sick building syndrome A circumstance in which many building occupants experience symptoms of illness (e.g., headaches, respiratory and eye irritation, nausea) for which no specific cause has been identified.

skimmer A device used to collect spilled oil from the water's surface.

slash-and-burn A common technique used in tropical countries to rapidly convert forestlands into temporary farms.

small-scale fishers Commercial fishers who use minimal gear and fish from small boats or nonmotorized canoes.

soil texture The relative fineness or coarseness of a soil, which is determined by its proportions of sand, silt, and clay.

solubility The amount of a substance capable of dissolving in a particular amount of solvent.

source reduction A waste management tactic aimed at reducing the amount of material that enters the waste stream.

speciation An evolutionary process by which new species arise.

species A group of interbreeding, or potentially interbreeding, populations, reproductively isolated from other populations.

species diversity A measure of diversity that combines the number of species in a community and their relative abundances.

species evenness How evenly individuals are apportioned among the species inhabiting a community; higher evenness increases species diversity.

species richness The number of species in a community or living in a local area or region; higher species richness increases species diversity.

sport (recreational) fishing The practice of fishing for pleasure (e.g., fly-fishing, hiring a tourist charter boat to catch trophy-sized fish).

S-shaped (logistic) growth Population growth in which the per capita rate of growth decreases with increasing population size as a result of predation or reduced availability of food, space, or other resources; eventually levels off at carrying capacity.

state property Property owned by federal, state, or local governments.

stratosphere The layer of Earth's atmosphere beginning at an elevation of 10 kilometers and extending outward to 50 kilometers (6.2 to 31.1 miles) above sea level.

strip mining A coal extraction technique in which overburden is removed from a long strip of land, exposing a coal seam; once the coal is removed, material from an adjacent strip is used to fill the excavation.

stock A discrete subpopulation of a species, which is reproductively isolated from other stocks.

stock assessment Estimated size of a fish stock, the rate at which the population is growing, and the rate of harvest.

subduction Process in which one tectonic plate moves under another, generally occurring where oceanic plates, which are of higher density, collide with continental plates.

subduction zone Zone where oceanic plates and continental plates collide, forming deep sea trenches and active volcanoes along the continental margin.

subsidence A settling or sudden sinking, in the case of sinkhole formation, of a land surface as a result of processes such as groundwater withdrawal or loss of organic matter in soil.

subsistence economy An economy in which individuals or groups produce or harvest enough resources to largely support themselves, with fewer resources gained through purchase or trade with other groups.

subsistence fishing The practice of catching enough fish for one's family plus a bit more for bartering or selling.

succession The gradual change in a community over time following a disturbance.

supply and demand An economic model stating that the price of a good (or service) will reach equilibrium when the consumer demand for it at a certain price equals the quantity supplied by producers.

sustainability The wise use of resources to ensure our ability to endure and live healthy lives, without compromising the welfare of future generations.

sustainable development A process of development that meets the needs of the present generation without reducing the ability of future generations to meet their needs. Development is sustainable when it does not, at a minimum, endanger Earth's natural life support system, including the atmosphere, waters, soils, and biological diversity.

sympatric speciation A process by which new species arise without geographic isolation.

synergistic effect An interaction of two toxic substances wherein their combined toxicity is greater than the sum of their individual effects.

take back laws State regulations that require manufacturers of various electronics to pay for e-waste recycling programs.

Taq polymerase An enzyme isolated from a bacterium discovered living in hot springs in Yellowstone National Park; used to amplify small quantities of DNA.

technology Practical application of scientific knowledge and methods to create products and processes.

teratogen A substance that causes abnormalities during embryonic growth and development, resulting in birth defects.

terra preta Dark, fertile soils high in charcoal and nutrient content, created by native populations in the Amazon River Basin before the arrival of Europeans.

terrestrial harvest systems Ways of extracting production from ecosystems, ranging from hunting and gathering in unmanaged natural ecosystems to nomadic herding and small-scale subsistence farming to industrialized agriculture.

tertiary treatment Advanced treatment of wastewater, which follows primary and secondary treatment, that removes dissolved organic chemicals, nitrogen, phosphorus, several other dissolved salts, and pathogens.

theory A scientific hypothesis that has withstood sufficient testing—through observation, experimentation, and modeling—so it has a high probability of being correct.

thermal (heat) energy A form of kinetic energy due to molecular motion in a mass of a substance, such as a mass of steam.

threshold dose The lowest dose (concentration) of a toxic substance that induces a toxicity response in an organism.

total fertility rate An estimate of the average number of children that a woman in a population gives birth to during her lifetime.

toxic Poisonous; a toxic substance is harmful to living organisms in relatively low amounts.

toxicant A toxic substance produced by humans or as a by-product of human activity. See *toxin*.

toxicology The science concerned with the effects of toxic substances on humans and other organisms.

toxin A poisonous substance produced by a living organism (e.g., a plant, animal, fungus, or bacterium) that can harm human health. See *toxicant*.

transboundary pollution The transport of pollutants by wind and water around the biosphere, across geographical and political borders.

transgenic organism A GM organism that contains genes from another species.

trophic level A step in the movement of materials or energy through an ecosystem or the position of a species in a food web.

ultraviolet (UV) light Shorter-wavelength, higher-energy rays from the Sun that can damage living tissue.

umbrella species A species whose protection provides protection for the entire ecosystem on which that species depends.

unsaturated zone The layers above the water table, which are not saturated with water.

upwelling The movement of cold subsurface water to the ocean's surface when warmer surface waters move offshore under the influence of prevailing or seasonal winds.

vector An organism that transmits a pathogen or parasite to other organisms (e.g., mosquitoes transmit malaria and other diseases to humans and other species).

virus A structurally simple disease-causing agent consisting of DNA or RNA encased in protein; viral diseases include common cold, flu, measles, mumps, chicken pox, smallpox, rabies, herpes, and human immunodeficiency virus (HIV, the virus responsible for AIDS).

waste stream The flow of discarded materials, especially municipal solid waste, from institutions, homes, and businesses.

water reclamation Any process of treating wastewater to make it safe for reuse or recycling.

water recycling Using treated wastewater for beneficial purposes, including industrial processes, irrigation, recharging groundwater supplies, restoring wetlands and aquatic ecosystems, and augmenting drinking-water supplies.

water table The uppermost level of groundwater, which forms the boundary between the saturated and unsaturated zones.

waterlogged soil A condition in which the water table is at or near the soil surface.

watershed (catchment, drainage basin) The land area from which an aquifer or river system acquires its water; also defined as the dividing line between catchments or drainage basins.

weather Atmospheric conditions, temperature, humidity, cloud cover, rainfall, etc. at a particular place and time (e.g., conditions during a particular day or month).

weathering The fragmentation and decomposition of mineral materials as a result of chemical, biological, and mechanical processes, resulting in the release of nitrogen, phosphorus, and other elements.

work A description of the transfer of energy; the work done on an object by a force is determined by the amount of force times the distance the object moves in the direction of the force. See *energy*.

zoonotic disease Any infectious disease that can spread from animals to humans.

Index

Note: Page numbers followed by f indicate figures; those followed by t indicate tables.

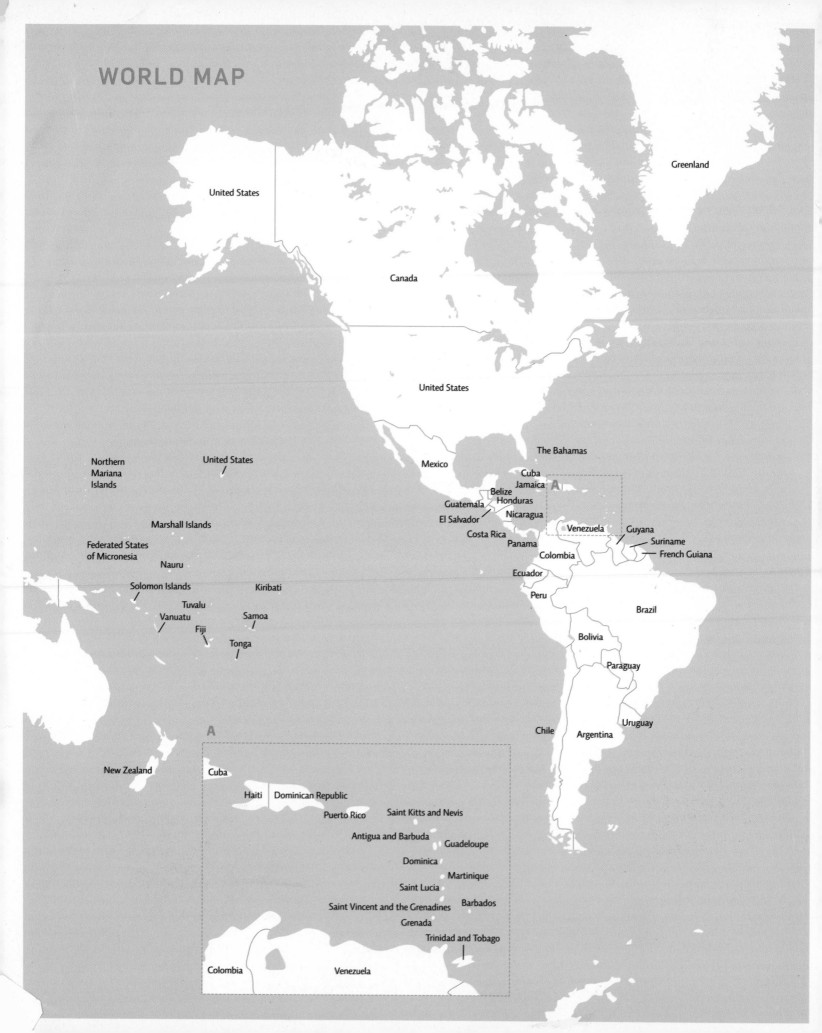

WORLD MAP

Greenland

United States

Canada

United States

Northern Mariana Islands

United States

The Bahamas

Mexico

Cuba

Jamaica

Belize

Honduras

Guatemala

El Salvador

Nicaragua

Costa Rica

Panama

Venezuela

Guyana

Suriname

French Guiana

Colombia

Ecuador

Peru

Brazil

Bolivia

Marshall Islands

Federated States of Micronesia

Nauru

Solomon Islands

Kiribati

Tuvalu

Vanuatu

Samoa

Fiji

Tonga

Paraguay

Chile

Argentina

Uruguay

New Zealand

A

Cuba

Haiti

Dominican Republic

Puerto Rico

Saint Kitts and Nevis

Antigua and Barbuda

Guadeloupe

Dominica

Martinique

Saint Lucia

Saint Vincent and the Grenadines

Barbados

Grenada

Trinidad and Tobago

Colombia

Venezuela